THIRD CENTURY OF BIOCHEMICAL OXYGEN DEMAND

Second Edition

Rodger B. Baird & Roy-Keith Smith, PhD

2022

Water Environment Federation
601 Wythe Street
Alexandria, VA 22314-1994 USA
https://www.wef.org

Library of Congress Cataloging-in-Publication Data

Baird, Rodger.
Third century of biochemical oxygen demand / by Rodger B. Baird and Roy-Keith Smith.
p. cm.
Includes bibliographical references and index.
1. Biochemical oxygen demand. I. Smith, Roy-Keith. II. Title.
TD737 .B35 2001
628.3—dc21 2001046825

First edition ISBN 9781572781719
Second edition ISBN 9781572784000

Printed in the USA

About WEF

The Water Environment Federation (WEF) is a not-for-profit technical and educational organization of 35,000 individual members and 75 affiliated Member Associations representing water quality professionals around the world. Since 1928, WEF and its members have protected public health and the environment. As a global water sector leader, our mission is to connect water professionals; enrich the expertise of water professionals; increase the awareness of the impact and value of water; and provide a platform for water sector innovation. To learn more, visit www.wef.org.

Text originally prepared by the **Third Century of Biochemical Oxygen Demand Task Force** of the **Water Environment Federation**

David S. Hasson, PhD, *Chair*

Principal Authors:

Peiffer Brandt
Kees Corssmit, PhD
Eric Rothstein, CPA

Reviewers and Contributors:

Rachel Stevens
Alan Cherubin
Rich Cortes, PE
Bob Decker
Myron Olstein
Karen Pallansch
Steven M. Ravel, PE, BCEE

Under the Direction of the **Utility Management Subcommittee** of the **Technical Practice Committee**

Special Publications of the Water Environment Federation

The WEF Technical Practice Committee (formerly the Committee on Sewage and Industrial Wastes Practice of the Federation of Sewage and Industrial Wastes Associations) was created by the Federation Board of Control on October 11, 1941. The primary function of the Committee is to originate and produce, through appropriate subcommittees, special publications dealing with technical aspects of the broad interests of the Federation. These publications are intended to provide background information through a review of technical practices and detailed procedures that research and experience have shown to be functional and practical.

Contents

List of Figures

List of Tables

About the Authors

Rodger B. Baird is Manager of Laboratories for the Los Angeles County Sanitation Districts, where he has worked in laboratory supervision and management for more than 30 years. His primary interests have been in wastewater reclamation and reuse, ocean monitoring, and environmental toxicology, and he has coauthored more than 80 publications and conference presentations. He holds a bachelor's degree in chemistry from Claremont Men's College and a master of science in inorganic chemistry from San Diego State University. He is a Part Coordinator for *Standard Methods for the Examination of Water and Wastewater* and is a member of ACS, WEF, AWWA, and SETAC.

Roy-Keith Smith is a seventh-generation Georgian on his father's side with roots in Warren, Taylor, and Talbot Counties. His great-great-grandfather was a captain in the War between the States, his grandfather was a state representative from Talbot County, and his father was a 1941 graduate in civil engineering from Georgia Tech, retiring as a commander in the Navy. He is a Floridian on his mother's side; his great-grandfather, Albert Keller, built the Saint Augustine water system and his grandfather, Sir Keith Keller, built the pipeline from the mainland to Key West as well as the Clearwater, Florida, water system. The S. K. Keller pumping plant in Pinelas County, Florida, is named after his grandfather.

Faced with a legacy of civil engineering and service to the water industry, this Vietnam veteran graduated from Georgia Tech with a BS in chemistry in 1976 and from Colorado State University with a PhD in chemistry in 1981 before moving on to the research faculty at the California Institute of Technology. After several years with the Georgia Department of Agriculture as a chemist, he was appointed to a position on the faculty at Southern Tech in 1990, then resigned in 1992 to join Analytical Services, Inc., as the Manager for Analytical Methods and Quality Assurance. He retired from Analytical Services in June 2000 to return to teaching, writing, and consulting. He taught chemistry at Pedro Menendez High School, St. Augustine, Florida, and St. Johns River Community College, Palatka, Florida, until his retirement in 2012.

Dr. Smith has written more than 60 published articles and is the author of 11 books (one in a second edition and one in the fourth edition) on environmental analysis. He has served as an expert witness for analytical environmental chemistry and quality control. He is a member of the

Standard Methods Committee, having been the Joint Task Group Chair for 4500-CN, Part Coordinator for Part 4000 (Inorganic Nonmetals) and Part 5000 (Aggregate Organic Constituents) at various times. He is a member of the Water Environment Federation since 1989, the Georgia WPCA, and the Florida WEA.

About the Reviewers

Marlene Patillo graduated from Austin Peay State University in 1981 with a BS in chemistry and biology. She started her environmental career as a laboratory director for the City of Memphis T. E. Maxon Wastewater Treatment Plant. Her career includes environmental research, developing manuals and training courses, pilot-plant studies, teaching environmental seminars, and developing pretreatment programs. She obtained wastewater operator certification in Tennessee and New Jersey. After working in the environmental field for city/county government and hazardous waste treatment facilities, she now is employed by the Maryland Department of the Environment conducting National Pollutant Discharge Elimination System inspections of municipal permittees and laboratories statewide. Marlene serves on the Awards Committee for CWEA, judges laboratory portions of Operations Challenge, is the state representative for NELAC for six years, is the state DMRQA Coordinator and has served on the Standard Methods Committee, specifically for BOD, Section 5210, for five years. She is also the author of Marlene's Mixture (MM), a synthetic wastewater, used in respirometry and oxygen-uptake studies. Marlene earned an Environmental Policy Management Certificate from the University of Maryland in 1998, was inducted to the International Who's Who of Public Service in 2001, and was the recipient of the 1999 WEF Laboratory Analyst Excellence Award from the CWEA.

Dr. Ray C. Whittemore has been with National Council for Air and Stream Improvement, Inc. (NCASI), for 30 years following undergraduate and graduate work in chemical engineering at the University of Maine. He holds the position of Principal Research Engineer at the Northeast Regional Center, now located in Lowell, Massachusetts. While at NCASI, he has worked on water quality issues and served as an instructor in U.S. Environmental Protection Agency (U.S. EPA) Qual2E workshops and now chairs the U.S. EPA MODEL 2000 Action Team devoted to modeling and monitoring. He also sits as an alternate member on the federal Advisory Committee for Water Information. In the 1980s, he developed the ultimate biochemical oxygen demand procedure for *Standard Methods for the Examination of Water and Wastewater* and advanced state-of-the-art technology in sediment oxygen demand measurement. He works with watershed-scale models and Total Maximum Daily Load (TMDL) issues for the forest products industry, and served on the planning committee for WEF's TMDL Science Workshop held in March 2001.

Preface

The protection of streams, lakes, rivers, and the oceans from pollution is one of the major tasks of the U.S. Environmental Protection Agency (U.S. EPA). Although there are many sources of pollution, man-made sources are overwhelmingly the most devastating. Under the authority of the Clean Water Act, the U.S. EPA has instituted a system of permits, the National Pollutant Discharge Elimination System (NPDES), for sources of waste entering into streams, rivers, and lakes (receiving waters). The permits generally list the maximum amount of specific pollutants that can be discharged per day or per month based on the volume of discharged water. The NPDES permit is normally issued to an industrial waste treatment facility or to a publicly owned treatment works (POTW). In addition to the NPDES permits, there are many specifically listed industries that have wastewater effluent guidelines and permits (40 CFR Subchapter N).

There are three classes of recognized pollutants:

1. Conventional pollutants: biochemical oxygen demand (BOD), chemical oxygen demand (COD), pH, total suspended solids (TSS), bacteria, oil and grease, and fecal coliform bacteria;
2. Nonconventional pollutants: nitrogen, phosphorus, ammonia, chloride, sulfate, and others that may endanger water quality; and
3. Toxic pollutants: the so-called 129 priority pollutants listed in 40 CFR 401.15.

There are requirements in the permits to periodically measure each of the pollutants mentioned above, and the approved measurement methods are listed in 40 CFR Part 136. All of the approved methods have several traits in common. They have a rigorously defined mechanism for the determination of the pollutant. There are well-recognized interferences in the method, with approved corrections. There is a specified endpoint (i.e., they give a result that generates a number that means there is no more pollutant beyond that number). They are amenable to specifying quality controls such as sampling, trip, storage and method blanks; laboratory control samples and laboratory fortified samples (matrix spikes); sample duplicates; calibration and calibration checks; determination of the method detection limit; instrument detection limit; operator demonstration of ability; and performance evaluation samples. There are specified sampling procedures and maximum holding time requirements. These traits are true for almost all of the listed approved methods, with the major exception of biochemical oxygen demand.

Biological oxygen demand is exceptional in a number of aspects; however, the most significant is that at the end of the 5-day incubation period the microbial utilization of the organic material in the sample is incomplete. Some estimates suggest up to 75% completion at 5 days; however, that is not a definitive percentage. In my education as an analytical chemist, a hard, definitive, reproducible result is an absolute characteristic of a defensible procedure. Complete oxidation can take up to 120 days according to many references; thus the uncertainty in the 5-day reported number is significant. Small alterations in the procedure can lead to dramatic changes in the results. That useful results can be obtained at all are largely due to the efforts of J. C. Young, who for many years has been the chair of the Standard Methods Joint Task Group for BOD. His work over many editions of *Standard Methods* has resulted in a procedure that gives as consistent results as can be expected given the limitations. However, problems still arise that are more suggestive of a procedural fault rather than any fundamental abnormality in a treatment works design or operation (Muirhead et al., 2006).

1.0 REFERENCE

Muirhead, W. M., Farmer, G., Walker, S., Robb, L., Elmendorf, H., Matthews, R., Butler, R., & Melcer, H. (2006). Study of raw wastewater BOD_5 and $cBOD_5$ relationship yields surprising results. *Proceedings of the Water Environment Federation*. WEF.

1

Introduction

Biochemical oxygen demand (BOD) is one of the fundamental concepts in wastewater treatment. The beginnings of the concept can be traced to Forchamer in 1849, who studied the oxidation of wastewater with potassium permanganate. Through the rest of the 1800s and the 1900s, BOD has been exhaustively studied and refined, both as a concept and as an analytical procedure. In the year 2022, in the third century of wastewater treatment and BOD, it is fitting that all of this previous work be reviewed and summarized. The writings on and about BOD are voluminous; however, much of the literature is relatively obscure or difficult to find. Earlier reviews include the work of Theriault (1927), with 175 references; Forster (1938), with 276 references; Hull (1955), with 302 references; and O'Brien & Clark (1962), with 816 references.

A National Technical Information Service bibliography (Brown, 1977), covering only a 2-year period, included 243 references to BOD. Many of these are reports that include BOD as an evaluation of a treatment process through an achieved percent BOD reduction, a metric introduced in the 1930s and 1940s (Sawyer & Romer, 1943). Assuming this to be an average number of publications, we can estimate that the literature as a whole contains more than 15 000 references in which BOD is mentioned in any context. Adding in the other topics covered in this book, the estimated citations rise to more than 20 000.

The current work has, in general, restricted the literature references to those from peer-reviewed sources, except in cases in which significant observations were published in symposium abstracts and then not followed up in the peer-reviewed literature. There has been no attempt to cite every reference to BOD in the literature. Many papers, especially in the last 40 years, have been reports of treatment process advancements, in which the efficiency of the process is evaluated in terms of BOD, total organic carbon, or chemical oxygen demand reduction. These types of papers are not included, as this work is not intended to be a review of engineering design and operation. Even with these self-imposed limitations, the current work contains more than 2100 citations, and it is almost certain that some relevant papers are overlooked.

Technicians, scientists, facility operators, regulators, and engineers have complained about the BOD test for many years, dating back to the 1930s, immediately after the test was first standardized. Specific complaints include 5 days being an excessive time, the performance of dilutions leading to technician errors, the static nature of the test not modeling for continuous-flow facility operations, and the test giving no indication of waste utilization rates (Banerji, 1971). Many wastewater professionals have predicted the imminent demise of the test (Masselli et al., 1972). But it is still here and is likely to remain so for the forseeable future, given how deeply it is imbedded in the regulatory framework that surrounds the industry.

This work is intended to serve three purposes. The first purpose is to describe BOD as a test procedure and biological phenomenon. The second purpose is to describe the place of BOD within the complex of testing that is used to evaluate treatment processes. The third purpose is to present the development of BOD and preserve, in one location, the literature citations that mark the road to the current test.

1.0 REFERENCES

Banerji, S. K. (1971). Laboratory tests for plant operation control and stream quality measurement. *J. Water Pollut. Control Fed., 43,* 399.

Brown, R. J. (1977). *Biochemical oxygen demand, 1975–May 1977: A bibliography with abstracts.* National Technical Information Service.

Forster, A. S. (1938). Bibliography on BOD. *Kleine Mitt. Mitglied. Ver. Wasser-, Boden-, u. Lufthyg., 14,* 227 (reproduced 1940 in *Chemical Abstracts, 34,* 5215).

Hull, C. H. J. (1955). Bibliography of biochemical oxygen demand. Low Flow Augmentation Project, Department of Sanitary Engineering, Water Resources, Rep. no. 5, Johns Hopkins University (revised 1961).

Masselli, J. W., et al. (1972). BOD? COD? TOD? TOC? *Text. Ind., 136*(9), 53.

O'Brien, W. J., & Clark, J. W. (1962). The historical development of the biochemical oxygen demand test. *Bulletin No. 20, Engineering Experimental Station, New Mexico State University.*

Sawyer, C. N., & Romer, H. (1943). Efficacy of chemical treatment as measured by BOD removal. *Sew. Works Eng.*, *14*, 351.

Theriault, E. J. (1927). The oxygen demand of polluted waters: I. A critical review; II. The rate of deoxygenation. *U.S. Public Health Service Bulletin*, 173.

2

Biochemical Oxygen Demand History

> Analysis of many problems is generally simplified by putting it into historical perspective. (Walker, 1971)

The standard 5-day biochemical oxygen demand (BOD_5) test as currently performed in treatment facility laboratories did not just appear full-blown one day. Nor was it even envisioned by one single person in a moment of brilliant insight. Instead, the BOD_5 test is the culmination of a tremendous number of small steps, both forward and backward, taken by many persons over a long period of time. Three major time periods characterize the history of BOD. The first, which covers the time up until the mid-1930s, is characterized by discovery and description of oxygen utilization in wastewater treatment. The second period, which overlaps with the first period, extends from about 1900 to the 1970s. It is characterized by development and refinement of the measurement of dissolved oxygen and BOD. The third period, which began in 1972 and extends to the present, is characterized by the use of the BOD_5 test as a regulatory compliance and process control tool.

The first small step was taken by Forchamer in 1849 (Foulds & Lunsford, 1968; O'Brien & Clark, 1962), when he determined the amount of potassium permanganate that a sample of polluted water consumes. Even though Forchamer was probably not thinking in terms of oxygen required for neutralization of the wastewater, he is the one who first implemented a chemical oxygen demand experiment. It is Brodie who in 1857 was credited with making the connection that pollutants could be removed from wastewater by oxidation (Brodie, 1865). Another 11 years passed before Frankland, in 1868, observes the absorption of gas by sealed containers of polluted water in an experimental setup that looks much like the modern multiple dilution BOD test in BOD bottles (Frankland, 1870).

The amount of chemical oxidant required for neutralization in a Forchamer experiment did not correlate well with the amount of gas absorbed in a Frankland procedure, and it was not until 1884 that Dupre proposed that microorganisms are the cause of the difference and that oxygen is the link (Dupre, 1885). Not everybody agreed with him (Drown, 1892). Still, the processes could not be fully elucidated until Winkler developed his titration procedure for determination of dissolved oxygen in 1888. Later studies (Muller, 1911; Sand & Trotman, 1912) firmly established the link between microorganisms and oxygen depletion. The next year, 1889, saw the first proposed standard method for the determination of the oxygen-consuming capacity of a polluted water.

Adeny, working with the Royal Commission on Sewage Disposal from 1898 to 1913, wass one of the principal early researchers on the oxygen demands of polluted waters (e.g., Adeny, 1896, 1897, 1905). He began in 1890 with respirometric studies of the oxygen uptake of wastewater and, by 1897, had separated oxygen demand into carbonaceous and nitrogenous components. Other giants on the Royal Commission included McGowan, Frye, and Kershaw. Kershaw's major lasting contribution is his 1915 guide that summarized the voluminous reports. Many abstracts were published from the works of the Royal Commission (Calvert, 1913; Letts, 1908; Letts & Richards, 1911; McGowan, 1908, 1910; McGowan & Frye, 1910; McGowan et al., 1913; Shenton & Easdale, 1914; Thompson, 1909), with much commentary (Rideal, 1913).

It is not just the British who were working on pollution abatement. There were ongoing efforts in most of the Western countries. In 1887, the American Association for the Advancement of Science released a summary of the effects of pollution. In 1899, the American Public Health Association (APHA) Committee on Standard Methods of Water Analysis was formed. In Germany, Spitta showed, in 1900, that oxidations in the light are performed preferentially by algae and diatoms, whereas bacterial oxidations predominate under dark conditions (Spitta, 1900a, 1900b).

The Winkler test was observed to exhibit erroneous results on some polluted water samples. In 1901, Rideal and Stewart described a permanganate modification of the Winkler titration that alleviated some of the problems. The Winkler test was used with an oxygen-consumed (chemical oxygen demand [COD]) test to show that there is an immediate oxygen demand of samples, followed by a slower biologically mediated process (Clark, 1901).

Whether dissolved oxygen (DO) levels or putrescibility results are the most appropriate measures of wastewater purification were not the only choices that were proposed. Brezinas, in 1906, preferred counting bacteria, while Kisskalt (1906) suggested that bacterial numbers are only related to infectivity rather than purification.

Papers detailing the individual steps that researchers took as understanding of a subject area made advances, and books summarized the available information on a topic and allowed students to come up to speed and quickly begin to contribute insight and data, overall providing the major spur to progress. By 1903, several texts described operation of treatment systems and analysis of effluents. These books include *Sanitary Examinations of Water, Air, and Food* by (Fox, 1886), *Handbuch der Wasser, 4th Edition* by Tiemann and Gartner (1895), *Sewage Works Analysis* by Fowler (1902), *The Purification of Sewage and Water, 3rd Edition* by Dibdin (1903), *Chemical Survey of Waters of Illinois* by Palmer (1903), *The Examination of Water and Water Supplies* by Thresh (1904), and *Sewage and Its Purification, 3rd Edition* by Rideal (1906).

The APHA Committee formally published its interim report in 1901 and the final report, on water analysis, in 1905 (APHA, 1901, 1905). This is regarded as the first edition of what is now titled *Standard Methods*. This edition contained an oxygen consumed test, the putrescibility test, and Winkler's original DO titration. Although the putrescibility test underwent many variations in attempts to obtain definitive information, in its original form, a portion of the wastewater was sealed in a glass bottle; then, after 24 hours, it was evaluated by smell and examined for evidence of purification (production of hydrogen sulfide or depletion of available oxygen, in modern terms).

In 1908, Adeny described the first dilution bottle test used for the express purpose of determining the oxygen demand of microorganisms as they utilize pollutants in wastewater (Adeny, 1908a, 1908b, 1908c, 1908d).

The theorists were hard at work also, beginning with Letts et al. (1900). Phelps published the first mathematical treatment of BOD of polluted wastewater in 1909 based on his earlier observations (Phelps & Farrell, 1905). Phelps made the assumption that oxygen utilization is a chemical process, thus

$$\text{Rate} = KL_t$$

where K is a constant and L_t can be interpreted as either the organic matter present at time t or the oxygen requirement of the sample at time t. This expression is integrated and terms are redefined to give

$$S = 1 - k^t$$

where S is the relative stability (available oxygen at time t divided by the total oxygen demand) and k is defined as 10^{-K}. Phelps's basic equations came to be known as the monomolecular model and were expanded, modified, and refined by many researchers over the ensuing years, and even find use to the current day.

Research is fine and dandy, but at some point there has to be a practical application for the research to be useful. In 1911, Hoover described the use of oxygen-consumed tests and 24-hour dilution bottle BOD tests for process control in the Columbus, Ohio, wastewater resource recovery facility (Hoover, 1911; and Hoover & McGuire, 1914). Up to this time, BOD had been used in the context of evaluating natural processes that occur in existing water bodies after the introduction of pollution (Hansen & Hilscher, 1916). Attention began to be paid to modeling the behavior of treatment facilities (Bartow & Mohlman, 1916; Tatham, 1916, 1921).

The British Isles are surrounded by seawater, and the effect of wastewater on the ocean was a concern from the beginning. Questions that needed answers centered on the mode of bacterial action in seawater and the measurement of pollution levels (Harvey, 1925; Lipman, 1922; Purvis & Coleman, 1906; Purvis & Walker, 1912).

The second edition of *Standard Methods* was published in 1912. Along with a reprint of the process tests of the first edition, it included the Rideal–Stewart modification of the Winkler titration (APHA, 1912).

Although one might hope otherwise, human beings perform science and sometimes their personal interests cloud their professional judgments. Such an incident occurred in the BOD history beginning in 1914. Lederer developed an oxygen-demand test that used added nitrate as the source of oxygen. When he came to chair the APHA *Standard Methods* Committee, he dismissed the "English Incubation Test," as Adeny's procedure was called, and proposed replacing it with his own nitrate-based procedure, although others were proponents of the English Test (Bruckmiller, 1916a, 1916b; Lederer, 1914a, 1914b; Mohlman, 1916). The publication of the third edition of *Standard Methods* in 1917 included BOD tests in both the excess oxygen and nitrate versions. The putrescibility test was renamed the relative stability test (APHA, 1917). Also in 1917, Mohlman was the first in a long series of efforts by both himself and many other researchers to describe the effect of germicides and other toxic chemicals on BOD and biologically based facility operations (Mohlman, 1917).

World War I (the Great War of our grandfathers) ended in 1918. Not everyone went to fight the war. Theriault stayed home and researched BOD for the U.S. Public Health Service. His wartime contribution to the development of BOD was the invention of quality control oriented at evaluating the dilution water quality and setting minimum and maximum oxygen-depletion readings for reliable results. Theriault also performed comparative testing of the different procedures of pollution measurement. In 1920, he publishes a strong recommendation for the dissolved oxygen BOD over Lederer's nitrate technique and the relative stability tests (Theriault, 1920).

The fourth edition of *Standard Methods* was published in 1920. There were no changes in either the BOD or DO procedures from the third edition (APHA, 1920). The fifth edition was published in 1923 and included Theriault's suggested quality controls on DO and dilution water. Lederer's nitrate-based procedure, although not removed, was de-emphasized (APHA, 1923). Also in 1923, Theriault proposed that monitoring BOD and total suspended solids (TSS) provided sufficient information for maintaining proper wastewater resource recovery facility operation. Warburg invented the respirometer, which is named after him, although it did not find use in wastewater studies until the 1940s.

The American Water Works Association (AWWA) joined forces with the APHA to publish the sixth edition of *Standard Methods* in 1925. The BOD and DO methods remain unchanged from the fifth edition (APHA & AWWA, 1925). In the same year, Streeter (1925) described the oxygen sag equation and Alsterberg (1925) invented the azide modification of the Winkler titration.

The first critical review of BOD was published in 1927 by Theriault. Reviews are essential in the evolution of science, as they summarize the state of the knowledge and inspire readers to new areas of research. Such was the effect of Theriault's review.

In 1928, Sierp invented the manometer, named after him, and applied it to wastewater samples for immediate measurement of oxygen utilization (Sierp, 1928). Manometers before Sierp's design were difficult to use, and this hampered adaptation by researchers. Respirometric determinations continue to this day.

In 1930, the proceedings of the Standard Methods for Examination of Sewage and Sewage Sludge Symposium were published and led to the "Uniform Procedure for Biochemical Oxygen Demand" by the Federation of Sewage Works Associations (FSWA, 1932). The FSWA changed its name to the Federation of Sewage and Industrial Waste Associations in 1950, Water Pollution Control Federation in 1960, and Water Environment Federation (WEF) in 1991.

In 1931, Eldridge and Mallman (1931) presented a comparison of dilution water that contains phosphate against one made up with bicarbonate.

The seventh edition of *Standard Methods* was published in 1933; the multiple dilution DO–BOD was the only biochemical demand test included.

A bicarbonate solution was recommended over tap water for dilutions, and recommendations for neutralization of alkaline or chlorinated water followed by reseeding were included as was a recommendation for a 5-day incubation time, or correction of the results to a 5-day incubation time (APHA & AWWA, 1933).

In 1936, the eighth edition of *Standard Methods* was published, including *Standard Methods of Sewage Analysis* as an appendix. The BOD procedure contained directions for preparation of separate phosphate-based and bicarbonate-based dilution waters. Adjustment of sample pH of 5 to 9 was recommended. The first modern-looking calculation of BOD results appeared (APHA & AWWA, 1936). In the same year, Wooldridge and Standfast (1936) described the use of the Barcroft manometer for BOD studies, and Lea and Nichols (1936) examined the effects of adding various mineral salts to dilution waters.

Ruchhoft reported on a comparison of the azide and Rideal–Stewart modifications of the Winkler titration in 1938. Petering and Daniels (1938) described the use of a dropping mercury electrode for DO determinations. This marks the first departure from use of visual indicators in the DO titration and eventually led to the direct-reading electrode-based measurements.

World War II began in Europe in 1939. C. N. Sawyer began a long career of publications on oxygen utilization in wastewater treatment. Ruchhoft (1941) reported on a detailed comparison of dilution water formulations, and the United States entered the war later that year. In 1942, Todt described the electrochemical determination of DO using metallic electrodes.

World War II ended in 1945. The paucity of wartime research in wastewater treatment and monitoring was made up by a tremendous infusion of federal funds administered by the U.S. Public Health Service. This resulted in an explosion in published research over the next 10 years.

The ninth edition of *Standard Methods* was released in 1946. The BOD method contained directions for preparation of dilution water with mineral salts, phosphates, and bicarbonates. A dilution BOD method for sludge and mud was included (APHA & AWWA, 1946). In 1947, the FSWA becomes a full partner with APHA and AWWA in the publication of *Standard Methods*, and the Joint Editorial Board was formed. Rhame (1947) described the use of dichromate COD values to estimate BOD values.

The year 1948 is an important year in this history. Abbott (1948) reported the use of methylene blue to inhibit nitrification in BOD. The use of the Warburg manometer for BOD was introduced (Caldwell & Langelier, 1948). Moore et al. (1948) described the use of a polarograph for DO determination. Finally, the Federal Water Pollution Control Act (FWPCA) was signed into law, and the federal and state governments began the long road to regulation of water pollution.

The use of pure chemicals to standardize the BOD procedure was reported by Sawyer et al. in 1950. This was not a new idea, as early work had been performed beginning with Jackson in 1900 and others. However, Sawyer approached the problem in a very methodical manner and developed the glucose–glutamic mixture, which is currently used.

Orford and Ingram, in 1953, used a logarithmic formula to describe oxygen utilization as the first major departure from monomolecular equation treatments (Orford & Ingram, 1953a, 1953b, 1953c). They also developed a formula for the temperature dependence of oxygen utilization by microorganisms. In a reworking of Lederer's 1914 idea, Bryan and Rohlich (1954) described the use of sodium chlorate as the oxygen source in a BOD-type test.

The 10th edition of *Standard Methods* (1955) contained a BOD test that is only slightly different from the modern procedure. Glucose–glutamic acid was used as the chemical standard. The dilution water contained a phosphate buffer, trace minerals, and bicarbonate. Nitrification control using acid or methylene blue was presented. The Warburg apparatus and the Sierp apparatus were described for manometric (respirometric) methods. The dichromate-based oxygen-consumed test (COD) appeared for the first time. Previous versions dating back to the first edition used permanganate.

Hull, in 1955, prepared the second major review of the BOD test since Theriault's review, containing 302 references. Unfortunately, it languished in the thesis archives of Johns Hopkins University in Maryland. Advances that year include Elmore's (1955) reaeration BOD technique. Wheatland and Smith described problems associated with storing dilution water for more than 24 hours (Wheatland and Smith, 1955). Values in oxygen-saturation tables that have been in use for more than 50 years for initial DO values were questioned by Truesdale et al. (1955).

Lee and Oswald (1958) performed experiments on seed preservation through freezing and lyophilization, leading to the commercial availability of standard seed mixtures. In 1959, Carritt and Kanwisher reported use of a membrane-covered polarographic electrode for DO determination, and L. C. Clark patented a similar device. J. W. Clark invented the electrolytic respirometer (O'Brien & Clark, 1962).

The 11th edition of *Standard Methods* appeared in 1960. The use of acclimated seed and soil preparations for seed were added to the BOD procedure. Values for oxygen saturation tables in the DO section were reduced to tenths of a milligram per liter from hundredths in previous editions (APHA et al., 1960).

O'Brien (with Clark) prepared the third major review of BOD for his master's thesis in 1962. It contained 816 references, but was only published as a *New Mexico State University Engineering Experiment Station Bulletin* and remains virtually unknown. The modern galvanic membrane-covered electrode was described by Mancy and Westgarth (1962) and Mancy et al. (1962), revolutionizing DO measurements. Van Hall et al. (1963) published a description of total organic carbon (TOC) determination.

The 12th edition of *Standard Methods* was published in 1965. A method (tentative) for DO measurement using a membrane-covered polarographic electrode was described. In the BOD section, the immediate dissolved oxygen demand (IDOD) was presented and the 10th edition's nitrification control techniques were deleted (APHA et al., 1965). The U.S. Congress passed the Water Quality Act (PL 89-234) that year, which required numeric water quality standards be developed for interstate waterways. The BOD_5 test was listed as a standard. The Federal Water Pollution Control Administration was established as part of the Department of the Interior. The other major advance in the late 1960s was the introduction of the Monod equation to describe bacterial growth in wastewater purification. It and the Haldane expression for inhibited growth replaced the monomolecular formula and became the standard equations that led to development of modern wastewater treatment models.

The U.S. Environmental Protection Agency (U.S. EPA) was established in 1970. The 13th edition of *Standard Methods* was released in 1971. More than 100 years had passed since Frankland's BOD test. The membrane electrode DO measurement became a standard method, and the combustion–infrared TOC technique was introduced as a tentative method (APHA et al., 1971).

Congress overrode a presidential veto, and the Federal Water Pollution Control Act Amendments (the so-called Clean Water Act) became law in 1972. The heart of the Act was the establishment of the National Pollutant Discharge Elimination System (NPDES) of permits. Five-day BOD and TSS, along with fecal coliform bacteria and pH, were designated as conventional pollutants. Almost all of the NPDES permits issued since this time have included effluent limitations for pH, BOD_5, fecal coliform, and TSS. This event moved the monitoring of BOD from a research or process control parameter into the realm of verifying NPDES permit compliance. The Act also required the U.S. EPA administrator to identify appropriate test methods for use in compliance monitoring. This requirement was fulfilled by the U.S. EPA in October 1973 with the promulgation of Title 40, Code of Federal Regulations, Part 136: *Guidelines Establishing Test Procedures for the Analysis of Pollutants*. The BOD_5 procedure from the 13th edition of *Standard Methods* was listed as the approved procedure (APHA et al., 1971).

Young (1973) reported on nitrification control using allyl thiourea, methylene blue, and tetrachloromethylpyridine. He later (1977) wrote a comprehensive book on BOD that was not published.

The 14th edition of *Standard Methods* became available in 1975. The entire book was reorganized to combine similar water and wastewater methods, which up to then have been kept in separate sections. The determination of DO using membrane-covered electrodes became a standard method. In the BOD method, reference was made to nitrification techniques but specific directions were not included. Sample preservation by cooling to less than 4 °C and a 24-hour maximum holding time was introduced (APHA et al., 1975).

Although the state of California began issuing BOD_5 performance evaluation samples earlier in the 1970s, U.S. EPA-sponsored nationwide laboratory proficiency evaluation samples began in 1978 with the issuance of the first *Water Pollution Quality Assurance Study* (WP 001). A BOD_5 test sample was included in the study. Two years later, in the fall of 1980, U.S EPA initiated the *Discharge Monitoring Report QA Studies* for holders of NPDES permits. The first such performance evaluation study, DMR QA 001, contains a BOD_5 test sample.

The 15th edition of *Standard Methods* (1980) described carbonaceous BOD with nitrification inhibition by 2-chloro-6 (trichloromethyl) pyridine (TCMP). The 16th edition of *Standard Methods* (1985) deleted the IDOD procedure from the BOD method. Total organic carbon determinations by the persulfate–UV oxidation and wet-oxidation were introduced along with miniaturized closed reflux methods using spectrophotometric or titration finishes for dichromate COD.

The 17th edition (1989) and the 18th edition (1992) of *Standard Methods* made no changes to the demand or DO methods. The 19th edition of *Standard Methods* (1995) finalized ultimate BOD and respirometric BOD as standard methods. The 20th edition (1998) made no changes to the demand or DO methods.

In 2002, Baird and Smith prepared the first edition of this book, a comprehensive text on BOD and related techniques, containing more than 2100 references. This was the first readily available in-depth treatment of the subject since the seminal work of Theriault in 1927, a period of almost 75 years.

The 21st edition of *Standard Methods* was published in 2005, and the 22nd edition in 2012. Quality control/quality assurance guidance (5020) was rewritten, and procedures for 5-day BOD (5210 B), ultimate BOD (5210 C), and respirometric BOD (5210 D) were revised to include the expanded quality assurance/quality control (QA/QC) measures. This edition was approved by the U.S. EPA for use by NPDES permit holders in 2017. The 23rd edition of *Standard Methods* was published in 2017, with enhanced quality control procedures throughout the edition.

Such is the official history of BOD (see Table 2.1). But the question that almost all persons ask has not been answered, and that is, "Why 5 days?"

TABLE 2.1. Timeline of biochemical oxygen demand and *Standard Methods*.

Year	Event
1849	Potassium permanganate method for determining oxidizable substances in wastewater
1857	Organic pollutants in wastewater can be removed by oxidation
1868	Frankland performs first BOD test
1884	Microorganisms use dissolved oxygen in metabolic processes explains differences between BOD and oxygen-consumed test
1888	Winkler titration method for dissolved oxygen
1889	"A Method, in Part, for the Sanitary Examination, and for the Statement of Results, Offered for General Adoption." Chemical Section of American Association for the Advancement of Science included a discussion of oxygen-consuming capacity
1897	Carbonaceous and nitrogenous oxygen demand recognized
1898–1913	Royal Commission on Sewage Disposal sponsors research on biochemical wastewater treatment and monitoring techniques
1899	APHA Committee on Standard Methods of Water Analysis formed
1900	Differences in oxidation of wastewater by algae and bacteria discovered
1901	Potassium permanganate modification of Winkler titration
1905	First edition of *Standard Methods* published—contains oxygen-consumed test, putrescibility test, and Winkler's original DO test
1908	Dilution bottle BOD technique
1909	Monomolecular equation
1911	Application of oxygen-consumed and 24-hour dilution BOD tests for wastewater treatment
1912	Second edition of *Standard Methods* published—Rideal–Stewart modified, Winkler test included
1914	APHA committee dismisses "English incubation test" in favor of nitrate determination of BOD
1917	Third edition of *Standard Methods* published—both excess oxygen and nitrate versions of BOD presented; putrescibility test renamed the relative stability test; effect of germicidal agents on oxygen demand tests described
1918	Quality control measures for dilution water and minimum/maximum DO residuals introduced
1920	Fourth edition of *Standard Methods* published—no changes in BOD or DO methods; DO-based BOD favored over nitrate-based BOD or relative stability test
1923	Fifth edition of *Standard Methods* published—nitrate method deemphasized; dilution water quality controls introduced to dissolved oxygen BOD as well as DO depletion and residual requirements; TSS and BOD proposed for process control; Warburg respirometer invented

(*continued*)

TABLE 2.1. Timeline of biochemical oxygen demand and *Standard Methods* (*continued*).

Year	Event
1925	Sixth edition of *Standard Methods* published—AWWA and APHA joint publication; no changes in BOD or DO methods; oxygen sag equation and azide modification of Winkler titration for DO published
1927	Theriault, *The Oxygen Demand of Polluted Waters,* Public Health Bulletin No. 173 published
1928	Sierp manometer invented; BOD Committee of Great Lakes and Ohio River Boards of Public Health Engineers appointed
1930	*Symposium: Standard Methods for the Examination of Sewage and Sewage Sludge* published
1931	Interactions between bacteria and plankton in BOD described; phosphate-based and bicarbonate-based dilution waters compared
1932	*Recommended Uniform Procedure for Biochemical Oxygen Demand* published
1933	Seventh edition of *Standard Methods* published—dilution BOD only test included, bicarbonate solution rather than tap water used for dilution water, and neutralization of alkaline or chlorinated waters followed by reseeding recommended
1935	*Standard Methods of Sewage Analysis* published, Federation of Sewage Works A sociations (now WEF) established
1936	Eighth edition of *Standard Methods* published—including *Standard Methods of Sewage Analysis:* separate phosphate and bicarbonate dilution waters and pH adjustment to 5–9 presented; modern-style BOD calculation appears; Barcroft differential manometer for BOD and initial studies of mineral salts in dilution waters reported
1938	Azide and Rideal–Stewart modifications of Winkler titration compared; dropping mercury electrode for DO described
1941	Comparative study of dilution waters
1942	Electrochemical determination of DO with metallic electrodes
1945	World War II ends and research in wastewater treatment fueled by copious federal funds distributed through U.S. Public Health Service
1946	Ninth edition of *Standard Methods* published—preparation of dilution water with trace minerals, phosphate and bicarbonate salts described; dilution method for BOD of sludges and muds presented
1947	Federation of Sewage Works Associations becomes full partner with AWWA and APHA in publishing *Standard Methods;* Joint Editorial Board formed; BOD values and dichromate COD values compared
1948	Nitrification inhibition with methylene blue, Warburg manometer for BOD, and polarograph for DO determinations reported; Federal Water Pollution Control Act established

(*continued*)

TABLE 2.1. Timeline of biochemical oxygen demand and *Standard Methods* (*continued*).

Year	Event
1950	Primary chemical standards for BOD evaluation
1953	Logarithmic formula for oxygen utilization and temperature dependence expression
1954	BOD test using sodium chlorate as the oxygen source
1955	10th edition of *Standard Methods* published—dilution test is the standard for BOD, glucose–glutamic acid standard is introduced, a phosphate buffering system added to the dilution water, nitrification control recommended by pasteurization or chemical (acid or methylene blue), dichromate COD appears, and manometric methods using the Warburg and Sierp apparatuses are presented; Hull's review published; reaeration BOD technique, studies on stored dilution water, and oxygen-saturation tables questioned
1956	Acclimated seeds and effect of molds on BOD determinations
1958	Seed preservation by freezing and lyophilization
1959	Polarographic membrane electrode for DO and electrolytic respirometer invented
1960	11th edition of *Standard Methods* published—use of acclimated seed or soil suspension introduced and oxygen-saturation table values shortened to tenths from hundredths
1962	O'Brien and Clark review history of BOD, galvanic membrane electrode for DO measurement
1963	Total organic carbon method
1965	12th edition of *Standard Methods* published—tentative polarographic method with membrane-covered electrode for DO measurement, immediate dissolved oxygen demand (IDOD) included; dichromate COD presented, nitrification control deleted; Monad equation introduced for bacterial growth in wastewater
1971	13th edition of *Standard Methods* published—membrane electrode measurement of DO introduced as a tentative method (method 218F), TOC by combustion-infrared introduced as tentative method
1972	Federal Water Pollution Control Act (FWPCA) reauthorized
1973	Nitrification control with allyl thiourea, TCMP; 40 CFR 136 promulgated
1975	14th edition of *Standard Methods* published—DO electrode fully approved, reference made to nitrification inhibition but no directions, separation of water and wastewater methods discontinued, sampling directions and holding times for BOD introduced
1977	Clean Water Act (reauthorization of the FWPCA)
1978	WP 001
1980	15th edition of *Standard Methods* published—TCMP included for nitrification control (CBOD); DMR QA Study 001

(*continued*)

TABLE 2.1. Timeline of biochemical oxygen demand and *Standard Methods* (*continued*).

Year	Event
1985	16th edition of *Standard Methods* published—IDOD procedure eliminated, TOC by persulfate–UV oxidation and wet-oxidation introduced, COD by closed reflux introduced
1987	Clean Water Act reauthorized
1989	17th edition of *Standard Methods* published—no changes to demand or DO methods
1992	18th edition of *Standard Methods* published—no changes to demand or DO methods
1995	19th edition of *Standard Methods* published—ultimate BOD and respirometric BOD become finalized methods
1998	20th edition of *Standard Methods* published—no changes to demand or DO methods
2001	*Third Century of BOD*, first edition, published
2005	21st edition of *Standard Methods* published
2012	22nd edition of *Standard Methods* published
2017	U.S. EPA approves 22nd edition of *Standard Methods*; 23rd edition of *Standard Methods* published

There are many stories about why 5 days is the standard incubation time. The most commonly encountered is that it takes 5 days for water to flow down the Thames River from London to the sea. A variation on this is that it takes a maximum of 5 days for water to flow from the most land-bound treatment facility in the British Isles to the ocean (LeBlanc, 1974a and 1974b), the idea being that once waste reaches the sea, DO content is no longer an issue. Another story is that nitrification does not begin until at least 5 days have passed. And another version is that all of the carbonaceous demand is completed by 5 days. Still another story dates back to when the standard work week was Monday until noon on Saturday; the BOD was set up on Monday morning, then read back on Saturday morning, 5 days later. Another story concerns the 5 days' average residence time of influent in a treatment facility.

A few facts may lay these ideas to rest. The distance from London to the coast is about 50 km (30 miles), and 5 days' travel time suggests a very slow-moving river (0.40 km/h) [0.25 mph]), which the Thames is not. As a comparison, the water in the lower Mississippi River, the "lazy" part of the Mississippi, moves at a velocity of approximately 2.2 km/h (1.4 mph) (Waldon, 1998). Most of the early work on BOD from the British described incubation times from 24 hours up to 20 or more days, 10 days being the most common. Although the 5-day maximum travel time from remote treatment facilities to the sea is plausible, there is no indication that at least up into the late 1920s there was any implementation of this idea. In fact, the selection of 5 days was frequently criticized in England in favor of a more useful 48-hour test (Cooper & Read, 1927). Neither of these reasons has any bearing on why 5 days is the standard incubation time for tests conducted in the United States, nor are they cited in any of the papers and discussions leading up to the publication of the seventh edition of *Standard Methods.*

Nitrification begins almost immediately if the appropriate organisms are present. That is why, if carbonaceous BOD is measured, a nitrification inhibitor is added at the beginning of the test. The ultimate demand of a wastewater is not substantially satisfied until the 20th day of incubation, with complete oxidation requiring up to 100 days.

The bacterial population in a dilution bottle maximizes within 30 hours, while the protozoan population can take 6 or more days to maximize, to head off other avenues of speculation. Oxygen utilization in a sample can reach a steady endogenous state in some cases as soon as 5 days, but it is more commonly longer.

The answer to "Why 5 days?" is found in the committee discussions that led up to the suggested standard procedure for BOD published by the FSWA in 1932. The joint BOD Committee of the Great Lakes and Ohio River Boards of Public Health Engineers was appointed in December 1928 and was composed of C. R. Cox, F. E. Daniels, R. D. Scott, E. H. Parks, H. A. Whittaker, E. J. Theriault, and L. F. Warrick (Chairman). An FSWA Committee on Methods of Sewage Analysis, appointed in January 1930, served as a review body and was composed of A. M. Buswell, W. D. Hatfield, J. J. Hinman, F. W. Mohlman, W. Rudolfs, and E .J. Theriault (Chairman).

The overall objective of a BOD test is to estimate the length of time that is required to completely neutralize (oxidize) polluted water, implying that retention time could be adjusted within the facility to achieve complete oxidation. In the 1930 symposium of the membership of both committees, Mohlman comments,

> It is now known that it is not correct to state, the complete demand is usually satisfied in 20 days at 20 °C. The first, or carbon-oxidation stage, is usually complete in less than 20 days at 20 °C, and the 5-day result bears a fairly constant relationship to the total first stage. Incubation for 5 days at 20 °C should be made standard. (FSWA, 1930)

And in the uniform procedure that appeared in 1932, in which the 5-day standard was published, it states,

> The 5-day demand, on the whole appears to bear the most definite relation to the values obtained (for complete oxidation) over other

> periods of incubation. For that reason it is strongly recommended that this period of incubation be adopted whenever the examinations are restricted to a single interval. In this connection it is suggested that valuable information may frequently be obtained by conducting the test over several periods of incubation and plotting the data so as to obtain a complete deoxygenation curve. (FSWA, 1932)

As a result of these developments, the seventh edition of *Standard Methods* (APHA & AWWA, 1933) is the earliest edition in which a 5-day incubation time is recommended.

So the answer to "Why 5 days?" is that total demand estimates calculated from the 5-day values are more consistent than other incubation time periods. The reason this answer has become obscure is that no one uses the 5-day value to estimate anything anymore. Instead, the BOD_5 measurement has become an endpoint in itself rather than being a step on the way to a process objective.

1.0 REFERENCES

Abbott, W. E. (1948). The bacteriostatic effect of methylene blue on the BOD test. *Water Sew. Works*, *95*, 424; *Public Health Eng. Abstr.*, *30*(3), 21 (1950).

Adeny, W. E. (1896). Recent advances in the bacterio-chemical study of sewage and other polluted waters. *Trans. Civ. Eng. Ireland*, *25*, 45; *Engineering* (London), *61*, 728, 762 (1908).

Adeny, W. E. (1897). The course and nature of fermentative changes in natural and polluted waters and in artificial solutions as indicated by the composition of the dissolved gases. Parts I–IV. *Trans. Royal Dublin Soc.*, 1895.

Adeny, W. E. (1905). Chemical changes attending the aerobic bacterial fermentation of simple organic substances. I. Urea, asparagin, albumose, and rochelle salt. *Proc. Royal Irish Acad.; Royal Commission Fifth Rep.*, App. VI, 13 (1908).

Adeny, W. E. (1908a). Nature and rate of the chemical changes which occur in polluted fresh and salt water. Addenum: Capacities of fresh and of sea water to bear pollution within limits of fouling. *Royal Commission Fifth Rep.*, App. I, 1, 433.

Adeny, W. E. (1908b). Reprint of papers by Adeny, 1895–1897. *Royal Commission Fifth Rep.*, App. VII, 5, 49.

Adeny, W. E. (1908c). Studies in the chemical analysis of fresh and salt waters. I. Applications of the aeration method of analysis to the study of river waters. *Royal Commission Fifth Rep.*, App. VII, 95.

Adeny, W. E. (1908d). Testimony by Adeny. *Royal Commission Fifth Rep.*, App. I, 665.

Alsterberg, G. (1925). Methods for the determination of elementary oxygen dissolved in water in the presence of nitrite. *Biochem. Z.*, *159*, 36.

American Association for the Advancement of Science. (1887). Preliminary committee report. *J. Anal. Chem.*, *3*, 398.

American Public Health Association. (1901). Report of the committee on standard methods of water analysis. *Public Health Papers Rep.*, *27*, 377.

American Public Health Association. (1905). *Standard methods for the examination of water and sewage. J. Infect. Dis.* Suppl. No. 1.

American Public Health Association. (1912). *Standard methods for the examination of water and sewage.* (2nd ed.).

American Public Health Association. (1917). *Standard methods for the examination of water and sewage.* (3rd ed.).

American Public Health Association. (1920). *standard methods for the examination of water and sewage* (4th ed.).

American Public Health Association (1923). *Standard methods for the examination of water and sewage.* (5th ed.).

American Public Health Association & American Water Works Association. (1925). *Standard methods for the examination of water and sewage.* (6th ed.). American Public Health Association.

American Public Health Association & American Water Works Association. (1933). *Standard methods for the examination of water and sewage* (7th ed.). American Public Health Association.

American Public Health Association & American Water Works Association. (1936). *Standard methods for the examination of water and sewage* (8th ed.). American Public Health Association.

American Public Health Association & American Water Works Association. (1946). *Standard methods for the examination of water and sewage* (9th ed.). American Public Health Association.

American Public Health Association, American Water Works Association, & Federation of Sewage and Industrial Wastes Associations. (1955). *Standard methods for the examination of water and wastewater* (10th ed.). American Public Health Association.

American Public Health Association, American Water Works Association, & Water Pollution Control Federation. (1960). *Standard methods for the examination of water and wastewater* (11th ed.). American Public Health Association.

American Public Health Association, American Water Works Association, & Water Pollution Control Federation (1965). *Standard methods for the examination of water and wastewater* (12th ed.). American Public Health Association.

American Public Health Association, American Water Works Association, & Water Pollution Control Federation. (1971). *Standard methods for the examination of water and wastewater* (13th ed.). American Public Health Association.

American Public Health Association, American Water Works Association, & Water Pollution Control Federation. (1975). *Standard methods for the examination of water and wastewater* (14th ed.). American Public Health Association.

American Public Health Association, American Water Works Association, & Water Pollution Control Federation. (1980). *Standard methods for the examination of water and wastewater* (15th ed.). American Public Health Association.

American Public Health Association, American Water Works Association, & Water Pollution Control Federation. (1985). *Standard methods for the examination of water and wastewater* (16th ed.). American Public Health Association.

American Public Health Association, American Water Works Association, & Water Pollution Control Federation. (1989). *Standard methods for the examination of water and wastewater* (17th ed.). American Public Health Association.

American Public Health Association, American Water Works Association, & Water Environment Federation. (1992). *Standard methods for the examination of water and wastewater* (18th ed.). American Public Health Association.

American Public Health Association, American Water Works Association, & Water Environment Federation. (1995). *Standard methods for the examination of water and wastewater* (19th ed.). American Public Health Association.

American Public Health Association, American Water Works Association, & Water Environment Federation. (1998). *Standard methods for the examination of water and wastewater* (20th ed.). American Public Health Association.

American Public Health Association, American Water Works Association, & Water Environment Federation. (2005). *Standard methods for the examination of water and wastewater* (21st ed.). American Public Health Association.

American Public Health Association, American Water Works Association, & Water Environment Federation. (2012). *Standard methods for the examination of water and wastewater* (22nd ed.). American Public Health Association.

American Public Health Association, American Water Works Association, & Water Environment Federation. (2017). *Standard methods for the examination of water and wastewater* (23rd ed.). American Public Health Association.

Bartow, E., & Mohlman, F. W. (1916). Purification of sewage by aeration in the presence of activated sludge. *Univ. Ill. Water Surv. Ser. Bull.* No. 13.

Brezinas, E. (1906). Die Donau von Leopoldberge bix Pressburg, die abwasser der stadt Wien und Deren Schicksal nach ihren einmundung in den strom. *Zeit. f. Hyg. 53*, 369.

Brodie, B. (1865). Minutes of evidence, Royal Sewage Commission of 1857. *Rep. Thames River*, 2, 49.

Bruckmiller, F. W. (1916a). Nitrates and oxygen demand. *J. Ind. Eng. Chem.*, *8*, 899; *Chemical Abstracts*, *10*, 3122 (1916).

Bruckmiller, F. W. (1916b). Oxygen demand of sewage. *J. Ind. Eng. Chem.*, *8*, 403.

Bryan, E. H., & Rohlich, G. A. (1954). Biological reduction of sodium chlorate as applied to measurement of sewage BOD. *Sew. Ind. Wastes*, *26*, 1315.

Caldwell, D. H., & Langelier, W. F. (1948). Manometric measurement of the biochemical oxygen demand of sewage. *Sew. Works J.*, *20*, 202.

Calvert, H. T. (1913). Eighth report of the Royal Commission on Sewage Disposal, 1898–1913. Standards and tests for sewage and sewage effluents discharging into rivers and streams. *J. Soc. Chem. Ind.*, *32*, 265; *Chemical Abstracts*, *7*, 2636 (1913).

Carritt, D. E., & Kanwisher, J. W. (1959). An electrode system for measuring dissolved oxygen. *Anal. Chem.*, *31*, 5.

Clark, H. W. (1901). *Massachusetts State Board Health Report, 1900–1901*, 373.

Clark, L. C. (1959) Electrochemical device for chemical analysis. U.S. Patent 2,913,386.

Cooper, E. A., & Read, W. H. (1927). Studies on the dissolved oxygen absorption test. *J. Soc. Chem. Ind.*, *46*, 154T.

Dibdin, W. J. (1903). *The purification of sewage and water* (3rd ed.). Van Nostrand Reinhold.

Drown, T. M. (1892). The effect of aeration of natural waters. *23rd Annual Report, Massachusetts State Board Health.*

Dupre, A. (1885). The estimation of dissolved oxygen in water. *Analyst, 10,* 156.

Eldridge, E. F., & Mallman, W. L. (1931). The influence of diluting waters on the biochemical oxygen demand. *Mich. State. Coll. Eng. Exp. Stn. Bull., 39,* 3.

Federation of Sewage Works Associations Committee Report. (1930). Symposium: Standard methods for the examination of sewage and sewage sludge. *Sew. Works J., 2,* 347.

Federation of Sewage Works Associations Committee Report: Joint Biochemical Oxygen Demand Committee of the Great Lakes and Ohio River Boards of Public Health Engineers. (1932). Progress report: Uniform procedure for biochemical oxygen demand. *Sew. Works J., 4,* 738.

Foulds, J. M., & Lundsford, J. V. (1968). An analysis of the COD method. *Water Sew. Works, 115,* 112.

Fowler, G. J. (1902). *Sewage works analysis.* Wiley & Sons.

Fox, C. B. (1886). *Sanitary examinations of water, air, and food.* J. and A. Churchill.

Frankland, E. (1870). *First report of the rivers pollution commission of 1868,* 18.

Hansen, P., & Hilscher, P. (1916). Pollution of Vermillion River at and below Streator. *Ill. State Water Surv. Bull.* Ser. B, 234; *Chemical Abstracts, 11,* 706 (1917).

Harvey, H. W. (1925). Oxidation in sea water. *J. Mar. Biol. Assoc., 13,* 953.

Hoover, C. B. (1911). A method for determining the parts per million of dissolved oxygen consumed by sewage and sewage effluents, Columbus Sewage Works. *Eng. News-Rec., 65,* 311.

Hoover, C. B., & McGuire, C. D. (1914). Experiments on the oxidation of sewage at Columbus, Ohio. *Eng. News-Rec., 71,* 1187.

Hull, C. H. J. (1955). Bibliography of biochemical oxygen demand. Low Flow Augmentation Project, Dep. Sanit. Eng. Water Resour., Rep. No. 5. Johns Hopkins University (revised 1961).

Kisskalt, K. (1906). Die verunreinigung der lahn und der wieseck durch die abwasser der stadt giesser, mit besonderen beruchtsichtigung der brauchbarkeit der ublichen methoden zur untersuchung von flussverunreinigung. *Zeit. f. Hyg., 53,* 305.

Lea, W. L., & Nichols, M. S. (1936). Influence of substrate on biochemical oxygen demand. *Sew. Works J.*, *8*, 435.

LeBlanc, P. J. (1974a). A review of the BOD_5 test. *Pulp Paper Mag. Can.*, *75*, 3, 79.

LeBlanc, P. J. (1974b). Review of rapid BOD test methods. *J. Water Pollut. Control. Fed.*, *46*, 2202.

Lederer, A. (1914a). A new method for determining the relative stability of sewage effluent, or polluted river water. *J. Infect. Dis.*, *14*, 482.

Lederer, A. (1914b). A serious fallacy of the "standard" methylene blue putrescibility test. *Am. J. Public Health*, *4*, 421.

Lee, E. W., & Oswald, W. J. (1958). Preserved inoculum for BOD determinations. Univ. Calif. Inst. Eng. Res.

Letts, E. A. (1908). The chemical changes which occur when potassium nitrate is added to an effluent from a septic tank. Royal Comm., Fifth Rep., App. VI, 182.

Letts, E. A., & Richards, F. H. (1911). The chemical changes occurring in sea water polluted with sewage, compared with those which take place in correspondingly polluted fresh water. Royal Comm., 7th Rep., 2, App. III, 110.

Letts, E. A., et al. (1900). On the nature and speed of the chemical changes occurring in mixtures of sewage and sea water. *Proc. Royal Dublin Soc.*, 9(3), 24.

Lipman, C. B. (1922). Does nitrification occur in sea water? *Science*, *56*, 501.

Mancy, K. H., Okun, D. A., & Reilley, C. N. (1962). A galvanic cell oxygen analyzer. *J. Electroanal. Chem.*, *4*, 65.

Mancy, K. H., & Westgarth, W. C. (1962). A galvanic cell oxygen analyzer. *J. Water Pollut. Control Fed.*, *34*, 1037.

McGowan, G. (1908). Standards and tests for sewage effluents. Royal Comm., 5th Rep., vol. 2, 217.

McGowan, G. (1910). Memorandum as to the best method of expressing or measuring strength of sewage as affecting purification. Royal Comm., 5th Rep., App. IV, 1.

McGowan, G., & Frye, C. C. (1910). Memorandum on the estimation of the work done by sewage filters. Royal Comm., 5th Rep., App. IV, 10.

McGowan, G., Frye, C. C., & Kershaw, G. B. (1913). Standards and tests for sewage and sewage effluents discharging into rivers and streams. Royal Comm., 8th Rep., vol. 2, 62.

Mills, E. J., Jr., & Stack, V. T., Jr. (1956). Biological oxydation parameter applied to industrial wastes. *Ind. Eng. Chem.*, *48*, 260.

Mohlman, F. W. (1916). The English incubation tests for the putrescibility of sewage and sewage effluents. *Univ. Ill. Water Surv. Ser. Bull.*, *13*, 315.

Mohlman, F. W. (1917). Copper in sewage at the New Haven sewage experiment station. *J. Ind. Eng. Chem.*, *9*, 1093.

Moore, E. W., Morris, J. C., & Okum, D. A. (1948). The polarographic determination of dissolved oxygen in water and sewage. *Sew. Works J.*, *20*, 1041.

Muller, A. (1911). The relation of oxygen loss in natural waters and artificial nutrient solutions to bacterial growth. *Arb. Kais. Gesundh.*, *38*, 294; *Chemical Abstracts*, *6*, 783 (1912).

O'Brien, W. J., & Clark, J. W. (1962). The historical development of the biochemical oxygen demand test. *Bulletin No. 20, Engineering Experimental Station, New Mexico State University*.

Orford, H. E., & Ingram, W. T. (1953a). Deoxygenation of sewage. I. Critical review of the monomolecular formula. *Sew. Ind. Wastes*, *25*, 419.

Orford, H. E., & Ingram, W. T. (1953b). Deoxygenation of sewage. II. The logarithmic formula as applied to sewage. *Sew. Ind. Wastes*, *25*, 424.

Orford, H. E., & Ingram, W. T. (1953c). Deoxygenation of sewage. III. The logarithmic formula applied to oxidations other than sewage. *Sew. Ind. Wastes*, *25*, 566.

Palmer, A. W. (1903). *Chemical Survey of Waters of Illinois, 1897–1903*. Univ. of Ill.

Petering, H. C., & Daniels, F. (1938). The determination of dissolved oxygen by means of the dropping mercury electrode with applications in biology. *J. Am. Chem. Soc.*, *60*, 2796.

Phelps, E. B. (1909). The disinfection of sewage and sewage filter effluents. *U.S. Geol. Surv. Water Supply Paper*, *229*, 74.

Phelps, E. B., & Farrell, F. W. (1905). The mode of action of the contact filter. *J. Infect. Dis. Suppl.*, *1*, 229.

Purvis, J. E., & Coleman, C. J. (1906). The influence of sea water on the decomposition of sewage. *J. Royal Sanit. Inst.*, *28*, 433.

Purvis, J. E., & Walker, G. R. (1912). The chemical and bacterial changes occurring in sewage when mixed with sea water. *J. Royal Sanit. Inst.*, *33*, 174.

Rhame, G. A. (1947). Determination of BOD values by chemical oxidation. *Water Sew. Works*, *94*, 192.

Rideal, S. (1906). *Sewage and its purification* (3rd ed.). Scott, Greenwood, & Sons.

Rideal, S. (1913). Review of the eighth report of the Royal Commission on Sewage Disposal. *Wasser u. Abwasser*, *7*, 397.

Rideal, S., & Stewart, G. G. (1901). The determination of dissolved oxygen in waters in the presence of nitrites and of organic matter. *Analyst*, *26*, 141.

Ruchhoft, C. C. (1938). The BOD test and oxidation in sewage treatment. *Sew. Works J.*, *10*, 264.

Ruchhoft, C. C. (1941). Report on the cooperative study of dilution waters made for the Standard Methods Committee of the Federation of the Sewage Works Association. *Sew. Works J.*, *13*, 669.

Sand, H. J. S., & Trotman, S. R. (1912). Measurement of absorption of oxygen by sewage effluents. *J. Soc. Chem. Ind.*, *31*, 1166.

Sawyer, C. N. (1941). BOD removal from waste sulfite liquor–sewage mixtures by activated sludge: factors governing the rate of removal. *Ind. Eng. Chem.*, *33*, 411.

Sawyer, C. N., Callejas, P., Moore, M., & Tom, A. Q. Y. (1950). Primary standards for BOD work. *Sew. Ind. Wastes*, *22*, 26.

Sawyer, C. N., Frame, J. D., & Wold, J. P. (1955). Revised concepts of biological treatment. *Sew. Ind. Wastes*, *27*, 929.

Sawyer, C. N., & Williams, A. W. (1942). The selection of a dilution water for the determination of the BOD of industrial wastes. *Sew. Works J.*, *14*, 1000.

Shenton, H. C. H., & Easdale, W. C. (1914). Royal Commission on Sewage Disposal: Extracts from the appendix to the eighth report. *Surveyor*, *44*, 581, 629.

Sierp, F. (1928). A new method for determining BOD. *J. Ind. Eng. Chem.*, *20*, 247.

Spitta, O. (1900a). Untersuchung uber die verunreinigung und selabstreinigung der flusse. II. Oxydative vorgunge im flusswasser. *Arch. f. Hyg.*, *38*, 215.

Spitta, O. (1900b). Untersuchung uber die verunreinigung und selbstreinigung der flusse. III Verhaltnisse des flussbodens. *Arch. f. Hyg.*, *38*, 266.

Streeter, H. W. (1925). The rate of atmospheric reaeration of sewage polluted streams. *Proc. Am. Soc. Civ. Eng.*, *51*, 1829.

Tatham, G. T. P. (1916). Some theoretical considerations bearing on the performance of biological sewage purification plants. *J. Soc. Chem. Ind., 35*, 711.

Tatham, G. T. P. (1921). Some practical applications of the theory of biological purification of waste liquors. *J. Soc. Chem. Ind., 40*, 201.

Theriault, E. J. (1920). A further study of the excess oxygen method for the determination of the biochemical oxygen demand of sewage and industrial wastes. *Public Health Rep., 35*, 1087.

Thompson, J. T. (1909). The fifth report of the Royal Commission on Sewage Disposal. *J. Soc. Chem. Ind., 28*, 72.

Thresh, J. C. (1904). *The examination of waters and water supplies*. P. Blakiston's Son & Co., Inc.

Tiemann, F., & Gartner, A. (1895). *Handbuch der wasser* (4th ed.). F. Vieweg und Sohn.

Todt, F. (1942). The electrochemical determination of dissolved oxygen in water. *Gesundh.-Ing.* (Germany), *65*, 76.

Truesdale, G. A., Downing, A. L., & Lowden, G. F. (1955). The solubility of oxygen in pure water and sea water. *J. Appl. Chem* (Great Britain), *5*, 53.

U.S. Environmental Protection Agency. (1973). Title 40, Code of Federal Regulations, Part 136, *Guidelines Establishing Test Procedures for the Analysis of Pollutants*.

U.S. Environmental Protection Agency. (2017). Federal Register 82 40846, August 28, 2017.

van Hall, C. E., Safranko, J., & Stenger, V. A. (1963). Rapid combustion method for the determination of organic substances in aqueous solutions. *Anal. Chem., 35*, 315.

Waldon, M. G. (1998). Time-of-travel in the lower Mississippi River: Model development, calibration, and application. *Water Environ. Res., 70*, 1132.

Walker, W. R. (1971). Law. *J. Water Pollut. Control Fed., 43*, 1376.

Wheatland, A. B., & Smith, R. G. (1955). BOD test: A note on variable results from the use of stored dilution water. *Analyst, 80*, 899.

Winkler, L. W. (1888). The determination of dissolved oxygen in water. *Berlin. Deutsch. Chem. Gellsch., 21*, 2843.

Wooldridge, W. R., & Standfast, A. F. B. (1936). Use of the Barcroft differential manometer in the estimation of the oxygen demand of sewage. *Biochem. J., 30*, 141.

Young, J. C. (1973). Chemical methods for nitrification control. *J. Water Pollut. Control Fed.*, *45*, 639.

Young, J. C. (1977). Biochemical oxygen demand: measurement and application. Unpublished manuscript.

3

Putrescibility and Relative Stability

The putrescibility, or relative stability, test arose in the late 1890s and early 1900s as an empirical procedure to determine if a wastewater effluent is satisfactorily neutralized or purified. It consists of sealing a small portion of the wastewater in a bottle and then incubating it for a period of time, 24 hours being common. After incubation, the bottle is opened and the contents are examined by eye and nose. Samples that have consumed all of the available oxygen and produced hydrogen sulfide, detected by the characteristic odor of rotten eggs, are judged not to be stable and need more time for purification.

Many modifications of the putrescibility test were presented in an effort to move away from a "feel-good" qualitative result. One direction was toward use of chemical reagents to detect the production of sulfide in the putrefying sample. Lead salts are added to the sample to more easily visualize the production of sulfide by the formation of black lead sulfide (Stoddard, 1901). Alternate proposals add reagents oriented toward detection of sulfide and including the use of Fischer's reagent, *p*-dimethylamino-aniline with iron(III) chloride, which in the presence of sulfide produces the brilliantly colored dye methylene blue (Weidert & Rohlich, 1908), and a more convoluted procedure called the Hamburg test (Korn & Kamman, 1907). The Hamburg test for organic sulfides consists of precipitation of the sulfides from the sample with barium oxide (baryta water) and barium chloride, then fusion of the washed and dried precipitate with potassium metal to form potassium sulfide. Potassium sulfide is detected by reaction with *p*-dimethylamino-aniline in the presence of ferrous chloride to form methylene blue. The Hamburg test was criticized as unreliable because of false-positive interferences caused by sulfate and nitrate (Guth, 1910), arising from reduction of these anions during the potassium fusion step. Stable drinking water was presented as one example of a sample giving a false-positive result.

The first edition of *Standard Methods* (APHA, 1905) included a putrescibility test for wastewater effluent. The test was primarily evaluated by means of sight, smell, or chemical detection of sulfide.

Another direction toward standardization of putrescibility was to approach the test as a measure of the use of oxygen by the sample. One approach is to add permanganate to the sample, then determine the amount of incubation time necessary for decolorization of the added reagent (Wyatt, 1893). Further work with permanganate led to the proposal of the "Manchester" stability procedure as a semiquantitative putrescibility test (Scudder, 1908). These variations of the oxygen-consumed test did not catch on as far as a putrescibility test is concerned; however, the idea that dissolved oxygen (DO) depletion is linked to sample purification remains and became a key concept in further test development.

Effluent stability was defined in a quantitative sense by the level of DO not falling below 50% of saturation (Dibdin & Thudichum, 1900), determined by measuring the DO level in the incubated sample. This oxygen-consumption-based definition forms a permanent link between the putrescibility test and biochemical oxygen demand (BOD), with the distinction being that the putrescibility test is a rapid full-volume procedure, whereas BOD evolved into the dilution test. Not all workers accepted this linkage (Eliassen, 1944), preferring to equate BOD with a measurement of load while using putrescibility to reveal oxygenation conditions.

Following up on the idea of oxygen depletion as the predominant characteristic, methylene blue was introduced as an indicator for the putrescibility test (Spitta, 1903). Methylene blue is a vivid blue dye that retains its color if the effluent is stable but becomes colorless if the oxygen is depleted in the sample. An improved methylene blue procedure quickly follows (Spitta & Weldert, 1906). The methylene blue putrescibility test was evaluated for use as a process control tool (Hale & Melia, 1915; Jackson & Horton, 1909; Johnson et al., 1905; Phelps & Winslow, 1907). The test was used to develop a mathematical description of stability and wastewater deoxygenation, which eventually led to development of the monomolecular equation for BOD (Phelps, 1909).

Other dyes were evaluated for use in the putrescibility test, and indigo carmine and methylene green were found to be usable. Results obtained from use of methylene blue were compared with those obtained from methylene green and indigo carmine in the putrescibility test (Clark & Adams, 1908). The latter two dyes were reported to become colorless earlier than methylene blue and led to quicker negative results. Another comparison of the three dyes found that methylene green gave the fastest results, whereas indigo carmine gave inconsistent results compared with the other two dyes (Jackson & Horton, 1909).

Another approach to putrescibility detection used the first two steps of the Winkler test to qualitatively determine DO depiction after incubation. If a white precipitate formed, it indicated the lack of oxygen and an unsuitable result, whereas a brown precipitate, indicating that DO is still present, revealed a stable effluent (Kimberly, 1908). This technique probably did not become popular for three reasons: (1) the need to add several reagents to the sample; (2) difficulty in the observation of a white or brown precipitate in a colored, particulate-laden effluent sample; and (3) the need for the complete depletion of oxygen from the sample to obtain a white precipitate.

The second edition of *Standard Methods* (APHA, 1912) includes a description of Phelps's (1909) work and a table (reproduced from third edition [APHA, 1917] as Table 3.1) for relative stability based on the methylene blue procedure derived from Phelps's equation. After the widespread availability of a standard procedure, users of the technique pointed out errors in the putrescibility test that can occur during its application. Incorrect results associated with adding either insufficient or too much methylene blue and problems with adsorption of the dye to clay particles in turbid samples were reported (Lederer, 1912, 1913, 1914a, 1914b, 1915). The detrimental effect of germicidals like copper, free chlorine, calcium hypochlorite, bisulfite, and

TABLE 3.1. Quantitative relative stability at 20 °C (APHA, 1917).

Days required for decolorization	Relative stability (%)	Days required for decolorization	Relative stability (%)
0.5	11	8.0	84
1.0	21	9.0	87
1.5	30	10.0	90
2.0	37	11.0	92
2.5	44	12.0	94
3.0	50	13.0	95
4.0	60	14.0	96
5.0	68	16.0	97
6.0	75	18.0	98
7.0	80	20.0	99

residual caustic alkalinity on relative stability results was discussed (Mohlman, 1917). The apparatus used for the tests had to ensure that the samples could not be reaerated during the incubation; otherwise, the methylene blue would not decolorize. A simple one-way gas release valve, consisting of a bottle stoppered with a gas-exit tube whose outlet is underneath the surface in a water-filled test tube, is described that ensures that no atmospheric reaeration occurs (Bachman, 1914; Buswell, 1914).

The third edition of *Standard Methods* (APHA, 1917) presented a technique to derive BOD values from putrescibility (renamed the relative stability test) with the following equation:

$$\text{Oxygen demand} = \frac{O[1 - p]}{Rp}$$

where O is the initial DO of the diluting water, p is the fraction of wastewater in the test, and R is the relative stability from the test. Later editions of *Standard Methods* presented only minor changes to the relative stability method.

Based on further work with indigo carmine and methylene blue, Clark reversed an earlier report (Clark & Adams, 1908) that indigo carmine gave faster results in the relative stability test. It was determined that indigo carmine requires a greater DO depiction in the sample than methylene blue to turn colorless (Clark, 1920, 1923; Clark & Cohen, 1922; Clark et al., 1925).

Calvery (FSWA, 1930) suggested relative stability be used for in-facility control and BOD for precise quantitative results and research work. However, by the time of the publication of the eighth edition of *Standard Methods* (APHA & AWWA, 1936), the relative stability test had been for the most part replaced in practice by the dilution bottle BOD. Subsequent editions of *Standard Methods*, while still carrying the relative stability test, included a statement that the popularity of the test is waning in favor of the dilution bottle BOD test.

Although in widespread use during the early parts of the 20th century, the exact reasons why the relative stability test works were unknown. Various mechanisms for the function of methylene blue in the test were proposed. Methylene blue, while well known as a reduction–oxidation (redox) indicator, is reduced only very slowly with sulfide. The reduction by sulfide becomes appreciable only in the presence of an elemental selenium catalyst (Feigl & Anger, 1972). Thus, the presence of sulfide, while in some cases easily detected by smell, is not the reason why the color of the dye fades. The reduction potential of methylene blue at pH 7 is +0.011 V, whereas the reduction potential of indigo carmine at pH 7 is −0.125 V (Clark, 1920, 1923; Clark & Cohen, 1922; Clark et al., 1925; Dean, 1992). The simple absence of oxygen from the sample, while causing a drop in reduction potential, is insufficient by itself to cause bleaching of the dyes, although the presence of oxygen is capable of oxidizing the leuco-dyes to the colored state, and this property was used to determine if samples are anaerobic (Hall, 1921). It was found that there must be a mediating agent for the bleaching to occur. Work on the effect of germicidals on the relative stability and dilution bottle tests suggests that microorganisms are the causitive agents (Mohlman, 1917).

Based on a series of experiments (Wooldridge, 1933; Wooldridge & Standfast, 1936a, 1936b, 1936c) it was found that microorganism dehydrogenases are the agents of the bleaching and, as oxygen is depleted from the closed system, the dyes serve as hydrogen acceptors from bacterial enzymes. These ideas are supported in more recent studies oriented toward comparison of methods for determination of enzyme activity in activated sludge (Jorgensen, 1984). Tests with methylene blue as a biological activity indicator revealed it to be easier to use than the triphenyl tetrazolium chloride or adenosine triphosphate procedures (further discussed in Chapter 9 on alternate BOD procedures). This later study also finds that there is some interference to the decolorization of methylene because of the presence of DO, and that samples have to be purged with an inert gas to remove DO before testing to obtain consistent results.

The dyes in the relative stability test have been reexamined several times. Methylene green was found in a series of controlled tests to give marginally faster results than methylene blue and, further, not to be affected by the presence of oxygen in the sample (Dickerson, 1939; Dickerson & Wein, 1940). Chemically, methylene green is 1-nitro-methylene blue. The presence of the nitro group gives methylene green increased ability to compete with oxygen as a terminal electron acceptor from the bacteria; however, similarity in molecular structure should not result in any fundamental difference in the mode of action (Figure 3.1).

Other dyes related to methylene blue have been tested. Dyes that have nitrogens replacing the sulfur in the ring are generally ineffective in the relative stability test. Thionin, which lacks any methyl groups on the nitrogens,

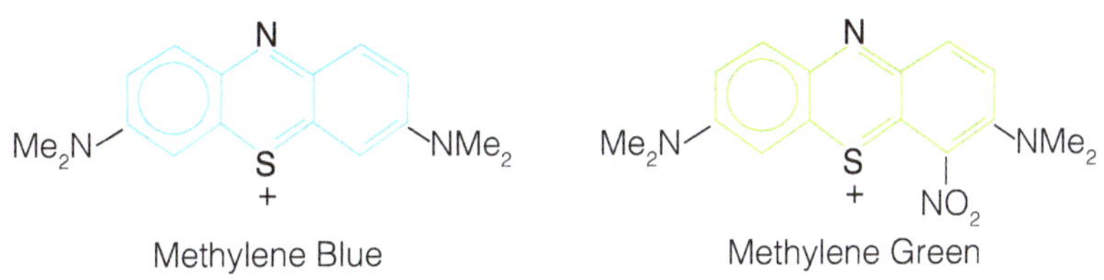

FIGURE 3.1. Structures of methylene blue and methylene green.

and brilliant cresyl blue, which has ethyl groups on one nitrogen and only hydrogens on the other, consistently function much faster than methylene blue (Hooper & Briscoe, 1944). Dyes that are structurally dissimilar to methylene blue, such as indigo carmine and dichlorophenolindophenol (Remy, 1937), were found to have erratic response times in the test compared with methylene blue, suggesting that they are not probing the same enzyme system in the bacteria.

The addition of thioglycolate medium to the sample, a supporting medium for the growth of faculative and strict anaerobes, further increased the speed of the determination such that a thionin-thioglycolate or brilliant cresyl blue-thioglycolate test takes about one-fifth to one-quarter the amount of time required for the standard methylene blue procedure (Hooper & Briscoe, 1944). These results suggested that the enzyme system reducing the methylene blue may be present in a faculatively anaerobic species or even a more or less strict anaerobe. This viewpoint was supported by results on synthetic wastewater (Butterfield et al., 1937) mixtures in which samples with low organic substrate, nitrate, and DO concentrations, but high bacterial counts, still exhibited quick bleaching of the methylene blue and thus low stability (Tidwell and Sorrels, 1956).

Although it still showed up in occasional papers (Gilcreas, 1952; Mortimer & Murray, 1957; Murray & Mortimer, 1957; Vezzi, 1970) reporting attempts to find a shortcut technique to monitor facility effluent, the putrescibility or relative stability test for wastewater has essentially disappeared from use, being withdrawn from *Standard Methods* by the 12th edition. That does not mean that the potential utility for the test has disappeared, or that reexamination of the biochemistry of the test may be completely valueless.

The concerns over responsible disposal and reuse of biosolids have given new life to the idea of relative stability; only this time, the samples are the solid and semisolid residue from treatment processes, rather than water. A paper in 1977 (Eikum & Paulsrud, 1977) summarized efforts to find reliable procedures for determining sludge stability, efforts that to a large degree duplicated those made on wastewater in the early 1900s. The use of the darkening of lead salts was reexamined (Ruffer, 1966). A strip of paper saturated with lead acetate was suspended over a 50 g sample of sludge in a sealed bottle. Production of volatile sulfides, indicating an unstable sludge, was indicated by the paper turning brown in color. Bleaching of various dyes, including methylene blue, was tested (Eikum & Paulsrud, 1977). Another study looks at the relation between 5-day BOD (BOD_5) of the sludge and volatile suspended solids (VSS). The sludge is considered to be odor free and stable if the BOD_5 is less than 0.2 g/g VSS (Koers & Mavinic, 1977). The final suite of methods presented in the biosolids regulations (Title 40

Code of Federal Regulations, Part 503; see Smith, 1999a, 1999b) contained no techniques even remotely similar to those investigated for putrescibility, but the idea of "stability" still lives on.

1.0 REFERENCES

American Public Health Association. (1905). Standard methods for the examination of water and sewage. *J. Infect. Dis.* Suppl. no. 1.

American Public Health Association. (1912). *Standard methods for the examination of water and sewage* (2nd ed.).

American Public Health Association. (1917). *Standard methods for the examination of water and sewage* (3rd ed.).

American Public Health Association & American Water Works Association. (1936). *Standard methods for the examination of water and sewage* (8th ed.). American Public Health Association.

Bachman, F. A. (1914). A new seal for the prevention of aeration in deaerated liquids. *J. Ind. Eng. Chem.*, *6*, 764.

Buswell, A. M. (1914). Modified apparatus for the putrescibility test. *J. Ind. Eng. Chem.*, *6*, 325.

Butterfield, C. T., Ruchhoft, C. C., & McNamee, R. D. (1937). Studies of sewage purification. VI. Biochemical oxidation by sludges developed by pure cultures of bacteria isolated from activated sludge. *Public Health Rep.*, *52*, 387; *Chemical Abstracts*, *31*, 5075 (1937).

Clark, H. W., & Adams, G. O. (1908). Studies on incubation tests. *J. Am. Chem. Soc.*, *30*, 1037.

Clark, W. M. (1920). Reduction potentials of mixtures of indigo and indigo white, and of mixtures of methylene blue and methylene white. *J. Wash. Acad. Sci.*, *10*, 255.

Clark, W. M. (1923). Studies on oxidation–reduction. I. Introduction. *Public Health Rep.*, *38*, 443.

Clark, W. M., & Cohen, B. (1922). Some elementary aspects of the putrescibility and similar tests. *Abstr. Bacteriol.*, *6*, 3.

Clark, W. M., Cohen, B., & Gibbs, H. D. (1925), Studies on oxidation–reduction. VIII. Methylene blue. *Public Health Rep.*, *40*, 1131.

Dean, J. A. (1992). *Lange's handbook of chemistry* (14th ed.). McGraw-Hill.

Dibdin, W. J., and Thudichum, G. (1900). Aeration as a test for the purity of sewage effluents. *J. Soc. Chem. Ind.*, *19*, 497.

Dickerson, D. (1939). The system activated sludge: oxygenated water. Part III. The adsorption and reduction of dyestuffs. *J. Soc. Chem. Ind.*, *58*, 199.

Dickerson, D., & Wein, E. U. (1940). Stability test for sewage effluents. *Analyst*, *65*, 289.

Eikum, A. S., & Paulsrud, B. (1977). Methods for measuring the degree of stability of aerobic stabilized sludges. *Water Res.*, *11*, 763.

Eliassen, R. (1944). Oxidants and reductants in sewage plant effluents. *Water Sew. Works*, *91*(6), 219.

Federation of Sewage Works Associations Committee Report. (1930). Symposium: Standard methods for the examination of sewage and sewage sludge. *Sew. Works J.*, *2*, 347.

Feigl, F., & Anger, V. (1972). *Spot tests in inorganic analysis* (6th ed.). Elsevier Science.

Gilcreas, F. W. (1952). Better plant operation with laboratory control. IV. Putrescibility and BOD. *Wastes Eng.*, *23*, 405.

Guth, F. (1910). The Hamburg test for putrescibility. *Gesundh. Ing.*, *33*, 586; *J. Soc. Chem. Ind.*, *29*, 1267 (1910).

Hale, F. E., & Melia, F. W. (1915). Comparison of methods for determining putresibility or oxygen demand. *Ind. Eng. Chem.*, *7*, 760.

Hall, I. C. (1921). Chemical criteria of anaerobiosis with special reference to methylene blue. *J. Bacteriol.*, *6*, 1.

Hooper, S. H., & Briscoe, O. W. (1944). Modifications of the relative stability test for sewage. *Sew. Works J.*, *16*, 558.

Jackson, D. D., & Horton, W. A. (1909). Experiments on the putrescibility test for sewage and sewage effluents. *J. Ind. Eng. Chem.*, *1*, 328.

Johnson, G. A., Copeland, W. R., & Kimberly, A. E. (1905). The relative applicability of current methods for the determination of putrescibility in sewage effluents. *Public Health*, *31*, 2, 80; *J. Inf. Dis. Suppl.*, *2*, 80 (1906).

Jorgensen, K. P. (1984). Determination of the enzyme activity of activated sludge by methylene blue reduction. *J. Water Pollut. Control Fed.*, *56*, 89.

Kimberly, A. E. (1908). Report on examination of sewage treatment plants. Ohio State Board of Health, Rep. Water Sew. Purif. Ohio 1906–1907, 371.

Koers, D. A., & Mavinic, D. S. (1977). Aerobic digestion of waste activated sludge at low temperatures. *J. Water Pollut. Control Fed.*, *49*, 460.

Korn, F., & Kamman, O. (1907). The Hamburg test for putrescibility in sewage effluents. *Gesund.-Ing.* (Germany), *30*, 165; *J. Soc. Chem. Ind.*, *26*, 888 (1907).

Lederer, A. (1912). The relation of the putrescibility of the settling and non-settling suspended matter in sewage. *Eng. News-Rec.*, *64*, 733.

Lederer, A. (1913). The relation of the nitrates to putrescibility of sewages. *J. Infect. Dis.*, *13*, 236.

Lederer, A. (1914a). A new method for determining the relative stability of sewage effluent, or polluted river water. *J. Infect. Dis.*, *14*, 482.

Lederer, A. (1914b). The biochemical oxygen demand of sewages. *J. Ind. Eng. Chem.*, *6*, 882.

Lederer, A. (1915). Relative stabilities in polluted waters carrying colloids. *Am. J. Public Health*, *5*, 740.

Mohlman, F. W. (1917). Copper in sewage at the new haven sewage experiment station. *J. Ind. Eng. Chem.*, *9*, 1093.

Mortimer, J. G., & Murray, K. A. (1957). Relative stability and oxygen uptake of effluents. *S. Afri. Ind. Chem.*, *11*, 118; *Chemical Abstracts*, *51*, 18404 (1957).

Murray, K. A., & Mortimer, J. G. (1957). Relative stability and oxygen uptake of effluents. *S. Afri. Ind. Chem.*, *11*, 118.

Phelps, E. B. (1909). The disinfection of sewage and sewage filter effluents. *U.S. Geol. Surv. Water Supply Paper*, *229*, 74.

Phelps, E. B., & Winslow, C. E. A. (1907). On the use of methylene blue in testing sewage effluents. *J. Infect. Dis. Suppl.*, *3*, 1.

Remy, E. (1937). Use of dichlorophenolindophenol as an especially sensitive indicator for the determination of the putrefiability of sewage. *Arch. Hyg. Bakt.*, *119*, 141; *Chemical Abstracts*, *32*, 2660 (1937).

Ruffer, H. (1966). Untersuchungen zur characterisierung aerob biologisch stabilizierter schlamme. *Vom Wasser* (Germany), *13*, 255.

Scudder, F. (1908). Testimony before the Royal Commission of Sewage Disposal. Royal Comm., 5th Rep., App. I, 684.

Smith, R.-K. (1999a). *Handbook of environmental analysis* (4th ed.). Genium Publishing.

Smith, R.-K. (1999b). *Lectures on wastewater analysis and interpretation.* Genium Publishing.

Spitta, O. (1903). Weiters untersuchungung uber flussverunreinigung. *Arch. f. Hyg.*, *46*, 113.

Spitta, O., & Weldert, R. (1906). Indicatoren fur die beurteilung biologisch gereinigter abwasser. *Mitt. Koenig. Pruf. Wasserversorg*, *6*, 160.

Stoddard, F. W. (1901). The incubator test for sewage effluents. *Analyst*, *26*, 281.

Tidwell, W. L., & Sorrels, J. H. (1956). Comparative evaluation of special 2-day and standard 5-day BOD tests. *Sew. Ind. Wastes*, *28*, 488.

Vezzi, S. (1970). Relationship between BOD, oxidizability, and stability tests as indexes of the degree of pollution of surface waters. *Rass. Chim.*, *22*(5), 149; *Chemical Abstracts*, *75*, 9670 (1971).

Weidert, R., & Rohlich, K. (1908). The determination of putrescibility in the effluents of biological sewage effluents. *Mitt. Kgl. Pruf. Wasser Abwasser*, 26; *Chemical Abstracts*, *4*, 1072 (1910).

Wooldridge, W. R. (1933). The stability test of sewage and its relation to enzyme activity. *Biochem J.*, *27*, 193.

Wooldridge, W. R., & Standfast, A. F. B. (1936a). Some experiments on the oxidation of sewage and sewage sludge systems. *Biochem. J.*, *30*, 149.

Wooldridge, W. R., & Standfast, A. F. B. (1936b). The role of enzymes in activated sludge and sewage oxidations. *Biochem. J.*, *30*, 1542.

Wooldridge, W. R., & Standfast, A. F. B. (1936c). Use of the Barcroft differential manometer in the estimation of the oxygen demand of sewage. *Biochem. J.*, *30*, 141.

Wyatt, F. (1893). Tests for purity of drinking water. *Eng. Mining J.*, *56*, 168.

4

Measurement of Dissolved Oxygen

Oxygen was discovered by Scheele in 1771 and independently by Priestly in 1774. The latter is generally credited in textbooks with the discovery. In 1857, Brodie advanced the opinion that oxidation of organic material in polluted water was the way to purify the water, but that oxygen dissolved in the water was not a significant source of oxidation (Brodie, 1865). Work by Frankland, Gerardin, and Dupre corrected Brodie's opinion and pointed out the tremendous importance of aqueous dissolved oxygen (DO) levels (Dupre, 1884; Frankland, 1870; Gerardin, 1875, 1876). Since then, a huge body of work, some of it cited in this book, has been devoted to exploring the ramifications and importance of DO in water on wastewater treatment and water quality (Krenkel et al., 1969). Receiving stream DO levels have even been proposed as legal measures of treatment acceptability (Beaujean, 1956).

Oxygen in the elemental form exists as the diatomic molecule O_2 and the triatomic allotrope O_3, called ozone. Diatomic oxygen is a diradical and very reactive as an oxidant. Life on earth has evolved to use oxygen as the major terminal electron acceptor in biological electron transport, but most organisms have concurrently evolved protective mechanisms to limit the contact of oxygen with sensitive enzyme systems. Oxygen is acutely toxic to life forms not adapted to its use. Ozone is even more reactive and toxic. Molecular oxygen is present in Earth's atmosphere to the extent of approximately 20%. Ozone, on the one hand, is a widely recognized contributor to and symptom of air pollution when present at ground level; on the other hand, the stratospheric ozone layer limits the amount of ultraviolet (UV) radiation that reaches the ground and helps protect life on Earth.

Early techniques for the analysis of oxygen required isolation of the gas, followed by measurement based on volumetric displacement of water in gas burets or manometers. Oxygen has very low solubility in water, the normal saturated concentration being around 9 mg/L at room temperature and sea level pressure (Elmore & Hayes, 1960, 1961; Fox, 1909; Hua, 1990; Klotts & Benson, 1963; Thresh, 1885; Truedale et al., 1955; Whipple & Whipple, 1911; Winkler, 1888). The presence of oxygen in only trace amounts in water hampered early work oriented toward quantitatively describing the role of oxygen in water purification. Improvements in micromanometric determinations led to a return to this approach for DO determination by some researchers (Scholander et al., 1955), and it is one of the primary tools in respirometric techniques for oxygen consumption. However, the evolution of easy chemical- and electrode-based measurement techniques for DO made possible most of the research into biochemical oxygen demand (BOD) and biological treatment of waste that has occurred over the last 100 years.

Much attention has been spent on the sampling of DO, dating back to the early years of the 1900s (Lederer, 1910). The current authors are of the opinion that immediate in situ measurements are the only reliable means of obtaining true DO values, regardless of any directions for preservation of samples that accompany the methods in recent editions of *Standard Methods.* There are too many reliable electrode-based field instruments available for immediate DO determination to be wasting time on preservation of samples.

This chapter is a review and discussion of the analysis of DO as it is applicable to waste treatment and BOD determinations. It is by no means a comprehensive discussion, as there have been many reviews (Heukelekian, 1938; Rose & Long, 1988; Rotshtein & Shemyakin, 1967) and books published on oxygen analysis. The book by Hitchman (1978) and the chapter by Grasshoff in *Marine Electrochemistry* (Grasshoff, 1981) can be singled out as particularly useful and informative.

1.0 WINKLER TITRATION AND MODIFICATIONS

Early research attempts concerning DO were hampered by a lack of good analytical techniques. Most depended on volumetric displacement of water or other indirect methods. In was not until the 1888 publication of Winkler's DO method that research could be performed in a truly quantitative sense (Theriault, 1925; Winkler, 1888, 1911, 1912). The Winkler titration allowed the development of DO measurements as a process control tool in wastewater treatment (Hoover, 1911).

The Winkler titration has, as the first step, the addition of a solution of manganese (Mn)(II) sulfate to the sample followed immediately by a strongly alkaline solution of potassium iodide. Though manganese(II) chloride has been used successfully (Clarke, 1910; Muller, 1889), the sulfate is the more common salt of manganese(II) and offers some stability advantages. Oxygen in the sample reacts with manganese(II) giving manganese(III) (Hitchman, 1978), which has a very fleeting existence, reacting with either another oxygen molecule to give manganese(IV) (Smith, 1999a, 1999b) or disproportionating to give manganese(II) and manganese(IV) (Boyles, 1997a, 1997b; Miller et al., 2001):

$$4Mn(II) + O_2 + 2H_2O + 8OH^- \rightarrow 4Mn(OH)_3$$

Further oxidation:

$$4Mn(OH)_3 + O_2 \rightarrow 4MnO(OH)_2 + 2H_2O$$

Disproportionation:

$$2Mn(OH)_3 \rightarrow Mn(II) + Mn(IV) + 6OH^-$$

Overall:

$$2Mn(II) + O_2 + 4OH^- \rightarrow 2MnO(OH)_2$$

The addition of strong chelating agents such as ethylenediaminetetraacetic acid (EDTA) or 1,2-diaminocyclohexane tetraacetic acid (DCTA) to the manganese(II) reagent allows stabilization of the initially formed manganese(III), as described below; however, in the normal Winkler procedure, the manganese(III) is unstable.

The appearance of the reaction supports the conclusion that the precipitate consists of manganese(IV). Solutions of manganese(III) salts are reported to appear purple, visibly similar to solutions of permanganate (Boyles, 1997a, 1997b; Miller et al., 2001), whereas solid forms of trivalent manganese are blue-purple. The presence of oxygen with addition of the manganese(II) and alkaline–iodide solutions generates a dirty brown precipitate, which is characteristic of manganese(IV) cruds that form on the sides of bottles of aged potassium permanganate solutions. Whether the oxidized manganese is in the 3+ or 4+ valence state is of interest in the sense that a correct description of the process is desired, but it is immaterial to the outcome of the standard Winkler titration, as the number of transferred electrons remains the same. As physical observation of the process favors the presence of manganese(IV), the rest of this discussion will assume that it is the active species unless special steps are taken to preserve the manganese(III) in the reaction.

The next step in the procedure is to acidify the sample by addition of acid. The acid dissolves the precipitated manganese and allows reaction with the iodide ion to form triiodide:

$$2MnO(OH)_2 + 6I^- + 8H \rightarrow 2Mn^{2+} + 2I_3^- + 6H_2O$$

The triiodide ion formally contains two atoms of iodine in the 0 valence state and one in the –1 valence state, although a more correct description regards the iodine atoms as covalently bonded to each other with the extra electron delocalized over the entire ion. The electron balance in the procedure has four electrons from six iodide ions transferred to each molecule of oxygen that was originally dissolved in the sample. Sulfuric acid is the preferred acid, although there are reports of the use of hydrochloric acid (Muller, 1889) or sulfamic acid (Nasbaum, 1958).

All that remains in the analysis is to determine how many electrons have been used to reduce the oxygen. This is performed by a redox titration of the triiodide with a standardized solution of a reducing agent, sodium thiosulfate being the agent chosen by Winkler. Muller chose a hyposulfite salt. The thiosulfate determination of iodine is

$$I_3^- + 2S_2O_3^{2-} \rightarrow 3I^- + S_4O_6^{2-}$$

The indicator for the titration is starch. Starch, depending on the source of the starch, forms an intensely colored blue-to-purple-to-black complex with iodine (from the triiodide ion), with the solution turning colorless at the endpoint of the titration.

Although most researchers and practitioners since Winkler have used thiosulfate as the reductant, it is not the most stable of reagents and requires frequent standardization (Smith, 1999a, 1999b). In addition to using a reagent known to present stability problems, the Winkler procedure has a serious shortcoming in that it relies solely on knowledge of the concentration of that titrant for quantitatively reliable results. There are few alternate checks for determining the accuracy of the measurement. The ideal would be to run a check of the Winkler procedure on a solution of known DO concentration; however, preparing solutions with known DO levels is quite challenging. There are tables of oxygen saturation levels available for different conditions of temperature and barometric pressure, but producing a saturated oxygen solution involves more than simply shaking the sample in the presence of the atmosphere and assuming that the book value has been achieved (Moggio & Burland, 1951). One proposal is to aerate a water sample, gasometrically determine the amount of oxygen in the water, then use the value as a check on the Winkler results (Wheatland & Smith, 1955). Another possibility is to titrate solutions containing known concentrations of hydrogen peroxide as a surrogate for oxygen. The lack of easily prepared and verified solutions of oxygen or an oxygen surrogate for the performance of the standard quality controls to demonstrate matrix spike recovery, laboratory control, and performance evaluation place undue reliance on the determinative procedure chosen.

There are numerous chemical interferences that were encountered in the application of the original Winkler procedure to wastewater. Nitrite ion, iron(II), reduced sulfur species such as sulfite and sulfide, and the presence of organic material that can react with triiodide ion can be named as a representative few (Clarke, 1910; Cost & Andrews, 1923; Elvove, 1914; FSWA, 1932). Several modifications of Winkler's original procedure have been incorporated to standard techniques to perform the DO determination, with the azide modification (Alsterberg, 1925) being the one most frequently used.

The azide modification (*Standard Methods*, 4500-O C [APHA et al., 1998]) adds sodium azide (NaN_3) to the alkaline potassium iodide solution to give alkali–iodide-azide (AIA) reagent (Alsterberg, 1925; Barnett & Hurwitz, 1939). On acidification of the sample, azide is converted to hydrazoic acid, which reacts with nitrite ion to remove it from interfering with the DO determination:

$$HN_3 + NO_2^- + H^+ \rightarrow N_2 + N_2O + H_2O$$

The azide modification will also correct for the presence of small (<1 mg/L) amounts of iron(II). If larger amounts of iron(II) are present, the permanganate modification is used (Rideal & Stewart, 1901; Theriault & McNamee, 1932), which can also compensate for the presence of nitrite, hypochlorite, and some interfering organic matter. The Winkler titration with AIA was miniaturized into a field-portable kit (Harper, 1953; Scott, 1964).

The permanganate modification (*Standard Methods*, 4500-O D [APHA et al., 1998]) consisted of pretreatment of the sample with potassium permanganate, sulfuric acid, and potassium fluoride. The permanganate oxidizes the interfering materials, then the excess of permanganate is removed by addition of a small amount of sodium or potassium oxalate. After the color of the excess permanganate fades, the rest of the azide modification of the Winkler titration is performed. Results of the permanganate modification have been evaluated (Placak & Ruchhoft, 1941; Ruchhoft et al., 1938).

The alum flocculation method (*Standard Methods*, 4500-O E [APHA et al., 1998]) was designed to remove interfering suspended organic matter from the sample that can present an iodine demand (Ruchhoft, 1941). In this pretreatment, a solution of alum [aluminum potassium sulfate, $AlK(SO_4)_2$] is added to the sample, followed by ammonium hydroxide. The sample is mixed, then the alum floc is allowed to form and settle. The clear supernatent is transferred to another container and the azide modification of the Winkler titration is performed.

The copper sulfate sulfamic acid method (*Standard Methods*, 4500-O F [APHA et al., 1998]) was designed to remove biologically active flocs from the sample. In this pretreatment, a solution of copper sulfate and sulfamic acid is added to the sample. The sample is then mixed and allowed to settle. The clear supernatent is transferred to another container and the azide modification of the Winkler titration is performed.

The alkali–hypochlorite method (Hitchman, 1978) is a pretreatment method to remove interferences caused by reduced sulfur species. A solution of sodium hypochlorite (NaOCl) is added to the sample and it is mixed. A solution of potassium iodide is then added, followed by acidification with sulfuric acid. The excess hypochlorite, indicated by the brown color of iodine, is neutralized with sodium sulfite solution. Starch can be added to more easily visualize the endpoint of the sodium sulfite addition. Potassium biiodate [$KH(IO_3)_2$] is added to just restore the starch color. Then, the azide modification of the Winkler titration is performed on the sample.

The titration in the Winkler procedure was recognized as one of the trickiest parts of the procedure. A joint Soviet–American interlaboratory study of the precision and accuracy of the procedure and some of the common variants was conducted and supported this observation (Chernyakova et al., 1983).

The first point of error is the determination of the endpoint of the titration. The endpoint, using the starch indicator, does not display an instantaneous transition from black to colorless (Smith, 1999b). The use of starch can be avoided by monitoring the near–UV absorbance of the triiodide ion and using its disappearance as the endpoint indicator. Used in conjunction with a screw-driven calibrated syringe for the titration, reproducible measurements exhibiting much less than 1% variation can be made (Bryan et al., 1976).

Iodine is subject to volatization into the atmosphere, giving a low bias to the determination when the solution is mixed too vigorously or the titration is performed too slowly. One answer has been to use an AIA solution that is up to 6 mol in potassium iodide (Carritt & Carpenter, 1966; Montgomery et al., 1964; Pomeroy & Kirchman, 1945). This solution is the one described as suitable for analysis of supersaturated oxygen solutions in *Standard Methods* 4500-O C (APHA et al., 1998). A photochemical oxidation of iodide to iodine resulting from exposure to bright sunlight was reported to contribute to a high bias in the titration (Carritt & Carpenter, 1966). The time that lapses between fixing the sample and the titration has been examined, and it was recommended that the sample be titrated as soon as possible (Toledo et al., 1982a).

Other chemical titration reagent and indicator combinations besides thiosulfate and starch have been used to conclude the procedure. Phenylarsine oxide (PAO) is the most common alternate reagent. The arsenic is in the 3+ valence state in PAO and is oxidized to the 5+ state by triiodide ion:

$$I_3^- + PAO + 2H_2O \rightarrow 2H^+ + 3I^- + PAO(OH)_2$$

Phenylarsine oxide can be used with starch indicator or several other indicator options. Although PAO is a versatile and stable reagent, arsenic is considered an environmentally hazardous material, and a number of practitioners have gone back to using thiosulfate, despite its stability problems (Smith, 1999a, 1999b). Another procedure used ascorbic acid for the titration with a vivid violet indicator (Erdey & Szabadvary, 1954; Lebska, 1977). A procedure using hydroquinone as the titrant with a diphenylamine indicator was reported (Berk & Hofmann, 1963). Another added excess hydroquinone, then performed a back titration with cerium(IV) and ferroin indicator to determine the excess (Reddy et al., 1978).

Electrode-based endpoint determinations for the Winkler titration have been frequently described. A platinum wire electrode with calomel reference was used to measure the voltage of the sample as it is titrated with thiosulfate in a potentiometric procedure (Freier, 1955; Hitchman, 1978). The potentiometric titration has been adapted to automated DO determinations (Griffiths & Jackman, 1959; Levi et al., 1977; Oudot et al., 1988). An amperometric titration that uses a rotating platinum wire electrode and a reference electrode to measure the current through the system at a constant voltage has been described (Jones & Mullen, 1973; Potter & Everitt, 1959; Toledo et al., 1982b). Plotting the current against the milliliters of titrant generates a line of sharply decreasing slope that abruptly levels off to a constant value at the end of the titration. If two platinum electrodes are used without a reference and a constant voltage is applied across the electrodes while the current is monitored, a dead-stop amperometric titration can be performed (Hitchman, 1978; Toledo et al., 1984). Techniques that use automatic titrators for the determination were reported (Ridley et al., 1960).

The Winkler procedure has been modified by a number of practitioners to avoid the use of a titration finish. The triiodide ion has a substantial absorbance, and it can be directly measured in the near–UV at 287 or 352 nm (Arnott & McPheat, 1953; Broenkow & Cline, 1969; Duval et al., 1974; Ivanoff, 1962; Resch, 1959). The iodine has also been extracted with organic solvents, like carbon tetrachloride (Hitchman, 1978) or chloroform (Elliot, 1963) and determined by colorimetry to improve sensitivity.

The chemistry of the Winkler procedure has been adapted for use in an autoanalyzer instrument (Saito & Uchiyama, 1980). Automated reagent dispenser systems were devised to make the manual method faster (Duedall et al., 1971).

When chelating agents are added to the sample, manganese(III) formed in the initial oxidation of manganese(II) can be trapped and stabilized. The original chelating agent used was pyrophosphate (Berka et al., 1968), and the pyrophosphate-Mn(III) complex was determined by hydroquinone titration with a ferroin indicator. Reagents such as EDTA or OCTA (1,2-cyclohexylenedinitrilo)-tetraacetic acid) form intensely colored complexes. The Mn(III)–EDTA complex has a bright violet-red color and is determined at 490 nm (de Carvalho et al., 1963; Rahim & Mohamed, 1978). It can also be titrated with ferrous ammonium sulfate (FAS) (Malik, 1964). A modification of the FAS titration involves addition of an excess of FAS, then backtitration with cerium(IV) and 1,10-phenanthroline indicator (Malik, 1966). The excess FAS has also been determined with an EDTA titration using a salicylate indicator (Fadrus & Maley, 1968a). The Mn(III)–DCTA has an absorbance maximum at 500 nm and can be quantitatively determined by colorimetry (Sastri & Maiaiyandi, 1983; Sastry et al., 1969). The presence of calcium ions was found to displace the manganese(III) from the OCTA complex and serves as a positive interference.

A great number of interferences to the Winkler procedure have been found. Iron(II) was identified early on as a problem (Buswell & Gallagher, 1923). Other metal ions can cause difficulties. Cobalt(III) and oxidized chromium salts were reported as interferences giving high bias to the test. One proposed solution was to run a blank analysis on the sample (Lakin, 1976; Potter & White, 1957). These cationic interferences can be removed by pretreating the sample with an ion-exchange resin before performing the Winkler titration (Potter & White, 1957).

Nitrite present in the sample has long been recognized as an interferent (Elvove, 1914). One tactic for nitrite correction involved a simultaneous determination of the iodine demand of the sample (Clarke, 1910). Two portions of sample were prepared; to one is added manganese(II) chloride and alkaline iodide, and to the other is added hydrochloric acid. They are allowed to sit overnight. The manganese(II)-treated sample is acidified with hydrochloric acid, then both are neutralized with magnesium carbonate and the liberated iodine titrated with sodium thiosulfate. The amount of thiosulfate needed to titrate the blank is subtracted from the manganese-treated sample. Blanks can also be used with the manganese(II) sulfate and sulfuric acid form of the Winkler procedure (Lakin, 1976; and Potter & White, 1957). An early technique for nitrite suppression added 1 mL of a 50% carbamide (urea) solution and sulfuric acid to the sample, then allowed it to sit for 4 hours before neutralization and addition of manganese(II) (Noll, 1917). Noll also added sodium bicarbonate to the reaction after addition of the alkaline iodide and acidification to precipitate any unreacted manganese(II) as the insoluble manganese(II) carbonate before titration. The azide modification of the Winkler titration was proposed as the solution to nitrites and has become the standard procedure (Alsterberg, 1925). Addition of sulfamic acid, H_2NSO_3H, as a replacement for sulfuric acid and simultaneous nitrite suppression, has been examined favorably in the azide procedure (Nasbaum, 1958). Sulfamic acid reacts with azide, but only under acidic conditions. The initially basic conditions existing after addition of manganese(II) and AIA allows the sulfamic acid and azide to remove nitrite, then slowly acidify the solution. Sulfamic acid, a solid, was suggested to be safer to work with than sulfuric acid, particularly in the field. Nitrite can be removed through initial addition of sulfamic acid, allowing the sample to sit, and then performing the Winkler titration (Fadrus & Maley, 1975).

Thiourea can serve as an interferent. The addition of mercury can sequester the thiourea; however, this makes use of the thiosulfate finish problematic because of the reaction of the sulfur in thiosulfate with mercury and release of the thiourea back into the solution. The answer is to perform the iodine determination with cerium(IV) (Fadrus & Maley, 1968c).

The problems associated with reduced sulfur species are well documented. Samples from pulp and paper mills with high levels of sulfite are a common occurrence, but other samples, such as those from swamp waters and others under the influence of reducing environments, also contain reduced sulfur species and cause difficulties (Beadle, 1958; Freier, 1952). These types of samples have led to development of several proposed solutions. One is to precipitate sulfide with zinc hydroxide, then determine the DO on the clear supernatant by the Winkler titration (Ingvorsen & Jorgensen, 1979). The amount of sulfide originally in the sample can be determined from analysis of the precipitate. The interference of sulfite can be eliminated using a rapid permanganate modification (Felicetta & Kendall, 1965). A different technique requires separate determination of the iodine demand of the sample, and then subtracting this value from the result of the Winkler titration (Bach, 1929, 1931; Baker, 1965).

Determination of the oxygen levels of actively respiring samples such as activated sludge often suffer from a lack of reproducibility because of shifting DO levels as organisms interact with the reagents (Theriault & McNamee, 1932, 1935). One technique is to stop the respiration of the sludge, then determine the DO. An inhibiting reagent containing sulfamic acid, acetic acid, and copper sulfate is added to the sludge sample and mixed by inversion in a sealed bottle. The sludge is allowed to coagulate and settle over 10 to 30 minutes, then an aliquot of the supernatant liquid is removed for determination of DO using the azide modified Winkler titration (Ruchhoft & Placak, 1942). This became the standard approach used in measurement of the BOD of raw and primary wastewater and sludge during the 1940 to 1960 period.

The application of the Winkler titration to determination of DO in seawater presents its own suite of problems. These have been reviewed in detail in a National Academy of Science study (Carritt & Carpenter, 1966).

2.0 OTHER CHEMICAL TECHNIQUES

An old method for oxygen determination, developed around the same time as the Winkler titration, used nitrite rather than manganese(II) as the reducing agent (Thresh, 1885). Nitrite and iodide ion were added to the sample generating iodine, which is determined by thiosulfate titration. Recent work on this reaction produced a modification that is easily adapted to automated DO determinations (Karlsson & Tortensson, 1974). Nitrite is mixed with iodide under acid conditions in a sealed reactor to form nitric oxide (NO) and iodine. The amount of iodine is determined coulometrically. The sample is added to the reactor. Oxygen in the sample reacts with the nitric oxide to transiently create nitrogen dioxide (NO_2), which further reacts with additional iodide to form more iodine, which is again determined coulometrically. The difference between the two determinations is directly related to the amount of oxygen in the added sample.

A number of direct colorimetric procedures were developed for DO determination. One added a solution of indigo carmine in the reduced form to the sample. Reduced indigo carmine, the so-called leuco form, is actually bright yellow-green. As it is oxidized by oxygen in the sample, the color changes from orange to red to purple to deep blue-green, depending on the amount of oxygen (Alcock & Coates, 1958; Buchoff et al., 1955), giving a visual estimation of the DO level. The decrease in absorbance at 410 nm because of the leuco form has been monitored in a quantitative spectrophotometric technique (Mizuniwa et al., 1976; St. John et al., 1964). Iron(II) chat can interfere with the results and can be removed from the sample by passing it through a mixed bed ion-exchange resin before addition of the reagent. The indigo carmine method has been compared with the results obtained using oxidation of cerium(III) (Ungureanu, 1964).

The putrescibility test has long been used as an indicator of the stability of a wastewater. Although most finishes for the test were oriented toward detection of sulfide or for the ability to decolorize a dye, the putrescibility test could be evaluated for the presence of DO after 24 hours, which would indicate that the waste is stable. The use of the first two steps of the Winkler test was suggested as being very sensitive (Kimberly, 1908); a white precipitate indicated that the waste was unsatisfactory, whereas a brown precipitate indicated that DO was still present.

Another method for DO determination used leuco phenothiazine dyes (Hamlin & Lambert, 1971), in which the amount of oxidized dye is proportional to the DO concentration. The leuco methylene blue determination of oxygen is another example of this procedure (Devdariani & Shmal'tsel, 1969; Schmiedt, 1934). Several direct colorimetric methods using leuco dyes were compared, with the safranine method being preferred (Ivanitskaya, 1969).

The direct colorimetric determination of DO with amidol, diaminophenol, has been examined several times (Ellis & Ellis, 1943; Gilcreas, 1935, 1952; Issacs, 1935). Addition of sodium citrate to buffer the sample and amidol generates red and yellow products from the DO. Spectrophotometric determination of the red product at 490 nm is linearly proportional to the amount of DO in the sample after a reaction time of 30 to 60 minutes. Extended reaction times skew the results. The test has been formatted in a color comparitor design for quick field determinations.

There have been several commercial kits on the market for direct colorimetric determination of DO levels. One such kit uses the proprietary dye, Rhodazine-D, as the active color reagent. The leuco form of this dye generates a red-purple color on reaction with oxygen, the intensity of the color being measured either with a colorimeter at 545 nm or a handheld color comparitor. A field test of the kit revealed it to give false-positive values in the presence of copper(II), iron(III), or chromiium(VI) (White et al., 1990). The prudent course to follow before depending on results obtained from this and similar kits is to compare readings with those generated by more definitive techniques such as the Winkler titration or a DO electrode.

Reversible colorimetric techniques for DO determination have been examined to make optical DO probes (optodes). An early effort measured the oxygen-sensitive fluorescence of acriflavine adsorbed onto silica gel (Zakharov & Grishaeva, 1980, 1982). A detection limit of 10 to 200 μg/L DO is reported for this system. Ruthenium complexes with 4,7-diphenyl-1,10-phenanthroline exhibit fluorescence in the absence of oxygen, which is quenched in the presence of oxygen. The complex is immobilized in a silica coating and excited with 450 nm light from a light emitting diode. The fluorescence is measured with a photodiode protected by a 550-nm cutoff filter. The detection limits for the optode are in the 0.015 mg/L range, and the linear calibration extends well beyond the oxygen saturation level in water. The device has a fast response and recovery time and is quite durable (McEvoy et al., 1996). Similar reagents have been used as fluorescent probes of DO concentration gradients (Wolff et al., 1991).

A qualitative colorimetric test for DO involves the addition of cuprous ammonium chloride to the sample. The copper(I) is oxidized to copper(II), which forms a vivid blue ammonia complex, indicating the presence of oxygen. This reaction has been adapted to use a catalytic amount of copper(I), with continuous electrolytic regeneration. The amount of current used in the coulometric electrolysis procedure is directly related to the amount of oxygen in the sample (Knapp, 1959).

A chemiluminescent technique for DO has been devised (Dubovenko et al., 1979). The test uses lucigenin, bis-*N*-methylacridinium nitrate, and luminol, 3-aminophthalhydrazide, with the luminol emitting light from oxidation and subsequent decomposition. The procedure is easily adapted to a continuous-flow autoanalyzer configuration.

The Miller titration was a direct titration of DO with iron(II), forming iron(III) in the presence of the redox indicator, phenosafranine (Miller, 1914). Optimum results are only achieved in this titration with precise pH control and rigorous exclusion of atmospheric oxygen (Ellis & Kanamori, 1973). A modification of the Miller procedure used salicylic acid as the indicator (Roskam & de Langen, 1963). The salicylic acid forms a violet complex with iron(III) produced from addition of an excess of iron(II) to the sample. The EDTA forms a stronger complex with iron(III) than salicylic acid, allowing the salicylic acid–iron(III) complex to be determined by titration with EDTA until the sample is colorless. A variation on this procedure first adds manganese(II), then iron(II) (Fadrus & Maley, 1968a, 1968b). A spectrophotometric modification that measures the bleaching of an ammonia solution of tris(4,7-dihydroxy-1, 10-phenanthroline) iron(II) complex by oxygen on addition to the sample was examined (Poe & Diehl, 1974). Interfering substances can be removed by addition of acidic iodide, then neutralization with an alkaline solution of tartrate, sodium thiosulfate, and methylene blue as an indicator, before iron(II) titration (Mori & Ishihara, 1971).

The oxidation of cerium(III) to cerium(IV) has been used in DO determinations. A titration procedure that uses *o*-toluidine as the indicator in a highly alkaline medium has been reported and the results compared with those using leuco indigo carmine (Ungureanu, 1964). Cerium(IV) exhibits an intense yellow color, and the determination can be made with a colorimeter or spectrophotometer (Golterman & Wisselo, 1981).

In a procedure that is widely used in quantitative microanalysis, oxygen is purged from the sample and trapped in an alkaline reaction solution containing pyrogallol (Flaschka et al., 1980). The amount of oxygen absorbed by the trap can be determined either gravimetrically or with a colorimeter based on the purple color of the oxidation product (Rezaeva, 1962). This technique is especially useful when samples rich in organic material are tested.

Thallium metal reacts rapidly with DO to form thallium hydroxide. The formation of thallium hydroxide is evidenced by the increase in conductivity of the sample, and this has been incorporated into DO analytical procedures (Doronin et al., 1969; Kabanova & Zalogina, 1965; Lueck, 1967, 1968, 1970). The dissolution of solid thallium electrodes has been used in an electrochemical technique (Chenouard et al., 1962). Thallium, although an important metal in the electronics industry, is very toxic. These methods for DO determination were not pursued because of the generation of hazardous and acutely toxic wastes.

The DO concentration of water samples has been directly determined using gas chromatography. A molecular sieve column with a thermal conductivity detector was reported on several times (Gaunt & Shanks, 1965; Sevcik & Tesarik, 1966). A Hersch coulometric oxygen detector has been used with gas chromatography (Hillman & Lightwood, 1966; Phillips et al., 1964). An exciting possibility is the use of a simple oxygen-selective detector (Kishi et al., 1998). Enriching the sample with a known amount of isotopically enriched oxygen ($^{18}O^{16}O$) led to a mass spectrometer isotopic dilution procedure (Mukhin & Aleskovskii, 1966). Oxygen and other gases in the sample were isolated from the sample using a gas–liquid separation (purging), followed by determination with gas chromatography (Li & Yang, 1982). A headspace sampling technique, in which the sample is allowed to equilibrate with the headspace in a sealed container, followed by gas chromatographic analysis, was published (Kampbell et al., 1989). This technique has been used as the determinative step in the headspace BOD procedure (Logan & Kohler, 2001; Logan & Patniak, 1997; Logan & Wagenseller, 1993).

Several biochemical techniques for determination of DO have been examined. Myoglobin is known for exhibiting color changes when complexed with oxygen. The chemical is immobilized in a silica matrix. Exposure to oxygen generates a measurable absorbance at 420 nm (Chung et al., 1995).

Glucose oxidase with flavin adenine dinucleotide (FAD) bound into the enzyme is immobilized on porous glass beads. On exposure to excess glucose, the limiting reagent for formation of γ-glucolactone by this enzyme complex is oxygen, which is necessary for the reoxidation of $FADH_2$ to FAD. Any oxygen in the sample is reduced to hydrogen peroxide. The flow from the oxidase reaction tube is subsequently mixed with *N,N*-dimethylaniline and 4-aminophenazone, which, in the presence of the hydrogen peroxide and immobilized peroxidase in a second reactor, combine to form an unidentified colored product. The colored product is determined at 555 nm, giving a linear correlation to the original DO concentration in the sample. The system is calibrated by injection of known amounts of hydrogen peroxide. Detection limits for the system are approximately 0.16 mg/L.

3.0 MAGNETIC MEASUREMENTS

Materials behave in two different fashions when placed in a magnetic field. Some materials (diamagnetic materials) disperse the field so that the intensity of the field is less in the material than the intensity in a vacuum. Other materials display paramagnetism, which is an increase in the strength of the magnetic field inside the material compared with a vacuum. Magnetic susceptibility is a weight or force measurement on a sample as a magnetic field is applied to the sample. It depends on the paramagnetic or diamagnetic character of the sample. For paramagnetic materials, the weight appears to increase when a magnetic field is applied. Paramagnetism is an extensive property in that some of the property is possessed by each molecule of oxygen and the total amount of paramagnetism is related to the concentration of the oxygen in the sample (Hoover et al., 1954). Oxygen is one of the few common gases that is paramagnetic because of the presence of unpaired electrons in the molecule's normally occurring triplet state. The other common gases that display paramagnetism are nitric oxide and NO_2, but they are not all that common in water, and exhibit only 40% of the paramagnetism that oxygen displays. Several commercial instruments are available for the direct determination of the magnetic susceptibility because of oxygen in the sample (Hitchman, 1978), and continuous monitoring of DO levels in fermentation tanks using a paramagnetic instrument has been described (Swartz & Cooney, 1978).

A detector that uses the principle of paramagnetism but does not directly measure it was developed for continuous determination of oxygen concentrations (Thiele, 1968; von Staden & Schlageter, 1965). A heated filament is contained in a permanent magnetic field. As a cool gas stream (purged from the water) enters the detector, the oxygen in the sample is attracted to the heated filament and serves to cool the filament, thereby changing its electrical resistance, which is the measured quantity. When heated, paramagnetic materials lose their paramagnetism. As the oxygen becomes heated, it loses its attraction for the filament, and the overall strength of the magnetic field remains constant. This allows the resistance of the filament to be quantitatively related to the concentration of oxygen in a simple fashion. The detector has demonstrated a rapid response time of approximately 1 second to changes in oxygen concentration.

4.0 ELECTROCHEMICAL AND ELECTRODE TECHNIQUES

Several electrode techniques have already been mentioned in relation to detection of the endpoint in reduction–oxidation titrations used with the Winkler and other chemical procedures. Electrodes occupy a more prominent position in the direct determination of oxygen than use in simple endpoint determinations might lead one to believe. This is because of the extreme versatility of electrode configurations and mode of operation. In some designs, such as potentiometric modes of operation, the electrode is used to measure the presence of an intensive property of the system. An intensive property is one that is displayed in the same amount regardless of the size of the sample. An example is the voltage (potential) exhibited by the redox reaction of oxygen under acidic and alkaline conditions:

$$O_2 + 4H^+ + 4e^- \leftrightarrow 2H_2O \qquad E^o = +1.229\ \text{V}$$

$$O_2 + 2H^2O + 4e^- \leftrightarrow 4OH^- \qquad E^o = +0.401\ \text{V}$$

although there is some small dependence of voltage on concentration, as expressed in the Nernst equation

$$E_{obs} = E^0 + \frac{RT}{nF}\ln\frac{[\text{oxidized species}]}{[\text{reduced species}]}$$

Where R is the gas constant, T is the temperature, n is the number of electrons transferred in the reaction, and F is the Faraday constant. This effect is slight in water until the concentration differences are relatively large, which is not to say that attempts to apply the theory to practice have not been made (see Laitinen et al., 1948; Smith & Delahay, 1965). It really does not matter how large the sample is—a milliliter to a full drum—the voltage will be close to the same. A potentiometric titration is based on monitoring the potential with a voltmeter, while a chemical procedure is performed on the sample to remove either the oxidized or reduced species. When the chemical procedure is completed, the observed potential changes rapidly to some other level. Thus, this technique is useful in titrations in which the quantitative information is derived from the amount of chemical reagent that must be added to the sample to bring about the change in voltage that is observed on the electrode monitoring device.

Relative quantitative potentiometric methods are applicable when the analysis is performed in the gas phase rather than in solution. The zirconium oxide (ZrO_2) cells mentioned as detectors in total oxygen demand instruments are examples. Zirconium oxide, commonly called zirconia, is a ceramic material that at temperatures in the range of 600 to 1600 °C becomes conductive to the oxygen ion. It is used to form a selectively permeable barrier between two platinum electrodes, one in an atmosphere of constant composition, while the other is in the analytical gas stream flowing from the oxidation chamber of the instrument. The potential difference between the two electrodes is related to the relative concentrations (partial pressures) of oxygen in the gas streams.

Electrochemical determinations can also be based on counting the number of electrons that are involved in oxidation and reduction reactions at the electrodes. Counting electrons translates to current measurements. This is an extensive property, in that the size of the current is related to the number of reacting molecules in the sample. Measurement of current is precise and sensitive; thus, this technique has evolved into the most widespread technology in use for DO determinations.

Monitoring the current in a sample is used in a qualitative sense in amperometric titrations, in which the electrodes are maintained at a constant voltage and an ammeter is placed in the circuit. As long as reactive species are present in the sample, current flows between the electrodes, as evidenced by a deflection on the ammeter. As titrant is added to the sample, the current level drops, going to zero as the analyte is exhausted from the system, signaling the end of the titration. A quantitative determination is achieved from the amount of chemical reagent that must be added to the sample to achieve zero current.

Theoretically, it is possible then to place some electrodes in a water sample, apply a voltage commensurate with the DO redox couple, and measure the resulting current to determine oxygen concentration. Although theoretically possible, this is largely unsuccessful unless electrodes are used that react specifically with DO, as, for example, a thallium electrode. The primary reason for the failure is that there are too many things in water besides oxygen that can contribute to or interfere with the observed current.

The second reason is that the reaction of oxygen to form water is not very reversible, largely being in the direction of water. Thus, there is only a tiny current resulting from oxygen that is generated in the system, which takes us back to the primary reason that it does not work well.

The commonly used solution to this problem is to separate the oxygen from the sample before electrochemical determination and then provide an alternate source of electrons to make up for the reverse reaction. An oxygen-permeable membrane isolates the electrodes from the sample and allows selective diffusion of oxygen through the membrane and into the electrolyte-filled cavity of the probe. A number of materials have been tested for use as membranes. Collodion was the first reported material (Davies & Brink, 1942) to be used on a DO probe. Many other materials, such as cellophane, silicone rubber, natural rubber, and polyvinyl chloride (PVC) have been examined. The development of fluorocarbon polymers such as polytetrafluoroethylene (PTFE, Teflon®) and the copolymer of tetrafluoroethylene-hexafluoropropylene (Teflon FEP®) as membranes have largely supplanted most of the other materials.

The membrane serves several functions other than isolating oxygen from the sample. It protects the electrodes from mechanical and chemical interferences. The membrane closes the probe compartment and maintains the exact geometric placement of the electrodes. The shape, consistency, and volume of the electrolyte solution that bathes the electrodes remain fixed. Although the mathematics of the operation of polarographic and galvanic DO probes are not being discussed here (an exhaustive discussion is presented in Hitchman, 1978), the rate of diffusion of oxygen from the membrane to the cathode is a key factor in the optimization of the probe. The membrane is an essential component in creating a thin layer of electrolyte solution between the membrane and the cathode and maximizing the mass of oxygen that reaches the cathode in a minimum time. Organic materials in the sample can coat and foul exposed electrodes, rendering them useless. The membrane keeps organics away from the electrodes; however, organics can coat the surface of the membrane and shut down oxygen transfer to the interior of the probe.

Sulfur-containing chemicals are one of the largest classes of substances that can react with the electrodes and cause irreparable damage. Few of these can pass the membrane, thus allowing the DO probe to be used to determine DO in samples that contain sulfites or complex sulfides, which present considerable problems to traditional chemical procedures like the Winkler titration and its variants. Hydrogen sulfide itself can affect the readings of the probe because of its ability to pass through the membrane and react with the anode, cathode, or both. There have been a number of design modifications to correct for the presence of hydrogen sulfide, including layering cadmium nitrate in a sandwiched membrane (Schmid & Mancy, 1969), using a nickel sulfide cathode (Huq & Rosenberg, 1964), or using a sulfide-based electrolyte (Amdur, 1970). But there are no simple ways to get around DO determination interferences in hydrogen sulfide environments with the standard commercial probe unless the sample is pretreated in one of the ways mentioned under the chemical methods for DO analysis.

Oxygen does not react in a single step with four electrons and four protons (hydrogen ions) to make water. As a first approximation of the overall complexity of the process, hydrogen peroxide is found to be a stable intermediate in the system (Hahn et al., 1975; Hitchman, 1978; Jacq & Blach, 1970; van Baalen & Marler, 1966). In recognition of this process, solutions of hydrogen peroxide of known concentration have been used to calibrate polarographic DO electrodes (Griffiths & Jackman, 1962).

$$O_2 + 2H^+ + 2e^- \leftrightarrow H_2O_2 \qquad E^o = +0.695$$

$$H_2O_2 + 2H^+ + 2e^- \leftrightarrow 2H_2O \qquad E^o = +1.776$$

More complexity is introduced through consideration that there are many intertwined routes to hydrogen peroxide with a variety of sequential reactions with single electrons and protons possible. At least four pathways operate to get to hydrogen peroxide, and which one is preferred depends on the exact conditions at that instant in time.

Electrode systems that are based on the reduction of oxygen to either hydrogen peroxide or all the way to water are called polarographic. These systems require that a voltage difference be applied to the two electrodes from an external source so that the cathode is significantly negative relative to the anode (reference) electrode. This allows the reduction reaction of oxygen to hydrogen peroxide or water to proceed and a current to flow. In these systems, the cathode is typically gold or platinum, while the reference electrode that provides an alternate source of electrons is typically Ag–AgCl. The silver anode functions corrosively, oxidizing the silver metal to silver cation and releasing electrons to the electrolyte so that the reduction of oxygen can occur at the cathode.

Electrodes can be chosen such that construction materials possess an intrinsic potential difference that is large enough for the oxygen reduction to occur. These types of cells are called galvanic cells, and function as batteries or fuel cells. The cathode is made of gold or platinum (rarely silver), and the anode is either lead/lead oxide or cadmium/cadmium oxide, the lead anode being more common. An aluminum/zinc combination of electrodes has been reported (Kutyurin, 1963; Smirnov et al., 1969) but did not become popular. The lead anode functions corrosively, generating lead oxide and electrons

$$Pb + 2OH^- \leftrightarrow PbO + H_2O + 2e^- \qquad E^o = +0.576\ V$$

$$O_2 + 2H_2O + 4e^- \leftrightarrow 4OH^- \qquad E^o = +0.401\ V$$

Net reaction is

$$O_2 + 2Pb \rightarrow 2PbO \qquad E^o = +0.977\ V$$

which, coupled with the oxygen reduction at the cathode, gives an overall cell potential of +0.977 V.

Both the polarographic and galvanic designs of the DO probe share the feature that the anode is consumed in the operation of the system. Although the mass of the anode is generally large enough to provide about 3 years of continuous operation, in practice, the working probe lasts a much shorter time. This is because of a buildup of the products of the electrolysis (lead oxide, lead sulfate, or silver chloride, depending on the electrolyte solution and anode material) on the surface of the anode, and a decreased ability to transport the needed neutralizing anions to the oxidized metal cations from the electrode. It is also possible for the anode salt to physically flake off and cause an electrical short within the probe. There are probes designed without anodes that are consumed during operation, generating oxygen from water and no net consumption of oxygen within the cell; however, these come with their own particular set of problems, including a slow response time.

Slow response time by an electrode to a change in DO concentration, also called hysteresis, is an often-encountered problem in electrometric DO measurement. It is particularly bothersome in continuous-flow devices and in situations in which a rapid DO change must be detected to quantify a kinetic event such as the diffusion rate of oxygen through water (Philichi & Stenstrom, 1989; Spanjers & Olsson, 1992). There is no easy way around this problem, and it does get worse as the electrode ages.

Electrochemical determinations of DO are subject to a salt effect (Todt, 1942). The observed salt effect is higher concentration readings from the meter than are actually present in the sample. The root cause for the high bias is that the response of the membrane probe is actually related to the activity times the concentration of the DO in the sample, $\alpha[O_2]$, rather than just the concentration. In samples with total salt concentration (ionic strength) less than 1%, the activity is approximately one, and there is little observable effect. However, at higher salt concentrations (more than 1%), there is a significant increase in the activity of the oxygen, and the observed response is higher than is expected based on the true oxygen concentration. For example, a 5% solution of sodium chloride gives a DO reading approximately 33% higher than the true concentration. Different salts that are dissolved in water affect the oxygen activity coefficient to differing degrees. Thus, for use in high-salt situations, electrodes need to be calibrated using similar solutions of known oxygen concentration. Hydrogen peroxide dosing can be effectively used to prepare known standards (Griffiths & Jackman, 1962), or the oxygen saturation level can be determined using the Winkler or another chemical procedure. Synthetic seawater solutions can be prepared (Smith, 1999a) for calibration of electrodes.

The Hersch cell is a coulometric detector consisting of a platinum or silver wire wound into a spiral cathode suspended in a strongly alkaline solution with a high electrode surface-area-to-electrolyte volume ratio. The anode is lead or lead amalgam (Hersch, 1952, 1960) running down the center of the spiral. Another Hersch cell design has the lead anode coating the walls of a long glass tube while the tube is packed with silver-plated glass beads (Kendall, 1971). The oxygen is reduced on the cathode and the amount of current generated is recorded. The technique requires that the reaction efficiency be close to 100%, which the Hirsch cell provides. The device has been miniaturized for use as a detector in gas chromatography and other instrumental applications.

Table 4.1 presents a summary of some of the significant papers that have been published concerning the construction and use of DO electrodes.

5.0 OXYGEN-FREE SAMPLES

An important quality control when using electrodes for DO determination is establishing the minimum level of response. As the electrode ages, the first symptom of performance degradation is the exhibition of false-positive signals in low-level solutions (Smith, 1999b).

The easiest way to check for degradation of electrode performance is to check the output from the electrode on an oxygen-free solution. Simply boiling a sample of water and then allowing it to cool under oxygen-free conditions is one way to create an oxygen-free test sample. Another is to chemically remove all of the oxygen from the sample. The first two steps of the Winkler titration will produce an oxygen-free sample.

The more commonly used technique has been to add sodium sulfite and a catalytic amount of cobalt to rapidly scrub a sample of all DO (Kikuchi, et al., 1954; Mattengly, 1977). Added cobalt(II) salts react almost instantaneously with oxygen to form the powerful oxidant cobalt (III) in the form of a cobalt–peroxide–cobalt bridged complex (Cotton et al., 1999;

TABLE 4.1. Dissolved oxygen electrode development.

Year	Relation and Comment	Reference
1938	Determination of DO with dropping mercury electrode	Petering & Daniels, 1938
1940	Continuous DO recordings	Manning, 1940
1941	Dropping mercury electrode for DO measurement	Ingols, 1941, 1942
1942	Polarographic DO determination	Todt, 1942
1945	Polarographic determination of DO in seawater	Giguere & Lauzier, 1945
1948	Polarographic determination of DO	Moore et al., 1948
1948	Dropping mercury electrode for continuous DO monitoring	Spoor, 1948
1949	Current reversal to clean electrodes in direct contact with sample	Olson et al., 1949
1950	Polarographic determination of DO	Seaman & Allen, 1950
1951	Compared Winkler and polarographic methods	Rand & Heukelekian, 1951
1952	Dropping mercury electrode	Busch & Sawyer, 1952a, 1952b
1952	Continuous recording of DO	Egalon & Vanhille, 1952
1952	Electrochemical determination of DO in water	Ohle, 1952
1954	Continuous recorder for DO	Briggs et al., 1954
1954	Continuous DO meter	Fitzpatrick & Harkeroad, 1954
1954	Polarographic DO measurement in presence of sulfite	Levine & Williams, 1954
1954	Polarographic determination	Schenk & Berg, 1954
1954	Polarographic determination of DO in water	Todt et al., 1954
1955	Solic platinum electrodes for DO measurement	Ingols, 1955
1955	Solic platinum electrodes for DO measurement	Lynn & Okun, 1955
1955	Polarographic determination of DO	Rosenthal & Conlon, 1955
1955	Oxygen plumet polarographic electrode	Todt & Petsch, 1955
1956	Continuous polarographic measurement of DO with meter and design	Levine et al., 1956
1957	Polarographic measurement of DO	Eckenfelder & Burris, 1957
1957	Continuous monitoring of DO	Husmann & Stracke, 1957
1957	Polarographic determination of DO	Klima, 1957
1957	Design problems of polarographic DO meters	Levine & Kelinschmidt, 1957

(continued)

TABLE 4.1. Dissolved oxygen electrode development (*continued*).

Year	Relation and Comment	Reference
1957	Rotating platinum disk electrode method	Okamoto et al., 1957
1957	Polarographic electrode	Todt, 1957
1958	DO measurement apparatus	Ebbinghaus, 1958
1959	Electric recorder for DO measurement	Bridges et al., 1959
1959	Electrode system for DO	Carritt & Kanwisher, 1959
1959	Performance testing of DO meters	Macklin et al., 1959
1959	Effect of stirring on dropping mercury electrode DO analysis	Potos, 1959
1959	Polarographic electrode	Todt, 1959
1960	Electrometric determination of DO	James & Stephen, 1960
1960	Polarographic measurements in the presence of surfactants	Mancy & Okun, 1960
1960	Electrochemical determination of DO	Okura et al., 1960
1960	Continuous electrode DO readings for process control	Stracke & Keefer, 1960
1962	Compared Winkler, polarographic, and colorimetric methods	Apte et al., 1962
1962	Thallium reactive electrodes	Chenouard et al., 1962
1962	Continuous amperometric determination of DO in seawater	Grasshoff, 1962a, 1962b
1962	Calibration of polarographic meter with hydrogen peroxide	Griffiths & Jackman, 1962
1962	Galvanic cell DO electrodes	Mancy et al., 1962; Mancy & Westgarth, 1962
1962	Cr(III) interference with polarographic DO electrode	Sunahara et al., 1962
1963	Electrically generated oxygen for electrode calibration	Jessop, 1963
1963	Wide bore dropping mercury electrode for continuous reading	Wood, 1963
1964	Polyethylene membrane over Au, Pt, or Ag/Pb galvanic cell	MacKereth, 1964, 1967
1964	Dropping mercury electrode	Malz & Bortlisz, 1964
1964	DO in spent sulfite waters determined with DO electrode	McKeown, 1964
1964	Mercury-coated platinum wire electrode	Ozaki et al., 1964
1964	Icing and storage in dark recommended for sample preservation	Porges, 1964

(*continued*)

TABLE 4.1. Dissolved oxygen electrode development (*continued*).

Year	Relation and Comment	Reference
1965	Pt/Ag galvanic cell with polyethylene membrane	Khaidarov, 1965
1965	Pb/Ag galvanic cell with polyethylene membrane	Kiss & Somloi, 1965
1966	Polarographic method for DO	Canuti, 1966
1966	Pb/Ag electrodes in a galvanic cell	Fleps & Farkas, 1966
1966	Membrane covered galvanic cell preferred for sulfite-containing samples	Petit, 1966
1966	A microelectrode sensor for DO	Pletcher & Sotiropoulos, 1966
1967	Polarographic method with current reversal to clean electrodes	Kamlyuk & Kamlyuk, 1967
1967	DO in swamp waters determined with DO electrode	McKeown et al., 1967
1967	DO determined in spent sulfite waters with electrode	Moore et al., 1967
1967	Tl/Al/Pb three-electrode sensor	Robinson, 1967
1967	Review of DO techniques	Rotshtein & Shemyakin, 1967
1968	Derivative polarographic method for DO in seawater	Parekh & Oza, 1968
1969	Sterilizable galvanic probe	Hemert et al., 1969
1969	Temperature compensation of galvanic DO sensors	Kuz'min & Smirnov, 1969
1969	Chronoamperometric determination of DO with membrane electrodes	Lilley et al., 1969
1969	Flowthrough respirometer with polarographic DO electrode	Olson et al., 1969
1969	Comparison of Winkler and DO sensor with temperature compensation	Reynolds, 1969
1969	Au/Zn membrane-covered galvanic electrode	Smirnov et al., 1969
1970	Sterilizable MacKereth electrode	Harrison & Melbourne, 1970
1970	DO sampler	Wells, 1970a
1971	DO integrator for long-term averaging	Slack, 1971
1973	DO depolarization effect on Zn/Ni electrodes in a cell	Klimenko 1972; Klimenko et al., 1973
1974	Tungsten bronze electrode for potentiometric system	Hahn et al., 1974
1974	Self-contained polarographic DO electrode	Kanwisher & Lawson, 1974
1975	Modification to minimize time response of polarographic electrodes	Kok & Zasic, 1975

(*continued*)

TABLE 4.1. Dissolved oxygen electrode development (*continued*).

Year	Relation and Comment	Reference
1977	Compared Winkler and polarographic electrode on seawater samples; found electrode needed constant recalibration	Atwood et al., 1977
1977	Sampling procedures for polarographic DO measurement	Davison, 1977
1977	A membrane covered Pt/Cu galvanic cell	Poole & Morrow, 1977
1979	Stability of membrane-covered polarographic probes	Heilmann, 1979
1979	Sampler for automated DO measurement	Mueller, 1979
1979	Air calibration of DO electrodes	Ruse, 1979
1979	Compared Winkler titration and DO electrode results	Smeaton, 1979
1979	Membrane-free DO electrodes	Zuellig, 1979
1980	Reducing flow sensitivity of a polarographic electrode	Ben-Yaakov & Ruth, 1980
1980	Review of electrochemical DO sensors	Bobeck, 1980
1980	Novel design of Clark electrode	Friese, 1980
1981	Sample container study for DO determinations	Pebler, 1981
1981	Interlaboratory study suggests DO electrodes are biased high at low DO concentrations	Wilcock et al., 1981
1982	Self-calibrating membrane electrode	Ben-Yaakov & Ben-Aher, 1982
1982	A maintenance-free DO electrode	Phelan et al., 1982
1982	Review of DO measurement	Schramm, 1982
1983	Maintenance requirements for DO meters	Kulin et al., 1983
1984	Compared several electrode systems for continuous monitoring	Barton, 1984
1984	Amperometric determination of DO	Calzolari, 1984
1984	Pulsed electrode design	Langdon, 1984
1984	Low-power DO meter design	Lastein, 1984
1984	Pulse-mode polarization in polarographic sensor	Serak, 1984
1984	U.S. Army Corps of Engineers study recommends polarographic electrodes for fieldwork	Wilhelms, 1984
1985	Examined probe with gelatinous electrolyte and another with a grinder mounted to continuously clean electrodes	Barton & McKeown, 1985
1985	Time-series analysis of electrode responses to monitor biofouling	Howell & Turner, 1985

(*continued*)

TABLE 4.1. Dissolved oxygen electrode development (*continued*).

Year	Relation and Comment	Reference
1985	Oxygen-sensitive membrane for automated DO analysis	Jourdan, 1985
1985	Calibration study for field use of DO electrodes	Prowse, 1985
1987	Polarographic stripping and analysis	Fayyad et al., 1987
1989	Oxygen electrode review	Orna, 1989
1989	Internal guard cathode to protect analytical cathode	Revsbech, 1989
1989	Micro DO sensor	Revsbech et al., 1989
1996	PVC membrane-covered electrode	Sanchez-Pedreno et al., 1996

Robertson, 1980). The initially formed cobalt–peroxide complex then serves to oxidize sulfite, SO_3^{2+}, to sulfate, SO_4^{2+}. This reaction is slower than the initial uptake of oxygen by cobalt. Sulfate is generally inert to further solution manipulations.

The cobalt (III)–peroxide complex is stable to a certain degree, particularly so when nitrogen ligands are present. The complex is found in biochemistry in the active site of vitamin B_{12}. In the presence of limited amounts of sulfite or other oxidizable materials, it can persist in solution. Iodide ion is easily oxidized to iodine by the complex, thus deoxygenations in which excessive amounts of cobalt are used can contribute to falsely high results in subsequent Winkler determinations. Most authors recommend that no more than 0.05 mg/L cobalt catalyst be used for deoxygenation (Kalinske, et al., 1973), and it has been demonstrated that as little as 0.01 mg/L cobalt is effective (Pye, 1947).

The amount of sulfite that needs to be added is based on having an excess in the sample beyond the amount of oxygen present. The most common source of sulfite is sodium sulfite, available as an anhydrous salt in high purity. Simple stoichiometry calculations indicate that at least 35.4 mg/L sodium sulfite is needed to completely react with the approximately 9 mg/L oxygen that is present in a saturated solution. An excess is ensured by adding at least 50 mg/L sodium sulfite.

6.0 CONCLUSION

Although there are many different techniques for measurement of DO, by far the most common method used is the membrane-covered DO electrode.

This is not to say that it is always the most appropriate method, and it is not free from certain limitations. A distinct area of research that is needed is not in the technical measurement of oxygen, but rather in the development of DO solutions of known concentration.

7.0 REFERENCES

Alcock, G. P., & Coates, K. B. (1958). Indigo carmine for the colorimetric determination of low concentrations of dissolved oxygen in water. *Chem. Ind.*, *10*, 554.

Alsterberg, G. (1925). Methods for the determination of elementary oxygen dissolved in water in the presence of nitrite. *Biochem. Z.*, *159*, 36.

Amdur, E. J. (1970). Electrochemical sensor. U.S. Patent 3515658.

American Public Health Association, American Water Works Association, & Water Environment Federation. (1998). *Standard methods for the examination of water and wastewater* (20th ed.). American Public Health Association.

Apte, V. R., Dabadghao, S. B., & Deshpande, W. M. (1962). Rapid methods for the determination of DO. *Environ. Health India*, *4*, 202; *Water Pollut. Abstr.*, *36*, 1442 (1963).

Arnott, J., & McPheat, J. (1953). Estimation of dissolved oxygen in de-aerated water. *J. Eng.*, *176*, 103.

Atwood, D. K., et al. (1977). Comparison of polarographic electrode and Winkler titration determinations of DO in oceanographic samples. *Deep-Sea Res.*, *24*, 311.

Bach, H. (1929). Determination of dissolved oxygen in water. *Gesundh. Ing.*, *52*, 36; *Chemical Abstracts*, *23*, 2773 (1929).

Bach, H. (1931). Sewage determinations. XIII. Determination of oxygen in the presence of sulfites. XIV. Practical limits in the determination of BOD. *Gesundh. Ing.*, *54*, 39; *Chemical Abstracts*, *25*, 3749 (1931).

Baker, M. C. (1965). Test for DO: Correction to the Winkler method. *Tappi J.*, *48*(4), 81.

Barnett, G. R., & Hurwitz, E. (1939). The use of sodium azide in the Winkler method for the determination of dissolved oxygen. *Sew. Works J.*, *11*, 781.

Barton, D. A. (1984). A study of recently developed continuous dissolved oxygen measurement systems based on their field performance. Natl. Council Air Stream Improvement Tech. Bull., No. 440.

Barton, D. A., & McKeown, J. J. (1985). Experience with recently developed dissolved oxygen instrumentation monitoring a paper industry wastewater treatment process. In R. A. R. Drake (Ed.), *Instrumentation and control of water and wastewater treatment and transport systems*. Pergamon Press.

Beadle, L. C. (1958). Measurement of dissolved oxygen in swamp waters: Further modification of the Winkler method. *J. Exp. Biol.* (Br.), *35*, 556. *Publ. Health Eng. Abstr.*, *39*(10), 30 (1959).

Beaujean, P. (1956). The consumption of dissolved oxygen a legal test of water. *Bull. Cent. Belge Etude Document Eaux*, *65*, 192; *Publ. Health Eng. Abstr.*, *37*(1), 18 (1957).

Ben-Yaakov, S., & Ben-Asher, Y. (1982). Continuous measurements of DO in water culture by a self-calibrating monitor. *Water Res.*, *16*, 169.

Ben-Yaakov, S., & Ruth, E. (1980). A method for reducing the flow sensitivity of a polarographic do sensor. *Talanta* (Great Britain), *27*, 391.

Berk, A., & Hofmann, P. (1963). Estimation of DO and determination of permanganate oxidizability in waters. *Chem. Prumysl.*, *13*, 287; *Chemical Abstracts*, *59*, 9675 (1963).

Berka, A., Glassi, H., & Hoffman, P. (1968). Die bestimmung gelosten sauerstoffes in gegenwart von zweiwertigen eisen und chlor in wasser. *Mikrochim. Acta*, *997*.

Bobeck, R. F. (1980). Electrochemical sensors for oxygen analysis. *Chem. Eng.*, *87*, 113.

Boyles, W. (1997a). Manganese III COD. Technical Paper 17057, Hach Company.

Boyles, W. (1997b). The science of chemical oxygen demand. Tech. Inf. Ser., Booklet No. 9, Hach Company.

Bridges, R., Dyke, G. V., & Knowles, G. (1959). Electric recorder for dissolved oxygen. *Water Waste Treatment J.*, *5*, 192; *Public Health Eng. Abstr.*, *39*(9), 27 (1959).

Briggs, R., Knowles, G., & Scragg, L. J. (1954). A continuous recorder for dissolved oxygen. *Water Sanit. Eng.*, *5*(5–6), 24; *Analyst*, *79*, 744 (1954); *Public Health Eng. Abstracts*, *34*(9), 19 (1954).

Brodie, B. (1865). Minutes of evidence, Royal Sewage Commission of 1857. *Rep. Thames River*, *2*, 49.

Broenkow, W. W., & Cline, J. D. (1969). Colorimetric determination of DO at low concentrations. *Limnol. Oceanogr.*, *14*, 450.

Bryan, J. R., Riley, J. P., & Williams, P. J. LeB. (1976). A Winkler procedure for making precise measurements of oxygen concentration for productivity and related studies. *J. Exp. Mar. Biol. Ecol., 21*, 191; *Chemical Abstracts, 85*, 166381 (1976).

Buchoff, L. S., Ingber, N. M., & Brady, J. H. (1955). Determination of low concentrations of dissolved oxygen in water. *Anal. Chem.*, 27, 1401.

Busch, A. W., & Sawyer, C. N. (1952a). Application of the dropping mercury electrode to BOD Determinations. U.S. Atomic Energy Comm. Rep., Unclassified, NYO 4436, Contract No. AT(30-1)-621; *Nuclr. Sci. Abstr.*, 6, 797; *Public Health Eng. Abstracts, 33*(3), 15 (1953).

Busch, A. W., & Sawyer, C. N. (1952b). Determination of the BOD of sewage and industrial wastes with the polarograph. *Anal. Chem., 24*, 1887.

Buswell, A. M., & Gallagher, W. W. (1923). The determination of dissolved oxygen in the presence of iron salts. *Ind. Eng. Chem., 15*, 1186.

Calzolari, S. G. (1984). Amperometric determination of DO and its application in wastewater treatment. *Inquinamento*, *26*(7–8), 47; *Chemical Abstracts, 101*, 197413 (1984).

Canuti, A. (1966). Rapid determination of DO and BOD in waters by a polarographic method. *Bell Lab. Chem. Proc., 17*, 93.

Carritt, D. E., & Carpenter, J. H. (1966). Comparison and evaluation of currently employed modifications of the Winkler method for determining dissolved oxygen in sea water. *J. Marine Res., 24*, 286.

Carritt, D. E., & Kanwisher, J. W. (1959). An electrode system for measuring dissolved oxygen. *Anal. Chem., 31*, 5.

Chenouard, J., Dirian, G., & Gabilly, R. (1962). Continuous determination of DO in water down to 0.002 mg/L. *Anal. Chim. Acta* (Netherlands), *44*, 334.

Chernyakova, A. M., et al. (1983). Comparison of the determination of oxygen dissolved in water using the Winkler method (Report of a 1981 Soviet-American experiment). *Okeanologiya, 23*, 681.

Chung, K. E., Lan, E. H., Davidson, M. S., Dunn, B. S., Valentine, J. S., & Zink, J. I. (1995). Measurement of DO in water glass-encapsulated myoglobin. *Anal. Chem., 67,* 1505.

Clarke, R. W. (1910). The determination of the amount of dissolved oxygen absorbed by sewage effluents containing nitrites and of the amount of nitrites in sewage effluents and water. *Analyst, 35*, 393.

Cost, J. H., & Andrews, E. R. (1923). A case in which Winkler's manganous process for dissolved oxygen is untrustworthy. *Analyst, 48*, 543; *J. Phys. Chem., 28*, 285 (1924).

Cotton, F. A., Wilkinson, G., Murillo, C. A., & Bachmann, M. (1999). *Advanced inorganic chemistry* (6th ed.). Wiley & Sons.

Davies, P. W., & Brink, F. (1942). Microelectrodes for measuring local oxygen tension in animal tissues. *Rev. Sci. Instr.*, *13*, 524.

Davison, W. (1977). Sampling and handling procedures for the polarographic measurement of oxygen in hypolimnetic waters. *Freshwater Biol.*, *7*, 393.

de Carvalho, A. H., Calado, J. G., & de Moura, M. L. (1963). Spectrophotometric determination of DO in demineralized water. *Rev. Port. Quim.*, *5*(1), 15; *Chemical Abstracts*, *60*, 10386 (1964).

Devdariani, I. V., & Shmal'tsel, G. N. (1969). Methods and apparatus for determining water-soluble oxygen with methylene blue. *Prib. Sisl. Upt.*, *9*, 42; *Chemical Abstracts*, *72*, 82834 (1970).

Doronin, A. N., Kabanova, O. L., & Timofeev, S. A. (1969). Thallium column for the determination of oxygen dissolved in water at 80 °C. *Zh. Anal. Khim.*, *24*, 274.

Dubovenko, L. I., Pilipenko, L. A., Sarylov, V. N., & Radchenko, E. M. (1979). Determination of oxygen dissolved in water by a chemiluminescent method. *Khim. Tekhnol. Vody* (Russia), *1*, 35; *Chemical Abstracts*, *93*, 209974 (1980).

Duedall, I. W., et al. (1971). Dispenser for Winkler's DO reagents. *J. Fish Res. Bd. Can.*, *28*, 1815.

Dupre, A. (1884). A report on the changes in the aeration of water, as indicating the nature of the impurities present in it. 14th Annual Report Local Gov. Board, 1883–1884, App. B., 11, 304.

Duval, W. S., et al. (1974). Spectrophotometric determination of DO concentration in water. *J. Fish Res. Bd. Can.*, *31*, 1529.

Ebbinghaus, E. (1958). Apparatus for recording the oxygen content of waste waters. *Wasserivertschaft*, *48*, 320; *Chemical Abstracts*, *53*, 10606 (1959).

Eckenfelder, W. W., & Burris, C. T. (1957). Polarographic measurement of dissolved oxygen. *Am. Soc. Testing Mater.*, Spec. Tech. Pub., *219*, 18; *Chemical Abstracts*, *52*, 9843 (1958).

Egalon, R., & Vanhille, H. (1952). Continuous analysis with recording of oxygen (or carbon dioxide). In *Boiler Feed Water* (IV Cong. Int. Chauffage Ind., reprint No. 224); *Chemical Abstracts*, *48*, 3601d (1954).

Elliot, J. W. (1963). A sensitive photometric procedure for DO with potential field applicability. *Prog. Fish-Cult.*, *25*, 42.

Ellis, M. M., & Ellis, M. D. (1943). Photoelectric determination of dissolved oxygen with amidol. *Sew. Works J.*, *15*, 1115.

Ellis, J., & Kanamori, S. (1973). An evaluation of the Miller method for do analysis. *Limnol. Oceanogr.*, *18*, 1002.

Elmore, H. A., & Hayes, T. W. (1960). Solubility of atmospheric oxygen in water. *J. Sanit. Eng. Div., Proc. Am. Soc. Civ. Eng.*, *86*(SA4), I, 41.

Elmore, H. A., & Hayes, T. W. (1961). Solubility of atmospheric oxygen in water. Closure by Comm. Sanit. Eng. Res., Sanit. Eng. Div., *Proc. Am. Soc. Civ. Eng.*, *87*(SA5), I, 73.

Elvove, E. (1914). A comparison of methods for the determination of oxygen in water in the presence of nitrite. *Hyg. Lab. Bull.*, *96*, 14; *Chemical Abstracts*, *9*, 2117 (1915).

Erdey, L., & Szabadvary, F. (1954). Ascorbinometric determination of oxygen dissolved in water. *Acta Chim. Hung.*, *4*, 325; *J. Appl. Chem.*, *5*(2), 161 (1955).

Fadrus, H., & Maley, J. (1968a). Modified determination of oxygen in water by titration with Fe(II). *Vom Wasser* (Germany), *34*, 132.

Fadrus, H., & Maley, J. (1968b). Oxygen determination in the presence of thiourea. *Staedtehygiene* (Germany), *19*, 93; *Chemical Abstracts*, *69*, 45866 (1968).

Fadrus, H., & Maley, J. (1968c). The indirect chelatometric determination of oxygen. *Stadtehygiene* (Germany), *19*, 53; *Chemical Abstracts*, *68*, 38670 (1968).

Fadrus, H., & Maley, J. (1975). Problems connected with the effect and removal of nitrites in oxygen determination in waters. *Acta Hydrochim. Hydrobiol.*, *3*, 333.

Fayyad, M., et al. (1987). Dissolved oxygen: Method comparison with potentiometric stripping analysis. *Anal. Lett.*, *20*, 529.

Federation of Sewage Works Associations Committee Report (1932). The determination of dissolved oxygen by the Winkler method. *Sew. Works J.*, *3*, 413.

Felicetta, V. F., & Kendall, D. R. (1965). DO determination in waters containing spent sulfite liquor. rapid permanganate modification of the Winkler test and galvanic cell methods. *Tappi J.*, *48*, 363.

Fitzpatrick, H. M., & Harkeroad, M. E. (1954). A meter for continuous indication of dissolved air in water. U.S. Atomic Energy Comm., Doc. No. DTMB-867; *Nucl. Sci. Abstr. 9*, 3068 (1955).

Flaschka, H. A., Barnard, A. J., & Sturrock, P. E. (1980). *Quantitative analytical chemistry* (2nd ed.). Willard Grant Press.

Fleps, W., & Farkas, P. (1966). Electrochemical method for DO determination in water. *Beszamolo Vizgazdalkodasi Tud. Kut. Intez. Munkajorol*, 228; *Chemical Abstracts*, *66*, 98381 (1967).

Fox, C. J. J. (1909). On the coefficients of absorption of nitrogen and oxygen in distilled water and sea water and of atmospheric carbonic acid in sea water. *Trans. Faraday Soc.*, *5*, 68.

Frankland, E. (1870). *First report of the Rivers Pollution Commission of 1868*, 18.

Freier, R. (1952). New methods for determination of dissolved oxygen in the presence of reducing substances. *Vom Wasser* (Germany), *19*, 124.

Freier, R. (1955). Potentiometric determination of small concentrations of oxygen by the Winkler method. *Mitt. Ver. Grosskesselbesitzer*, *36*, 62; *Chemical Abstracts*, *50*, 727 (1956).

Friese, P. (1980). A new type of Clark oxygen electrode. *Electroanal. Chem. Interfacial Electrochem.*, *106*, 409.

Gaunt, H., & Shanks, C. (1965). Determination of oxygen in boiler feed water. *Chem. Ind.*, *8*, 328; *Chemical Abstracts*, *62*, 11533 (1965).

Gerardin, A. (1875). Alterations de la Seine aux abords de Paris depuis Novembre, 1874, Justua Mai, 1875. *Compt. Rend.*, *80*, 1326.

Gerardin, A. (1876). Examen des eaux pluviales relevees aux eudiometres de l'Observatoire de Paris. *Compt. Rend.*, *81*, 989.

Giguere, P. A., & Lauzier, L. (1945). Polarographic determination of oxygen dissolved in sea water. *Can. J. Res.*, *23B*, 76.

Gilcreas, F. W. (1935). A colorimetric method for the determination of dissolved oxygen. *J. Am. Water Works Assoc.*, *27*, 1166.

Gilcreas, F. W. (1952). Better plant operation with laboratory control. III. The oxygen requirements of sewage. *Wastes Eng.*, *23*, 348.

Golterman, H. L., & Wisselo, A. G. (1981). Ceriometry, a combined method for COD and DO (with a discussion on the precision of the Winkler technique). *Hydrobiologia*, *77*, 37.

Grasshoff, K. (1962a). Determination of DO in sea-water. II. Continuous amperometric determination of the vertical distribution of oxygen in seawater. *Kiel. Meeresforsch.* (Germany), *18*, 2151; *Chemical Abstracts*, *58*, 11096 (1963).

Grasshoff, K. (1962b). Determination of oxygen in sea-water. *Kiel. Meeresforsch.* (Germany), *18*, 42.

Grasshoff, K. (1981). The electrochemical determination of oxygen. In M. Whitfield and D. Jagner (Eds.), *Marine electrochemistry.* Wiley & Sons.

Griffiths, V. S., & Jackman, M. I. (1959). Automatic potentiometric titration of dissolved oxygen. *Anal. Chem.*, *31*, 161.

Griffiths, V. S., & Jackman, M. I. (1962). A polarographic study of DO. *Talanta*, *9*, 205.

Hahn, C. E. W., Davis, A. H., & Albery, W. J. (1975). Electrochemical improvement of the performance of pO_2 electrodes. *Respir. Physiol.*, *25*, 109.

Hahn, P. B., et al. (1974). Measuring DO in aqueous solution using tungsten bronzes as a potentiometric indicating electrode. U.S. Patent 3856634.

Hamlin, P. A., & Lambert, J. L. (1971). Determination of dissolved oxygen using photoreduced leuco phenothiazine dyes. *Anal. Chem.*, *43*, 618.

Harper, E. L. (1953). Semi-microdetermination of dissolved oxygen. *Anal. Chem.*, *25*, 187.

Harrison, D. E. F., & Melbourne, K. V. (1970). Autoclavable version of the Mackereth oxygen probe. *Biotechnol. Bioeng.*, *12*, 633.

Heilmann, J. (1979). Stability of membrane-coated polarographic oxygen electrodes. *Gewaesserschutz, Wasser. Abwass.* (Germany), *39*, 67; *Chemical Abstracts*, *92*, 99347 (1980).

Hemert, P. V., et al. (1969). A steam-sterilizable electrode of the galvanic type for the measurement of DO. *Biotechnol. Bioeng.*, *11*, 549.

Hersch, P. A. (1952). Galvanic determination of traces of oxygen in gases. *Nature*, *169*, 792.

Hersch, P. A. (1960). Trace monitoring in gases using galvanic systems. *Anal. Chem.*, *32*, 1030.

Heukelekian, H. (1938). Dissolved oxygen: Its determination and importance in the activated sludge process. *Water Works Sewerage*, *85*, 715.

Hillman, G. E., & Lightwood, J. (1966). Determination of small amounts of oxygen using a Hersch cell as a gas chromatographic detector. *Anal. Chem.*, *38*, 1430.

Hitchman, M. L. (1978). *Chemical analysis, Vol 49: Measurement of dissolved oxygen.* Wiley & Sons.

Hoover, C. B. (1911). A method for determining the parts per million of dissolved oxygen consumed by sewage and sewage effluents, Columbus Sewage Works. *Eng. News-Rec.*, *65*, 311.

Hoover, S. R., Jaswicz, L., & Porges, N. (1954). The magnetic oxygen analyzer in studies of oxygen uptake: A new instrument for measuring aerobic oxidation rates. *Water Sew. Works*, *101*, 84.

Howell, J. A., & Turner, C. (1985). On-line estimation of the time constant of oxygen electrodes by time series analysis. *Inst. Chem. Eng. Symp. Ser.*, *94*, 309.

Hua, H. (1990). Accurate method for calculation of saturation DO. *J. Environ. Eng., 116*, 988.

Husmann, W., & Stracke, G. (1957). Continuous measurement of oxygen in river water and sewage. *Wasserwirtsch.-Wassertech.* (Germany), *48*, 13; *Chemical Abstracts, 52*, 19679 (1958).

Huq, A. K. M. S., & Rosenberg, A. J. (1964). Electrochemical behavior of nickel compounds. *J. Electrochem. Soc., 111*, 270.

Ingols, R. S. (1941). Determination of dissolved oxygen by the dropping mercury electrode. *Sew. Works J., 13*, 1097.

Ingols, R. S. (1942). Dissolved oxygen recordings with the dropping mercury electrode. *Ind. Eng. Chem., Anal. Ed., 14*, 256.

Ingols, R. S. (1955). Experience with solid platinum electrodes in the determination of dissolved oxygen. II. Georgia Institute of Technology studies. *Sew. Ind. Wastes, 27*, 7.

Ingvorsen, K., & Jorgensen, B. B. (1979). Combined measurement of oxygen and sulfide in water samples. *Limnol. Oceanogr., 24*, 390.

Issacs, M. L. (1935). A colorimetric method for the determination of dissolved oxygen. *Sew. Works J., 7*, 435.

Ivanitskaya, A. S. (1969). Selection of the optimum method for determining small concentrations of oxygen in water at operating conditions. *Vodopodgot. Vodn. Rezhim Khimkontrol Parosilovykh Ustanovkakh* (Russia), *3*, 174; *Chemical Abstracts, 71*, 84421 (1969).

Ivanoff, A. (1962). On a colorimetric method of determining DO in sea water. *C. R. Acad. Sci., 254*, 4493.

Jacq, J., & Blach, O. (1970). Electrochemical reduction of oxygen in hydrogen peroxide solution. *Electrochim. Acta, 15*, 1945.

James, G., & Stephen, M. (1960). Electrometric determination of dissolved oxygen in aqueous solution. *Analyst, 85*, 35.

Jessop, G. (1963). Improvements in and relating to methods of and apparatus for measuring the concentration of oxygen dissolved in liquids. British patent 858677. *Water Pollut. Abstr., 36*, 432 (1963).

Jones, M. M., & Mullen, M. W. (1973). Some aspects of the Winkler determination of oxygen in water. *Talanta, 20*, 327.

Jourdan, J. (1985). Analysis of residual concentration of dissolved oxygen in boiler water. *Eau. Ind. Nuisances, 92*, 46.

Kabanova, O. L., & Zalogina, E. A. (1965). The determination of trace amounts of dissolved oxygen in water by the use of a thallium column at 32 and 40 °C. *Zh. Anal. Chim., 20*, 608.

Kalinske, A. A., Lash, L. D., & Shell, G. L. (1973). Cobalt interferences in non-steady state clean water test. *Water Sew. Works*, *120*, 7, 54.

Kamlyuk, L. V., & Kamlyuk, A. K. (1967). Determination of oxygen dissolved in water by a polarographic method with the aid of a solid electrode with electrochemical depolarization. *Dokl. Akad. Nauk. SSSR* (Russia), *11*, 368; *Chemical Abstracts*, *67*, 76179 (1967).

Kampbell, D. H., et al. (1989). Dissolved oxygen and methane in water by a GC headspace equilibration technique. *Int. J. Environ. Anal. Chem.*, *36*, 249.

Kanwisher, J. W., & Lawson, K. D. (1974). An improved, self-contained polarographic DO probe. *Limnol. Oceanogr.*, *19*, 700.

Karlsson, R., & Torstensson, L. G. (1974). Coulometricdetermination of DO in water. *Talanta*, *21*, 957.

Kendall, D. R. (1971). Improved Hersch cell system for rapid determination of oxygen in gases. *Anal. Chem.*, *43*, 944.

Khaidarov, I. S. (1965). Measuring cell for the determination of DO. *Vestnik Moskov. Univ. Ser. VI Biol., Pochnvoved*, *20*(4), 59; *Chemical Abstracts*, *64*, 7867 (1966).

Kikuchi, S., Honda, K., & Kim, S. (1954). Removal of dissolved oxygen by sodium sulfite: use in polarographic measurements. *Bull. Chem. Soc. Japan*, *27*, 65; *Public Health Eng. Abstr.*, *38*(12), 26 (1958).

Kimberly, A. E. (1908). Report on examination of sewage treatment plants. Ohio State Board of Health, Rep. Water Sew. Purif. Ohio 1906–1907, 371.

Kishi, H., Arimoto, H., & Fujii, T. (1998). Analysis of alcohols and phenols with a newly designed gas chromatographic detector. *Anal. Chem.*, *70*, 3488.

Kiss, P., & Somloi, J. (1965). The electrochemical determination of oxygen dissolved in water. *Kozp. Elelmiszeripari Kutato Intez. Kozlem. N.*, *1*, 23; *Chemical Abstracts*, *67*, 36305 (1967).

Klima, D. (1957). Simple arrangement for the polarographic determination of dissolved oxygen in industrial wastes. *Voda*, *36*, 148.

Klimenko, A. M. (1972). Theory of electrochemical devices for continuous measurement of concentration of oxygen dissolved in water. *Zb. Nauk. Rob. Aspir., L'vov Politekh. Inst.*, *6*, 113; *Chemical Abstracts*, *78*, 143394 (1973).

Klimenko, A. M., et al. (1973). Industrial testing of an automatic oxygen meter. *Energ. Elek.*, *1*, 13; *Chemical Abstracts*, *78*, 163888 (1973).

Klotts, C. E., & Benson, B. B. (1963). Solubilities of nitrogen, oxygen and argon in distilled water. *J. Mar. Res.*, *21*, 48.

Knapp, W. G. (1959). Internal electrolysis colorimetric method for determination of small quantities of oxygen. *Anal. Chem.*, *31*, 1463.

Kok, R., & Zasic, J. E. (1975). Dynamic response of a polarographic oxygen probe. *Biotechnol. Bioeng.*, *17*, 527.

Krenkel, P. A., Thackston, E. L., & Parker, F. L. (1969). Impoundment and temperature effects on waste assimilation. *J. Sanit. Eng. Div., Proc. Am. Soc. Civ. Eng.*, *95*, 37.

Kulin, G., et al. (1983). Evaluation of a dissolved oxygen field test protocol. *J. Water Pollut. Control Fed.*, *55*, 178.

Kutyurin, V. M. (1963). An amperometric method for the determination of oxygen in water. *Zh. Anal. Khim.* (Russia), *18*, 765; *Chemical Abstracts*, *59*, 12507 (1964).

Kuz'min, A. A., & Smirnov, D.N. (1969). Modern methods and apparatus for the automatic control of the concentration of DO in wastewaters. *Ochistka Proiz-ed Stochnykh Vod.*, *4*, 200; *Chemical Abstracts*, *73*, 91032 (1970).

Laitinen, H. A., Higuchi, T., & Czuha, M. (1948). Potentiometric determination of oxygen using the dropping mercury electrode. *J. Am. Chem. Soc.*, *70*, 561.

Lakin, M. B. (1976). Chemical catalyst interference in the Winkler titration determinations of DO. A Method for Correction. *Water Res.*, *10*, 961.

Langdon, C. (1984). Dissolved oxygen monitoring system using a pulsed electrode: Design, performance, and evaluation. *Deep-Sea Res., Part A.*, *31*, 1357.

Lastein, E. (1984). Automatic systems for measurements in shallow lakes. *Arch. Hydrobiol.*, *96*, 245.

Lebska, B. (1977). Ascorbimetric determination of oxygen content of water. *Przegl. Papier.* (Poland), *33*(4), 123; *Abstr. Bull. Inst. Paper Chem.*, *48*, 4933 (1977).

Lederer, A. (1910). Influence of storage and various preservatives upon the dissolved oxygen in water. Sewage Disposal Investigations, Chicago Sanit. Dist.

Levi, E. M., et al. (1977). A titration apparatus for the determination of DO in seawater. *J. Fish Res. Bd. Can.*, *34*, 2218.

Levine, H. S., & Kelinschmidt, R. S. (1957). Principles and problems in development of a dissolved oxygen analyzer. *Sew. Ind. Wastes*, *29*, 856.

Levine, H. S., Warren, W. V., Tsivoglou, E. C., & Walker, W. W. (1956). Continuous measurement of dissolved oxygen in water. *Anal. Chem.*, *28*, 343.

Levine, H. S., & Williams, O. J. (1954). Polarographic determinations of dissolved oxygen in dilute sulfite waste liquor. *Anal. Chem.*, *26*, 1297.

Li, Y., & Yang, Y. (1982). Determination of dissolved oxygen, nitrogen, and total carbon dioxide in water by gas chromatography after gas–liquid extraction. *Fenxi Huaxue*, *10*, 762; *Chemical Abstracts*, *98*, 185267 (1984).

Lilley, M. D., Story, J. B., & Raible, R. W. (1969). Chronoamperometric determination of DO using membrane electrodes. *J. Electroanal. Chem. Interfacial Electrochem.*, *23*, 425.

Logan, B. E., & Kohler, D. (2001). Oxygen mass-transfer coefficients for different sample containers used in the headspace BOD test. *Water Environ. Res.*, *73*, 58.

Logan, B. E., & Patniak, R. (1997). A gas chromatographic-based Headspace BOD test. *Water Environ. Res.*, *69*, 206.

Logan, B. E., & Wagenseller, G. A. (1993). The HBOD test: A new method for determining BOD. *Water Environ. Res.*, *65*, 862.

Lueck, W. (1967). Oxygen trace measurements in water by the thallium method. *Energie*, *19*, 203; *Chemical Abstracts*, *68*, 71997 (1968).

Lueck, W. (1968). Oxygen trace measurements in water by the thallium method. *Deut. Ges. Chem. Apparatewesen Monogr.*, *61*, 155; *Chemical Abstracts*, *70*, 50366(1969).

Lueck, W. (1970). Industrial measurements of traces of oxygen in light and heavy water by the thallium method. *Messtechnik*, *78*, 181.

Lynn, W. R., & Okun, D. A. (1955). Experience with solid platinum electrodes in the determination of dissolved oxygen. *Sew. Ind. Wastes*, *27*, 4.

MacKereth, F. J. H. (1964). An improved galvanic cell for determination of oxygen concentration in fluids. *J. Sci. Instr.*, *41*, 1, 38; *Chemical Abstracts*, *60*, 2307 (1964).

MacKereth, F. J. H. (1967). Electrolytic oxygen sensor. U.S. Patent 3322662; *Chemical Abstracts*, *67*, 78556 (1967).

Macklin, M. O., Baumgartner, D. J., & Ettinger, M. B. (1959). Performance test of continuous recording dissolved oxygen analyzer. *Sew. Ind. Wastes*, *31*, 804.

Malik, A. V. (1964). On the complexiometric determination of DO in water. *Vom Wasser* (Germany), *31*, 301.

Malik, A. V. (1966). Modification of Winkler's method for the determination of DO in water. *Indian J. Tech.*, *4*, 32.

Malz, F., & Bortlisz, H. (1964). Application of the controlled dropping mercury electrode to the continuous measurement of dissolved oxygen in sewage or sewage-activated sludge mixtures. *Zeits. Anal. Chem., 206*, 409; *Chemical Abstracts, 62*, 2611 (1965).

Mancy, K. H., & Okun, D. (1960). Automatic recording of dissolved oxygen in aqueous systems containing surface-active agents. *Anal. Chem., 32*, 108.

Mancy, K. H., Okun, D. A., & Reilley, C. N. (1962). A galvanic cell oxygen analyzer. *J. Electroanal. Chem., 4*, 65.

Mancy, K. H., & Westgarth, W. C. (1962). A galvanic cell oxygen analyzer. *J. Water Pollut. Control Fed., 34*, 1037.

Manning, M. (1940). A method of obtaining continuous records of dissolved oxygen in lake waters. *Ecology, 21*, 509.

Mattengly, G. E. (1977). Experimental study of wind effects on aeration. *J. Hydraul. Div., Proc. Am. Soc. Civ. Eng., 103*, 311.

McEvoy, A. K., McDonagh, C. M., & MacCraith, B. D. (1996). Dissolved oxygen sensor based on fluorescence quenching of oxygen sensitive ruthenium complexes immobilized in sol-gel derived porous silica coatings. *Analyst, 121*, 785.

McKeown, J. J. (1964). Analytical instrumentation and stream ecology investigations. *Pulp Paper Mag. Can., 65*, 251; *Chemical Abstracts, 61*, 15823 (1964).

McKeown, J. J., Brown, L. C., & Gove, G. W. (1967). Comparative studies of DO analysis methods. *J. Water Pollut. Control Fed., 39*, 1323.

Miller, D. G., Brayton, S. V., & Boyles, W. T. (2001). COD analysis of wastewater using trivalent manganese oxidant with chloride removal by sodium bismuthate pretreatment. *Water Environ. Res., 73*, 63.

Miller, J. (1914). A field method for determining dissolved oxygen in water. *J. Soc. Chem. Ind., 33*, 185.

Mizuniwa, F., et al. (1976). Spectrophotometric determination of DO in water with indigo carmine. *Bunseki Kagaku, Shimpo Sosetsu, 25*, 89; *Chemical Abstracts, 85*, 182165 (1976).

Moggio, W. A., & Burland, P. D. (1951). Standardization of *n*/40 thiosulfate solutions with saturated solutions of dissolved oxygen. *Sew. Ind. Wastes, 23*, 1243.

Montgomery, H. A. C., Thom, N. S., & Cockburn, A. (1964). Determination of dissolved oxygen by the Winkler method, and the solubility of oxygen in pure water and sea water. *J. Appl. Chem., 14*, 280.

Moore, E. W., Morris, J. C., & Okum, D. A. (1948). The polarographic determination of dissolved oxygen in water and sewage. *Sew. Works J., 20*, 1041.

Moore, M. G., Rieck, R. H., & Moore, W. A. (1967). Determination of DO in the presence of sulfite waste liquor. *J. Water Pollut. Control Fed.*, *39*, 64.

Mori, M., & Ishihara, Y. (1971). Determination of dissolved oxygen. III. Modification of Shibata–Miller method. *Kogyo Yousi*, *149*, 32; *Chemical Abstracts*, *75*, 40176 (1971).

Mueller, D. (1979). Sampler for automatic measurement of the oxygen content of water samples. *Z. Wasser Abwasser Forsch.* (Germany), *12*, 179; *Chemical Abstracts*, *92*, 28299 (1980).

Mukhin, N. N., & Aleskovskii, V. B. (1966). Determination of oxygen dissolved in water by the isotope dilution method. *Zavod. Lab.* (Russia), *32*, 110; *Chemical Abstracts*, *66*, 5654 (1967).

Muller, M. (1889). Determination of oxygen dissolved in water. *Chem. Zeit.*, *13*, 1188.

Nasbaum, I. (1958). A modified Winkler method for dissolved oxygen determination. *Water Sew. Works*, *105*, 469.

Noll, H. (1917). Estimation of dissolved oxygen in water in the presence of nitrites and organic substances. *Z. Angew. Chem.*, *30*, 105; *J. Chem. Soc.*, *112*, 502 (1917).

Ogren, L., & Johansson, G. (1983). An enzymatic flow injection method for the determination of oxygen. *Anal. Chem. Acta*, *145*, 101.

Ohle, W. (1952). Investigation and application of electrochemical determination of dissolved oxygen for examination of waters. *Vom Wasser* (Germany), *19*, 99.

Okamoto, G., Okura, T., & Goto, K. (1957). Determination of dissolved oxygen in waste waters with a rotating platinum electrode. *J. Appl. Chem.*, *7*, 338.

Okura, T., Goto, K., & Furukawa, T. (1960). Determination of dissolved oxygen in polluted waters. *Mem. Fac. Eng. Hokkaido Univ.*, *2*, 2.

Olson, R. A., Brackett, F. S., & Crickart, R. G. (1949). Oxygen tension measurement by a method of time selection using the static platinum electrode with an alternating potential. *J. Gen. Physiol.*, *32*, 681.

Olson, T. A., et al. (1969). Flow-through polarographic respirometer for aquatic animals. *Hydrobiologia*, *34*, 322.

Orna, M. V. (1989). Oxygen electrode. History, design, and applications. *Am. Chem. Soc. Symp. Ser.*, *390*, 196.

Oudot, C., Gerard, R., Morin, P., & Gringue, I. (1988). Precise shipboard determination of dissolved oxygen (Winkler procedure) for productivity studies with a commercial system. *Limnol. Oceanogr.*, *33*, 146.

Ozaki, T., Suzuki, J., & Izawa, K. (1964). Analysis with convection electrodes. VI. Determination of dissolved oxygen. *Bunseki Kagaku, Shimpo Sosetsu, 13*, 107; *Chemical Abstracts, 67*, 926 (1967).

Parekh, K. J. M., & Oza, P. M. (1968). Determination of oxygen and nitrate in sea water by derivative polarographic method. *Salt. Res. Ind., 5*, 44; *Chemical Abstracts, 69*, 99250 (1968).

Pebler, A. (1981). Sample containers for trace analysis of dissolved oxygen, hydrogen, and carbon dioxide. *Anal. Chem., 53*, 361.

Petering, H. C., & Daniels, F. (1938). The determination of dissolved oxygen by means of the dropping mercury electrode with applications in biology. *J. Am. Chem. Soc., 60*, 2796.

Petit, P. J. (1966). DO analysis in the presence of spent sulfite liquor. *Water Wastes Eng., 3*, 8, 36.

Phelan, D. M., et al. (1982). A maintenance-free DO monitor. *Int. Lab., 12*, 60.

Philichi, T. L., & Stenstrom, M. K. (1989). Effects of dissolved oxygen probe lag on oxygen transfer parameter estimation. *J. Water Pollut. Control Fed., 61*, 183.

Phillips, T. R., Johnson, E. G., & Woodword, H. (1964). The use of a Hersch cell as a detector in gas chromatography. *Anal. Chem., 36*, 450.

Placak, O. R., & Ruchhoft, C. C. (1941). Determination of BOD: Comparative study of azide and Rideal–Stewart modifications of the Winkler method. *Chemical Abstracts, 35*, 4887.

Pletcher, D., & Sotiropoulos, S. (1966). Towards a microelectrode sensor for the determination of oxygen in waters. *Anal. Chim. Acta* (Netherlands), *322*, 83.

Poe, D. P., & Diehl, H. (1974). Spectrophotometric determination of DO with tris(4,7-dihydroxy-1,10-phenanthroline). Iron(II). *Talanta, 21*, 1065.

Pomeroy, R., & Kirchman, H. D. (1945). Determination of dissolved oxygen: Proposed modification of the Winkler method. *Anal. Chem., 17*, 715.

Poole, R., & Morrow, J. (1977). Improved galvanic DO sensor for activated sludge. *J. Water Pollut. Control Fed., 49*, 422.

Porges, R. (1964). Dissolved oxygen determinations for field surveys. *J. Water Pollut. Control Fed., 36*, 1247.

Potos, C. (1959). Effects of magnetic stirring on the dropping mercury electrode. *Sew. Ind. Wastes, 31*, 254.

Potter, E. C., & Everitt, G. E. (1959). Further advances in dissolved oxygen micro-analysis. I. Small-scale water sampling vessels and amperometric titrations. *J. Appl. Chem., 9*, 642.

Potter, E. C., & White, J. F. (1957). Micro determination of dissolved oxygen in water. V. Development of self-testing and fully compensating analytical method. *J. Appl. Chem.*, *7*, 459.

Prowse, C. W. (1985). Field-use of dissolved oxygen electrodes. *J. Inst. Water Eng. Sci.*, *39*, 179.

Pye, D. J. (1947). Chemical fixation of oxygen. *J. Am. Water Works Assoc.*, *39*, 1121.

Rahim, S. A., & Mohamed, S. H. (1978). Determination of DO in water. *Talanta*, *25*, 519.

Rand, M. C., & Heukelekian, H. (1951). Determination of dissolved oxygen in industrial wastes by the Winkler and polarographic methods. *Sew. Ind. Wastes*, *23*, 1141.

Reddy, G. S., Soundar Rajan, S. C., & Reddy, Y. K. (1978). Titrimetric determination of DO in waste water. *Talanta*, *25*, 480.

Resch, G. (1959). Photometric determination of dissolved oxygen on the basis of the Winkler method. *Combustion*, *31*(1), 61.

Revsbech, N. P. (1989). An oxygen microsensor with a guard cathode. *Limnol. Oceanogr.*, *34*, 474.

Revsbech, N. P., et al. (1989). Denitrification in a trickling filter biofilm studied by a microsensor for oxygen and nitrous oxide. *Water Res.*, *23*, 867.

Reynolds, J. F. (1969). Comparison studies of Winkler vs. oxygen sensor. *J. Water Pollut. Control Fed.*, *41*, 2002.

Rezaeva, L. T. (1962). Photometric micro-determination of oxygen in liquids rich in organic substances. *Zh. Anal. Khim.* (Russia), *17*, 874; *Chemical Abstracts*, *58*, 8403 (1963).

Rideal, S., & Stewart, G. G. (1901). The determination of dissolved oxygen in waters in the presence of nitrites and of organic matter. *Analyst*, *26*, 141.

Ridley, J., Elliot, D., & Oaten, A. (1960). An automatic titrator as applied to Winkler's method for dissolved oxygen. *Analyst*, *85*, 508.

Robertson, W. S. (1980). The role of the cobalt catalyst in the unsteady-state test. *Water Pollut. Control*, *79*, 487.

Robinson, R. H. (1967). Apparatus for measuring DO in water. U.S. Patent 3313720; *Chemical Abstracts*, *66*, 118754 (1967).

Rose, S., & Long, A. (1988). Monitoring dissolved oxygen in ground water: Some basic considerations. *Ground Water Monit. Rev.*, *8*(1), 93.

Rosenthal, B. L., & Conlon, R. L. (1955). Polarographic determination of BOD: A progress report. *Sanitalk*, *3*(7), 1.

Roskam, R. T., & de Langen, D. (1963). A compleximetric method for the determination of DO in water. *Anal. Chim. Acta*, *28*, 78.

Rotshtein, V. P., & Shemyakin, V. N. (1967). Determination of oxygen in natural waters. *Dokl. Otd. Kom., Geogr. Ovo. SSSR* (Russia), *2*, 168; *Chemical Abstracts*, *70*, 109013 (1969).

Ruchhoft, C. C. (1941). Dissolved oxygen and BOD determinations: Their application and interpretation. *Sew. Works J.*, *13*, 542.

Ruchhoft, C. C., Moore, W. A., & Placak, O. R. (1938). Determination of dissolved oxygen by the Rideal–Stewart and Alsterberg modifications of the Winkler method. *Ind. Eng. Chem. Anal. Educ.*, *10*, 701.

Ruchhoft, C. C., & Placak, O. R. (1942). Studies of sewage purification. XVI. Determination of dissolved oxygen in activated sludge sewage mixtures. *Sew. Works J.*, *14*, 638.

Ruse, A. (1979). Standardization of membrane-coated oxygen electrodes with air. *Gewaesserschutz, Wasser, Abwasser*, *39*, 77; *Chemical Abstracts*, *92*, 99172 (1980).

Saito, H., & Uchiyama, H. (1980). Automatic analyzer for DO in water. *Keiryo Kenkyusho Hokolu*, *29*, 29.

Sanchez-Pedreno, C., Ortuno, J. A., & Hernandez, J. (1996). Determination of chlorine and dissolved oxygen in waters and of ascorbic acid in pharmaceuticals by iodometric potentiometric titration using an elasticized poly(vinyl chloride): Membrane electrode. *Anal. Chim. Acta* (Netherlands), *333*, 107.

Sastri, V. S., & Maiaiyandi, M. (1983). Sudies on the determination of dissolved oxygen in water. *Talanta*, *30*, 983.

Sastry, G. S., Hamm, R. E., & Poole, K. H. (1969). Spectrophotometric determination of DO in water. *Anal. Chem.*, *41*, 857.

Schenk, G., & Berg, G. (1954). Electrochemical determination of dissolved oxygen according to F. Todt's method. *Naturwissenschaften* (Germany), *41*, 425.

Schmid, M., & Mancy, K. H. (1969). The electrochemical determination of dissolved oxygen in water in the presence of hydrogen sulfide. *Chimia* (Switzerland), *23*, 398.

Schmiedt, F. (1934). A practical biochemical method for the determination of oxygen in sewage. *Tech. Gemeindeblatt*, *37*, 181; *Chemical Abstracts*, *28*, 7389 (1934).

Scholander, G. F., Van Dam, L., Claff, C. L., & Kanwisher, T. W. (1955). Micro-gasometric determination of dissolved oxygen and nitrogen. *Biol. Bull.*, *109*, 328.

Schramm, H. (1982). Oxygen measurement in the operation of wastewater treatment plants and sewers. *Ber. Abwassertech. Ver.* (Germany), *34*, 407; *Chemical Abstracts*, *97*, 222644 (1982).

Scott, R. H. (1964). Miniaturized DO field kit. *J. Water Pollut. Control Fed.*, *36*, 342.

Seaman, W., & Allen, W. (1950). Polarographic determination of dissolved oxygen in water containing industrial effluent. *Sew. Ind. Wastes*, *22*, 912.

Serak, L. (1984). On the electrochemical measurement of dissolved oxygen in water. *Sci. Total Environ.*, *37*, 107.

Sevcik, A., & Tesarik, K. (1966). Direct chromatographic determination of gases dissolved in water. *Prace Vyzkum. Ustavu Cs. Naftovych Dolu. Publ.*, *23*, 93; *Chemical Abstracts*, *66*, 5646 (1967).

Slack, K. V. (1971). Average dissolved oxygen measurement and water quality significance. *J. Water Pollut. Control Fed.*, *43*, 433.

Smeaton, G. (1979). DO probe wrinkles Winkler. *Water Sew. Works*, *126*, 8, 52.

Smirnov, D. N., Kuz'min, A. A., & Saifi, R. N. (1969). Automatic control of the amount of DO in waste water. *Vodosnabzh. Sanit. Tekh.* (Russia), *5*, 18; *Chemical Abstracts*, *71*, 104988 (1969).

Smith, F. R., & Delahay, P. (1965). Coulostatic study of residual oxygen in electrochemical investigations. *J. Electroanal. Chem.*, *10*, 435.

Smith, R.-K. (1999a). *Handbook of environmental analysis* (4th ed.). Genium Publishing.

Smith, R.-K. (1999b). *Lectures on wastewater analysis and interpretation*. Genium Publishing.

Spanjers, H., & Olsson, G. (1992). Modelling of the dissolved oxygen probe response in the improvement of a continuous respiration meter. *Water Res.*, *26*, 945.

Spoor, W. A. (1948). Use of the dropping mercury electrode for the continuous measurement of dissolved oxygen in flowing water. *Science*, *108*, 421.

St. John, P. A., Winefordner, J. D., & Silver, W. S. (1964). Microdetermination of DO in water by a rapid spectrophotometric method. *Anal. Chim. Acta* (Netherlands), *30*, 49.

Stracke, G., & Keefer, C. (1960). Continuous measurement of oxygen to control and regulate aeration. *Water Sew. Works*, *107*, 388.

Sunahara, H., Nakashima, R., Sasaki, S., & Matsumae, T. (1962). Rapid determination of DO in industrial wastes. I. Polarographic determination of the DO in waste from chromium plating. *Rep. Gov. Ind. Res. Inst., Nagoya*, *11*, 402; *Anal. Abstr.*, 4438 (1963).

Swartz, J. R., & Cooney, C. L. (1978). Instrumentation in computer-aided fermentation. *Process Biochem.*, *13*(2), 3.

Theriault, E. J. (1925). The rate of deoxygenation of polluted waters. *Proc. Am. Soc. Civ. Eng.*, *51*, 1819.

Theriault, E. J., & McNamee, P. D. (1932). Dissolved oxygen in the presence of organic matter, hypochlorites, and sulfite wastes. *Ind. Eng. Chem. Anal. Educ.*, *4*, 59.

Theriault, E. J., & McNamee, P. D. (1935). Studies on sewage purification. I. Apparatus for the determination of dissolved oxygen in sludge–sewage mixtures. *Public Health Rep.*, *50*, 480.

Thiele, H. (1968). Possibility of continuous measurement of oxygen content in water and sewage as well as oxygen consumption in systems with high consumption rate. *Industrieabwasser* (Germany), *22*, 6; *Chemical Abstracts*, *72*, 103479 (1970).

Thresh, J. C. (1885). A new method of estimating the oxygen dissolved in water. *J. Chem. Soc.*, *5*, 185.

Todt, F. (1942). The electrochemical determination of dissolved oxygen in water. *Gesundh.-Ing.* (Germany), *65*, 76.

Todt, F. (1957). The importance and use of an instantaneous electrochemical determination of oxygen for the control of polluted waters. *Bull. Cent. Belge Etude et Document Eaux*, *36*, 70.

Todt, F. (1959). Experience with electrochemical oxygen determinations in liquids in recent years. *Chem. Ztg.*, *83*, 483; *Chemical Abstracts*, *53*, 22632 (1959).

Todt, F., & Petsch, G. (1955). The further development of the oxygen plumet for direct electrochemical determination of dissolved oxygen in natural waters. *Gesundh.-Ing.* (Germany), *76*, 104.

Todt, F., Schwarz, W., & Todt, H. G. (1954). Electrochemical determination of dissolved oxygen in surface waters. *Gesundh.-Ing.* (Germany), *75*, 224.

Toledo, A. P. P., et al. (1982a). Amperometric determination of DO in small volume samples. *An. Simp. Bras. Eletroquim. Eletroanal.*, *1*, 93; *Chemical Abstracts*, *98*, 95358 (1983).

Toledo, A. P. P., et al. (1982b). Determination of DO: A comparison of some already proposed modifications of the Winkler method. *J. Environ. Stud.*, *16*, 219.

Toledo, A. P. P., et al. (1984). A semi-micro cell for dead stop determination of DO. *Cienc. Cult.*, *36*, 98; *Chemical Abstracts*, *100*, 126635 (1984).

Truedale, G. A., Downing, A. L., & Lowden, G. F. (1955). The solubility of oxygen in pure water and sea water. *J. Appl. Chem* (Great Britain), *5*, 53.

Ungureanu, C. (1964). Methods for determination of oxygen in water by using indigo carmine and cerium salts, by the photometric method. *Energetica, 12*(4), 182; *Chemical Abstracts, 60*, 11755 (1964).

van Baalen, C., & Marler, J. E. (1966). Occurrence of hydrogen peroxide in sea water. *Nature, 211*, 951.

von Staden, J., & Schlageter, H. (1965). The chlorator phase exchanger, an instrument for continuous measurement and control of oxygen content in strongly contaminated waters. *Oesterr. Abwasser-Rundsch.* (Austria), *10*, 90; *Chemical Abstracts, 67*, 25251 (1967).

Wells, W. N. (1970). Dissolved oxygen sampler. *Water Sew. Works, 117*, 282.

Wheatland, A. B., & Smith, L. J. (1955). Gasometric determination of dissolved oxygen as a check of titrimetric methods. *J. Appl. Chem., 5*, 144.

Whipple, G. C., & Whipple, M. C. (1911). Solubility of oxygen in sea water. *J. Am. Chem. Soc., 33*, 362.

White, A. F., Peterson, M. L., & Solbau, R. D. (1990). Measurement and interpretation of low levels of dissolved oxygen in ground water. *Ground Water, 28*, 584.

Wilcock, R. J., et al. (1981). An interlaboratory study of dissolved oxygen in water. *Water Res., 15*, 321.

Wilhelms, S. C. (1984). Measurement of dissolved gases at corps of engineers projects. CORPS Rep. WES/TR/HL-E-84-6. U.S. Army Eng. Waterways Experimental Station, Vicksburg, MS.

Winkler, L. W. (1888). The determination of dissolved oxygen in water. *Berlin. Deutsch. Chem. Gellsch., 21*, 2843.

Winkler, L. W. (1911). Determination of dissolved oxygen (addition). *Z. Angew. Chem., 24*, 831.

Winkler, L. W. (1912). Sauerstoff-flasche. *Z. Angew. Chem., 25*, 1563.

Wolff, L. M., et al. (1991). A fluorescence technique to measure concentration gradients hear the interface. In S. C. Wilhelms and J. S. Gulliver (Eds.), *Air-water mass transfer*. Am. Soc. Civ. Eng.

Wood, A. A. (1963). Operation of a wide-bore polarograph continuous DO recorder. *Effluent Water Treat., J3*, 23; *Chemical Abstracts, 59*, 312 (1963).

Zakharov, I. A., & Grishaeva, T. I. (1980). Determination of microconcentrations of water-dissolved oxygen by the quenching of the adsorbate afterglow. *Zh. Anal. Khim., 35*, 481; *Chemical Abstracts, 93*, 225379 (1980).

Zakharov, I. A., & Grishaeva, T. I. (1982). Determination of trace concentrations of DO in water by using quenching of the afterglow of its hydrophobic adsorbates. *Zh. Anal. Khim.*, *37*, 1735; *Chemical Abstracts*, *98*, 59589 (1983).

Zuellig, H. (1979). Membrane-free electrodes for the measurement of dissolved oxygen. *Gewaesser, Wasser, Abwasser* (Germany), *39*, 37; *Chemical Abstracts*, *92*, 116140 (1980).

5

Process of Biochemical Oxygen Demand

Many people perform the biochemical oxygen demand (BOD) test as a cookbook exercise without having an understanding of the processes that are taking place during the test. Fundamental research into the mechanics of the test and its relation to biological waste treatment processes have been occurring since the late 1800s; however, even more than 100 years later, it cannot be said that a complete understanding is available. Much of the research has been reviewed numerous times in varying degrees of depth (Camp, 1963; Eckenfelder, 1989; Gaudy & Gaudy, 1980; Hull, 1955; Kamhawy, 1996; Lea, 1940; LeBlanc, 1974a, 1974b; O'Brien & Clark, 1962; Sawyer et al., 1994; Theriault, 1927; Young, 1977). Each year has brought new insights to the process and, with each telling, the story has become richer. This chapter is a summary of the current level of understanding.

The short version of the BOD process is that a portion of waste, diluted with water containing mineral nutrients, is seeded with a small number of microorganisms. The dissolved oxygen (DO) concentration of the sample is determined, then the diluted sample is incubated in the dark at 20 °C for a period of time, 5 days being the nominal standard. The microorganisms use organic substances in the waste as food to grow and reproduce. At the same time, the organisms consume oxygen. At the end of the incubation period, the residual DO is determined. The amount of oxygen consumed by the organisms feeding on the diluted sample is extrapolated back to the amount of oxygen that would be consumed by similar organisms utilizating an undiluted sample.

Obviously, there are a lot of important procedural details left out of this short description. However, the major concepts are present; they are:

- Microorganisms consume organic substrates in the waste,
- Oxygen is used by the microorganisms,
- The metric of the test is the difference between the initial and final values of DO,
- A diluted portion of the sample is tested, and
- The result is extrapolated back to the undiluted sample.

The interpretation of the test results depends on how these major concepts are understood by the user of the results and exactly how these are applied in the performance of the test.

There are two additional important factors in the BOD test: (1) interactions of mixed populations of microorganisms, and (2) toxicity and acclimation of organisms. These factors may need to be considered to understand certain BOD test results.

This chapter presents the status of the understanding of these concepts and factors as it has evolved through the work of many investigators over more than 100 years. It must be stressed that this is not by any means a complete understanding of the BOD process. At the same time, what is already known about the BOD process is impressive and can serve as a useful platform for further research and understanding.

1.0 MICROORGANISMS CONSUME ORGANIC SUBSTRATES IN THE WASTE

There are many types of waste that are produced by humans. Some arise from the basic needs of the human animal to acquire, prepare, and consume food and then eliminate the wastes generated by the body from the utilization of the food. The term "animal" is stressed here because the characteristics of this type of waste are little different from similar wastes from any animal, with the exception that humans often spend a considerable amount of time in the preparation of the food before consumption and that the preparation itself generates waste. The animal functioning of people is the source of the major portion of what is termed domestic waste and has historically been the driving force for societies to develop techniques for waste collection and disposal. Domestic waste has been used by many cultures as an agricultural fertilizer in an early form of biological treatment. Although believed for thousands of years by some to be a source of illness, it was not until the middle of the 1800s that scientific techniques demonstrated domestic waste to be a health threat to public drinking water supplies, spurring even greater efforts to safely dispose of these wastes.

The human animal portion of domestic waste chemically consists of carbohydrates, proteins, simple and combined lipids, complex nucleic acids, and a host of other minor components such as natural pigments and alkaloids. These are all biologically derived organic substances and are subject to utilization by microorganisms as a food source. Over eons, microorganisms have evolved abilities to utilize these substances because, to a large degree, they are the same substances that all animals and plants generate as wastes. These wastes are an essential component in the cycling of food and minerals in the biosphere of the Earth.

The animal functioning of humans accounts for a significant portion of wastes generated by humans, but by no means is it the only source of wastes. Humans differ from animals in the ability to use materials from the environment to manufacture tools. Beginning with simple tools made of wood and stone, humans have progressed to making very complex tools such as microwave ovens, fiber extruders, books, automatic rifles, table saws, and millions of other items. These tools are then used to manufacture things to fulfill the basic needs of the human population for food, clothing, sanitation, protection, and shelter. Tools are also used to manufacture items of comfort and entertainment for humans. Together, all of these manufacturing efforts comprise the industrial society of humankind, and it all generates waste—considerably more waste than the human animal component. The industrial society originated only about 10 000 years ago, with most of the development occurring within the last 300 years. Microorganisms did not evolve in the presence of the wastes of the industrial society and, as a first approximation, have difficulty utilizing these wastes as food sources. Even the organic chemical components of industrial society wastes are largely inert to microbial utilization.

Bacteria are the smallest organisms that participate in microbial degradation of wastes, yet are the fundamental agents of that degradation. Neither plant nor animal, bacteria lack a distinct DNA-filled nucleus and are classified in a group, procaryotes, all by themselves. Modern classification schemes based on molecular biology principles of evolution list most of the bacteria encountered in wastewater treatment in the domain of Eubacteria, which serves as a superclassification over several kingdoms of bacteria; the Gram-positive and Gram-negative bacteria being included in this grouping. Older classification schemes, such as that used in *Bergey's Manual of Systematic Bacteriology* (Sneath, 1986), present four phyla of bacteria in one kingdom, with relationships among individual species defined by the reaction to the Gram stain and metabolic similarities.

Regardless of the classification scheme used, bacteria are single-celled organisms that perform all of the essential functions of life. They are isolated from the environment; assimilate organic substrates from their environment; use organic materials for energy production, growth, and reproduction; and, finally, excrete wastes back to the environment. Although most bacteria are smaller than can be discerned by the unaided human eye, they are exceedingly complex organisms. *Escherichia coli* (*E. coli*) is considered by some to be one of the simpler bacteria, yet even it is estimated to contain more than 5000 different organic molecules, including approximately 3000 proteins and 1000 different nucleic acids (Beveridge & Doyle, 1989).

Bacteria are separated from their surroundings by an outermost cell wall, called a sacculus, and an inner cytoplasmic membrane. Some bacteria secrete a slime coat or capsule surrounding the cell wall. These structures isolate the bacteria from the environment while allowing interaction with the environment. The cell wall is found in most bacteria and is a porous, more or less rigid structure composed of polysaccharides, teichoic acids, and cross-linked peptidoglycans (mucopeptides) that give shape to the bacteria. Bacteria display considerable variety in the composition of the cell wall. The Gram stain is a probe of the cell wall chemistry and is used to differentiate bacteria. Gram-positive bacteria are found to contain 50% to 90% of peptidiglycans in the cell wall, making it quite rigid, while Gram-negative bacteria contain 10% or less peptidoglycans, rendering the wall more membrane-like. Gram-negative bacteria also exhibit proteins and layers of lipopolysaccharide and lipoprotein in the cell wall, which are largely lacking in the Gram-positive bacteria.

The cytoplasmic membrane of bacteria consists of a bilayer of phospholipids in which proteins are embedded. In many bacteria, the membrane is not shaped like a balloon, pressed up rigidly against the underside of the cell wall; rather, it is highly folded, which tremendously increases the surface area. The membrane is semipermeable and provides an active interaction with the environment. Proteins in the membrane serve to transport needed ions and food across the membrane to the internal parts of the cell. Dehydrogenases and many of the enzymes associated with the electron-transport system are located in the cytoplasmic membrane. Synthesis systems for preparation of proteins, carbohydrates, phospholipids, and other needed cell components are located inside the cell.

Bacteria, because of the rigid cell wall and cytoplasmic membrane, are unable to take particles or large molecules, such as proteins or polysaccharides, directly into the cell for use as food. Pores are formed in the wall of the bacteria from porin polypeptides. Food intake is restricted to small molecules with a maximum size of trisaccharides, or short peptides up to five or six amino acids in length. Some molecules and ions can pass through the pores easily; however, others are actively transported into the cell by specific porin proteins (Beveridge & Doyle, 1989). Volumetric size is a more important limitation than strict molecular weight, although most researchers do not make a distinction between them and tend to equate the two. Examination of molecular size distributions of readily usable food (soluble 5-day biochemical oxygen demand [BOD_5]) in an influent waste stream finds that 76% to 88% are in the less than 1000 molecular weight range, while only 4% to 13% are in the greater than 10 000 molecular weight range. Further, the bulk of the material in the greater than 10 000 molecular weight range passes through the biological treatment unchanged or even increases as measured by total organic carbon (TOC) (Levine et al., 1985; Logan & Wagenseller, 2000; Rickert & Hunter, 1971). Results from gel permeation chromatography (GPC), a molecular-sizing analysis, have been found to be highly correlated to BOD removal efficiencies for fish-processing waste streams (Onari et al., 1981). Components of landfill leachate less than 5000 molecular weight as determined by GPC are found to be biologically treatable, whereas the greater-than-5000 molecular weight fractions require physical and chemical treatment for remediation (Millot et al., 1987).

Many bacteria release extracellular enzymes to the surrounding media. The extracellular enzymes degrade large polymeric food molecules into smaller pieces outside the cell wall. Bacteria then transport the solubilized pieces into the cell. Complex carbohydrates (polysaccharides) are attacked by extracellular glucanase, glucosidase, amylase, and lysozyme, each being specific for a particular linkage in the polysaccharide (Goel et al., 1999). One of the more difficult carbohydrate substrates for bacteria to use is cellulose, a common material found in waste streams from paper and pulp mills. A few members of the myxobacteria, the cytophagas, actinomycetes, and *Clostridium*, have been found to have the genetic ability to prepare extracellular enzymes for the hydrolysis of cellulose; however, most of the successful degradation is by fungi. *Chaetomium globosum, Myrothecium verrucaria, Trichderma reesi*, and *Stachybotrys atra* are specific fungi that thrive on cellulose through the ability to synthesize required cellulases (Gaudy & Gaudy, 1980).

Proteins are degraded extracellularly by proteases to amino acids and small peptides. Bacterial proteases are much less specific than the proteases of higher animals and will cleave amide bonds in proteins almost at random. This allows bacteria to be more flexible in their ability to acquire needed amino acids. Once the small peptides are transported into the cell, they are further degraded to amino acids by internal peptidases. Amino acids, after appropriate activation, are directly used as building blocks for needed proteins and enzymes.

Bacteria find lipids generally in the form of triglycerides and phospholipids, with occasional encounters with more complex structures. Lipases and phospholipases, such as alkaline and acid phosphatase (Goel et al., 1999), are secreted to hydrolyze the ester bonds of lipids extracellularly. Resulting fatty acids and glycerol are brought into the cell, where they are catabolized to recover energy and make acetyl-coenzyme A (acetyl-CoA) for synthesis of other needed materials.

The DNA and RNA are hydrolyzed extracellularly by deoxyribonucleases (DNA-ases) and ribonucleases (RNA-ases), respectively. Appended phosphate groups on molecules and initial cleavage products are hydrolyzed externally by phosphatases. Not all bacteria have the ability to secrete DNA-ases or RNA-ases, with members of *Bacillus* and *Streptococcus* being the most active.

Straight-chain hydrocarbons are oxidized, with direct participation of molecular oxygen and a variety of dehydrogenases, to fatty acids that can be transported into the cell. This requires an aerobic environment for most bacteria. Further utilization of the fatty acids is intercellular as energy sources.

The extracellular hydrolysis does not result in preparation of small food molecules for the exclusive use of the individual cell, although the presence of a slime coating or capsule secreted by the single cell can help concentrate and retain the hydrolysis products. For the most part, products are available for many different cells. Further, the individual cell generally does not have the ability to secrete all of the enzymes necessary for complete hydrolysis of all substances that are potential food sources. Often, specific enzymes are required for breaking each of the different types of chemical bonds found in large food molecules. Thus, hydrolysis and subsequent use of the components in a complex waste are, in many cases, an exercise in mutualism by several bacterial species.

Slimes and capsules are composed primarily of polysaccharides that form a gel with water around the bacterial cell. Slimes and capsules are similar in composition, but differ in the amount of exopolymer in them, although both are composed of approximately 99% water. Slimes are often regarded as simply more watery and less rigid forms of capsules, as there are some bacteria that can produce one or the other depending on conditions

(Beveridge & Doyle, 1989). Capsules often contain molecular attachments to the cell wall. Both slimes and capsules serve to retain extracellular hydrolysis enzymes in the immediate vicinity of the cell, although they may also contain other functional enzymes.

The transport process across the cellular membrane is an active and specific process mediated by proteins, called permeases. For each compound to be transported across the membrane, there is a specific protein. If a protein is lacking, the bacteria cannot transport the compound. For example, *E. coli* lacks a protein to transport citrate, and citrate cannot be assimilated by this bacteria, even though citrate is an important component of the tricarboxylic acid (or Krebs) cycle. *E. coli* must synthesize the citrate that it needs because of a lack of transport capability (Gaudy & Gaudy, 1980). This specificity in transport has been found to apply to individual amino acids, although it is less applicable to small peptides. Several aurotrophs have been found that lack the ability to transport a specific amino acid but can bring it into the cell if it is part of a dipeptide (Gaudy & Gaudy, 1980). Transport is active and does require using chemical energy in the form of adenosine tripolyphosphate (ATP) or a proton gradient.

The inability to directly assimilate large molecules is one of the reasons why some waste components, such as humic acids, cellulose, and lignins, take extremely long times to degrade even though they are composed of what would nominally be considered easily degradable subunits (Heukelekian, 1950; Ponce, 1974; Raabe, 1968; ZoBell & Stadler, 1940b). Even in acclimated cultures, the time required for extracellular breakdown of lignins and subsequent assimilation of the digestable pieces results in an initiation time lag of 12 to 20 hours as compared with a readily usable food like glucose (Lomova & Ryabkova, 1976). Waste treatment processes that include oxidation of waste by chemicals such as chlorine or ozone before biological treatment help in the breakdown of the larger molecules into usable-sized pieces. This results in the frequently observed phenomenon of BOD increasing, sometimes even doubling, after chemical treatment with oxidants, while TOC and chemical oxygen demand (COD) are unaffected or decrease (Bauman & Lutz, 1974; Beaman et al., 1998; Bekbolet et al., 1996; Carmichael, 1947; Groff & Ridenour, 1943; Ingols & Jacobs, 1957; Karrer et al., 1997; Matsui et al., 1984; McGuire, 1927; Mebley & Stover, 1983; Snyder, 1947; Stover & Kincannon, 1983; Stover et al., 1983). Treatments that are nonoxidative, such as exposure to beta, gamma, or UV radiation, are largely inconsequential to the ability of the bacteria to assimilate food (Grune, 1951, 1953; Grune & Eliassen, 1951; Lee et al., 1982; Skrindle & Sawyer, 1952; and Yamazaki et al., 1983).

The range of substances that bacteria can use as food is fairly wide. For the most part, naturally occurring simple carbohydrates and sugars, peptides and amino acids, nucleic acids, fatty acids, and a number of oxidized hydrocarbons are readily assimilated, although there are rate differences generally noted between the classes of substrates (Carter & Malaney, 1971). Polymers of these materials take time to digest extracellularly but eventually can be assimilated. Manufactured substances can present problems to bacteria, particularly substances that are chemically distinct from the normal range of biologically synthesized compounds. As an example, nitrilotriacetic acid, a chelator used in detergents, is an inert material to all but a few bacteria and most often gives a BOD of zero (Pfeil & Lee, 1968).

Many persons in the wastewater industry equate BOD with pollutant load. This is not correct. The BOD measures a biological response to the presence of food, and the food content of a waste may comprise only a small fraction of the total organic content (Clark, 1974; Hoskins, 1933; Van Beneden, 1956). Confusion over this point has often led to derision of the BOD result as not being suitable for material balance calculations (Hoffman et al., 1973). The derision is unwarranted, as BOD is fundamentally inappropriate for use in this regard (Mohlman, 1936; Ruchhoft, 1941). Even the approximation of organic load being the sum of BOD and an oil and grease determination (Smith, 1999a, 1999b) is not suitable, as the sum tends to double count some of the constituents (Schaeffer et al., 1977). If material balances are the objective of an investigation, suitable chemically based techniques such as COD, TOC, total oxygen demand (TOD), or other more selective analytical methodology must be used to generate the necessary data.

Bacteria and other organisms found in waste treatment are divided into broad metabolic groups. Autotrophs use inorganic carbon as the primary carbon source. Heterotrophs use organic carbon as the primary carbon source. Organisms that derive their energy from chemical (food) sources are called chemotrophs, while those that derive energy from light are called phototrophs. Organisms that use inorganic electron donators are called lithotrophs. Those that use organic foodstuffs as electron donators are called organotrophs. Thus, a chemolithotropic organism is one that uses a substance like ammonia, nitrite, iron(II), sulfide, elemental sulfur, sulfite, or hydrogen as the energy source and electron donator, and most commonly uses carbon dioxide as the major carbon source. Finally, organisms that use oxygen as the electron acceptor in respiration are called aerobic, while those that use other substances such as sulfur chemicals or carbon dioxide are called anaerobic. Some organisms are flexible in their respiration and, while they function most efficiently when using oxygen, they are capable of using other terminal electron acceptors when oxygen is not present. These are called facultative anaerobes. Other organisms are poisoned by the presence of oxygen and are called obligate (strict) anaerobes.

Bacteria are not the only organisms present in the BOD bottle, although they frequently are responsible for the majority of the substrate utilization. Other organisms present include fungi (yeasts), algae, protozoa, and higher animal and plant forms. The interactions among these organisms are discussed below.

2.0 OXYGEN IS USED BY MICROORGANISMS

Microorganisms take compounds that are assimilated and break them down into smaller usable pieces, collectively termed central pathway intermediates, predominantly through the enzyme-catalyzed oxidation reactions of glycolysis, tricarboxylic acid, and hexose monophosphate pathways. Some of the intermediates are oxidized further to carbon dioxide, which is passed back out of the cell to the environment while, at the same time, chemical energy is stored in the form of ATP. The oxidants used in these reactions are, for the most part, nicotinamide adenine dinucleotide (NAD), nicotinamide adenine dinucleotide phosphate (NADP), flavin adenine mononucleotide (FMN), and flavin adenine dinucleotide (FAD). Once oxidants have been reduced in catabolic reactions, they are reoxidized for reuse by ubiquinone and the cytochromes in the electron-transport system. Other intermediates are converted to carbon skeletons such as acetyl-CoA. These breakdown processes are called catabolism. Carbon skeletons are used as the building blocks in the synthesis of the needed protein, carbohydrate, lipid, and nucleic acid cellular components. Excess building blocks are used for synthesis of storage polymers such as glycogen, polyhydroxybutyrate (PHB), and starch. The synthesis functions are collectively termed anabolism and result in an increase in size of the organism and allow reproduction, both leading to an increase in biomass. Oxidation reactions that occur during anabolism are linked with the reduction of one of the electron carriers: NAD, NADP, FMN, or FAD. Again, the reduced carriers are reoxidized by ubiquinone and cytochromes, with the final electron acceptor being oxygen.

Oxygen is the major terminal electron acceptor in aerobic systems, and its use is termed respiration. Although the respiration reaction is generally limited to the actual oxidation of cytochromes, cytochromes are intimately linked to all cellular functions involving food assimilation and metabolism (Clifton & Logan, 1939; Greig & Hoogerheide, 1941). Respiration is the prime example of the "tail wagging the dog," as nothing else can function in the cell without dumping excess electrons into oxygen.

In eukaryotic organisms, the electron-transfer system is located in the mitochondria, and the system is almost identical in all mitochondria. The electron-transport system links catabolism of organic substrates, oxidative phosphorylation, and anabolic synthesis. The final step in the system is the oxidation of reduced cytochromes to oxidized cytochromes, coupled with simultaneous reduction of oxygen to water, all catalyzed by oxidases. There are three types of cytochromes that have been identified in mitochondria: cytochromes *a, b*, and *c*.

Bacteria lack mitochondria and, in fact, one school of thought holds that mitochondria arose in higher life-forms when a Gram-negative heterotrophic bacterial cell was engulfed by the first ancestral eukaryotic organism and, rather than being consumed as food, developed a symbiosis with the eucaryote (Gaudy & Gaudy, 1980; Ingraham & Ingraham, 1995; Tortora et al., 1998). This is termed endosymbiotic cellular evolution, and the theory is bolstered by the observation that most existing bacteria and all mitochondria have 70S ribosomes, while the ribosomes found elsewhere in eukaryotic cells are 80S. Further, the DNA in mitochondria and bacteria is a single strand fused into a circle, rather than the linear form characteristic of higher cells. Chloroplasts—organelles that perform photosynthesis—are believed to have arisen in eukaryotic cells by a similar pathway.

Lacking mitochondria, bacteria contain the electron-transport components in the cytoplasmic membrane. There is a great diversity of electron-transfer mechanisms in bacteria, although cytochromes linked with an oxidase reducing oxygen to peroxide or water tends to always be the final step. At least eight different cytochromes have been identified in various species of bacteria in addition to the three found in mitochondria. Further, bacteria capable of existence in aerobic environments are found to possess generally more than one path through different cytochromes to oxygen reduction. These additional paths give the organisms more survival flexibility.

Bacteria and higher organisms do not store the oxidized NAD, NADP, FMN, FAD, quinones, or cytochromes of the electron-transport system. Instead, they are oxidized as immediately needed for metabolism in the cell. Also, the cell does not store oxygen. Although the direct participation of oxygen is limited for the most part to the oxidation of the cytochromes, this is the key exit point for electrons from the cellular metabolism and allows every other system in the cell to function. Altogether, this means that the cell must have a continual supply of oxygen, and that the amount of oxygen used by the cell at any one instant of time is going to change depending on the status and rate of metabolism that the cell needs at that instant.

Neither does a cell store all the enzymes that it needs to perform all the functions of which the cell is capable. The information needed to prepare each enzyme is stored in the DNA of the cell; however if, for example, lipase is not needed at the moment, it is not sitting in a storage location waiting for potential use. When a need for the lipase arises, the cell responds first by synthesizing the lipase. This generally requires internal catabolism of stored carbohydrates, such as glycogen, PHB, or starch, and existing proteins to obtain the building blocks for the needed enzyme.

When a bacterial cell is at a metabolic idle, it still is consuming oxygen (respiring). Such a situation arises when there is little to no food in the external environment, as would be the case when a technician prepares seed for a BOD test. This could involve making a dilute suspension of bacteria from a dried commercial preparation of seed or preparing a dilution of seed acquired from a location downstream of the effluent point from a treatment facility.

Actually, these two situations are quite different from the point of view of the bacteria. The commercial seed is in a dormant state, with minimal oxygen use and metabolic rate, and it is awakening (recovering) to living in an aqueous environment. For the bacteria that go dormant as spores, this mean hatching out from a spore state to a free-living state, which requires a substantial reorganization of the cellular components to re-form the cell wall and cellular membrane. The reorganization consists of catabolism of the spore coat and elaboration of the pieces of the coat to new structures. This requires a lot of oxygen and is a very busy time for the cell.

In the other situation, the preparation of seed from a downstream location, bacteria are experiencing a change of environment from one of relative plenty to one of relative scarcity. The enzymes required for active food acquisition and use are no longer needed and are shut down and catabolized while the cell prepares to enter a resting state. For bacteria that exist as spores in dormancy, this may mean beginning to construct the spore coat within the cell, depending on the severity of the environment. These also are activities that require oxygen.

In either case, the cell is using internal resources to synthesize structures of immediate need. The term *endogeneous respiration* is used to describe situations in which the bacterial cells are not actively acquiring external food. Both of the above situations, although oriented toward different ends, are characterized by significant levels of metabolic activity, and both require consumption of oxygen as the cell changes from one state to another.

When the seed is added to the diluted waste, bacteria enter an environment rich in food, at least rich compared with the environment of the seed suspension. At first the bacteria were still in a state of endogeneous respiration, but are now oriented toward catabolizing internal components and synthesizing enzymes and proteins that can acquire food. This includes enzymes to be secreted for external hydrolysis of food and transport proteins for the cellular membrane to bring the food into the cell (Rao & Gaudy, 1966). As soon as food begins to enter the cell, enzymes for internal catabolism are synthesized. Oxygen consumption of the cell is still at the endogeneous respiration rate until the food actually enters the cell. (Although this is a very reasonable portrayal of the lag-time process, it is not held by all [see Sykes, 1976]). Once food is inside the cell, the oxygen consumption rate increases dramatically (Martin, 1932). As food is catabolized, the cell begins to manufacture enzymes for anabolic metabolism so that the cell can grow in size, resulting in a further increase in oxygen consumption.

After achieving a certain increase in size, and assuming that food acquisition is continuing, the cell synthesizes enzymes that prepare the cell for reproduction (Rao & Gaudy, 1966). The preparation for reproduction represents the maximum rate of oxygen use by the individual cell. Once reproduction occurs, the original cell returns to the growth phase and related lower oxygen consumption rate. The overall oxygen consumption in the culture increases because, where originally there was only one cell, now there are two. As long as food and oxygen are available for the bacterial cells in the culture to grow and reproduce, the rate of oxygen consumption will generally increase logarithmically in parallel with the reproduction rate.

Two factors lead to a slowing of the rate of oxygen consumption. One occurs when either the food or oxygen supply for supporting the population runs out. When the food runs out, bacteria change to the state of endogeneous respiration, which is characterized by a lower oxygen consumption rate. The other arises because individual bacterial cells do not live forever, even in the presence of adequate food. The death of individual cells releases metabolic wastes into the environment. When the population becomes so large that the metabolic wastes produced by dead bacteria accumulate to a level that is detrimental through a toxic effect to the living cells, it results in the population entering a steady state of death and reproduction and an overall leveling of the oxygen consumption rate.

The overall shape of the oxygen-consumption curve for the BOD test describes a double-S shape. It begins with a very slow rate of increase as the bacteria adjust to the new environment. After the metabolic adjustment is complete, a rapidly increasing rate in consumption is exhibited until the food runs out, when the consumption levels (the plateau region) briefly as the bacteria adjust to endogenous respiration. After metabolic adjustment to endogenous respiration, the oxygen consumption rate again increases briefly, then the rate decreases as a steady state is approached.

The double-S shape for a readily usable food source can be placed in a time reference. The first phase (lag) depends on how long the seed was maintained in the suspension (i.e., how long the bacteria were starved before use in the BOD test), but is only of short duration, 1 or 2 hours at the most. The rapid growth and reproduction phase (assimilation) lasts for 24 to 48 hours, depending on the amount of food present. The plateau region of metabolic adjustment is also brief, on the order of fewer than several hours. The remainder of the time of the test has bacteria in the endogeneous respiration phase. Overall, although the assimilation phase is fast with a high rate of oxygen use, the slower rate associated with endogeneous respiration lasts considerably longer. It is generally agreed that the synthesis phase of oxidation is more rapid but the endogeneous phase accounts for a larger fraction of the total oxidative destruction (Ludzack et al., 1964). Current estimates hold that at least 50% of total oxygen utilization occurs through endogeneous respiration (van Loosdrecht & Henze, 1999).

Examination of the BOD for pure substrates for a given seed generates a time axis for the double-S curve that is characteristic of the substrate. Substances that are readily assimilated, such as glucose, succinate, glutamic acid, and sorbitol, each gave separate curves. Mixtures of these compounds resulted in curves that are superpositions (mathematical averages) of the individual curves (Parisod & Schroeder, 1978). The amount of oxygen consumed by the individual cell during active assimilation has been examined (Kolaczkowski et al., 1965; Matsunaga et al., 1981). The amount required depends on the substrate and the species, with values ranging from 0.1 to 1.0×10^{-9} mg O^2/cell.

A food source that is less readily assimilated will stretch these events out in time. Long-term BOD studies (up to 200 days) of kraft mill effluents that are rich in slowly degradable materials, such as cellulose and lignin, reveal a rapid phase of assimilation within the first several days, and then another assimilation phase beginning around day 28 (Heukelekian, 1957; Hiidenheimo & Wilson, 1974). In BOD studies on receiving streams extending up to 100 days, a second assimilation phase occurred at 13 days (Williams, 1951). The BOD of algae biomass is sluggish, with 20-day BOD experiments accounting for only one-half of the theoretical BOD (Milojevic, 1967). The presence of both easily and less easily utilized food sources can give superimposed oxygen-use curves that are quite complex (Lomova & Ryabkova, 1976). The ratio of 20-day to 5-day biochemical oxygen demand (BOD_{20}:BOD_5) on filtered raw wastewater is used as a rough guide to the suitability of different types of waste treatment. Biological treatment processes are recommended if the ratio is less than 1.35, while physical–chemical rectification techniques are more likely to be time efficient when the ratio is greater than 1.35 (Iwugo, 1980).

Changing states can be viewed as the overriding characteristic of bacterial life, and even of life in general. Regardless of the immediate situation, the bacterial cell is always preparing to do something else. In rich environments, it is preparing to grow and possibly reproduce. In scarce environments, it is preparing to enter dormancy. In dormancy, most metabolic functions are shut down but some systems still are necessarily operating as the bacterial cell strives to maintain a minimum definition of life and avoid death, which is the final change. For aerobic organisms, all of these require continuous consumption of oxygen.

3.0 THE METRIC OF THE TEST IS THE DIFFERENCE BETWEEN THE INITIAL AND FINAL VALUES OF DISSOLVED OXYGEN

The saturation concentration of oxygen in water is not a constant. Rather, it changes with barometric pressure and temperature. Other factors that affect the measured level of oxygen in a sample are the presence of salts, as described in the chapter on DO. In most situations, the room temperature saturation concentration of freshwater is approximately 9 mg/L.

As the test proceeds, bacteria respire oxygen, and the DO concentration decreases in the bottle. Individual bacterial species have an optimum level of DO for maximum growth (Elsworth et al., 1957; ZoBell & Stadler, 1940a). Aerobic bacteria require a critical minimum level of oxygen, generally in the 0.1 to 1.0 mg/L level, for proper functioning, and most aerobic treatment units are operated at approximately 2 mg/L. Receiving streams generally are not saturated with DO, and 4 to 5 mg/L or less is a normal level for natural freshwater streams with good reaeration. Lakes tend to be stratified with higher levels of DO on the surface and anaerobic conditions at the bottom of the lake. Thus, the bottle BOD test requires the bacteria to respire in conditions not mirrored in any environment (Pomeroy, 1938). Increasing the DO level to the saturation point for the BOD test is the first nonrepresentative factor, although there are some data showing that an increased oxygen tension is directly related to increases in bacterial population (Heukelekian, 1936). Some BOD techniques that have been proposed to extend the range of DO depletion observed actually start with a solution that is supersaturated in oxygen (Bohmeke, 1975; Viehl, 1940, 1953, 1955), which is inhibitory to many microorganisms. As the DO is depleted in the bottle, it may be at a level commensurate with the normal situation for the seed and waste for only a brief period of time. The final DO is generally well below the normal operating concentration, thus leading to a second nonrepresentative factor.

The DO criteria within the performance of the BOD test are to have at least a 2.0 mg/L decrease from the initial level of DO and to have at least 1.0 mg/L DO remaining at the end of the test. Good analytical practice leads technicians to strive for a reasonable depletion in the DO of 5 to 7 mg/L to maximize the reliability and reproducibility of the measurement. Different species of bacteria exhibit maximum efficiency at different DO concentrations. Bacterial populations acclimate to DO levels just as they do to any environmental factor. As the DO level drops in the bottle, there is a succession of maximally performing bacterial species that are consuming the waste, and these may not be the same bacteria that are optimally functioning in the treatment unit or receiving stream. Although a depletion of 5 to 7 mg/L does not seem like a great difference, it represents a 2.7- to 8-fold decrease in concentration. A relatively similar reduction in the atmospheric concentration of oxygen from 20% to 2.5% would lead to the death of most air-breathing organisms.

Different populations of organisms utilize different substrates, as expressed by oxygen uptake, at different rates. Even when efforts are taken to provide the same seed population for each test, such as use of a commercial seed, variances in utilization rates are seen when comparing different substrates. Especially evident in studies in which substances are compared that exhibit similar ultimate (20-day or greater incubation times) BOD values, comparison of the 5-day values indicates wide variations. With the exception of cases in which a complete oxygen utilization curve has been obtained for a particular waste and the specific relation of the 5-day value to the ultimate BOD value has been determined, there is absolutely no justification for use of a 5-day incubation period as a standard, and the practice has been frequently derided (Eldridge, 1938; Gameson & Wheatland, 1958; Lenov, 1974; Lomova & Ryabkova, 1976; Mills & Stack, 1955; Mohlman, 1936; Rudolfs et al., 1950). For example, a 7-day incubation period is the standard in Norway (Broch-due et al., 1994), with no more justification than that which supports a 5-day incubation.

Once a complete oxygen-utilization curve has been established for a particular waste and seed, any incubation time period can be suitable. Thus, 2-day incubations are quite popular for process control of well-characterized waste streams.

4.0 A DILUTED PORTION OF THE SAMPLE IS TESTED

The DO criteria within the performance of the BOD test are to have at least a 2.0 mg/L decrease from the initial level of DO and to have at least 1.0 mg/L DO remaining at the end of the test. The very narrow range of acceptable DO depletions means that most wastes must be diluted to provide for the minimum residual DO at the end of the test period. These dilutions range from simple factors of 2 to 5 for most surface waters that are largely devoid of organic material, up to factors of 5000 or greater for strong wastes with high organic content. Aside from the analytical considerations of how accurately these dilutions are prepared, there are other considerations that lead to the test generating unrepresentative results (Heulelekian, 1933; Ruchhoft et al., 1948). Significant sample characteristics that change as a result of dilution are pH, ion concentration, and organic activity.

A 1000-fold dilution can change the pH of the sample from acidic (pH 5) or basic (pH 9) to neutrality (pH 7), and thereby change the sample's character physically, chemically, and biologically. The pH of the sample has a great effect on the solubility of key waste components. Hydrated metal ions are generally insoluble at basic pH but become more soluble as the pH is lowered, making the ions available for interaction with microorganisms. Frequently, interaction with soluble metals is biologically inhibitory, resulting in an underestimation of the BOD of the diluted sample compared with the natural sample. The reverse is true for organic acids in the sample. Organic acids are generally insoluble at acid pH and increase their availability as the pH is raised. The result can be an overestimation of the BOD of the natural sample.

The dominant ions that are present in the sample are replaced by those present in the dilution water, and the overall ionic activity of the waste is replaced by that of the dilution water. Both of these changes represent significant alterations in the character of the waste. Biologically toxic metals in the sample can have the toxic effect diminished by the dilution. The degree of solubility of organic materials in water is affected by both the dominant ions and the overall ionic activity. In general, organic solubility decreases as ionic activity increases; however, if a specific organic component in the waste can interact with a specific metal to form a soluble complex, this can lead to an increase in the bioavailability of both the metal and the organic substance. The source of the needed specific metal can be either the original sample or the nutrients in the dilution water, and the effect is difficult to predict in any given situation.

The organic activity of a sample is affected by the level of dilution. Most organic molecules in the sample, particularly in high-strength wastes, do not exist as discrete, individually solvated molecules. Instead, they exist as clusters of molecules, which are solvated as a group, even forming micelles when the concentration reaches critical levels. At greater concentrations, the micelles coalesce and an organic layer floating on the surface of the sample can be observed. As solvated clusters, and especially when a micelle forms, the organic molecules become essentially inert to bacterial usage. All bacterial extracellular enzymatic activity takes place in water solution, and none occurs in organic solvent, which is the closest analogy to the cluster or micelle of organic molecules in the sample. The case of the common industrial chemical bis(2-ethylhexyl)phthalate (DEHP) is described in more detail in the chapter on biodegradation and toxicity. However, in brief, at levels below 0.38 mg/L, DEHP is soluble in water and readily degradable by bacteria, while at greater concentrations it is found to be inert because of the formation of clusters. Thus, dilution can resolubilize organic substrates and give an overestimate of the BOD compared to that expressed by the undiluted sample.

Another effect of dilution is to change the population dynamics of the microorganisms compared with the species distribution in the unmodified sample (Chin et al., 1972; Lester et al., 1979; Liang et al., 1979; Stephanopoulos & Fredrickson, 1979). Of lesser concern when a commercial seed is used, results obtained by using a seed from the sample source on a highly diluted portion are reflective of microorganisms having to acclimate to an environment of scarcity rather than plenty, in addition to overall utilization of the waste as food (Hartmann & Wilderer, 1968).

Dilution results in an overall change in the kinetics and mechanism of the BOD and oxygen utilization (Jennelle & Gaudy, 1970) such that growth curves obtained in dilute systems bear little relation to those from the more concentrated state in the treatment facility. For this reason, many efforts have been made to perform the BOD on a sample with a single concentration that is representative of the conditions of treatment. Often this requires reaeration of the sample. This is cumbersome with the dilution bottle technique, but can be accomplished (Orford et al., 1953). More frequently, respirometer techniques using electrolytic instruments provide the needed information.

5.0 THE RESULT IS EXTRAPOLATED BACK TO THE UNDILUTED SAMPLE

The BOD result that is obtained on the diluted sample is converted to a BOD result on the undiluted sample, taking into account only the dilution factor. As described above, the native and diluted samples may be significantly different in chemical, biological, and physical characteristics. In extreme situations, they may have no relation to each other. However, the simple correction for the dilution factor forces the user of the data into assuming that they are identical.

The calculation of BOD presented in *Standard Methods* (APHA et al., 1998), although it accounts for the seed contribution, does not factor in the depletion by the dilution water. This is a technical characteristic of the method rather than being a strictly process consideration, and is normally not a big issue. However, BOD results reported from diluted samples that exhibit only the minimum level of depletion—that is, 2.0 mg/L—can have up to a 10% variance in the value due solely to the dilution water. In situations in which a 10% data variance can influence a design or regulatory compliance decision, it behooves the data user to carefully check the raw data of the analysis. There are calculation techniques available than can include this depletion, and they are discussed in detail in the theory discussion.

6.0 INTERACTIONS OF MIXED POPULATIONS OF MICROORGANISMS

The above overall description of the BOD test is presented under the implied conditions of a pure culture of bacteria as a seed. The BOD test can be performed with a single organism; however, with the exception of research situations, there is not much reason for using a monoculture. Growth curves developed from observations of pure cultures bear little resemblance to those obtained from mixed cultures (Coackley, 1969). Treatment facility and effluent-receiving streams are definitely not monocultures. The normal complement contains many different types of bacteria, protozoa, algae, fungi, rotifers, and other higher forms of plant and animal life, as suggested in Table 5.1, any of which can be a significant population in the treatment of the waste (Campbell & Rudolfs, 1923; Cox, 1921).

The bacteria present in wastewater come from fecal, soil, and water inputs to the waste. The first studies of the bacterial fauna of wastewater encountered the tremendous diversity of organisms present in these samples (Buxton, 1902; Kofoid, 1903). Faced with the taxonomic complexity of the task, researchers resorted to enumerating the isolates by function (Hotchkiss, 1924). Later practitioners approached the task by grouping similar bacteria by morphology and biochemical responses. In one effort, more than 450 pure bacterial cultures were isolated from the various stages in a treatment facility (Lighthart & Loew, 1972). These were broken down into nine groups based on morphological and chemical similarities. The nine groups were identified as *Alkaligenes–Achromobacter, Pseudomonas*, Enterobacteriaceae, *Arthrobacter–Bacillus, Cytophaga–Flavobacterium, Pseudomonas–Vibrio–Aeromonas, Achromobacter* and, finally, two miscellaneous groups. Fecally derived bacteria, such as *E. coli*, can be found in influent and effluent (Martin, 1932) but, by the end of the treatment train, were of minor abundance, in the 1% to 10% range, compared with *Pseudomonas* spp., *Flavobacterium* spp., and *Achromobacter* spp. (Harkness, 1966). Other known important species include the nitrification bacteria *Nitrosomonas* and *Nitrobacter*, although the characterization battery designed for the above study did not include tests of nitrification ability. Even to this day, modelers of waste treatment resort to describing the bacterial influences by function, that is, polyphosphorus-accumulating heterotrophs, anoxic rapidly biodegradable heterotrophic COD, anoxic slowly biodegradable heterotrophic COD, and aerobic slowly biodegradable heterotrophic COD (Griffiths, 1994), rather than tracking each species.

There were early attempts to develop bacterial seeds for innoculants in BOD tests from pure cultures of bacteria. In one such attempt, pure and mixed seeds prepared from *E. coli, Pseudomonas aeruginosa, Bacillus cereus, Serratia marcescens, Proteus vulgaris, Alcaligenes fecalis, Alcaligenes metalcaligenes, Achromobacter liquefaciens, Achromobacter recti*, and *Achromobacter bookeri* were found to produce very consistent results from one test to the next; however, the overall BOD expressed by the seeds were substantially lower than that of a mixed fauna wastewater seed (Tidwell & Sorrels, 1955; Zehnpfennig & Nichols, 1953). Current commercial seeds are prepared from pure cultures, with 13 to 14 microbial strains commonly being used (Paixao et al., 2000).

Bacteria are the primary consumers of soluble substrate in the media, the soluble materials being defined as sorbable by the bacteria with or without extracellular hydrolysis. Other classes of organisms can also directly utilize substrate, but the amount consumed by them is most often minor compared with that of bacteria. A study of the BOD of pure and mixed cultures of fungi showed that these organisms could, in general, over a 10-day period, accomplish approximately 0.4 to 0.6 of the BOD exhibited by bacterial mixed seeds. Further, the fungi were slower to accomplish full substrate assimilation, taking approximately 4 days, whereas bacteria generally did the task in 1 to 2 days (Cooke et al., 1956). Commercial BOD seeds do not normally contain fungi; however, for selected substrates, these organisms have been shown to have particularly useful food utilization abilities (Butcher, 1932). As an example, dairy wastes are 62% reduced in 72 hours by *Saccharomyces fragilis*, a lactose fermenter (Porges et al., 1950a, 1950b). One group of fungi that has received considerable attention is the white rot fungi, *Polystictus* spp. and *Armillaria* spp., which thrive on diverse foods like explosives residues and lignins, substrates virtually inert to bacterial decomposition. The term "white rot" comes from the ability of these fungi to decompose dark-colored lignins in wood, leaving the white cellulose untouched. Other useful fungi, such as *Pleurotus* spp. and *Polyporus* spp., can consume both lignins and cellulose. The presence of *Aphanomyces* and *Allomyces* fungi in water samples has been linked to the presence of pollution and can serve as indicator organisms for river systems affected by

TABLE 5.1. Organisms other than bacteria found in waste treatment.

Type	Organisms	Reference
Fungi	*Aspergillus* spp.	Cooke, 1954a, 1954b, 1954c; Harvey, 1952
Fungi	*Beggiatoa* spp.	Cooke, 1954a, 1954b, 1954c; Harvey, 1952
Fungi	*Fusarium* spp.	Cooke, 1954a, 1954b, 1954c; Harvey, 1952
Fungi	*Geotrichun* spp.	Cooke, 1954a, 1954b, 1954c; Harvey, 1952
Fungi	*Leptomitus* spp.	Cooke, 1954a, 1954b, 1954c; Harvey, 1952
Fungi	*Margarinomyces* spp.	Cooke, 1954a, 1954b, 1954c; Harvey, 1952
Fungi	*Oospora* spp.	Cooke, 1954a, 1954b, 1954c; Harvey, 1952
Fungi	*Penicillium* spp.	Cooke, 1954a, 1954b, 1954c; Harvey, 1952
Fungi	*Sepedonium* spp.	Cooke, 1954a, 1954b, 1954c; Harvey, 1952
Fungi	*Sphaerotilus* spp.	Cooke, 1954a, 1954b, 1954c; Harvey, 1952
Fungi	*Trichoderma* spp.	Cooke, 1954a, 1954b, 1954c; Harvey, 1952
Fungi	*Zoogloea* spp.	Cooke, 1954a, 1954b, 1954c; Harvey, 1952
Yeast		Heukelekian, 1942
Heterotrich ciliates	*Blepharisma* spp.	Poole, 1984
Heterotrich ciliates	*Climacostomum* spp.	Poole, 1984
Heterotrich ciliates	*Spirostomum* spp.	Poole, 1984
Heterotrich ciliates	*Stentor* spp.	Poole, 1984
Holotrich ciliates	*Trachelophyllum* spp.	Curds & Cockburn, 1970; Klimowicz, 1972; Poole, 1984; Sydenham, 1971
Holotrich ciliates	*Litonotus* spp.	Poole, 1984
Holotrich ciliates	*Hemiophyrus* spp.	Poole, 1984
Holotrich ciliates	*Porodon* spp.	Poole, 1984
Holotrich ciliates	*Chilodonella* spp.	Klimowicz, 1972; Poole, 1984; Sydenham, 1971
Holotrich ciliates	*Dysteriidae* spp.	Poole, 1984
Holotrich ciliates	*Trochilia* spp.	Sydenham, 1971
Holotrich ciliates	*Depanomonas* spp.	Poole, 1984
Holotrich ciliates	*Uronema* spp.	Poole, 1984
Holotrich ciliates	*Colpidium* spp.	Poole, 1984
Holotrich ciliates	*Lembadion* spp.	Poole, 1984
Holotrich ciliates	*Paramecium* spp.	Poole, 1984
Holotrich ciliates	*Cinetochilum* spp.	Poole, 1984
Hypotrich ciliates	*Aspidisca* spp.	Klimowicz, 1972; Poole, 1984; Sydenham, 1971
Hypotrich ciliates	*Oxytricha* spp.	Klimowicz, 1972; Poole, 1984

(*continued*)

TABLE 5.1. Organisms other than bacteria found in waste treatment (*continued*).

Type	Organisms	Reference
Hypotrich ciliates	*Stylonychia* spp.	Poole, 1984
Hypotrich ciliates	*Euplotes* spp.	Klimowicz, 1972; Poole, 1984
Peritrich ciliates	*Epistylis* spp.	Cossignanni et al., 1992; Klimowicz, 1972; Poole, 1984; Sydenham, 1971
Peritrich ciliates	*Vorticella* spp.	Klimowicz, 1972; Pillai & Subramanyan, 1944; Poole, 1984; Sydenham, 1971
Peritrich ciliates	*Zoothamnium* spp.	Poole, 1984
Peritrich ciliates	*Carchesium* spp.	Poole, 1984
Peritrich ciliates	*Opercularia* spp.	Klimowicz, 1972; Poole, 1984; Sydenham, 1971
Suctorian ciliates	*Acineta* spp.	Poole, 1984
Suctorian ciliates	*Metacineta* spp.	Poole, 1984
Suctorian ciliates	*Podophyra* spp.	Poole, 1984
Suctorian ciliates	*Sphaerophrya* spp.	Poole, 1984
Flagellate protozoa	*Peranema* spp.	Poole, 1984; Sydenham, 1971
Flagellate protozoa	*Anisonema* spp.	Sydenham, 1971
Rhizopod protozoa	*Amoeba* spp.	Klimowicz, 1972; Poole, 1984
Rhizopod protozoa	*Euglypha* spp.	Poole, 1984; Sydenham, 1971
Rhizopod protozoa	*Flabellula* spp.	Sydenham, 1971
Rhizopod protozoa	*Mayorella* spp.	Sydenham, 1971
Rhizopod protozoa	*Arcelia* spp.	Sydenham, 1971
Rhizopod protozoa	*Cochliopodium* spp.	Barker, 1943, 1946, 1948, 1949; Biernacka, 1958; Cutler et al., 1932; Klimowicz, 1972; Sydenham, 1971
Rotifers	*Monogononta* spp.	Poole, 1984
Rotifers	*Bdelloidea* spp.	Poole, 1984
Rotifers	*Lecane* spp.	Klimowicz, 1972; Sydenham, 1971
Rotifers	*Mniobia* spp.	Sydenham, 1971
Rotifers	*Philodina* spp.	Klimowicz, 1972; Sydenham, 1971
Rotifers	*Habrotrocha* spp.	Klimowicz, 1972
Rotifers	*Adineta* spp.	Klimowicz, 1972
Rotifers	*Epiphanes* spp.	Klimowicz, 1972
Rotifers	*Brachionus* spp.	Klimowicz, 1972
Rotifers	*Colurella* spp.	Klimowicz, 1972
Rotifers	*Lepadella* spp.	Klimowicz, 1972

(*continued*)

TABLE 5.1. Organisms other than bacteria found in waste treatment (*continued*).

Type	Organisms	Reference
Rotifers	*Menostyla* spp.	Klimowicz, 1972
Rotifers	*Proales* spp.	Klimowicz, 1972
Rotifers	*Notommata* spp.	Klimowicz, 1972
Rotifers	*Cephalodella* spp.	Klimowicz, 1972
Tardigrades	*Rotaria tardigrada*	Klimowicz, 1972; Poole, 1984
Nematodes	Unspecified species	Poole, 1984; Sydenham, 1971
Annelids	*Aeolosoma* spp.	Sydenham, 1971
Gastrotrichs	*Chaetonotus* spp.	Klimowicz, 1972; Poole, 1984
Oligochaetes	*Aeolosomatidae* spp.	Poole, 1984
Oligochaetes	*Naididae* spp.	Poole, 1984
Copepod larvae		Poole, 1984

wastes other than the fecally derived pollution suggested by the presence of *E. coli* (Harvey, 1952).

Competition among species of bacteria for food is most frequently based on whichever species is the most efficient in assimilating available food, although there are examples of a bacterial species functioning as a predator on other bacteria, such as *Bdellovibrio bacteriovorous* (Stolp & Starr, 1963). For the most part, there is a general lack of active obstruction of one species by another in acquiring food (Bustamonte & Gaudy, 1968). Despite the exceptions, such as antibiotic-producing microorganisms, bacteria found in waste treatment generally do not produce offensive chemicals to prevent other cells from approaching or feeding. The functioning of the extracellular enzymes to hydrolyze polymeric food substrates is often beneficial to all nearby bacteria, as the cell that releases the enzymes is incapable of harvesting 100% of the products of the hydrolysis, even in the presence of a slime coating around the cell. And the extracellular enzymes themselves can serve as food for other organisms.

Individual species and even genotypes (called strains) of bacteria within a species do not have the same ability to acquire individual food substrates. Bacteria are often characterized by their ability to utilize specific substrates. On one hand, there are the generalists that are capable of utilizing many different substrates, having a genetic makeup that allows production of many types of enzymes. On the other hand, there are the specialists that tend to utilize one substrate, preferentially and in some cases exclusively, over other substrates. As a rule of thumb, specialists tend to be more efficient than generalists and, in competition for limited food, a specialist bacteria will be more successful than a generalist (Lu & Grady, 1988). Thus, one type of acclimation mechanism that occurs in mixed populations under a limited range of food resources consists of specialist organisms becoming numerically dominant.

Bacteria function as the lowest link on the food chain. Most higher organisms, including almost all animals, consume other creatures. Within a waste treatment environment, eukaryotic organisms are the major predators. These organisms have either a mouth or can mold the cell membrane around the bacteria, as in phagocytosis, to bring the bacterial cell and its nutrients into the organism. The effect of predators on bacteria has been studied (Butterfield & Purdy, 1931; Butterfield et al., 1931; Gaudy & Gaudy, 1980; Javornicky & Prokesova, 1963; Purdy & Butterfield, 1918), but in the view of most researchers, "[predators] are always clearly visible in the microscope, seldom visible in the literature, and never explicitly visible in the models" (van Loosdrecht & Henze, 1999) as significant factors in waste treatment. Even in BOD tests conducted with commercial seed that intentionally avoid inclusion of predators, unless the waste has been disinfected or sterilized before the test, predators will be present.

The existence of the plateau in oxygen consumption by mixed organism populations in the BOD test was originally ascribed to the effect of developing predator populations on the numbers of bacteria in the sample. This model initially has the bacterial population growing and reproducing while feeding on the soluble substrate. As the slower-growing predator population increases in number, at some point these begin to consume more bacteria than bacteria are reproducing, leading to a leveling, generally within 2 days of the initiation of the test, and then a drop in the numbers of bacteria present (Gaudy, 1972; Gaudy & Gaudy, 1980; Gaudy et al., 1965). A tacit assumption in this model is that the use of oxygen by predators is at a lower rate than that caused bt the growth of the bacteria. Overall oxygen consumption is an additive function of the numbers of bacteria and predators and, depending on how fast the predator population outstrips the bacterial population, a plateau in oxygen consumption is observed. The difference in total oxygen consumption rate by the culture before and after the plateau is then reflective of the differences between bacterial use of oxygen and predator use of oxygen, with the later stages of oxygen use determined solely by predator endogeneous respiration, as the vast majority of bacteria have been consumed.

Other explanations for the plateau, especially as observed in cultures that are devoid of predators such as occurs in commercial seed preparations, hold that it is a manifestation of the bacterial population increasing to the point where available food becomes the limiting factor in continued growth. When all the soluble food is gone, the bacteria switch over to an endogeneous respiration state and the oxygen consumption rate changes to the slower rate reflective of endogeneous metabolism and population reduction because of death. The effect of predators is to put a break on the bacterial growth and consumption of food, thereby prolonging the time needed to reach the maximum bacterial population and the plateau. A corollary of this view is that after the soluble food is gone, bacteria are performing endogeneous respiration, but the bacterial population never reaches zero because of predation. Thus, the rate of oxygen consumption after the plateau is primarily attributed to bacterial respiration. One set of authors has described the BOD test as a three-step process, the sequence being bacterial utilization of substrate, protozoan consumption of bacteria, and finally autodestruction (Gates & Ghosh, 1971). Autodestruction in this model contains elements of bacterial and protozoan endogeneous metabolism, and death and lysis of bacterial cells that return some consumable materials to the medium.

Experiments have demonstrated that the presence of predators does serve to dampen bacterial growth. Experimental separation of bacteria from protozoa in the laboratory is easily achieved by filtration of the liquid culture through a 5- to 8-μm pore-size membrane filter (Tezuka, 1968). The half-life of the bacteria *E. coli* drops from a normal 16 hours to 1.8 hours in the presence of protozoan predators (Curds & Fey, 1969). One discussion presented a scenario in which a bacterial species that is highly efficient at utilizing the food substrate for growth is decimated by predation, allowing a less efficient species to thrive and perform most of the food assimilation (Curds, 1974). This view holds that predation is detrimental to the waste purification process and goes so far as to recommend separation of bacteria and predators into sequential treatment stages, thereby maximizing the rate of treatment (Pirt & Bazin, 1972). This idea has been implemented in an actual treatment facility design (Lee & Welander, 1996).

The actual situation is a combination of these extreme views. Application of Monod growth models to both bacteria and protozoa reflects an interactive approach. The double Monod model presents the growth of the bacteria as dependent on the numbers of bacteria present, the effect of predation, and available food, while predator growth depends on the numbers of bacteria present (Canale et al., 1973; Curds, 1971; Curds & Cockburn, 1968; Proper & Garver, 1966; Sudo et al., 1975). Feedback effects of the double Monod model suggest that neither the bacterial or predator populations ever reach zero, nor is the available food completely consumed. Instead, relative equilibrium states tend to be achieved between bacteria and predators that are determined by the amount of food present (Nisbet et al., 1983). The determination of the relative abundance of a key organism in the ecological equilibrium, such as the protozoan *Epistylis plicatilis*, has been proposed as a quantitative metric of the microorganism balance in the treatment system and can be used as process control for facility operation (Cossignanni et al., 1992).

Protozoan predators do not limit their menu of acceptable fare to simply bacteria. The diurnal interactions of carnivorous and bacteriophagous ciliated protozoa has been examined and modeled (Curds, 1973). The predation of protozoa on other protozoa adds a significant braking factor into the overall ecology of the protozoa–bacteria interaction.

An aspect of microorganism interdependence observed in mixed cultures that is for the most part absent in monocultures is the formation of flocs (Butterfield, 1940; Butterfield & Wattie, 1938). Flocs are groups of organisms held together by filaments and slime. Most floc particles are started by bacteria and fungi and require quiescent conditions. Inert colloids are caught in the filaments of the clustered microorganisms (Bhargava, 1983). As the floc grows, other organisms in the mixed culture accumulate onto the particle. For sessile organisms, such as stalked ciliates, the floc particle serves as an anchor point and provides conditions favorable to population increase beyond that possible in microbially dispersed samples. For motile predators, it provides a concentration of food. A cycling of food, nutrients, and waste products develops within the floc, and the particles are viewed as the smallest ecosystem that exists.

The development of floc particles, although beneficial for the most part to the participating organism and highly desirable in treatment facilities, was detrimental to the progress of the BOD test (Butterfield & Wattie, 1938). Because of inhibited diffusion of food, nutrients, and oxygen into the bulk of the particle, there was an overall slowing of oxygen and food consumption and thus slower bacterial reproduction in the bottle compared with a situation with dispersed organisms (Pasveer, 1954). The accumulation of bacterial waste products within the particle further served to inhibit bacterial growth (Bogan, 1958). The presence of bacterial flocs also favors more rapid development of the predator population, as there is a significant decrease in the amount of time and energy that predators have to spend searching for food.

A rapidly stirred BOD test has often been observed to exhibit a higher result than that obtained under quiescent conditions (Morrissette & Mavinic, 1978; Shrivastava & Dixit, 1970). Using a mixed population seed, the BOD_5 of glucose–glutamic acid standards has been observed to increase by 7%, the BOD_5 of raw and primary effluent increases 17%, and the BOD_5 of final effluent increases 44% (Ali & Bewtra, 1972; McGhee & Walsh, 1974). The monomolecular rate constant, k, is larger in all samples except the final effluent; however, the value of the ultimate carbonaceous BOD increased only for the final effluent samples. A contributing factor to these observations may be the prevention of floc formation.

7.0 TOXICITY AND ACCLIMATION OF ORGANISMS

As introduced in the next chapter on theory, acclimation of cultures to substrates can take many mechanistic forms. The substrate is not the only driving force for acclimation. Other challenges include incubation temperature, salinity of the medium, DO level, and presence of competing organisms or predators.

On the individual species level, acclimation can be a result of metabolic or structural adjustment by the individual cell in response to a change in substrate composition or environmental conditions. This is a temporary adjustment, with the individual organism able to revert back to the standard state when the challenge is removed. One common example of this type of acclimation is observed when starved organisms are given food. The metabolism of the individual changes from endogeneous to active food acquisition. Removal of food has the individual return to a state of endogeneous metabolism. A rather extreme example of individual cell acclimation is seen in *Pseudomonas aeruginosa* (Miguez et al., 1986). When forced to use hexadecane as the sole carbon source, this bacteria reworks its outermost lipopolysaccharide layer so that it is hydrophobic rather than hydrophilic and thus is able to attract the nonpolar food substrate.

More permanent acclimation by an individual species can be a result of a resistance-like response, in which a limited number of organisms (even one) possesses a genotype mutation that confers an advantage to the individual organism under the specific condition of the challenge. Susceptible organisms in the species die off under the challenge, while the mutant genotype survives and reproduces. Eventually, the species is dominated and characterized by individuals possessing the mutant genotype. Under this mechanism of acclimation, removal of the challenge may or may not allow the individuals to revert to the previous standard state, but now all the individuals permanently possess the mutant genotype. If the mutation is detrimental to the individual under the standard conditions, the species may again go through the resistance response and lose the mutation.

More frequently, the acclimation response takes place in mixed cultures containing a variety of species, and somewhat different mechanisms can take place in addition to those mentioned. The acclimation from starvation to feeding is a common observation; however, the response of the mixed culture is not the same for all of the species that are present. Generally one, or at most a few species, will be able to effectively avail themselves of the resources and achieve optimum growth, thus becoming the dominant species in the culture. Slight differences in the medium or interspecies interactions can favor one species over another, and even seemingly identical environments generated from aliquots of the same wastewater seeded with the same population of microorganisms can lead to development of different dominant species (Klimowicz, 1972). Sometimes, it is a mutualistic effect between two species that results in optimum survival for both and dominance in the mixed culture; whereas in isolation, neither species thrives.

Factors that have been observed to change population dynamics in a mixed culture have been DO levels (Amberson, 1928; Myrick & Busch, 1960), salinity (Gotaas, 1949; Kincannon & Gaudy, 1966; Purvis, 1926), pH (Mukherjee et al., 1968), dilution of the medium (Chin et al., 1972; Hoskins & Butterfield, 1933; Lester et al., 1979), changes in substrate load (Duggan & Cleasby, 1976; Liang et al., 1979; Ramanathan & Gaudy, 1969; Stephanopoulos & Fredrickson, 1979), nitrogen-phosphorus and nitrogen-carbon nutrient levels (Gehm, 1947, 1950, 1953; Helmers et al., 1950; Holderby & Lea, 1935; Lea & Nichols, 1936, 1937; Van Esseltime & Rahn, 1939), available surface-to-volume ratio (Heukelekian & Heller, 1940; ZoBell, 1943), and especially temperature (Carpenter & Berger, 1978; Clarke & Bonnet, 1939; Gotaas, 1948; Greenfield & Elder, 1926; Greenfield et al., 1926; Heilbrunn, 1925; Hiidenheimo & Wilson, 1974; Howland, 1953; Lee, 1975; Liagina & Kusnetznow, 1937; Moore, 1941; Phelps, 1930; Shih & Stack, 1969; Ware, 1958). Although other factors might be found, these tend to be the major ones.

The performance of BOD tests on samples exhibiting high salinity levels arising from certain industrial effluents or seawater offers considerable challenges (Grindby & Wheatland, 1956; Wheatland & Wheatland, 1956). Dissolved oxygen levels determined on dilution water can be strongly affected when used to dilute hypersaline solutions (Davis et al., 1978). This observation may, in part, be caused by the salt effect noted for DO determinations using electrode-based procedures as discussed in Chapter 4 on DO. A recommended dilution water for ocean samples, consisting of seawater that has been aged for 20 to 30 days to remove inherent demand, avoided the DO-measurement problem (Kuftarkova, 1972). Microorganisms in most seeds are generally intolerant of saline conditions, and a lengthy acclimation period is needed before satisfactory BOD results can be obtained.

Microorganisms can function best within a very limited temperature range; however, the range that they can tolerate can be quite wide, especially on the side of lower temperatures (Crozier, 1924). The major effect of low temperature is a lengthening of the lag time by a factor of three to four at 4 °C and a 33% decrease of the BOD_5 compared with results obtained at 20 °C (Nylander, 1964). Low temperatures seem to have a minor effect on the achievement of the plateau once the lag phase is passed; however, a significant effect has been noted on the rate of endogeneous respiration beyond the plateau (Benedict, 1977; Moore, 1941). Increasing the temperature above 20 °C does not increase the total BOD appreciably; however, it does increase the rate at which it is achieved. Particularly applicable to tropical areas where the ambient temperature is greater than 20 °C, performing the BOD test at an elevated temperature is found to exhibit a better relationship to conditions found in a treatment facility. In one study in the tropics, duplicate samples were incubated at 20 and 27 °C, and it was found that a 3-day BOD (BOD_3) at 27 °C achieved the same results as a BOD_5 at 20 °C (Chaudhuri et al., 1992). Increasing the incubation temperature to 30 °C allows a utilization equivalent to the BOD_5 at 20 °C to be achieved in 2.6 days (Varma & Nepal, 1972). An increase in incubation temperature can allow performance of 1-day BOD (BOD_1) that gave results linearly related to those obtained in 5 days at 20 °C (Carpenter & Berger, 1978). The ultimate BOD was only slightly affected by an increase in incubation temperature, with 35 °C only giving up to a 7% increase for some samples (Nylander, 1964). The maximum tolerable temperature tended to be only slightly greater than the optimum temperature, with the major effect being inhibition of the enzymes that perform respiration (Edwards & Rettger, 1937; Elder, 1925). Thus, a 37 °C incubation temperature was found to produce erratic results (Leszczynski, 1969). The time for return to optimum performance by a mixed culture at room temperature was found to be 2 weeks for cultures acclimated to 4 °C and months for 32 °C acclimated cultures. Rapidly changing the temperature to another level resulted in inhibited respiration (Benedict & Carlson, 1973). Others expanded the observation range from 2 to 52 °C and calculated deoxygenation rates (Carpenter et al., 1968) with essentially the same observations. Protozoa and fungi have been found to be more detrimentally affected than bacteria to raised temperatures (Lee, 1975).

In some cases, microorganisms cannot adjust their metabolism sufficiently to overcome a challenge, and the organisms die. This effect is termed toxicity, and it can be limited to observations of the effect on a single species or be exhibited by all the subpopulations in a mixed culture.

The quantitative determination of toxicity on a pure culture of organisms is a well-defined concept, and the concept has been put into practice in procedures that attempted to explicitly define and rigidly control the experimental protocols (Bernheim, 1942). Even with quantitative quality control limits and continuous sensitivity evaluations, toxicity tests on pure cultures have been found to generate data that are best described as semiquantitative, as the 95% confidence intervals range up to an order of magnitude (Smith, 1999a, 1999b). It should be understood that assignment of quantitative toxicity evaluations to mixed populations is tenuous, if not completely inappropriate. This inability to define a toxicity threshold for a mixed culture is largely caused by the differing response characteristics exhibited by any two mixed populations that are compared. This problem also exists when quantitative biodegradation studies are attempted with mixed cultures. One proposed solution was to try several different sources of inocula before the decision is made that the substrate is nonbiodegradable (Vernimen et al., 1967).

These observations have not prevented toxicity evaluations on mixed cultures from being performed and reported many times. An example is the depression of poultry wastewater BOD by 33% in the presence of 1 mg/L copper and 16% by 1 mg/L zinc (Andemoroti, 1988). It is probably true that the authors did observe that 1 mg/L zinc did reduce their BOD by 16%; however, the data are meaningless for prediction of the effect of zinc on other poultry wastewaters. This is not to say that determination of percentage reduction in respiration attributed to toxic substrates is valueless. Indeed, using a Warburg respirometer to generate more precise respiration data than is possible through the use of the traditional dilution bottle BOD technique, it has been shown many times that toxicity can be traced in a quantitative fashion through collection systems back to the source. However, these data are not transportable to another system regardless of measurement technique. Thus, the reporting of quantitative toxicity values on mixed cultures is a largely futile gesture. Other reported examples of quantitative toxicity determinations on mixed cultures are presented in Table 5.2.

Microorganisms produce waste just like all forms of life. For the most part, bacteria accumulate the waste products inside the cells because they lack specific mechanisms to expel waste. One exception is the bacterial production of polymeric extracellular slime, which can be considered to be a waste expulsion mechanism (Dean, 1969). Upon cellular death and lysis, all the components of the cell, including the waste products, are released into the medium (Daigger & Grady, 1977). Predation by protozoa, which use vacuoles for active waste expulsion of materials that cannot be utilized, can also lead to the accumulation of bacterial waste products in the medium.

Organisms do not thrive on their own waste products, and the accumulation of waste in the media, particularly in pure cultures, elicits a toxicity response from the organisms. In mixed culture, the effect is moderated by the ability of the different species to utilize each other's waste.

8.0 CONCLUSION

The dilution bottle BOD result relies on a complex interplay among many factors, including the composition and history of the microorganisms, the physical and chemical characteristics of the medium, the pretreatment and handling of the sample, and the incubation environment. Minute variations in any of these factors can drastically change the relationships among the

TABLE 5.2. Some examples of quantitative toxicity evaluations of mixed bacterial cultures.

Toxicant	Result	Reference
6 mg/L $HgCl_2$	100% inhibition of glucose BOD	Berkun, 1982
9 mg/L $HgSO_4$	100% inhibition of glucose BOD	Berkun, 1982
2 ppm $HgCl_2$	100% inhibition	FSIWARC, 1954
1 mg/L Hg	100% inhibition of glucose BOD	Mowat, 1976
30 mg/L $K_2Cr_20_7$	100% inhibition of glucose BOD	Berkun, 1982
1 mg/L Cr(VI)	20% inhibition of glucose BOD	Ingols & Kirkpatrick, 1952a, 1952b; Placak et al., 1948, 1949
10 mg/L Cr(VI)	50% inhibition of glucose BOD	Mowat, 1976
10 mg/L Cr(III)	75% inhibition of glucose BOD	Mowat, 1976
Ethylenediaminetetra-acetic acid (EDTA)–Cr	No BOD inhibition	Morgan & Lackey, 1958
40 mg/L $CuSO_4$	100% inhibition of glucose BOD	Berkun, 1982
1 mg/L Cu	33%–66% inhibition of glucose BOD	FSIWARC, 1956; Placak et al., 1948, 1949; Mowat, 1976
EDTA–Cu	No BOD inhibition	Morgan & Lackey, 1958
200 mg/L $Al_2(SO_4)_3$	25% inhibition of glucose BOD	Berkun, 1982
1 mg/L Zn	17% reduction in BOD	FSIWARC, 1956
200 mg/L $ZnSO_4$	64% inhibition of glucose BOD	Berkun, 1982
1 mg/L Ag	30% inhibition of glucose BOD	Mowat, 1976
25 ppm cyanide	96% reduction of BOD_5	Ludzack et al., 1951, 1959
2.5 mg/L cyanide	45% inhibition of glucose BOD	Mowat, 1976
0.2% chlorine	43% inhibition of BOD	Baity & Bell, 1929; Crain & Gottlieb, 1935; Gaunt & Abbott, 1926, 1928
Pharmaceutical waste	Inhibition of BOD	Ajaml et al., 1980
Metals	Supression of BOD	Ajaml et al., 1983
Acetonitrile	0%–2% of ThOD[a]	Ludzack et al., 1959
Acrylonitrile	0% of ThOD	Ludzack et al., 1959
Adiponitrile	40% of ThOD	Ludzack et al., 1959
Benzonitrile	40%–60% of ThOD	Ludzack et al., 1959
Lactonitrile	67%–70% of ThOD	Ludzack et al., 1959
Oxydipropionitrile	0% of ThOD	Ludzack et al., 1959

[a]ThOD = theoretical oxygen demand.

microorganisms and the interactions between the microorganisms and the consumable substrate in the sample. The lack of ability to exercise significant analytical control over these divergent factors means that the test is, even under the best of conditions, only semiquantitative.

Consideration of all these factors leads to the conclusion that time-defined BOD studies using dilution bottle techniques have only a tenuous connection to the conditions existing in real-life waste treatment situations. Available biological test procedures, such as continuous culture or batch respirometry studies, produce more clearly defined results that have inherently better interpretational value. Chemical procedures such as COD, TOC, or TOD give more rigorous and reproducible chemical characterizations of wastes.

The format and practice of the 5-day dilution bottle BOD test as a regulatory compliance tool should be closely examined with regard to what is intended to be measured and controlled. The confusion and complexity of this issue cannot be understated. Even the two authors of this book cannot agree on a statement of the regulatory intent of the BOD_5 test. One believes that the intent of listing BOD_5 limits on final effluent permits is to provide a standard to monitor treatment facility performance, while the other believes it better serves as a water quality standard in the receiving water system. Regulatory intent to measure the ultimate biodegradability of the final effluent is better served by extending the incubation time beyond 5 days to at least 20 days. If the immediate effect of the effluent on receiving stream oxygen levels is the key intent, then respirometry kinetic data obtained from batch reactors maintained at the ambient conditions of the stream provide better guidance. Other regulatory intents, once explicitly identified, can perhaps be better served by more precisely oriented test procedures.

The 5-day dilution bottle BOD test developed as scientists and engineers struggled to understand the processes of biological waste treatment and pollution effects on water environments. The knowledge and technological innovations that are the result of numerous studies have, for the most part, been retained within the offices of academics and consulting engineers, leaving most practitioners in the dark about the intricacies of the test. We intended this chapter to illuminate the BOD test from a mechanistic viewpoint. The next chapter examines the theoretical aspects of the test. We feel that understanding the test thoroughly is a necessary prerequisite to successful performance and interpretation of the test.

9.0 REFERENCES

Ajmal, M., et al. (1980). Detrimental effects of pharmaceutical industrial waste on microorganisms. *Water, Air Soil Pollut.*, *13*, 447.

Ajmal, M., et al. (1983). Influence of toxic metals on the repression of CBOD. *Water Res.*, *17*, 799.

Ali, H. I., & Bewtra, J. K. (1972). Effect of turbulence on BOD. *J. Water Pollut. Control Fed.*, *44*, 1798.

Amberson, W. R. (1928). The influence of oxygen tension upon the respiration of unicellular organisms. *Biol. Bull.*, *55*, 79.

American Public Health Association, American Water Works Association, & Water Environment Federation. (1998). *Standard methods for the examination of water and wastewater* (20th ed.). American Public Health Association.

Andemoroti, C. M. A. (1988). The effect of metallic toxicants on BOD measurements. *Biol. Wastes*, *24*, 259.

Baity, H. G., & Bell, E. M. (1929). Reduction of the BOD of sewage chlorination. *Sew. Works J.*, *1*, 279.

Barker, A. N. (1943). The protozoan fauna of sewage disposal plants. *Naturalist*, July–Sept., 64.

Barker, A. N. (1946). The ecology and function of protozoa in sewage. *Ann. Appl. Biol.*, *33*, 314.

Barker, A. N. (1948). The protozoan fauna of sewage. *J. Gen. Microbiol Camb.*, *2*, 21.

Barker, A. N. (1949). Some microbial aspects of sewage purification. *J. Inst. Sew. Purif.*, *7*.

Bauman, H. D., & Lutz, L. R. (1974). Ozonation of a kraft mill effluent. *Tappi J.*, *57*(5), 116.

Beaman, M., Lambert, S., Graham, N., & Anderson, R. (1998). Role of ozone and recirculation in the stabilization of landfills and leachates. *Ozone: Sci. Eng.*, *20*(2), 121.

Bekbolet, M., Cecen, F., & Ozkosemen, G. (1996). Photocatalytic oxidation and subsequent adsorption characteristics of humic acids. *Water Sci. Technol.*, *34*, 65.

Benedict, A. H. (1977). Discussion of temperature effects on BOD stoichiometry and oxygen uptake rate. *J. Water Pollut. Control Fed.*, *49*, 864.

Benedict, A. H., & Carlson, D.A. (1973). Temperature acclimation in aerobic bio-oxidation systems. *J. Water Pollut. Control Fed.*, *45*, 10.

Berkun, M. (1982). Effects of inorganic metal toxicity on BOD—I. Methods for the estimation of BOD parameters—II. *Water Res.*, *16*, 559.

Bernheim, F. (1942). The effects of various substances on the oxygen uptake of *Blastomyces dermatiditis*. *J. Bacteriol.*, *44*, 533.

Beveridge, T. J., & Doyle, R. J. (1989). *Metal ions and bacteria.* Wiley.

Bhargava, D. S. (1983). Most rapid BOD assimilation in Ganga and Yamuna Rivers. *J. Environ. Eng.*, *109*, 102.

Biernacka, I. (1958). Seasonal changes in the protozoan fauna in the activated sludge of the purification plant in Danzig. *Pol. Arch. Hydrobiol.*, *5*, 51.

Bogan, R. H. (1958). Biochemical degradation products: A new dimension in stream pollution. *Sew. Ind. Wastes*, *30*, 208.

Bohmeke, S. (1975). Experiences with BOD determinations with special considerations for the modified Viehl dilution method. *Papier*, *29*(8), 358.

Broch-due, A., Anderson, R., & Kristoffersen, O. (1994). Pilot plant experience with an aerobic moving bed biofilm reactor for treatment of NSSC wastewater. *Water Sci. Technol.*, *29*(5–6), 283.

Bustamonte, R. B., & Gaudy, A. F., Jr. (1968). Comparison of bacterial growth in single and mixed populations of sewage origin. *Proc. Oklahoma Academy Sci.*, *4*, 441.

Butcher, R. W. (1932). Contributions to our knowledge of the ecology of sewage fungus. *Brit. Myc. Soc. Trans.*, *17*, 112.

Butterfield, C. T. (1940). Some functions of bacteria in the purification of polluted water. *J. Bacteriol.*, *39*, 527.

Butterfield, C. T., & Purdy, W. C. (1931). Some interrelationships of plankton and bacteria in natural purification of polluted water. *Ind. Eng. Chem.*, *23*, 213.

Butterfield, C. T., Purdy, W. C., & Theriault, E. J. (1931). Studies of natural purification in polluted water. IV. The influence of plankton on the biochemical oxidation of organic matter. *Public Health Rep.*, *46*, 393; *Chemical Abstracts*, *25*, 1929.

Butterfield, C. T., & Wattie, E. (1938). Studies of sewage purification. VIII. Observations on the effect of variations in the initial numbers of bacteria and the dispersion of sludge flocs on the course of oxidation of organic matter by bacteria in pure culture. *Public Health Rep.*, *53*, 1912; *Sew. Works J.*, *10*, 815 (1938).

Buxton, B. H. (1902). Bacterial purification of sewage. *Philadelphia Med. J.*, *9*, 621; *Rev. Am. Chem. Res.*, *8*, 365 (1902).

Camp, T. R. (1963). *Water and its impurities.* Reinhold.

Campbell, F. L., & Rudolfs, W. (1923). Studies on the biology of sewage disposal. *New Jersey Agr. Exp. Stn. Bull.*, *390*, 5; *Am. J. Public Health*, *14*, 398 (1924).

Canale, R. P., Lustig, T. D., Kehrberger, P. M., & Salo, J. E. (1973). Experimental and mathematical modeling studies of protozoan predation on bacteria. *Biotechnol. Bioeng.*, *15*, 707.

Carmichael, J. C. (1947). Correlation between BOD and chlorine consumed. Unpublished master's thesis, Dep. Sanit. Eng., Johns Hopkins University.

Carpenter, W. L., & Berger, H. F. (1978). Investigation of factors affecting BOD measurement and experience with the 1-day BOD test. *Natl. Council Air Stream Improvement Tech. Bull.*, 305.

Carpenter, W. L., Vamvakais, J. G., & Gellman, I. (1968). Temperature relationships in aerobic treatment and disposal of pulp and paper wastes. *J. Water Pollut. Control Fed.*, *40*, 733.

Carter, J. A., Jr., & Malaney, G. W. (1971). Dual substrate utilization by activated sludge microflora. *Water Sew. Works*, *118*, 411.

Chaudhuri, N., Tyagi, P. C., Niyogi, N., Thergaonkar, V. P., & Hanna, P. K. (1992). BOD test for tropical countries. *J. Environ. Eng.*, *118*, 298.

Chin, S. Y., et al. (1972). Kinetic behavior of M populations of activated sludge. *Biotechnol. Bioeng.*, *14*, 179.

Clark, D. W. (1974). BOD: A re-evaluation. *Water Sew. Works*, *121*, 68.

Clarke, G. L., & Bonnet, D. D. (1939). The influence of temperature on the survival, growth, and respiration of *Calamus firmachicus*. *Biol. Bull.*, *76*, 371.

Clifton, C. E., & Logan, W. A. (1939). On the relation between assimilation and respiration in suspensions and in cultures of *Escherichia coli*. *J. Bacteriol.*, *37*, 523.

Coackley, P. (1969). Aspects of activated sludge. *Process Biochem.*, *4*, 35.

Cooke, W. B. (1954a). Fungi in polluted water and sewage. I. Literature review. *Sew. Ind. Wastes*, *26*, 539.

Cooke, W. B. (1954b). Fungi in polluted water and sewage. II. Isolation technique. *Sew. Ind. Wastes*, *26*, 661.

Cooke, W. B. (1954c). Fungi in polluted water and sewage. III. Fungi in a small polluted stream. *Sew. Ind. Wastes*, *26*, 790.

Cooke, W. B., Moore, W. A., & Kabler, P. (1956). Comparative satisfaction of BOD by fungi in pure culture and in competition with domestic sewage seed. *Sew. Ind. Wastes*, *28*, 1075.

Cossignanni, M., Cingolani, L., Giacche, L., & Miliani, R. (1992). Investigations on activated sludge microfauna and treatment plant disfunctions. *Water Sci. Technol.*, *26*, 2519.

Cox, R. R. (1921). Non-bacterial populations of sewage trickling filters. *Eng. News-Rec.*, *87*, 720.

Crain, F. T., & Gottlieb, S. (1935). A study of the mechanism of the effect of chlorine on the BOD. *Trans. Kans. Acad. Sci.*, *38*, 145; *Chemical Abstracts*, *31*, 8088 (1937).

Crozier, W. J. (1924). On biological oxidations as function of T. *J. Gen. Physiol.*, *7*, 189.

Curds, C. R. (1971). Computer simulations of microbial population dynamics in the activated sludge process. *Water Res.*, *5*, 1049.

Curds, C. R. (1973). A theoretical study of factors influencing the microbial population dynamics of the activated-sludge process I. The effects of diurnal variations of sewage and carnivorous ciliated protozoa. *Water Res.*, *7*, 1269.

Curds, C. R. (1974). Computer simulations of some complex microbial food chains. *Water Res.*, *8*, 769.

Curds, C. R., & Cockburn, A. (1968). Studies on the growth and feeding of *Tetrahymena pyriformis* in axemic and monoxenic culture. *J. Gen. Microbiol.*, 54, 343.

Curds, C. R., & Cockburn, A. (1970). Protozoa in biological sewage-treatment processes. I. A survey of the protozoan fauna in british percolating filters and activated sludge plants. *Water Res.*, *4*, 225.

Curds, C. R., & Fey, G. J. (1969). The effect of ciliated protozoa on the fate of *Escherichia coli* in the activated sludge process. *Water Res.*, *3*, 853.

Cutler, D. W., Crump, L. M., & Dixon, A. (1932). Some factors influencing the distribution of certain protozoa in biological filters. *J. Anim. Ecol.*, *1*, 143.

Daigger, G. T., & Grady, C. P. L., Jr. (1977). A model for the bio-oxidation process based on product formation concepts. *Water Res.*, *11*, 1049.

Davis, E. M., et al. (1978). Probable causes for reduced BOD test results in hypersaline wastewaters. *Water Res.*, *12*, 917.

Dean, R. B. (1969). Colloids complicate treatment processes. *Environ. Sci. Technol.*, *3*, 820.

Duggan, J. B., & Cleasby, J. L. (1976). Effect of variable loading on oxygen uptake rate. *J. Water Pollut. Control Fed.*, *48*, 540.

Eckenfelder, W. W., Jr. (1989). *Industrial water pollution control*. McGraw-Hill.

Edwards, O. F., & Rettger, L. F. (1937). The relation of certain respiratory enzymes to the maximum growth temperature of bacteria. *J. Bacteriol.*, *34*, 489.

Elder, A. L. (1925). The effect of temperature on the rate of deoxygenation of diluted sewage. Unpublished master's thesis, University of Illinois.

Eldridge, E. F. (1938). The BOD of industrial wastes. *Michigan State Coll. Eng. Exp. Stn. Bull.*, *78*, 16.

Elsworth, R., Williams, V., & Smith, R. H. (1957). The effect of oxygen supply on the rate of growth of *Aerobacter cloacae*. *J. Appl. Chem.*, *7*(5), 269.

Federation of Sewage and Industrial Waste Associations Research Committee. (1954). Toxicity of mercuric chloride, chromic sulfate, and sodium chromate in the dilution BOD test. *Sew. Ind. Wastes*, *26*, 536.

Federation of Sewage and Industrial Waste Associations Research Committee. (1956). Toxicity of copper and zinc ions in the dilution BOD test. *Sew. Ind. Wastes*, *28*, 1168.

Gameson, A. L. H., & Wheatland, A. B. (1958). The ultimate oxygen demand and course of oxidation of sewage effluents. *J. Proc. Inst. Sew. Purif.*, 106.

Gates, W. E., & Ghosh, S. (1971). Biokinetic evaluation of BOD concepts and data. *J. Sanit. Eng. Div., Proc. Am. Soc. Civ. Eng.*, *97*, 287.

Gaudy, A. F., Jr. (1972). Biochemical oxygen demand. In *Water pollution microbiology*. Wiley & Sons.

Gaudy, A. F., Jr., Bhatla, M. N., Follett, R. H., & Abu-Niaaj, F. (1965). Factors affecting the existence of the plateau during the exertion of BOD. *J. Water Pollut. Control Fed.*, *37*, 444.

Gaudy, A. F., Jr., & Gaudy, E. T. (1980). *Microbiology for environmental scientists and engineers*. McGraw-Hill.

Gaunt, P., & Abbott, W. E. (1926). The effect of chlorine on the absorption of dissolved oxygen by polluted waters. *J. Soc. Chem. Ind.*, *45*, 323.

Gaunt, P., & Abbott, W. E. (1928). Dissolved oxygen absorption time relation of activated sludge effluents. *J. Soc. Chem. Ind.*, *47*, 14.

Gehm, H. W. (1947). Problems in the application of the BOD test to pulp and paper wastes. *Sew. Works J.*, *19*, 865.

Gehm, H. W. (1950). Adaption of the BOD test to pulp and paper mill wastes. *Tappi J.*, *33*,124.

Gehm, H. W. (1953). Aerobic treatment of wastes high in BOD concentration. *Proc. 8th Ind. Waste Conf., Purdue University*

Goel, R., Mino, T., Satoh, H., & Matsuo, T. (1999). Modeling hydrolysis processes considering intracellular storage. *Water Sci. Technol.*, *39*, 1, 97.

Gotaas, H. B. (1948). Effect of temperature on biochemical oxidation of sewage. *Sew. Works J.*, *20*, 441.

Gotaas, H. B. (1949). Effect of sea water on biochemical oxidation of sewage. *Sew. Works J.*, *21*, 818.

Greenfield, R. E., & Elder, A. L. (1926). The effect of temperature on the rate of deoxygenation of diluted sewage. *Ind. Eng. Chem.*, *18*, 291.

Greenfield, R. E., Elder, A. L., & McMurray, R. E. (1926). Further studies on the BOD test. *Ind. Eng. Chem.*, *18*, 1276.

Greig, M., & Hoogerheide, J. C. (1941). The correlation of bacterial growth with oxygen consumption. *J. Bacteriol.*, *41*, 549.

Griffiths, P. (1994). Modifications to the IAWPRC Task Group activated sludge model. *Water Res.*, *28*, 657.

Grindby, J., & Wheatland, A. B. (1956). Salinity and the BOD of estuary water. *Water Sanit. Eng.*, *6*(1), 10.

Groff, G., & Ridenour, G. M. (1943). Effect of breakpoint chlorination on the BOD of sewage. *Sew. Works J.*, *15*, 847.

Grune, W. N. (1951). The effect of radioactivity on the bod of sewage. Unpublished thesis, Coll. Eng., New York University.

Grune, W. N. (1953). Radioactive effects on the BOD of sewage. *Sew. Ind. Wastes*, *25*, 882.

Grune, W. N., & Eliassen, R. (1951). Studies on the effect of radioactive phosphorus (P^{32}) on the BOD of sewage. *Sew. Ind. Wastes*, *23*, 141.

Harkness, N. (1966). Bacteria in sewage treatment processes. *J. Proc. Inst. Sew. Purif.*, 542.

Hartmann, L., & Wilderer, P. (1968). Physical and biological aspects of BOD kinetics. *Vodni Hospod. A or B* (Czech.), 6, 258; *Chemical Abstracts*, *69*, 8391 (1968).

Harvey, J. V. (1952). Relationship of aquatic fungi to water pollution. *Sew. Ind. Wastes*, *24*, 1159.

Heilbrunn, L. V. (1925). The use of temperature coefficients in the interpretation of biological processes. *Science*, *62*, 268.

Helmers, E. N., Anderson, E. J., Kilgore, H. D., Weinberger, L. W., & Sawyer, C. N. (1950). Nutritional requirements in the biological stabilization of industrial wastes. I. Experimental method. *Sew. Ind. Wastes*, *22*, 1200.

Heukelekian, H. (1933). Effect of dilution of sewage on bacterial numbers. *Sew. Works J.*, *5*, 774.

Heukelekian, H. (1936). Oxygen tension and bacterial numbers. *Sew. Works J.*, *8*, 415.

Heukelekian, H. (1942). Effect of yeast on the stabilization of sewage and sludge digestion. *Public Works*, *73*(9), 17.

Heukelekian, H. (1950). Biological oxidation of industrial wastes. *Sew. Ind. Wastes*, *22*, 87.

Heukelekian, H. (1957). The BOD test in textile mill waste problems. *Textile Ind.*, *121*(7), 74.

Heukelekian, H., & Heller, A. (1940). Relation between food concentration and surface for bacterial growth. *J. Bacteriol.*, *40*, 547.

Hiidenheimo, H., & Wilson, M. F. (1974). Long term BOD determinations on biologically treated and untreated kraft mill effluents discharged to Lake Saimaa, Finland. *Water Res.*, *8*, 535.

Hoffman, T. W., et al. (1973). Simulation of a petroleum refinery waste treatment process. *J. Water Pollut. Control Fed.*, *45*, 2321.

Holderby, J. M., & Lea, W. L. (1935). The BOD test, as influenced by the ratio of organic carbon to total nitrogen. *Sew. Works J.*, *7*, 36.

Hoskins, J. D. (1933). The oxygen demand test and its application to sewage treatment. *Sew. Works J.*, *5*, 923.

Hoskins, J. K., & Butterfield, C. T. (1933). Some observed effects of dilution on the bacterial changes in polluted water. *Sew. Works J.*, *5*, 763.

Hotchkiss, M. (1924). Studies on the biology of sewage disposal: A survey of the bacteriological flora of a sewage treatment plant. *J. Bacteriol.*, *9*, 437.

Howland, W. E. (1953). Effect of temperature on sewage treatment processes. *Sew. Ind. Wastes*, *25*, 161.

Hull, C. H. J. (1955). Bibliography of biochemical oxygen demand. Low Flow Augmentation Project, Dep. Sanit. Eng. Water Resour. Rep. No. 5, Johns Hopkins University (revised 1961).

Ingols, R. S., & Jacobs, G. M. (1957). BOD reduction by chlorination of phenol and amino acids. *Sew. Ind. Wastes*, *29*, 258.

Ingols, R. S., & Kirkpatrick, E. S. (1952a). A study of chromium toxicity with various oxygen demand tests. *Waste Eng.*, *23*, 256.

Ingols, R. S., & Kirkpatrick, E. S. (1952b). Study of chromium toxicity by several oxygen demand tests. *Anal. Chem.*, *24*, 1881.

Ingraham, J. L., & Ingraham, C. A. (1995). *Introduction to Microbiology*. Wadsworth Publishing.

Iwugo, K. O. (1980). Testing points the way to proper organics treatment. *Water Sew. Works*, *127*(4), 42.

Javornicky, P., & Prokesova, V. (1963). The influence of protozoa and bacteria upon the oxidation of organic substances in water. *Int. Rev. Ges. Hydrobiol.*, *48*, 335.

Jennelle, E. M., & Gaudy, A. F., Jr. (1970). Studies on the kinetics and mechanism of bod exertion in dilute systems. *Biotechnol. Bioeng.*, *12*, 519.

Kamhawy, S. M. (1996). Biochemical oxygen demand (BOD) revisited. 78th Annual Conf., Iowa Water Pollut. Control Assoc.

Karrer, N. J., Ryhiner, G., & Heinzle, E. (1997). Applicability test for combined biological–chemical treatment of wastewaters containing biorefractory compounds. *Water Res.*, *31*, 1013.

Kincannon, D. F., & Gaudy, A. F., Jr. (1966). Some effects of high salt concentrations on activated sludge. *J. Water Pollut. Control Fed.*, *38*, 1148.

Klimowicz, H. (1972). The microfauna of activated sludge: II. Assemblages of microfauna in block aeration tanks. *Acta Hydrobiol.* (Poland), *14*, 19.

Kofoid, C. A. (1903). Plankton studies. IV. The plankton of the Illinois River (temperature as a factor in the environment). *Illinois State Lab. Natl. History Bull.*, *6*, 175.

Kolaczkowski, S. T., Mejboum, Z., & Spandowska, S. (1965). The BOD of some organic substances. *Fortschr. wasserchem. ihrer grenzgeb.* (Germany), *3*, 113; *Chemical Abstracts*, *68*, 5156 (1968).

Kuftarkova, E. A. (1972). Determination of BOD in sea water polluted with industrial and every day wastes. *Biol. Morya*, *27*, 162; *Chemical Abstracts*, *77*, 156130 (1972).

Lea, W. L. (1940). Some factors affecting the results of five-day BOD determinations. Unpublished thesis, Univ. Wisconsin–Madison.

Lea, W. L., & Nichols, M.S. (1936). Influence of substrate on biochemical oxygen demand. *Sew. Works J.*, *8*, 435.

Lea, W. L., & Nichols, M.S. (1937). Influence of phosphorus and nitrogen on BOD. *Sew. Works J.*, *9*, 34.

LeBlanc, P. J. (1974a). A review of the BOD_5 test. *Pulp Paper Mag. Can.*, *75*, 3, 79.

LeBlanc, P. J. (1974b). Review of rapid BOD test methods. *J. Water Pollut. Control. Fed.*, *46*, 2202.

Lee, E. G.-H. (1975). Effect of temperature and sludge loading on BOD_5 removal and sludge setteability in activated sludge systems treating bleached kraft effluents. *Tappi J.*, *58*, 6, 100.

Lee, N. E., et al. (1982). Ultraviolet irradiation of municipal wastewater: Evaluation of effects on organic constituents. *Environ. Int.*, *7*, 403.

Lee, N. M., & Welander, T. (1996). Reducing sludge production in aerobic wastewater treatment through manipulation of the ecosystem. *Water Res.*, *30*, 1781.

Lenov, A. V. (1974). Generalization, typification, and kinetic analysis of biochemical consumption of oxygen curves based on BOD experiments. *Okeanologiia*, *14*, 82.

Lester, J. N., et al. (1979). Cultivation of a mixed bacterial population of sewage origin in the chemostat. *Water Res.*, *13*, 545.

Leszczynski, C. (1969). Attempts at replacing the standard 5-day BOD of pulp mill wastewaters by a 1-day BOD determination. *Przegl. Papie.* (Poland), *25*(5), 155; *Chemical Abstracts*, *71*(12), 269 (1969).

Levine, A. D., Tchobanoglous, G., & Asano, T. (1985). Characterization of the size distribution of contaminants in wastewater: treatment and reuse implications. *J. Water Pollut. Control Fed.*, *57*, 805.

Liagina, N. M., & Kusnetzow, S. I. (1937). The determination of the intensity of respiration of some species of water bacteria at various temperatures under laboratory conditions. *Microbiology*, *6*, 21.

Liang, D. H., et al. (1979). Bipopulation approach to CMAS simulation. *J. Environ. Eng.*, *105*, 905.

Lighthart, B., & Loew, G. A. (1972). Identification key for bacteria clusters from an activated sludge plant. *J. Water Pollut. Control Fed.*, *44*, 2078.

Logan, B. E., & Wagenseller, G. A. (2000). Molecular size distribution in a wastewater transformed by treatment in a full-scale trickling filter. *Water Environ. Res.*, *72*, 277.

Lomova, M. A., & Ryabkova, R. M. (1976). Biological degradation of difficult-to-oxidize effluent components. *Bum. Prom.* (Russia), *11*, 24; *Abstr. Bull. Inst. Pap. Chem.*, *47*, 11602 (1977).

Lu, Y.-T., & Grady, C. P. L., Jr. (1988). Effects of interactions within a three-membered microbial community on the kinetics of dual substrate removal in continuous culture. *Water Sci. Technol.*, *20*(11–12), 11.

Ludzack, F. J., Krieger, H. L., & Ettinger, M. B. (1964). Interactions of waste feed, activated sludge, and oxygen as traced by radioactive carbon. *J. Water Pollut. Control Fed.*, *36*, 782.

Ludzack, F. J., Moore, W. A., Krieger, H. L., & Ruchhoft, C. C. (1951). Effect of cyanide on biochemical oxidation in sewage and polluted water. *Sew. Ind. Wastes*, *23*, 1298.

Ludzack, F. J., Schaffer, R. B., Bloomhoff, R. N., & Ettinger, M. B. (1959). Biochemical oxidation of some commercially important organic cyanides. *Sew. Ind. Wastes*, *31*, 33.

Martin, D. C. (1932). The oxygen consumption of *Escherichia coli* during the lag and logarithmic phases of growth. *J. Gen. Physiol.*, *15*, 691.

Matsui, M., et al. (1984). Reaction of water-soluble dyes with ozone. *J. Soc. Dyers Colourists*, 10, 350.

Matsunaga, T., et al. (1981). Amperometric determination of viable cell numbers based on sensing microbial respiration. *Eur. J. Appl. Microbiol.. Biotechnol.*, *12*, 97.

McGhee, T. J., & Walsh, R. D. (1974). The effect of mixing on the bod determination. *Water Sew. Works*, *121*, 46.

McGuire, C. D. (1927). Effect of chlorine treatment on the BOD of sewage filter effluents. Columbus, Ohio, Div. Sew. Disposal Report, *20*, 36.

Mebley, D. R., & Stover, E. L. (1983). Effects of ozone on the biodegradability of biorefractory pollutants. *J. Water Pollut. Control Fed.*, *55*, 489.

Miguez, C. B., Beveridge, T. J., & Ingram, J. M. (1986). Lipopolysaccharide changes and cytoplasmic polyphosphate granule accumulation in *Pseudomonas aeruginosa* during growth on hexadecane. *Can. J. Microbiol.*, *32*, 248.

Millot, N., et al. (1987). Application of GPC processing system to landfill leachates. *Water Res.*, *21*, 709.

Mills, E. J., Jr., & Stack, V. T., Jr. (1955). Suggested procedure for evaluation of biological oxidation of organic chemicals. *Sew. Ind. Wastes*, *27*, 1061.

Milojevic, M. (1967). Contribution of steam-sterilized algae to BOD. *J. Water Pollut. Control Fed.*, *39*, 1201.

Mohlman, F. W. (1936). Use and abuse of the BOD test. *Sew. Works J.*, *8*, 499.

Moore, E. W. (1941). Long time BOD's at low temperatures. *Sew. Works J.*, *13*, 561.

Morgan, G. B., & Lackey, J. B. (1958). BOD determinations in wastes containing chelated copper or chromium. *Sew. Ind. Wastes*, *30*, 283.

Morrissette, D. G., & Mavinic, D. S. (1978). BOD test variables. *J. Environ. Eng.*, *104*, 1213.

Mowat, A. (1976). Measurement of metal toxicity by BOD. *J. Water Pollut. Control Fed.*, *48*, 853.

Mukherjee, S. K., Chatterji, A. K., & Saraswat, I. P. (1968). Effect of pH on the rate of BOD of wastewater. *J. Water Pollut. Control Fed.*, *40*, 1934.

Myrick, N., & Busch, A. W. (1960). The selective stimulation of respiration in mixed cultures of bacteria and protozoa. *J. Water Pollut. Control Fed.*, *32*, 741.

Nisbet, R. M., Cunningham, A., & Gurney, W. S. C. (1983). Endogenous metabolism and the stability of microbial prey–predator systems. *Biotechnol. Bioeng.*, *25*, 301.

Nylander, G. (1964). A study of waste waters from forest product industry. 5. A study of the course of biochemical oxidation. *Sven. Papperstidn.* (Sweden), *67*, 562.

O'Brien, W. J., & Clark, J. W. (1962). The historical development of the biochemical oxygen demand test. *Bulletin No. 20, Engineering Experimental Station, New Mexico State University.*

Onari, Y., et al. (1981). Change of gel-filtration chromatography of pollutants by aerobic microbiological treatment. *Mizu Shori Gijutsu*, *22*, 1083; *Chemical Abstracts*, *96*, 91054 (1982).

Orford, H. E., Rand, M. C., & Gellman, I. (1953). A single dilution technique for BOD studies. *Sew. Ind. Wastes*, *25*, 284.

Paixao, S. M., Baeta-Hall, L., & Anselmo, A. M. (2000). Evaluation of two commercial microbial inocula as seed in a 5-day BOD test. *Water Environ. Res.*, *72*, 282.

Parisod, J. P., & Schroeder, E. D. (1978). BOD progression in mixed substrates. *J. Water Pollut. Control Fed.*, *50*, 1827.

Pasveer, A. (1954). Distribution of oxygen in activated sludge floc. *Sew. Ind. Wastes*, *26*, 28.

Pfeil, B. H., & Lee, G. F. (1968). Biodegradation of nitrilotriacetic acid in aerobic systems. *Environ. Sci. Technol.*, *2*, 543.

Phelps, E. B. (1930). Stream pollution from the operator's point of view. *Sew. Works J.*, *2*, *555*.

Pillai, S. C., & Subramanyan, V. (1944). The role of protozoa in the aerobic purification of sewage. *Nature*, *154*, 179.

Pirt, S. J., & Bazin, M. J. (1972). Possible adverse effect of protozoa on effluent purification systems. *Nature*, *239*, 290.

Placak, O. R., Ruchhoft, C. C., & Sapp, R. G. (1948). Effect of copper and chromate ions in the BOD of sewage dilutions. *Sew. Works Eng.*, *19*, 300.

Placak, O. R., Ruchhoft, C. C., & Sapp, R. G. (1949). Copper and chromate ions in sewage dilutions. effect on BOD. *Ind. Eng. Chem.*, *41*, 2238.

Pomeroy, R. (1938). Influence of oxygen concentration on BOD. *Sew. Works J.*, *10*, 465.

Ponce, S. L. (1974). BOD of finely divided logging debris in stream water. *Water Resour. Res.*, *10*, 983.

Poole, J. E. P. (1984). A study of the relationship between the mixed liquor fauna and plant performance for a variety of activated sludge sewage treatment works. *Water Res.*, *18*, 281.

Porges, N., Pepinsky, J. B., Hendler, N. C., & Hoover, S. R. (1950a). Biochemical oxidation of dairy wastes. I. methods of study. *Sew. Ind. Wastes*, 22, 318.

Porges, N., Pepinsky, J. B., Hendler, N. C., & Hoover, S. R. (1950b). Biochemical oxidation of dairy wastes. II. Comparative study of yeasts. *Sew. Ind. Wastes*, *22*, 888.

Proper, G., & Garver, J. C. (1966). Mass culture of the protozoa *Colpoda steinii*. *Biotechnol. Bioeng.*, *8*, 287.

Purdy, W. C., & Butterfield, C. T. (1918). The effect of plankton animals upon bacterial death rates. *Am. J. Public Health*, 499.

Purvis, J. E. (1926). The influence of soft and of hard water on the decomposition of sewage. *Surveyor*, *69*, 277.

Raabe, E. W. (1968). BOD and degradation of lignin in natural waters. *J. Water Pollut. Control Fed.*, *40*, 145.

Ramanathan, M., & Gaudy, A. F., Jr. (1969). Effect of high substrate concentration and cell feedback on kinetic behavior of heterogeneous populations in completely mixed systems. *Biotechnol. Bioeng.*, *11*, 207.

Rao, B. S., & Gaudy, A. F., Jr. (1966). Effect of sludge concentration on various aspects of biological activity in activated sludge. *J. Water Pollut. Control Fed.*, *38*, 794.

Rickert, D. A., & Hunter, J. V. (1971). General nature of soluble and particulate organics in sewage and secondary effluent. *Water Res.*, *5*, 421.

Ruchhoft, C. C. (1941). Dissolved oxygen and BOD determinations: Their application and interpretation. *Sew. Works J.*, *13*, 542.

Ruchhoft, C. C., Placak, O. R., & Ettinger, M. B. (1948). Variations in BOD velocity constants of sewage dilutions. *Ind. Eng. Chem.*, *40*, 1290.

Rudolfs, W., Barnes, G. E., Edwards, G. P., Heukelekian, H., Hurwitz, E., Renn, C. E., Steinberg, S., & Vaughan, W. F. (1950). Review of literature on toxic materials affecting sewage treatment processes, streams, and BOD determinations. *Sew. Ind. Wastes*, *22*, 1157.

Sawyer, C. N., McCarty, P. L., & Parkin, G. F. (1994). *Chemistry for environmental engineering* (4th ed.). McGraw-Hill.

Schaeffer, D. J., et al. (1977). Relationship between BOD_5 and fats, oils, and greases. *Water Sew. Works*, *124*(3), 82.

Shih, C. S., & Stack, V. T., Jr. (1969). Temperature effects on energy oxygen requirements in biological oxidation. *J. Water Pollut. Control Fed.*, *41*, 461.

Shrivastava, S. K., & Dixit, R. C. (1970). Effect of mixing on the BOD test. *Environ. Health*, *12*, 334.

Skrindle, R., & Sawyer, C. N. (1952). Application of the Warburg respirometer to industrial waste analysis, special studies on radioactive wastes. *Proc. 7th Ind. Waste Conf., Purdue University*, 481.

Smith, R.-K. (1999a). *Handbook of environmental analysis* (4th ed.). Genium Publishing.

Smith, R.-K. (1999b). *Lectures on wastewater analysis and interpretation.* Genium Publishing.

Sneath, P. H. A. (Ed.). (1986). *Bergey's manual of systematic bacteriology.* Lippincott, Williams and Wilkins.

Snyder, C. W. (1947). Chlorine absorption as an indicator of BOD. Unpublished thesis, Dep. Sanit. Eng., Johns Hopkins University.

Stephanopoulos, G., & Fredrickson, A. G. (1979). Effect of spatial inhomogeneities on the coexistence of competing microbial populations. *Biotechnol. Bioeng.*, *21*, 1491.

Stolp, H., & Starr, M. P. (1963). *Bdellovibrio bacteriovorous* Gen. and Sp. N., A predatory ectoparasitic and bacteriolytic microorganism. *Antonie van Leeuwenhoek*, *29*, 217.

Stover, E. L., & Kincannon, D. F. (1983). Biological treatability of specific organic compounds found in chemical industry wastewater. *J. Water Pollut. Control Fed.*, *55*, 97.

Stover, E. L., et al. (1983). Ozone assisted biological treatment of industrial wastewaters containing biorefractory compounds. *Ozone Sci. Eng.*, *4*, 177.

Sudo, R., Kobayashi, K., & Aiba, S. (1975). Some experiments and analysis of a predator–prey model: interaction between *Colibium campylum* and *Alcaligenes faecalis* in continuous and mixed culture. *Biotechnol. Bioeng.*, *17*, 167.

Sydenham, D. H. J. (1971). A re-assessment of the importance of ciliates, rhizopods, and rotatorians in the ecology of activated sludge. *Hydrobiologia*, *38*, 553.

Sykes, R. M. (1976). Microbial product formation and variable yield. *J. Water Pollut. Control Fed.*, *48*, 2046.

Tezuka, Y. (1968). A method for estimating bacterial respiration in natural water. *Japan. J. Ecol.*, *18*, 60.

Theriault, E. J. (1927). The oxygen demand of polluted waters: I. A critical review; II. The rate of deoxygenation. *U.S. Public Health Service Bulletin*, 173.

Tidwell, W. L., & Sorrels, J. H. (1955). Effect of bacterial flora on deoxygenation. *Sew. Ind. Wastes*, *27*, 167.

Tortora, G. J., Funke, B. R., and Case, C. L. (1998). *Microbiology, an introduction* (6th ed.), Addison-Wesley.

Van Beneden, B. (1956). Is BOD a measure of the load or of bacterial activity? *Bull. Cent. Belge Etude et Document Eaux*, *68–69*, 313.

Van Esseltime, W. P., & Rahn, O. (1939). The effect of temperature on bacteriostasis. *J. Bacteriol.*, *57*, 547.

van Loosdrecht, M. C. M., & Henze, M. (1999). Maintenance, endogeneous respiration, lysis, decay, and predation. *Water Sci. Technol.*, *39*, 1, 107.

Varma, M. M., & Nepal, J. K. (1972). Kinetics of soluble substrate assimilation. *J. Water Pollut. Control Fed.*, *44*, 2316.

Vernimen, A. P., Henken, E. R., & Lamb, J. C., III. (1967). A short-term BOD test. *J. Water Pollut. Control Fed.*, *39*, 1006.

Viehl, K. (1940). The effect of the content of oxygen on BOD and on the natural self-purification on water as well as on the activated sludge process. *Zeit. Hyg. Infektr.*, *122*, 533; *Chemical Abstracts*, *35*, 5613 (1941).

Viehl, K. (1953). Determination of oxygen consumption and BOD using dissolved oxygen. *Gesundh.-Ing.* (Germany), *74*, 123.

Viehl, K. (1955). Effect of temperature on biological water purification. *Gesundh.-Ing.* (Germany), *76*, 555.

Ware, G. C. (1958). Effect of temperature on the biological destruction of cyanide. *Water Waste Treatment J.*, *6*, 537.

Wheatland, A. B., & Wheatland, J. (1956). Salinity and the BOD of estuary water. *Water Sanit. Eng.*, *6*(5–6), 10.

Williams, W. E. (1951). Relation of BOD to stream deoxygenation. *Sew. Ind. Wastes*, *23*, 1161.

Yamazaki, M., et al. (1983). Radiation treatment of gamma radiation and activated sludge treatment. *Yousi to Haisui*, *25*, 2, 159.

Young, J. C. (1977). Biochemical oxygen demand: Measurement and application. Unpublished manuscript.

Zehnpfennig, R., & Nichols, M. S. (1953). Inoculation studies in BOD determination on sewage and pulp mill waste. *Sew. Ind. Wastes*, *25*, 61.

ZoBell, C. E. (1943). The effect of solids surfaces upon bacterial activity. *J. Bacteriol.*, *46*, 39.

ZoBell, C. E., & Stadler, J. (1940a). The effect of oxygen tension on the oxygen uptake of lake bacteria. *J. Bacteriol.*, *40*, 307.

ZoBell, C. E., & Stadler, J. (1940b). The oxidation of lignin by lake bacteria. *Arch. f. Hydrobiol.*, *37*, 163.

6

Theory of Biochemical Oxygen Demand

The theory of biochemical oxygen demand (BOD) is still evolving. Perhaps a more descriptive title for this chapter could be "Mathematical Methods for BOD Modeling and Prediction." However, the word "theory" is preferred as it connotes that the subjects touched on here lie more often in the realm of conjecture than experience, and that there may be a somewhat tenuous connection to observation. In no case has there been any mathematical treatment devised for description of a biological system that is completely accurate or applicable; at best, there are approximations.

The approach taken here is to initially progress from a consideration of the oxygen demand of a pure substance to the biological oxygen demand of pure substances. Approximations of mixed substrates as pure substances and the related oxygen demand are then described before moving on to consider how BOD has been modeled as a process. The application of biological process modeling as a factor in predicting the behavior of treatment systems and natural waters is considered next. The chapter concludes with a return to a description of the explicit modeling of BOD as it has been pursued over the last 50 years; unfortunately, this area of interest has been sorely neglected despite the great advances that have been made in understanding the fundamentals of biological processes.

The treatment in this chapter depends necessarily on mathematics, but a conscious attempt is made to limit the mathematical presentation to descriptions of results rather than lengthy derivations. Those interested in derivations should consult the original literature that is cited.

1.0 THEORETICAL OXYGEN DEMAND OF PURE SUBSTANCES

The theoretical oxygen demand (ThOD) is used in assessing the biodegradability of pure compounds (OECD, 1998, Method 30, Annex 4). The ThOD is calculated based on the products of complete oxidation of each of the elements in the compound listed in Table 6.1 (Ettinger, 1958). For example, ethanol is consumed in a combustion reaction giving carbon dioxide and water as the products

$$CH_3CH_2OH + 3O_2 \rightarrow 2CO_2 + 3H_2O$$

For a generalized compound of the formula

$$C_cH_hCl_{cl}N_nNa_{na}O_oP_pS_s$$

TABLE 6.1. Theoretical oxidation products of commonly occurring elements.

Element	Oxidation product
C	CO_2
H	H_2O
N	N (CBOD[a])
N	N_2O_5 (BOD)
Na	Na_2O
S	SO_3
X (Halogen)	X
P	P_2O_5

[a]CBOD = Carbonaceous BOD.

the ThOD without nitrification, likened to the carbonaceous BOD, can be calculated as

$$\text{ThOD}_{\text{NH3}} \frac{16[2c+0.5(h-cl-3n)+3s+2.5p+0.5na-o]}{\text{mol wt}} \text{mg } O_2/\text{mg}$$

Where mol wt is the molecular or formula weight of the compound. The ThOD with nitrification, similar to an uninhibited BOD, is

$$\text{ThOD}_{\text{NH3}} \frac{16[2c+0.5(h-cl)+2.5n+3s+2.5p+0.5na-o]}{\text{mol wt}} \text{mg } O_2/\text{mg}$$

Using the formula for glucose ($C_6H_{12}O_6$, mol wt 180), ThOD without nitrification is calculated as

$$\text{ThOD}_{\text{NH3}} \frac{16[2\times 6+0.5(12)-6]}{180} \text{mg } O_2/\text{mg}$$

$$\text{ThOD}_{\text{NH3}} = 1.07 \text{ mg } O_2/\text{mg glucose}$$

The idea of ThOD can be applied to the organic contents of domestic wastewater influents by using a generalized formula, $C_5H_7NO_2$, for bacteria (biomass) (Porges et al., 1956). Other studies have generated biomass formulas ranging from C_4H_8ON to $C_8H_{15}O_5N$, depending on the state of the biomass and the relative proportions of carbohydrate to protein in the substrate used to feed the microorganisms (Burkhead & Waddell, 1971).

A generalized formula for algae (Milojevic, 1967) is

$$C_5H_8NO_2 \text{ or } C_{5.6}H_{9.8}O_{2.3}N$$

However, a more complete formula for algae is

$$C_{106}H_{263}O_{110}N_{16}P$$

which is presented in the context of the Redfield stoichiometry

$$C_{106}H_{263}O_{110}N_{16}P \leftrightarrow 106CO_2 + 16HNO_3 + H_3PO_4 + 122\ H_2O$$

where the forward reaction is characteristic of heterotrophic respiration and the reverse reaction is photosynthetic production (Stumm & Morgan, 1996).

As described below and elsewhere in this work, BOD is not equivalent to a chemical combustion reaction. The closest analogy to theoretical oxygen demand is either chemical oxygen demand (COD) or total oxygen demand (TOD). For many biodegradation studies in which the substrate is a mixture of substances or a nonstoichiometric material, COD is used in place of ThOD.

2.0 THEORETICAL APPROXIMATIONS OF BIOCHEMICAL OXYGEN DEMAND

Limiting our thinking to only oxidation of the carbon content of the molecule, an approximation of the oxygen demand to form carbon dioxide can be calculated based on the percent carbon in the substance. Using algae as an example, the carbon content is 53%; thus, the oxygen demand is approximately 0.53 × 32/12 = 1.4 mg O_2/mg algae (Milojevic, 1967), where 32 and 12 are the molecular weights of oxygen and carbon, respectively. Applying this approximation to glucose ($C_6H_{12}O_6$, mol wt 180) gives 0.40 × 32/12 = 1.07 mg O_2/mg glucose; however, this procedure ignores any oxygen already present in the substrate. If the amount of oxygen already present in the substrate (0.53% for glucose) is accounted for, the estimate becomes 0.54 mg O_2/mg glucose, which is lower than the experimentally derived value (0.68 mg O_2/mg glucose) but a closer estimate.

From the viewpoint that COD is an invariant and calculable characterization of a substrate, thermodynamics have been used to relate COD and BOD (Servizi & Bogan, 1964). The derived formula is

$$\text{Theoretical BOD g } O_2/\text{mol} = \left[1 - A\frac{COD}{C}\right]\text{COD}$$

Where COD is the calculated (32 × moles O_2 to oxidize 1 mole substrate in balanced combustion reaction) or experimentally derived COD of the substrate, C is moles of carbon per moles substrate, and A is a constant ranging from 0.015 to 0.017, depending on substrate. Carbohydrates have a value of A = 0.017. Applying the formula to glucose ($C_6H_{12}O_6$, mol wt 180) gives

$$\text{Theoretical BOD} = \left[1 - 0.017\frac{32 \times 6}{6}\right]32 \times 6 = 87.2 \text{ g } O_2/\text{mol glucose}$$

$$= 0.48 \text{ mg } O_2/\text{mg glucose}$$

This calculated BOD is a low estimate, an average BOD_5 for glucose being 0.68 g/g (Heukelekian & Rand, 1955).

A total BOD (L_o) calculation for use in biodegradation studies based on determination of BOD_{20}, COD, and TOC has been reported (Mao & Smith, 1995)

$$L_O = 87.5\left[\text{BOD}_{20}\frac{k_3}{k_1}\right]^{0.72}\left[\frac{TOC}{COD}\right]^{1.60}$$

where the constants k_1 and k_3 are derived from the plotted BOD versus time curve. The BOD has been calculated from an empirical formula called the equivalent BOD (EBOD) (Goldsmith & Watson, 1980) based on the percentage composition of the waste:

$$\text{EBOD} = 1.301\ \text{protein} + 0.891\ \text{fat} + 0.691\ \text{carbohydrates}$$

Applying this estimate to glucose, which is a carbohydrate, gives a value of 0.691 mg O_2/mg glucose, which is fairly close to the average experimentally derived value.

The amount of oxygen required for conversion of ammonia to nitrate is called nitrification biochemical oxygen demand (NBOD or NOD). It is frequently determined as the difference between BOD and carbonaceous BOD (CBOD) (Deai et al., 1991):

$$\text{NBOD} = \text{BOD} - \text{CBOD}$$

Another approach for determination of NBOD is to calculate it based on initial and final values of ammonia and nitrite in the incubated sample. There are two formulas proposed for the calculation (Cooper, 1986; Wezernak & Gannon, 1967). The second is thought to be more reliable.

$$\begin{aligned}\text{NBOD} = {} & 3.43(\text{NH}_4^+ - \text{N}_{\text{Initial}} - \text{NH}_4^+ - \text{N}_{\text{Final}}) + \\ & 1.14(\text{NO}_2^- - \text{N}_{\text{Initial}} - \text{NO}_2^- - \text{N}_{\text{Final}})\end{aligned}$$

$$\begin{aligned}\text{NBOD} = {} & 3.22(\text{NH}_4^+ - \text{N}_{\text{Initial}} - \text{NH}_4^+ - \text{N}_{\text{Final}}) + \\ & 1.11(\text{NO}_2^- - \text{N}_{\text{Initial}} - \text{NO}_2^- - \text{N}_{\text{Final}})\end{aligned}$$

3.0 MODELING BIOCHEMICAL OXYGEN DEMAND OF NATURAL AND WASTE WATERS

The first major attempt to model the course of deoxygenation of wastewater was performed by Phelps (1909) in a study of the relative stability test. Deoxygenation was considered to be similar to a chemical reaction, and chemical reaction kinetics were applied. Chemical reaction theory holds that the rate of the reaction is related to the products of the concentrations of the consumable reactants, each raised to some exponential power, times a constant. The sum of the exponents is called the order of the reaction. In the deoxygenation process, the consumable reactants are the organic pollutant substrate and oxygen; however, the rate of deoxygenation was assumed to be equivalent to the rate of substrate use, thus the rate of reaction is

$$\text{Rate of deoxygenation} = -\frac{dL}{dt} = K_1 L$$

where t is the time, L is the concentration of the pollutant(s), and K_1 is the reaction rate constant. As there is only one identifiable reactant, this is a monomolecular reaction rate expression, similar to a chemical decomposition. The sum of the exponents is 1; thus, it is a first-order reaction. The terms monomolecular and first-order have stuck as identifiers of this rate expression.

The differential expression can be integrated to give

$$L_t = L_0 \exp[-K_1 t]$$

which relates the remaining pollutant at time $t(L_t)$ to the original amount of pollutant (L_0). Phelps was interested in developing an expression for relative stability, s, and went on to derive the expression:

$$\text{Relative stability} = \frac{L - L_t}{L} = 1 - k^t = 1 - 0.794^t$$

where k is equal to 10^{-K}, with K being estimated as 0.1 (Spitta & Weldert, 1906).

Pleissner (1910) presented an equation for a 48-hour incubation of the form:

$$\text{Oxygen loss per hour} = \frac{\text{Oxygen demand}}{17 \log^2 t}$$

where t is time.

An early equation for the deoxygenation rate of wastewater that included consideration of the oxygen concentration takes the form:

$$\log \frac{[O_2]_{initial}}{[O_2]_{final}} = A L_0 t$$

where A is a constant, L_0 is the initial amount of organic matter, and t is time (Black & Phelps, 1911).

The amount of organic content in the sample (L_0) was somewhat indeterminate until it was equated with the ultimate BOD (Streeter, 1925; Streeter & Phelps, 1925; Theriault, 1925, 1927). Theriault rearranged Phelp's equation to give

$$y_t = L_0\,(1 - 10^{-kt}) \text{ or } L_0\,(1 - e^{-Kt})$$

where k is the base 10 constant, and K is the base e constant. This equation is called the monomolecular equation. Theriault also presented a temperature dependence of the velocity constant k

$$\frac{k_T}{k_{20}} = \theta^{T-20}$$

where T is the temperature and the temperature constant θ was found to be 1.049.

A plot of the monomolecular equation (Figure 6.1) describes a curve opening along the time axis with the origin at time zero. From the time zero

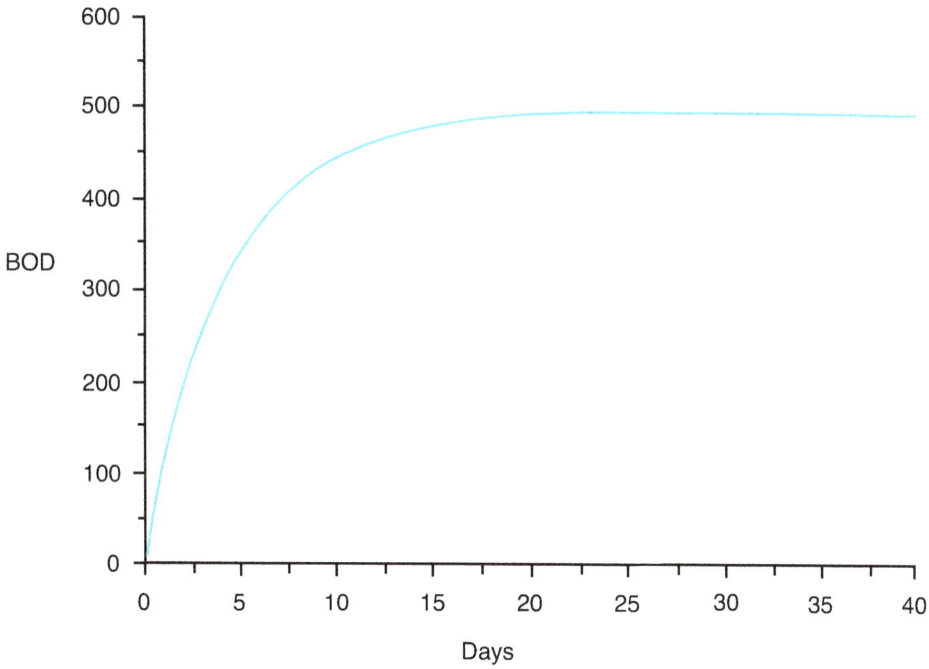

FIGURE 6.1. A plot of a first-order monomolecular equation with an ultimate BOD of 500.

point, the curve displays an initial rapid increase in measured BOD that tends to exponentially level off with time and asymptotically approach the ultimate BOD. This type of plot approximates experimental BOD results when they are determined over extended periods of time (up to 20 or more days). A key characteristic of this model, and a direct consequence to equating a biological process with a chemical process, is that the rate of reaction is assumed to be at the maximum rate possible at the instant of mixing the reactants and that the rate decreases from then on.

Theriault originally obtained values for the constants L and k from least-squares analysis of long-term BOD plots. Other techniques for determination of these constants included the log difference method (Fair, 1936), the slope method (Thomas, 1937, 1940), the method of moments (Lee, 1951; Ludzack et al., 1953; Moore et al., 1950; Ruchhoft, 1950), graphical approximations (Bagchi & Chaudhuri, 1970; Herman, 1956; Sletten, 1966; Thomas, 1950), statistical method (Grune, 1950), simultaneous equations (Navone, 1960), slide rule (Sheehy, 1960), and empirical method (Rhame, 1956). These approaches have been compared with each other in both the ease of use and ability to predict ultimate BOD (Hewitt & Hunter, 1975; Marske & Polkowski, 1972). A method for determining k as a function of loading, retention time, temperature, and hydraulic flow pattern has been reported (Uhlmann, 1979).

In attempting to use the monomolecular equation, it is found that it does not describe the early parts of the BOD exertion curve. This early deviation from the equation is called the "lag" phase, meaning that the BOD rate is lagging behind that predicted from the monomolecular equation. Some attempts were made to modify the monomolecular expression to account for the lag. One (Thomas, 1940) takes the form

$$y_t = L(1 - c10^{-kt})$$

where the added constant c was approximately equal to 10^{-kt}, t being at the end of the lag phase. Another modification (Fair et al., 1941) is

$$y_t = L(1 - [1 + at]^{-k/a})$$

where a is called the constant of retardation. Still another (Wiegand, 1954) is

$$\frac{L-y}{L-y_0} = 10^{-kt}$$

where L is the ultimate first-stage oxygen demand, y is the BOD at t days, and y_0 is the BOD at zero days (no actual physical significance). One report

(Rhame, 1956) resorted to using data from 7- and 14-day BOD to determine *k* and *L*, but mentioned that 3- and 6-day BOD data could be used if precautions were taken to prevent any initial lag in the BOD bottles.

Other work (Moore, 1941) found that the monomolecular equation needed to have the constants reevaluated when the temperature was below 5 °C and the curve exhibited a secondary oxidation peak. Attempts to apply the monomolecular equation to 5-day BOD results from a set of similar treatment facilities found a wide variation in the value of *k* (National Research Council, 1946), which in most earlier work had been found or was assumed to be 0.1. Values averaging 0.18 with a range of 0.10 to 0.30 were reported in the study. This led to the stated conclusion that 5-day BOD was not proportional to wastewater strength and that evaluation of facility performance based on it was less significant than commonly assumed. Another evaluation of the velocity constant *k* gave 0.15 (Ruchhoft et al., 1948a and 1948b), and values of *k* and *L* were found to vary with wastewater concentration. Manometric determination of *k* and *L* gave values significantly higher than the dilution bottle techniques (Caldwell & Langelier, 1948). Considerable variation in *k* was found when BOD was determined on samples containing seawater (Gotaas, 1949). Radiation was found to affect *k* values (Eliassen et al., 1950). Still other workers (Berthouex & Hunter, 1971a, 1971b; Constable & McBean, 1977; Gannon, 1966) stressed that the methods and errors associated with determination of rate constants greatly influence the prediction of BOD results. When statistical techniques are applied to the determination of *k*, the distribution of predicted CBOD is reported to be log-normal (Kothandaraman, 1970).

The velocity constant, *k*, changes with different substrates, seed populations, and conditions. Under constant conditions and microorganism populations, it is related to the ability of the seed to use the substrate as food. Easily degradable materials such as alcohols, carbohydrates, and hydroxyacids produce *k* values in the 0.12 to 0.20 range. Slowly degradable materials such as lignins exhibit values for *k* as low as 0.0016 (Bosko, 1965). When applied as a metric for characterization of waste streams, *k* determinations can be difficult when there are both rapidly and slowly degradable materials present, as is the case in effluent from pulp mills (Bouveng & Solyom, 1973).

Stochastic methods use mathematics to quantitatively define variations in processes. A stochastic model features an equation that predicts a result under a given set of conditions, termed the expectation, and another equation that represents the normal variance of the prediction, called the variance. The expectation and variance are written as functions or operators on the variables such as $\mathbf{E}[L]$, $\mathbf{E}[y]$, $\mathbf{V}[L]$, and $\mathbf{V}[y]$. Stochastic methods have been applied to first-order BOD kinetics (LeDuc et al., 1986). The stochastic form of the first-order BOD equation is

$$\frac{dL}{dt} = k_s L + S$$

where S is the randomness function. Application of suitable transforms gives the expectation equation

$$\mathrm{E}[y] = L_o\,[1 - \exp(-k_1 t)]$$

which is the same as the integrated form of the first-order BOD equation. The variance equation is

$$\mathrm{V}[y] = \frac{\sigma^2}{2k_1}[1 - \exp(-2k_1 t)]$$

where σ^2 represents the statistical variance in the process expressed as an average variance obtained from replicate measurements of BOD at a variety of time t. Plotting the functions $\mathbf{E}(y)$ and $\mathbf{E}[y] \pm 2\sigma$ generates the familiar first-order BOD curve, bounded by two additional lines that represent the 95% confidence interval of the measurement.

The basic assumption of the monomolecular theory as a first-order reaction is that the rate of utilization of substrate is similar to a single-molecule decomposition reaction. In pure chemical systems, the rate of decomposition is determined primarily by the concentration of the chemical. The rate constant is subject to temperature (Arrhenius, 1915), pressure, volume, and initiator effects; however, for a given set of environmental conditions, the concentration is the main determinant. In this model, if the concentration is doubled, the rate is doubled. Further, the maximum rate for any given experiment is generally achieved at the instant the experiment is initiated. Thus, there is considerable interest in chemical kinetics in closely studying the first moments of reaction.

The exertion of BOD as a process in a BOD bottle or stream or treatment facility does not bear up under this scrutiny, and the analogy to a single-molecule decomposition breaks down. In particular, many researchers observed that the first 24 to 48 hours of the BOD experiment does not conform to the first-order monomolecular equation (e.g., Gram, 1956; Hagawara, 1956). In most cases, the dissolved oxygen (DO) uptake is found to start off significantly slower than predicted; become faster than predicted; then after 24 to 48 hours, falls along the consumption indicated by the equation, overall describing an S-shape. This early slow increase in the DO consumption curve, the lower tail of the S, is called the lag phase. The duration of the lag phase is reported to be temperature dependent. At 4 °C, it is three to four times longer than at 20 °C (Nylander, 1964).

The first major complication with the monomolecular description is that the BOD process is not a decomposition reaction. Instead, it is likened overall to a combustion reaction, and oxygen is one of the required reactants. Methane (CH_4) is one of the simplest organic compounds. Methane reacts with oxygen in a balanced equation as follows:

$$CH_4 + 2O_2 \rightarrow CO_2 + 2H_2O$$

An approximation of the overall rate equation for this combustion is

$$\frac{dCH_4}{dt} = k[CH_4]^a[O_2]^b$$

which indicates that the rate of reaction is determined by not only the concentration of methane but also the concentration of oxygen, both raised to some power. It is a fundamental tenet in chemical kinetics that the observed reaction rate is governed by the slowest step, and it is more than somewhat unreasonable to make the assumption that this combustion process takes place in a single step with all molecules participating. The overall reaction order is generally equal to the sum of the exponents of the participating chemicals in the slow step. Pollutant substrates are much more complicated than methane and, while an exact empirical composition is not available, using one for biomass (Porges et al., 1956) can serve as a further example:

$$C_5H_7NO_2 + 5O_2 \rightarrow 5CO_2 + 2H_2O + NH_3$$

An approximation of an overall rate law for the complete combustion of biomass is

$$\frac{d\ \text{Biomass}}{dt} = k[\text{Biomass}]^a[O_2]^b$$

where the concentrations of biomass and oxygen are raised to some power and the number of molecules participating in the slow step is unknown (Castellan, 1972). In fact, there may be little contribution of oxygen to the slow step. However, the important point is that the model should at least be bimolecular and include some participation by oxygen. Although the need for incorporation of oxygen concentration to a rate expression for BOD had been proposed very early (Pleisnner, 1910), it was not until many years later that it was actually done. Instead, more complex equations that depended only on substrate concentration were investigated.

A second-order equation for BOD was presented in 1953 by Woodward. He presented an equation of the form

$$-\frac{dL}{dt} = KL^2$$

which was integrated to give

$$y = \frac{L_o^2 Kt}{1 + L_o Kt}$$

A slightly different form of a second-order BOD equation was presented in 1965 by Young and Clark:

$$y_t = \frac{t}{a + bt}$$

and was derived based on curve-fitting procedures. The constants a and b were described as being the inverse of the reaction rate constant and the inverse of the ultimate BOD, respectively. First- and second-order BOD equations were compared (Landine, 1971; Marske & Polkowski, 1972; Zanoni, 1967) using both respirometry and dilution bottle techniques. Depending on the application, some practitioners found the monomolecular formulation a better fit to their data than a second-order equation (Butts & Kothandaraman, 1970). Others found that second-order kinetics operated at low substrate concentrations but not at higher levels (Hartmann & Wilderer, 1968).

Higher-order kinetic equations, up to fourth, have been used in efforts to fit BOD curves using regression techniques (Argamam et al., 2000; Lenov, 1974):

$$-\frac{dL}{dt} = KL^2$$

which integrates, giving

$$y = L_o - [(n - 1)Kt - L_o^{1-n}]^{1/(1-n)}$$

which is a generic representation of the higher-order reactions. This equation, with $n = 2$ and higher values, was used to predict ultimate BOD

(Hewitt et al., 1979) from real data sets. These types of equations gave increasing values of ultimate BOD with increasing orders, with orders of three and higher giving values in excess of measured COD in some cases.

Trigonometric polynomial analysis of the diurnal fluctuations of BOD data results in an equation of the form (Papadopoulos et al., 1991; Sampathkumaran, 1953)

$$y = \text{mean BOD} + \alpha \cos(\beta t) + \beta_1 \sin(\beta t) + \sigma$$

where α is the amplitude, β is the periodicity of the fluctuations, and σ is an error function. This equation provides a somewhat better mathematical fit to the data but no real improvement in understanding BOD.

In studies examining data from batch-wise biochemical oxidations, it is found that, as each experiment progressed, the order of the descriptive equation changed. A second-order equation fit the initial data; however, by day 10, a one-tenth order equation was needed (Stones, 1963, 1979, 1984). Regardless of the form, the monomolecular and subsequent equations based solely on the utilization of substrate in general do not account for the lag phase at the beginning of the BOD curve, although they have been used quite successfully to predict ultimate BOD.

These models all lack contributions from other components of the process. A second-order equation that includes contributions from both oxygen and substrate has been described (Stones, 1981):

$$\frac{\mathrm{d}y}{\mathrm{d}t} = k_2(a - y)(b - y)$$

where a is the total BOD and b is the initial DO.

This is not to say that n-order equations based strictly on substrate concentration have no use in treatment design and control. Particularly in situations of continuous-flow processing, in which the overall concentration of oxygen and biomass are constants, n-order equations of the form

$$-\frac{\mathrm{d}L}{\mathrm{d}t} = K_n X_v \left[\frac{L}{L_o}\right]^n$$

where K_n is a curve-fitting coefficient and X_v is an active biomass concentration, have found great utility (Adams et al., 1975; Argamam et al., 2000).

Chemical oxidation reactions take place in flames (combustion processes) or with strong chemical oxidants such as permanganate, peroxide, ozone, or dichromate. Chemical oxidations with molecular oxygen are generally slow unless there is some facilitating mechanism. In the BOD process, the oxidation of the substrate with molecular oxygen is facilitated by microorganisms. Microorganisms serve a much more intimate role in the process than merely speeding the reactions through a catalytic effect. The microorganisms initially convert the substrate to individual cellular growth, then reproduction, and eventually more oxidized products that include carbon dioxide, water, and organic and cellular decomposition byproducts. A chemical reaction for this process cannot be balanced in the general sense but would look something like

$$\text{Substrate} + \text{microorganism} + xO_2 \rightarrow yCO_2 + zH_2O + \text{bioproducts}$$

And an analogous overall rate expression would start with assumption that

$$\frac{\text{dSubstrate}}{\text{d}t} = k[\text{Substrate}]^a[\text{Microorganism}]^b[O_2]^c$$

The key point to be made here is that the microorganisms play an essential role in the rate of the reaction. For any given initial concentration of oxygen and substrate, the reaction rate is going to depend on the concentration of functioning microorganisms (ReVelle et al., 1965; Van Beneden, 1956). If there are abundant active microorganisms, the rate is high, but if there are few active microorganisms, the rate is low. An examination (Buswell et al., 1954) of the rate of oxygen consumption found no justification for the calculation of L_0 from a 5-day BOD value. Instead, a direct relationship between maximum bacterial population and pollutant load was found. Most of the achievements in BOD theory that were real advancements beyond the monomolecular theory have included either the effects of oxygen concentration or microorganisms.

Logarithmic equations have been proposed as models of BOD reactions. The first was by Gaunt and Abbott (1928), who based their choice of function on the binary fission and subsequent logarithmic growth of nitrifying bacteria. Orford and Ingram (1953a, 1953b, 1953c) described the use of a logarithmic equation of the form

$$y_t = s(M \log[t] + B)$$

which was developed as an empirical best straight-line fit of semilog plots of BOD data. When applied to any BOD curve

$$y = S(0.85 \log[at] + 0.41)$$

where S is the 5-day BOD and a is the x-axis intercept of the sample/x-axis intercept of normal domestic wastewater. A temperature dependence of a was reported of the form

$$a_T = a20\theta^{(T-20)}$$

and θ was found to have a value of 1.072. The logarithmic model was criticized as unrealistic (Woodward, 1953); however, further work (Gameson & Wheatland, 1958) produced a related equation of the form

$$y = \text{Ultimate BOD}[1 - (Ae^{-k't} + Be^{-k''i} + Ce^{k'''t})]$$

where the velocity constants k', k'', and k''' are derived for each system by regression techniques, and the coefficients are restrained by the condition $A + B + C = 1$.

The early parts of the BOD curve, as a deviation from the monomolecular model, have been the subject of much study and debate. Detailed examinations of the early parts of the curve reveal that the slowly increasing rate of oxygen uptake in the lag phase beginning at time zero is observed to give way to a very rapid increase in exerted BOD, followed again by a slowing of the oxygen utilization and rate of approach to the ultimate BOD, an S-shaped behavior. A second-order reaction based on the amount of substrate and the number of bacteria was presented in 1968 (Simpson, 1968). Starting from the premise that

$$\frac{dy}{dt} = \gamma lq$$

where y is the BOD at time t, l is the substrate remaining at time t, q is the total number of bacteria, and γ is reaction rate constant, an expression for the BOD is derived

$$y = L_r/[1+K] - \frac{L_r/[1+K]}{1 + \frac{q_0}{L_r}\frac{1+K}{K}\exp[L_r K\gamma t]}$$

where K is a constant, L_r is the total removable oxygen demand, and q_0 is the initial quantity of bacteria. This equation is noteworthy in that the constants K and γ are readily derived from log plots of the BOD data and that the graph of the equation accurately reproduces the lag phase and the S-shape of the BOD curve. However, interest in the Simpson equation has been minimal because of the concurrent development of biological growth models.

There have been two essentially different explanations for the S-shape to the BOD curve. The first explanation described a rapid growth of cells with assimilation of soluble (available) nutrients, followed by a slower endogenous cellular respiration (Edeline & Lambert, 1979; Hoover et al., 1953). Endogenous respiration has been described as the catabolism of cellular enzymes. Another viewpoint attributed the latter stages of the respiration to catabolism of absorbed substrate rather than internal cellular enzymes (Sykes, 1976). Both effects are probably at work.

The rapid growth period after the initial lag is observed to be completed within the first 24 to 36 hours of the experiment and is often called the assimilation phase. It is generally agreed that the assimilation phase of oxidation is more rapid, but the endogenous phase accounts for a larger fraction of the total oxidative destruction (Hoover & Porges, 1951; Hoover et al., 1952, 1953; Ingram, 1939a, 1939b, 1940; Ludzack, et al., 1964; Van Loosdrecht & Henze, 1999).

The second explanation for the S-shape in the curve holds that it is to the result of the predation of protozoa on the bacteria (Bhatla & Gaudy, 1965a, 1965b, 1966; Gates & Ghosh, 1971). The leveling of the oxygen consumption within 24 to 48 hours of the initiation of the experiment has been ascribed to the decrease in the number of bacteria resulting from predation. However, even pure cultures of bacteria can exhibit a pause in respiration (Bhatia & Gaudy, 1965c; Wilson & Harrison, 1960). This observed pause in the overall respiration, regardless of cause, is termed the plateau. The assimilation period and plateau in respiration were summarized (Busch, 1958; Busch & Myrick 1960, 1961; Busch et al., 1960; Canale & Cheng, 1974; Gaudy, 1972) as the initial active bacterial respiration associated with food uptake and reproduction, followed by a declining bacterial respiration (endogenous respiration), and finally by increasingly active protozoal respiration. The S-shaped curve of the early stages of BOD exertion has evolved into a double S-curve, as in Figure 6.2.

An initial lag in oxygen utilization before the beginning of the bacterial exponential growth phase is frequently observed, particularly in dilution bottle and less so in respirometer studies. This effect has been ascribed to starvation of the microorganisms. In the preparation for the experiment, the seed organisms enter a state of endogenous respiration, where oxygen is being utilized to metabolize cellular mass. The cellular mass metabolized is predominantly those enzymes and proteins that are involved in reproduction and uptake of new food. When the starved organisms are exposed to food, it takes a short period of time to regenerate the needed biochemicals to acquire the food and resume replication (Apoteker & Thevenot, 1983; Rao & Gaudy, 1966).

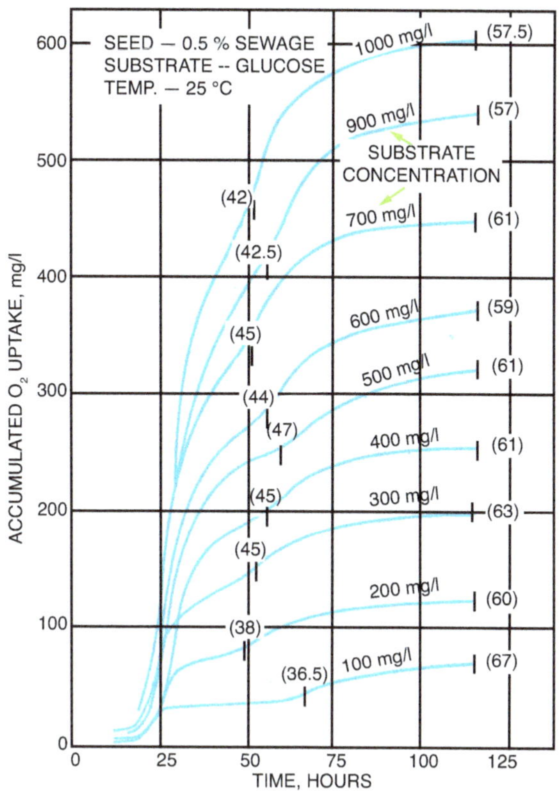

FIGURE 6.2. Plateau region in the early phase of BOD (Bhatia & Gaudy, 1966).

Many practitioners have described the plateau and studied various aspects of it. Examination of more than 50 long-term BOD curves indicated that a lag phase, assimilation phase, and plateau were found in almost all of them (Gaudy et al., 1965). Benedict (1977) held that the plateau region is universal in BOD experiments and is not affected by seed or temperature. He also observed considerable variation in oxygen uptake rates beyond the plateau region. The temperature coefficient (2) is determined to be greater (1.065) before the plateau than after the plateau (1.051) (Flegal & Schroeder, 1976). The plateau BOD for pure compounds has been compared with the plateau BOD for mixtures of the compounds. The plateau BOD for the mixtures can be approximated from the mean of the plateau values for the pure compounds and, in all cases, the measured rates of achieving the plateau were much higher than that predicted by the monomolecular equation (Parisod & Schroeder, 1978). An online instrument has been described that measures the plateau BOD and uses the value as an important process control parameter (Lenk et al., 1990).

The plateau region of the BOD curve has been the sticking point in most of the attempted modeling. A fairly advanced model has been developed, called the autocatalytic model. The autocatalytic model (Maoyu, 1990) is derived from the idea that the microorganisms in the test use the substrate to produce more microorganisms that subsequently also use substrate. It takes the form

$$\frac{dy}{dt} = k_3(a-y)(b-y)(y+y_0)$$

or

$$\frac{dy}{dt} = k_3y(a-y)(b-y) + k_3y_0(a-y)(b-y)$$

where a is the total CBOD, b is the initial DO concentration, and y_0 is related to the size of the initial microorganism inocula and microbial activity. The model includes all of the abovementioned influences: substrate, DO, and organisms. The integrated form of the equation is

$$k_3t = \frac{1}{[b-a][a+y_0]}\ln\frac{a}{a-y} + \frac{1}{[b-a][b+y_0]}\ln\frac{b-y}{b} + \frac{1}{[a+y_0][b+y_0]}\ln\frac{y_0+y}{y_0}$$

At time zero, the initial rate of the process is described by k_3aby_0. Plots of this equation for various values of y_0 reproduce the characteristic shape of the BOD lag when y_0 is small and the first-order curve when y_0 is large. However, in no case is there any indication of a plateau (double S) in the plots.

A series of equations have been derived from empirical examination of double-log plots of BOD test data (Swamee & Ojha, 1991). These equations were derived from consideration of increasingly complex BOD curves. For the simplest type of carbonaceous BOD curve, similar to a first-order curve, the derived expression is

$$y = L\,[(t_c/t)^{mc/n} + 1]^{-n}$$

where L is the ultimate first-stage BOD, t_c is the time for reaching the ultimate first stage CBOD, and mc is the carbonaceous BOD rate exponent; n is a transition exponent such that

$$n = 1.4427 \ln \frac{y_c}{L}$$

where y_c is the observed BOD exerted at time t_c. The presence of a lag phase in this model has been attributed to the use of unacclimated seed or the presence of toxic substrate. The presence of a lag phase is included by expanding the simplest equation to

$$y = L\,[(t_L/t)^{mL/n} + (t_c/t)^{mc/n} + 1]^{-n}$$

where terms for the lag time (t_L) and the double-log slope (mL) of the lag portion of the BOD curve ($t < t_L$) are added.

The presence of a second stage in the BOD curve has been attributed to the presence of nitrification. The second-stage behavior is added as follows:

$$\begin{aligned} y = {} & L\,[(t_c/t_L)^{mc/n}\,(t_L/t)^{mL/n} + (t_c/t)^{mc/n} + 1]^{-n} \\ & + (w_n - 1)\,L[(t_n/t_s)^{mn/n}\,(t_s/t)^{ms/n} + (t_n/t)^{mn/n} + 1]^{-n} \end{aligned}$$

where w_nL is the ultimate BOD exerted at the end of the second stage, t_s is the conceptual lag time, t_n is the ultimate time for the second stage, and mn and ms are the nitrogenous rate exponents.

The final variation includes a description of the plateau in the carbonaceous stage, and the equation becomes very unwieldy:

$$\begin{aligned} y = {} & w_pL\,[(t_p/t_L)^{mca/n}\,(t_L/t)^{mL/n} + (t_p/t)^{mca/n} + 1]^{-n} \\ & + (1 - w_p)\,L\,[(t_c/t_{cL})^{mcb/n}\,(t_{cL}/t)^{mcL/n} + (t_c/t)^{mcb/n} + 1]^{-n} \\ & + (w_n - 1)\,L\,[(t_n/t_s)^{mn/n}\,(t_s/t)^{ms/n} + (t_n/t)^{mn/n} + 1]^{-n} \end{aligned}$$

where w_pL is the plateau BOD, mca is the slope of the BOD curve before the plateau, and t_p is the time of the start of the plateau.

Although the plateau was successfully reproduced in an equation in this latter case, the equation itself has no foundation in recognized biological or chemical theory and principles. Thus, it really represents a dead end in the advancement of knowledge. Successful description of the plateau requires a completely different approach.

The major weak point in the monomolecular equation is the assumption that the microorganism population in the seed is constant, despite the widely known fact that the number of organisms increases in the BOD bottle, reaching a maximum density approximately 22 hours after the beginning of the experiment (Butterfield, 1929a & b, 1933). The rapidly expanding population in some cases is directly responsible in—as few as 5 hours–for up to 38% to 62% of the oxidation of the BOD exhibited at 5 days (Butterfield, 1935, 1937a, 1937b, 1940). Substrate utilization rates must factor in the effect of an expanding population (Buswell et al., 1954). Biological growth models examine the development of populations of organisms as they are affected by the substrate concentration. The most frequently used model has been the Monod equation (Monod, 1942, 1950), which describes the relationship between a logarithmic growth rate constant (μ) and substrate concentration (S)

$$\mu = \frac{\mu_m S}{k_s + S}$$

where k_s is a constant similar to the Michaelis–Menten constant (Michaelis & Menten, 1913) and is the substrate concentration at the point where $\mu = \frac{\mu_m}{2}$; μ_m is the maximum growth rate constant. A plot of μ against S is presented in Figure 6.3.

The growth rate (Gaudy, 1972; Gaudy et al., 1967; Ramanathan & Gaudy, 1969) for heterogeneous wastewater-derived populations and for

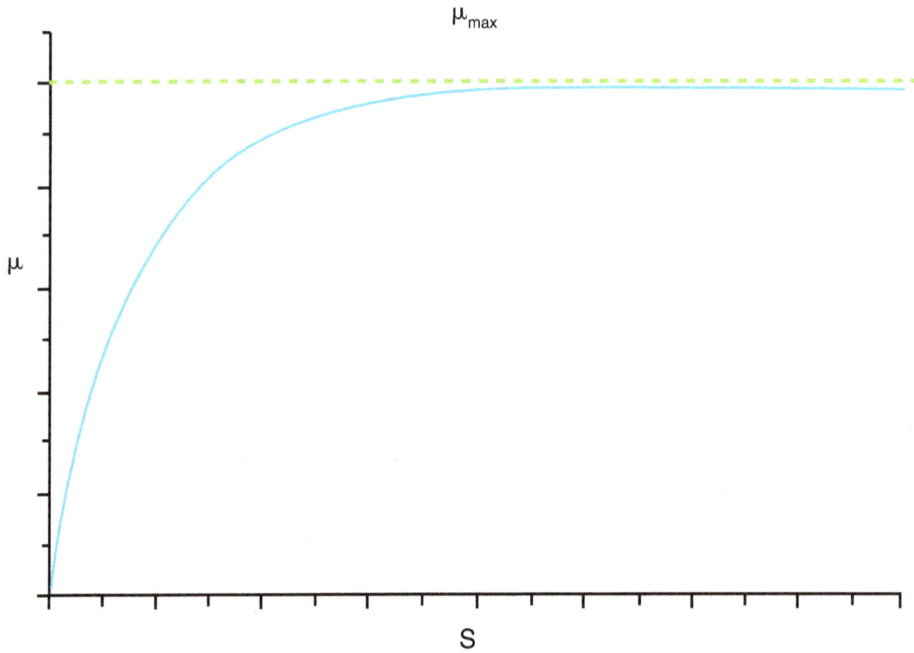

FIGURE 6.3. Growth curve with uninhibited (Monod) kinetics.

natural stream-derived populations is related to the size of the bacterial population by

$$\frac{dN}{dt} = \mu N$$

where N is the number of bacteria. Because the growth rate is related to the food concentration, this means that the growth rate and oxygen-utilization rate change depending on the substrate concentration. And it is decidedly not linear. A 1:10 dilution of the sample does not give a 1:10 reduction in the oxygen-utilization rate.

The assimilation phase of the BOD curve before and including the plateau has been described using the Monod equation (Braun & Berthouex, 1970; Gates et al., 1969; Gaudy et al., 1967). Application to a BOD process, or disappearance of substrate (L), gives

$$-\frac{dL}{dt} = \frac{dX/dt}{Y} = \mu_m \frac{LX}{Y[k_s + L]}$$

where X is the microorganism concentration and Y is the yield factor.

The first-order monomolecular equation and its derivatives are not a simplified case of the Monod expression. The Monod equation becomes equivalent to the first-order monomolecular BOD equation if, and only if, the relationship holds as follows:

$$\frac{\mu_{\max} X}{Yk_s} = \text{constant}$$

which means that the microorganism population cannot increase or decrease during the BOD test, a situation that is obviously almost physically impossible.

The Monod differential equation for BOD is integrated to give

$$-\frac{1}{Y}\mu_{\max} t = \frac{k_s}{X_a + YL_a} \ln \frac{L_a - y}{L_a} + \frac{1}{Y} \frac{X_a + YL_a + K_s Y}{X_a + YL_a} \ln \frac{X_a}{X_a + Y_y}$$

where the subscript a is used to indicate the initial conditions and the variable y is the exerted BOD and is equal to $L_a - L$. Needless to say, this is a cumbersome expression, and two simplified equations for this general expression can be derived. In the first situation, which applies to most dilution bottle BOD experiments, the constant k_s is much larger than the substrate load (L), and the differential equation reduces to the form

$$-\frac{dL}{dt} = \mu_{max}\frac{XL}{Yk_s}$$

This is integrated to give

$$y = L_a - \frac{L_a + X_a/Y}{1 + \frac{X_a}{YL_a}\exp[\mu_{max}/k_s][X_a/Y + L_a]t}$$

In the second case, which occurs during undiluted respirometer tests using pure cultures, substrates, or both, the constant k_s is much smaller than the substrate load (L) and the description reduces to

$$-\frac{dL}{dt} = \mu_{max}\frac{X}{Y}$$

which integrates to give

$$y = \frac{X_a}{Y}(\exp[\mu_{max}t] - 1)$$

The feeding of the protozoa on the bacteria is amenable to description by the Monad equation (Bonomi & Frederickson, 1976). The interactions of the predator protozoa and the bacteria prey in continuous culture bioreactors have been described using a protozoan Monad equation that contains a bacterial Monad equation, a double Monad relationship (Nisbet et al., 1983) of the form

$$\frac{dX_1}{dt} = (\mu_1 - D)X_1 - \frac{\mu_2 X_2}{Y_2} - M_1$$

$$\frac{dX_1}{dt} = (\mu_2 - D)X_2 - M_s$$

$$\frac{dS_1}{dt} = D(S_r - S) - \frac{\mu_1 X_1}{Y_1}$$

$$\mu_1 = \mu_{1_{max}} \frac{S}{K_1 + S}$$

$$\mu_2 = \mu_{2_{max}} \frac{X_1}{K_2 + X_1}$$

where X_1 is the prey density, X_2 is the predator density, S is the substrate concentration, S_r is the reservoir substrate, μ_1 is the prey-specific growth rate, μ_2 is the predator-specific growth rate, D is the dilution rate, Y_1 and Y_2 are the prey and predator yield coefficients, M_1 and M_2 are the endogenous metabolism losses of prey and predators, and K_1 and K_2 are the half-saturation constants for the prey and predators, respectively. Models of predation that do not use Monad kinetics have provided only semiquantitative results (Canale, 1969).

The overall progress of BOD is described as consisting of three phases. The initial bacterial utilization of substrate and subsequent protozoan predation on the bacteria are described by Monad models as presented above. The last phase is termed autodestruction or endogenous respiration (Gaudy & Engelbrecht, 1960). This also is amenable to modeling using a Monad equation of the form

$$\frac{dX^b}{dt} = \frac{k^{EM} K''}{K'' - X^b}$$

where X^b is the biomass of the organisms (Gates & Ghosh, 1971).

4.0 BIOCHEMICAL OXYGEN DEMAND IN WASTEWATER TREATMENT MODELING

There have been many attempts over time to mathematically describe the operational aspects of wastewater treatment. One of the early attempts at modeling focuses on activated sludge (Garrett, 1951; Garrett & Sawyer, 1952). Hindsight from 50 years later reveals this selection of problem to model was overly ambitious and beyond the abilities of that time, the primary shortcoming being a lack of a usable expression for the biological activity. It is currently recognized that BOD is not the only needed element in a description of facility operation or even an adequate model for the incoming waste load to the facility. One early model for waste load considered it as a function of BOD, COD, total organic carbon (TOC), ammonia, and *Nitrosomonas* count (Buswell, 1955).

The Monod equation is a key feature in most successful models of wastewater treatment facility operations. Biological rotating disc reactors have been modeled using Monod kinetics (Paolini, 1979; Stover & Kincannon, 1982), although other efforts have used first-order bacterial growth kinetics and had to neglect oxygen limitations (Hansford et al., 1978). Trickling filters have been modeled using Monod kinetics (Moodie & Greenfield, 1979). Great strides have been made toward modeling the activated sludge process using Monod kinetics, but it is still a work in progress (Gujer et al., 1999).

A key to the use of the Monod or any other equation for the description of a process is the estimation of the values of the constants in the equation. The literature often shows that values and estimation methods initially proposed for constants by the original authors tend to become fixed in usage. It is only after extensive use of the equations that alternate techniques used for parameter estimation are investigated, and it may be found that the initial estimation methods actually incorporate hidden interdependencies between the constants. This is true for the Monod constants. The use of nonlinear estimation techniques was able to satisfactorily separate the various Monad constants (Simkins & Alexander, 1985).

It was initially found that apparently not all BOD curves could be described by application of the Monod equation. One study even went so far to say that, if the Monod equation described the behavior of the system, then the system is stable; otherwise, it is not stable (Koga & Humphrey, 1967). Systems were observed to exhibit hysteresis (slow response) or lag in achieving performance predicted by simple application of Monod descriptions. One cause of hysteresis can be attributed to starvation, which is an artifact of the experiment. Another cause arises from a shift in predominant populations of bacteria as the initial load concentration changed, requiring a Monod description for each population, with the overall result being an interaction of the individual equations (Liang et al., 1979). The interactions of protozoan predators and bacterial consumers of soluble substrate have been described with a double Monod expression (Nisbet et al., 1983).

Bacterial acclimation is a catchall term used to describe a variety of causes of hysteresis. It can be observed in individual bacterial species, where one cause is the time required for activation of latent ability to utilize the substrate, that is, synthesis of needed enzymes by the individual bacterium. Another cause in the bacterial species is an expression of resistance (Smith, 1999). Some individuals in the species possess a genetic quirk that enables the individual to utilize the substrate, while other individuals lack the genetic ability. The possessors of the quirk are able to thrive on the substrate and reproduce, while those that lack it die. Over time, the individuals exhibiting the genetic quirk become the dominant genotypes in the species, and the species is said to have become acclimated to the substrate.

An example is the hydrolysis of the α-glucosido bond in disaccharides (Varma & Lindley, 1966).

Acclimation is also applied to mixed populations containing several bacterial species. The bacteria able to use the substrate become the dominant species in the mixture, and the time required for the readjustment of the relative population numbers is called acclimation.

Regardless of the cause of the acclimation, different kinetics are observed in non-acclimated and acclimated cultures. The term *extant kinetics* has been coined to describe culture characteristics at any time under any conditions, whereas *intrinsic kinetics* is used to describe the fully optimized, acclimated behavior of a culture (Grady et al., 1966). For continuous-culture experiments, the substrate-to-microorganism ratio (S_0/X_0) is often used as a guide to which set of kinetics is operating (Ellis et al., 1996a), with a ratio of greater than 20 indicating that intrinsic results may be obtained. An S_0/X_0 of 20 is frequently exceeded in the BOD dilution bottle; however, intrinsic kinetics are frequently non-operational at any time in the bottle.

Application of Monod kinetics assumes that the substrate is (eventually) a perfect food for the organisms. That is not always the case. Some bacteria are completely unable to utilize certain substrates. Other bacteria are able to utilize, even after acclimation, only small amounts of the substrate; then, at higher concentrations, the substrate becomes toxic to the bacteria. This effect is called inhibition and is frequently encountered in wastewater treatment. One specific example of inhibition is the effect of *n*-propyl to *n-octyl* alcohols on bacteria (Therien et al., 1984). At low concentrations these compounds are metabolized by the bacteria. Alcohols can act as surfactants, with a hydrophobic tail and a hydrophilic head (the −OH group), and insinuate themselves to the hydrophobic interior of the cell membrane. The longer the carbon chain on the alcohol, the more efficiently they function as a surfactant to disrupt the cell membrane. After significant alcohol incorporation, the functioning of the cell membrane is compromised, resulting in decreased respiration and food assimilation by the bacteria. The concentration of alcohol in the media that initiates cell membrane disruption is called the critical level and is found to decrease with increasing carbon-chain length, reflecting the alcohol's surfactant capabilities.

Inhibition can be modeled (Ramkrishna et al., 1967). There have been alternate models of inhibition proposed, particularly for the case of high feed concentration inhibition (Bachanek, 1971); however, the most frequently used equation for inhibited growth uses the one proposed by Haldane (1930), called Haldane kinetics:

$$\mu = \mu_{max} S/[S + K_s + (S^2/K_i)]$$

where μ is the growth rate and K_i is the inhibitory constant. Figure 6.4 is a plot of the growth rate under the Haldane model, and this should be compared with Figure 6.3, which represents uninhibited growth. The growth rate is related to biomass (X) or bacterial numbers (N) as follows:

$$\mu = \frac{1}{X}\frac{dX}{dt} = \frac{1}{N}\frac{dN}{dt}$$

and

$$X_t = X_o e^{\mu t} \quad \text{or} \quad N_t = N_o e^{\mu t}$$

There are many different possible modes of inhibition of bacteria by substrate. Some are (Edwards, 1970) as follows:

- Modify chemical potential of substrates, intermediates, or products;
- Alter cell's permeability;
- Change activity of one or more enzymes;

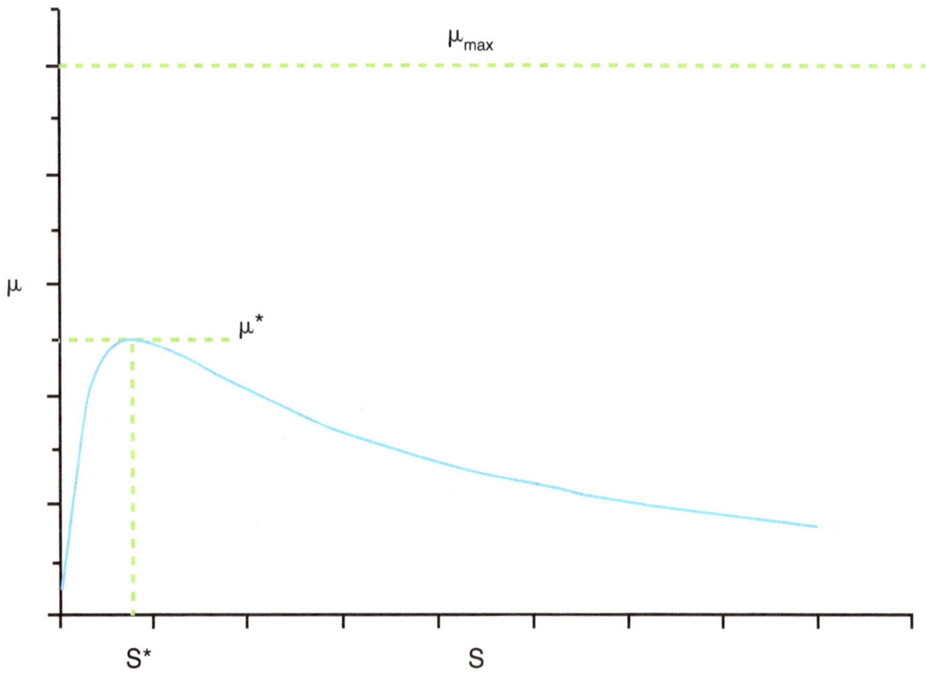

FIGURE 6.4. Growth rate curve with inhibited (Haldane) kinetics.

- Prevent dissociation of an enzyme or metabolic aggregate;
- Interfere with genome or transcription process in synthesis of enzymes; and
- Influence functional activity of the cell.

The chemical mechanisms that can cause one or more of these effects (Edwards, 1970) include:

- Irreversible chemical reaction with a cellular component;
- Preferential complexation with an enzyme or coenzyme;
- Substitution of substrate for an intermediate in a metabolic sequence;
- Dissociation of an enzyme aggregate;
- Altering pH, ionic strength, dielectric constant, or salvation in the bacterial environment; and
- Interference with a feedback control loop in the cell.

There are a number of different expressions other than the Haldane equation that have been proposed to describe inhibition (Edwards, 1970; Pawlowsky & Howell, 1973), including

$$\mu = \mu_{max} S\,[1 + (S/K\ )]\,/\,[S + K_s + (S^2/K_i)]$$

$$\mu = \mu_{max} S\,/\{S + K_s + (S^2/K_i)[1 + (S/K_s)]\}$$

$$\mu = \mu_{max}\,[S \exp(-\,S/K_i)]\,/\,(S + K_s)$$

$$\mu = \mu_{max}\,[\exp(-\,S/K_i) - \exp(-\,S/K_s)]$$

These more involved equations have found less use in treatment modeling and are infrequently encountered.

Haldane kinetics have been expressed in a form for substrate utilization (Andrews, 1968) that is frequently used:

$$\text{Rate of substrate depletion} = \frac{X}{Y}\frac{\mu_{max} S}{S + K_s + S^2/K_i}$$

where Y is the true growth yield. This form of the Haldane equation is called Andrews kinetics.

A key point in the graph of inhibited growth is the point where the concentration of the substrate changes the growth rate from positive to negative, the highest point in the curve in Figure 6.4. This is called the critical specific growth rate (μ^*) and can be calculated as

$$\mu^* = \frac{\mu_{max}}{1 + 2\sqrt{K_s / K_i}}$$

The concentration at which this occurs is the critical concentration (S^*) and

$$S^* = \sqrt{K_s K_i}$$

There have been numerous techniques applied to the estimation of the parameters of the Haldane equation (Lewandowski et al., 1985; Oner & Erickson, 1983; Ong, 1983; Rozich, 1992; Rozich & Gaudy, 1992). Most depended on the determination of the critical concentration (Rozich & Gaudy, 1984).

The widespread use of increasingly powerful computers since the 1960s led to an increased ability to model finer details of interactions in complex systems, and the microbial interactions in wastewater rectification are indeed a complex system. The use of computers has led to an increased interest in attempting to develop models. In recognition of this important aspect of wastewater research, in 1968 the Water Environment Federation added kinetic models to the list of topics in the annual literature review issue of *Water Environment Research*, and it has since become one of the more voluminous subjects in each year's review.

An example of the degree of detail that computer models have been constructed to consider is presented in Table 6.2 (from Barker & Dold, 1997a). This system of equations was designed to model the removal of biological nutrients from activated sludge systems. It was designed as an improvement of the ASM1 and ASM2 models (Henze et al., 1987; Henze et al., 1994).

Without going into an exhaustive description of the model components presented in Table 6.2, there are several important observations that can be drawn from its examination. The first is the extensive use of Monad and Haldane kinetic equations to describe the contribution of each component. Of all of the mathematical models that have been developed, these two equations have produced the most durable results (Boyden et al., 1994; Corman & Pave, 1983; Daigger & Grady, 1977, 1982; Dochain et al., 1995; Ellis et al., 1996a, 1996b; Fujie et al., 1997; Gujer et al., 1995; Marais & Ekama, 1976, 1977; Simkins et al., 1986; Sollfrank & Gujer, 1991; Tabak et al., 1992; Vanrolleghem et al., 1995; Watkin & Eckenfelder, 1984). They are vital components of the ASM1 (Henze et al., 1987), ASM2 (Henze et al., 1994), ASM2D (Henze et al., 1999), and ASM3 (Gujer et al., 1999) models

TABLE 6.2. Example equations used in advanced computer models of treatment system operations (Barker & Dold, 1997a).

Switching function	Equivalent	Definition	Switching function	Equivalent	Definition
AirYesHet	1 – AirNoHet	$\frac{S_O}{K_{O\,HET} + S_O}$	PO_4GroYes	1 – PO_4GroNo	$\frac{P_{O4}}{K_{LP\text{-}GRO} + P_{O4}}$
AirNoHet	1 – AirYesHet	$\frac{K_{OHET}}{K_{O\,HET} + S_O}$	PO_4GroNo	1 – PO_4GroYes	$\frac{K_{LP\text{-}GRO}}{K_{LP\text{-}GRO} + P_{O4}}$
AirYesAut	1 – AirNoAut	$\frac{S_O}{K_{O\,AUT} + S_O}$	PO_4UptYes	1 – PO_4UptNo	$\frac{P_{O4}}{K_{LP\text{-}UPT} + P_{O4}}$
AirNoAut	1 – AirYesAut	$\frac{K_{OAUT}}{K_{O\,AUT} + S_O}$	PO_4UptNo	1 – PO_4UptYes	$\frac{K_{LP\text{-}UPT}}{K_{LP\text{-}UPT} + P_{O4}}$
NH_3Yes	1 – NH_3No	$\frac{N_{H3}}{K_{NA} + N_{H3}}$	S_{BSA}Yes	1 – S_{BSA}No	$\frac{S_{BSA}}{K_{SSEQ} + S_{BSA}}$
NH_3No	1 – NH_3Yes	$\frac{K_{NA}}{K_{NA} + N_{H3}}$	S_{BSA}No	1 – S_{BSA}Yes	$\frac{K_{SSEQ}}{K_{SSEQ} + S_{BSA}}$
NO_3Yes	1 – NO_3No	$\frac{N_{O3}}{K_{NO} + N_{O3}}$	PolyPYes	1 – PolyPNo	$\frac{P_{PP\text{-}LO}}{K_{XP} + P_{PP\text{-}LO}}$
NO_3No	1 – NO_3Yes	$\frac{K_{NO}}{K_{NO} + N_{O3}}$	PolyPNo	1 – PolyPYes	$\frac{K_{XP}}{K_{XP} + P_{PP\text{-}LO}}$

(continued)

TABLE 6.2. Example equations used in advanced computer models of treatment system operations (*continued*).

		1	2	3	4	5	6	7	8	9	10	11	12
	Component Process	Z_H	Z_A	Z_P	Z_E	S_{ENM}	S_{BSC}	S_{BSA}	S_{PHB}	S_{UP}	S_{US}	$P_{PP\text{-}LO}$	$P_{PP\text{-}HI}$
1	Aerobic growth of heterotrophs on S_{BSC} with N_{H3}	1					$\frac{-1}{Y_{H,AER}}$						
2	Anoxic growth of heterotrophs on S_{BSC} with N_{H3}	1					$\frac{-1}{Y_{H,ANOX}}$						
3	Aerobic growth of heterotrophs on S_{BSC} with N_{O3}	1					$\frac{-1}{Y_{H,AER}}$						
4	Anoxic growth of heterotrophs on S_{BSC} with N_{O3}	1					$\frac{-1}{Y_{H,ANOX}}$						
5	Aerobic growth of heterotrophs on S_{BSA} with N_{H3}	1						$\frac{-1}{Y_{H,AER}}$					
6	Anoxic growth of heterotrophs on S_{BSA} with N_{H3}	1						$\frac{-1}{Y_{H,ANOX}}$					

7	Aerobic growth of heterotrophs on S_{BSA} with N_{O3}	1					$\frac{-1}{Y_{H,AER}}$
8	Anoxic growth of heterotrophs on S_{BSA} with N_{O3}	1					$\frac{-1}{Y_{H,ANOX}}$
9	Decay of heterotrophs	−1		$f_{EP.H}$	$1 - f_{EP.H}$		
10	Aerobic hydrolysis of stored/enmeshed COD				−1	1	
11	Anoxic hydrolysis of stored/enmeshed COD				−1	E_{ANOX}	
12	Anaerobic hydrolysis of stored/enmeshed COD				−1	E_{ANA}	
13	Hydrolysis of organic nitrogen						
14	Ammonification						
15	Fermentation of S_{BSC} to S_{BSA}	$Y_{H,ANA}$				−1	$\frac{(1 - Y_{H,ANA})}{Y_{AC}}$
16	Autotroph growth		1				
17	Autotroph decay		−1	$f_{EP,A}$	$1 - f_{EP,A}$		

(*continued*)

TABLE 6.2. Example equations used in advanced computer models of treatment system operations (*continued*).

	Component Process	13 P_{O4}	14 N_{BP}	15 N_{BS}	16 N_{OS}	17 N_{H3}	18 N_{US}	19 S_O	Rage equation, ρi
1	Aerobic growth of heterotrophs on S_{BSC} with N_{H3}	$-f_{P,ZH}$				$-f_{N,ZH}$		$-\frac{(1-Y_{H,AER})}{Y_{H,AER}}$	$\mu_H \cdot \frac{S_{BSC}}{K_{SH}+S_{BSC}} \cdot Z_H \cdot AirYesHet \cdot NH_3Yes \cdot PO_4\ GroYes$
2	Anoxic growth of heterotrophs on S_{BSC} with N_{H3}	$-f_{P,ZH}$			$-\frac{(1-Y_{H,ANOX})}{2.86Y_{H,ANOX}}$	$-f_{N,ZH}$			$\eta_{GRO} \cdot \mu_H \cdot \frac{S_{BSC}}{K_{SH}+S_{BSC}} \cdot Z_H \cdot AirNoHet \cdot NO_3Yes \cdot NH_3Yes \cdot PO_4GroYes$
3	Aerobic growth of heterotrophs on S_{BSC} with N_{O3}	$-f_{P,ZH}$			$-f_{N,ZH}$			$-\frac{(1-Y_{H,AER})}{Y_{H,AER}}$	$\mu_H \cdot \frac{S_{BSC}}{K_{SH}+S_{BSC}} \cdot Z_H \cdot AirYesHet \cdot NO_3Yes \cdot NH_3No \cdot PO_4GroYes$
4	Anoxic growth of heterotrophs on S_{BSC} with N_{O3}	$-f_{P,ZH}$			$-f_{N,ZH} - \frac{(1-Y_{H,ANOX})}{2.86Y_{H,ANOX}}$				$\eta_{GRO} \cdot \mu_H \cdot \frac{S_{BSC}}{K_{SH}+S_{BSC}} \cdot Z_H \cdot AirNoHet \cdot NH_3Yes \cdot NH_3No \cdot PO_4GroYes$
5	Aerobic growth of heterotrophs on S_{BSA} with N_{H3}	$-f_{P,ZH}$				$-f_{N,ZH}$		$-\frac{(1-Y_{H,AER})}{Y_{H,AER}}$	$\mu_H \cdot \frac{S_{BSC}}{K_{SH}+S_{BSC}} \cdot Z_H \cdot AirYesHet \cdot NH_3Yes \cdot PO_4GroYes$

6	Anoxic growth of heterotrophs on S_{BSA} with N_{H3}	$-f_{P,ZH}$		$\frac{(1-Y_{H,ANOX})}{2.86Y_{H,ANOX}}$	$-f_{N,ZH}$		$\eta_{GRO} \cdot \mu_H \cdot \frac{S_{BSA}}{K_{SH} + S_{BSA}} \cdot Z_H \cdot AirNoHet \cdot NO_3Yes \cdot NH_3Yes \cdot PO_4GroYes$
7	Aerobic growth of heterotrophs on S_{BSA} with N_{O3}	$-f_{P,ZH}$		$-f_{N,ZH}$		$-\frac{(1-Y_{H,AER})}{Y_{H,AER}}$	$\mu_H \cdot \frac{S_{BSA}}{K_{SH} + S_{BSA}} \cdot Z_H \cdot AirYesHet \cdot NO_3Yes \cdot NH_3No \cdot PO_4GroYes$
8	Anoxic growth of heterotrophs on S_{BSA} with N_{O3}	$-f_{P,ZH}$		$-f_{N.ZH} - \frac{(1-Y_{H,ANOX})}{2.86Y_{H,ANOX}}$			$\eta_{GRO} \cdot \mu_H \cdot \frac{S_{BSA}}{K_{SH} + S_{BSA}} \cdot Z_H \cdot AirNoHet \cdot NO_3Yes \cdot NH_3No \cdot PO_4GroYes$
9	Decay of heterotrophs	$-f_{P,ZH} - f_{EP,H} \cdot f_{P,Z,EH}$	$-f_{N,ZH} - f_{E,P,H} \cdot f_{N,ZEH}$				$B_H \cdot Z_H$
10	Aerobic hydrolysis of stored/enmeshed COD						$K_H \cdot \frac{S_{ENM}/Z_H}{K_X + S_{ENM}/Z_H} \cdot Z_H \cdot AirYesHet$
11	Anoxic hydrolysis of stored/ enmeshed COD						$\eta_{S,ANOX} \cdot K_H \cdot \frac{S_{ENM}/Z_H}{K_X + S_{ENM}/Z_H} \cdot Z_H \cdot AirYesHet \cdot NO_3Yes$

(continued)

TABLE 6.2. Example equations used in advanced computer models of treatment system operations (*continued*).

	Component Process	13 P_{O4}	14 N_{BP}	15 N_{BS}	16 N_{OS}	17 N_{H3}	18 N_{US}	19 S_O	Rage equation, ρi
12	Anaerobic hydrolysis of stored/enmeshed COD								$\eta_{S,ANA} \cdot K_H \cdot \frac{S_{ENM}/Z_H}{K_X + S_{ENM}/Z_H} \cdot Z_H \cdot$ AirNoHet $\cdot\, NO_3No$
13	Hydrolysis of organic nitrogen		–1	1					$\rho_{10} \cdot \frac{N_{BP}}{S_{ENM}}$ or $\rho_{11} \cdot \frac{N_{BP}}{S_{ENM}}$ or $\rho_{12} \cdot \frac{N_{BP}}{S_{ENM}}$
14	Ammonification			–1		1			$K_R \cdot N_{BS} \cdot (Z_H + Z_P)$
15	Fermentation of S_{BSC} to S_{BSA} – $f_{P,ZH}$, $Y_{H,ANA}$					$-f_{N,ZH} \cdot Y_{H,NA}$			$K_C \cdot \frac{S_{BSC}}{K_{S,ANA} + S_{BSC}} \cdot Z_H \cdot$ AirNoHet $\cdot\, NO_3No$
16	Autograph growth	$-f_{P,ZA}$			$\frac{1}{Y_A}$	$-f_{N,ZA} \frac{1}{Y_A}$		$\frac{(4.57 - Y_A)}{Y_A}$	$\mu_A \cdot \frac{N_{H3}}{K_{NH} + H_{H3}} \cdot Z_H \cdot$ AirYesAut
17	Autotroph decay	$-f_{P,ZA} - f_{EP,A} \cdot f_{P,ZEA}$	$-f_{N,ZA} - f_{EP,A} \cdot f_{N,ZEA}$						$b_A \cdot Z_A$

of activated sludge treatment. See Patry and Chapman (1989) for an in-depth discussion of all aspects of modeling. Alternate kinetic expressions, such as first-order kinetics (Logan et al., 1987a, 1987b; Mona et al., 1979) or higher-order equations, have seen only limited application.

The second observation to make is that, in the modeling of uptake, utilization, and decay by each microorganism subpopulation, the subpopulations are defined by their overall effect on nutrient concentration. The most common microorganism population that is characterized is the effect of heterotrophs on the various nutrient concentrations (Griffiths, 1994), although in the example in Table 6.2, the effects of autotrophs and polyP organisms are also included. This is not the only modeling style that is possible. High-power computers made it possible to look at each subpopulation of microorganisms, defined by species, and describe how the individual species utilize each essential element of growth and their interactions with competing organisms and predators. Models of specific predator-prey interactions have been examined as one example (Marsili-Libelli, 1980). Although these alternate views may lead to a more realistic model, they lack general applicability, as each treatment facility has its own unique microbiological ecology.

The third observation of the model parameters in Table 6.2 is the treatment of the influent "food." Early models viewed the influent pollutant load as a lump-sum parameter, equated with COD, BOD, and TOC, even though these are not interchangeable measures (Aziz & Tebbutt, 1980; Hao & Li, 1987; Ibrahim & Abasaeed, 1995; Wu et al., 1980). The model illustrated in Table 6.2 breaks the influent pollutant load down into subcategories based on how the microorganisms use the particular component. As presented in Table 6.2, an influent can be characterized by how it will interact with the treatment process. Various authors have presented a classification similar to that in Figure 6.5. Some concentrated on wastewater organics (Droste & Sanchez, 1986; Eckenfelder, 1989; Ros, 1993; Tyteca, 1981; Watkin & Eckenfelder, 1985), whereas others defined "pollutants" in terms of COD fractions, meaning the carbonaceous material (Barker & Dold, 1997a, 1997b; Ekama et al., 1986), even though COD can measure contributions from inorganics and other organics than strictly carbonaceous. There have been some efforts toward using model inputs that relate to specific test results, such as CBOD (Nolasco et al., 1998) and a soluble, readily biodegradable hydrolysis product component of COD (Novak et al., 1995). Still others split out influent heterotrophic biomass as an additional subcategory of the influent COD load (Kappeler & Gujer, 1992). A standard notation has been suggested for model influent characteristics (Grau et al., 1982; Henze et al., 1982, 1995). It is noteworthy that, reflecting the research preference for COD over BOD, the trend in modeling

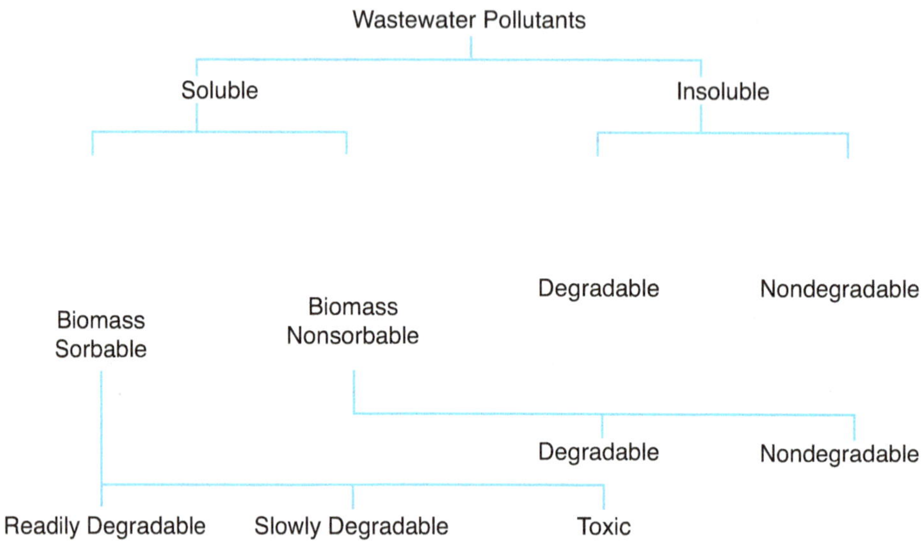

FIGURE 6.5. Classification of pollutants in wastewater according to biological utilization characteristics.

since the 1970s has predominately been the use of COD as a quantifiable variable, rather than BOD.

5.0 OXYGEN SAG AND WATER QUALITY EQUATIONS

When considering streams that are recipients of pollution, the DO level is of prime importance. When pollutants are added to the stream, the DO is observed to drop for a while and then recover slowly toward the saturation level (Gerardin, 1875). This phenomenon is called the oxygen sag. The measured DO level at any time is a function of the rate of removal through the oxidation of any pollutant load in the stream and added oxygen from other sources.

These ideas of depletion and replenishment are combined in the Streeter–Phelps oxygen sag equations (Streeter, 1925; Streeter & Phelps, 1925). The rate expression for oxygen depletion is

$$\frac{\mathrm{d}D}{\mathrm{d}t} = K_1 L - K_2 D$$

where D is oxygen deficit (difference between DO saturation and observed DO); t is time; L is BOD load; K_1 is deoxygenation rate coefficient, base e

natural logarithms (k_1 base 10 logarithms); and K_2 is reaeration rate coefficient, base e natural logarithms (k_2 base 10 logarithms). This rate expression is integrated to give

$$D = \frac{K_1 L_o}{K_2 - K_1}(\exp[-K_1 t] - \exp[-K_2 t] + D_o \exp[-K_2 t])$$

A generalized plot of the DO and remaining BOD over time in a stream that is subject to a pollutant load is illustrated in Figure 6.6.

Since 1927, the oxygen sag equations have been subject to many modifications, most commonly to the very difficult reaeration term (U.S. Public Health Service, 1958). Reaeration from the atmosphere has long been considered as one of the most significant terms in any water quality equation, yet an adequate description and measurement technique for the oxygen reaeration constant, $K_L a$, was not produced until the discovery that any two nonreactive gases behaved according to the following relationship:

$$\frac{k_1}{k_2} = \frac{D_1}{D_2} = \frac{d_2}{d_1}$$

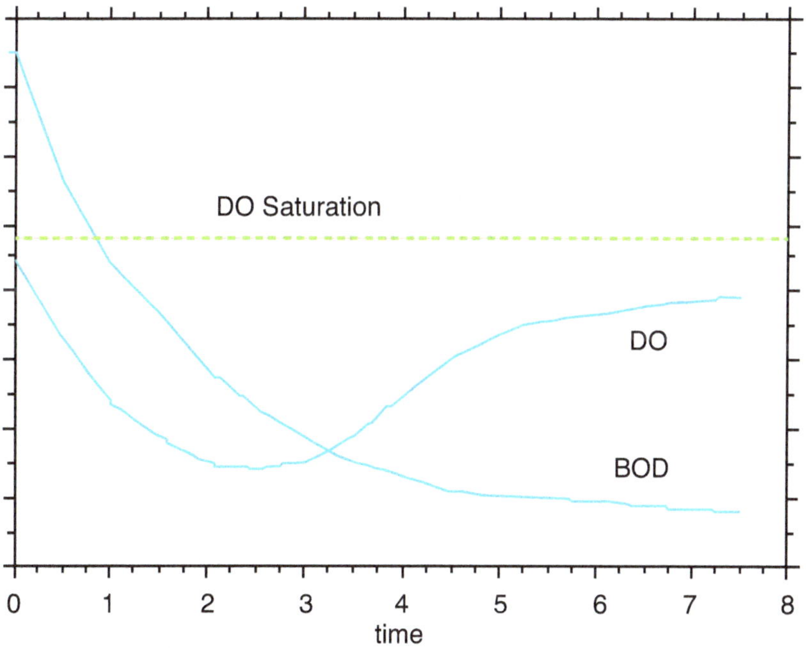

FIGURE 6.6. Generalized plot of remaining BOD and DO behavior in a stream by the Streeter–Phelps oxygen sag equations.

where k is the transfer or reaeration coefficient, D is the molecular diffusivity, and d is the molecular diameter (Tsivoglou, 1967; Tsivoglou et al., 1968). The kinetics of gas transfer is further described as

$$k_2 = an\frac{A}{V}$$

where a is a constant related to the molecular properties of the gas, n is the rate of surface replacement, A is the surface area, and V is the volume (if the sample is homogeneously mixed). These relationships allow the estimation of oxygen-reaeration rates from studies on other, more easily measured tracer gases, krypton being the most commonly employed. Rathburn (1977) reviewed early work on the subject of tracer gas measurements. Reaeration rates are also directly determined, especially with the availability of oxygen-sensing electrodes. These determinations have their own particular set of problems, the response time of the electrode being one of the more prominent (Koizumi & Aiba, 1984). Online techniques have been developed for continuous reassessment of reaeration rates (Sollfrank & Gujer, 1990).

Further studies found other factors that influence reaeration rates that are necessary to consider for application to actual water bodies. The degree of agitation is an important variable, and laboratory studies showed that, as the stirring speed changed, the steady-state DO in the sample changed (Mignone & Ertola, 1984). The microorganisms in the sample affect the oxygen-transfer rates, and a two-film model was modified to describe this effect (Mines & Sherrard, 1987). A significant study compared the predictions of 36 different equations found in the literature for reaeration rate calculations with actual measurements and concluded that reaeration rates should be actually measured to obtain reliable predictions from models (Thyssen & Erlandsen, 1987). Another model included the effects of dynamic and static surface tension in addition to airflow rates (Masutani & Stenstrom, 1991).

Even though the equations describing the influences and effects have been subject to extensive modification, the basic idea that DO is a function of the DO entering the system and the DO leaving the system has remained a constant. Considerable effort has been expended toward discovering the many entry and exit mechanisms for the DO. As they are discovered, these mechanisms have been modeled. Terms describing the effects of reaeration by photosynthesis, sediment composition, rates and type of turbulence, and temperature have been added. Other factors added to various model modifications have included coefficient of organic settling, rate of organic addition from runoff and bottom sediments, sediment oxygen demand (Dobbins, 1964), and longitudinal dispersion coefficient. The Camp–Dobbins equation

(Camp, 1963; Dobbins, 1964) is representative of the degree to which the Streeter–Phelps model has been modified:

$$D_t = \frac{K_1}{K_2 - [K_3]}\left\{\left[L_a - \frac{R}{K_1 + K_3}\right][\exp(-K_1 t - K_3 t) - \exp(-K_2 t)]\right\}$$
$$+ \frac{K_1}{K_2}\left\{\left[\frac{R}{K_1 + K_3} - \frac{A}{K_1}\right][1 - \exp(-K_2 t)]\right\} + D_a \exp(-K_2 t)$$

where D_t is the DO deficit at time t, D_a is the initial DO deficit, L_a is the initial BOD, K_1 is the deoxygenation rate constant, K_2 is the reaeration rate constant, K_3 is the sedimentation and absorption rate constant, A is the oxygen production or reduction caused by plants and benthal demand, and R is the BOD addition rate caused by runoff and scour (after Loucks, 1967). The use of this model was criticized (DiToro et al., 1967) in favor of models that included space and time boundary conditions. Another modification included the effects of ice coverage of the stream (for one example see Ranjie & Huimin, 1987). Beginning in the 1960s, models were computerized (Worley et al., 1965), which led to many possible modifications of the basic Streeter–Phelps equation, as summarized in Table 6.3.

Many efforts have been made to examine alternate equations for the description of the DO–BOD interactions in streams, rivers, and estuaries. The major spur for these efforts has been the often-made observation that the monomolecular form of the oxygen sag equation is not effective as a model (Beck & Young, 1976; Edeline & Lambert, 1974; Whipple & Hunter, 1981). An example of this approach is embodied in the Churchill–Buckingham model, in which first-order or higher explicit modeling of BOD was abandoned in favor of an empirical fitting of a descriptive equation after data from the stream had been acquired (Churchill & Buckingham, 1956; Wang, Kao et al., 1979; Wang, Wang et al., 1979). The empirical equations are

$$Y = a_0 + a_1 X_1 + a_2 X_2 + a_3 X_3$$

$$W = b_0 + b_1 Z_1 + b_2 Z_2 + b_3 Z_3$$

where Y is the DO drop, X_1 is the BOD at sag, X_2 and Z_2 are the stream temperature at sag, X_3 and Z_3 are the stream flow at sag, W is the BOD remaining at sag, Z_1 is the combined BOD loads at the upstream stations, and the constants a and b are derived from fitting the equations to the actual stream data. The constants are optimized from the solution of multiple

TABLE 6.3. Modifications and expansions of the Streeter–Phelps oxygen sag equation.

Year	Citation	Comments
1930	Streeter, 1930	Compared rates of natural and controlled conditions
1934	Streeter, 1934	Formulated bacterial changes
1964	Dobbins, 1964	Added additional terms to Streeter–Phelps equation
	Harley, 1964	Nomograph solutions to Streeter–Phelps equation
1965	Worley et al., 1965	Computer modeling of Streeter–Phelps equation
1966	Liebman & Loucks, 1966	Presented a method for calculating the critical time in Streeter–Phelps equation
1969	Lofy, 1969	Used a nomograph for the Streeter–Phelps equation
1972	Koivo & Phillips, 1972	Worked backwards through Streeter–Phelps model to determine causative BOD loads and BOD reaction rates that generate observed DO values downstream
1973	Schmied, 1973	Used Streeter–Phelps equation to select mill sites for minimum impact on receiving waters
1974	Arbabi et al., 1974	Critical deficit = $E + R$(BOD load) + A(DO deficit), where E, R, and A are stream self-purification constants
1975	Arbabi & Elzinga, 1975	Used linear approximation to solve the Streeter–Phelps oxygen sag equation for the critical deficit
	Brutsaert, 1975	Tested Streeter–Phelps equation in a hypothetical situation using probability theory; output presented as probability distributions of water quality parameters
	Burges & Lettenmaier, 1975	Discussed need for stochastic models to simulate the probabilistic nature of water quality. A first-order analysis was used to determine the uncertainty in the Streeter–Phelps deterministic model parameters and Monte Carlo simulation used to evaluate the accuracy of the results. Areas of greatest uncertainty were travel times, BOD decay constant, and reaeration coefficient
	Rinaldi & Soncini-Sessa, 1978	Variation of predicted DO and BOD because of variation in estimates of parameters
1979	Bathala et al., 1979	Calibrated Streeter–Phelps model with NBOD in small streams and found that model is more sensitive to reaeration coefficient than to CBOD and NBOD decay rate constants
	Rinaldi et al., 1979a	Parameter estimation of Streeter–Phelps model
	Shen, 1979	Modeling of rivers (Book)
	Wang, Kao et al., 1979; Wang, Wang et al., 1979	Multiple linear correlation based on the Streeter–Phelps and Churchill and Buckingham models were used to predict DO drops as a function of BOD loading, stream temperature, and stream flow
1981	Wen & Kao, 1981; Wen et al., 1980	Method of moments applied to Streeter–Phelps equation
1982	McBride, 1982	Nomographs for rapid solutions for the Streeter–Phelps equation
1984	Dewey, 1984	Streeter–Phelps DO model modified with stochastic variables to compare alternatives of flow augmentation and more efficient wastewater treatment

simultaneous equations and require a minimum of six sets of observational data. Note that the approach contains no mechanistic relation between BOD and DO. Solution and use of these type models have been aided significantly by extensive use of computers.

Water quality modeling has advanced from the Churchill–Buckingham design to incorporate more sophisticated terms; however, the basic empirical approach has remained. A nonlinear decay term for BOD gave noted improvement in predictions (Shasty et al., 1973). Nontidal river systems have been examined with a lumped parameter model, where all possible factors that could affect the DO and BOD were added to the linear equations; then the constants for those factors were manipulated to fit the observational data (Beck & Young, 1975, 1976). Extended Kalman filters have been applied as one form of manipulation (Beck & Young, 1976; Constable & McBean, 1979) to identify the important parameters. Random differential equations have been used in river quality models (Finney et al., 1982). Stochastic optimization techniques have been applied to linear and random differential expressions many times to river models in order to find the significant factors that cause the observed DO and BOD fluctuations and to present those fluctuations as probability distributions (Beck & Young, 1976; Burn & McBean, 1986; Custer & Krutchkoff, 1969; Damaskos & Papadopoulos, 1983; Lohani, 1982; Padgett & Rao, 1979; Padgett et al., 1977; Sornberger & Keshavan, 1973; Thayer & Krutchoff, 1967). Examples of the differential equations that have been employed for solution of stochastic modifications include moment equations, the Fokker–Planck equation, and the Riccati equation (LeDuc et al., 1986; Zielinski, 1988). Several computerized packages have been developed, including QUAL-II (Willis et al., 1976), the Dissolved Oxygen Routing Model (DORM) (Fullen et al., 1982), and DOSAG-1 (Novotny & Krenkel, 1975).

The sediment oxygen demand (SOD) has long been recognized as a significant contributing factor to DO–BOD behavior in rivers (Jansa & Akerlindh, 1941). The BOD from sediments has been most commonly modeled using first-order equations (Chen et al., 1983). Experiments have shown that SOD can contribute up to 25% of the DO–BOD interactions in the overlying water (Snodgrass & Ng, 1985a, 1985b). The oxidation of methane and ammonia in sediments has been modeled and is an important interaction between the anaerobic and aerobic layers (DiToro et al., 1990). These interactions are important factors in the overall contribution of sediment stabilization to water quality.

A modern expression for sediment stabilization that includes contributions from SOD takes the form of the Klapwijk–Snodgrass model

$$J_{\mathrm{COD}} = \mathrm{SOD} - 4.57\, RN \times H_1 + 2.87\, J_{\mathrm{N2}}' + J_{\mathrm{COD}}'$$

where J_{COD} is the sediment stabilization rate, RN is the zero-order rate of nitrification, H_1 is the thickness of the aerobic layer, J_{N2}' is the flux of nitrogen from the sediment to the overlying water column, and J_{COD}' is the flux of COD (total of the organic and chemically amenable materials) between the sediment and the overlying water column (Hatcher, 1986). Sediment oxygen demand in this model is defined as

$$\mathrm{SOD} = \sqrt{2COX\ ROX\ DOX}$$

where COX is the concentration of oxygen in the water column, ROX is the zero-order rate of oxygen consumption in the aerobic zone, and DOX is the diffusion coefficient for oxygen. If there is substantial disturbance of the sediments attributed to burrowing by worms, mollusks, and other creatures (biotubation), this factor is included in the DOX term. Further examination of the oxygen dependence of biological SOD (BSOD) found that a Michaelis–Menton type expression

$$\mathrm{BSOD} = \mu(T)\frac{[\mathrm{O_2}]}{K_o + [\mathrm{O_2}]}$$

where $\mu(T)$ is a temperature-dependent Michaelis–Menton rate constant and K_o is the half-saturation constant (about 1.4 mg/L) described the strong dependence of BSOD on oxygen up to 2 mg/L in the overlying water column, which leveled off at greater concentrations (Walker & Snodgrass, 1986).

Other efforts have concentrated on finding the significant mechanistic factors in water quality. Factors that have been considered include dispersion (Koussis et al., 1990), growth and decomposition of phytoplankton (Beck & Young, 1975; Cohen, 1990; Frisk, 1982; Frisk & Kyla-Harakka, 1982), energy dissipation (Kawashima & Suzuki, 1989), surface area for attached biomass (Kawashima & Suzuki, 1989), benthic oxygen demand (Frisk, 1982; Frisk & Kyla-Harakka, 1982), and slowly decaying organic matter (Frisk, 1982; Frisk & Kyla-Harakka, 1982). Numerous reviews (see Firth, 1984) and books (James, 1978; Loucks et al., 1981; Reckhow & Chapra, 1983; Rinaldi et al., 1979b; Shen, 1979; Zison et al., 1978) have been published on this topic.

Detailed examination of the biodegradation of pollutants by bacteria in riverine situations reveals that the process is little different from that found in the BOD bottle (Apotecker & Thevenot, 1983). Upon addition of substrate to the water, a lag phase of 10 to 20 hours occurs, during which time the bacteria adjust their metabolism to assimilate the substrate. This is followed by a subsequent rapid growth phase, where DO depletion, cell growth and reproduction, and BOD uptake rates are proportional to and exponentially increase with time. These observations suggest that a Monod model for uninhibited uptake and a Haldane model for inhibited assimilation may be appropriate for use in water quality models.

Overall, the practice of water quality modeling depends on the input of BOD data to the model; however, it does not depend to any appreciable degree on the mechanistic interpretation of BOD utilization. This lies in stark contrast to the modeling of treatment facility operations, where a great deal of effort has gone into a detailed description of the processes that contribute to BOD; however, researchers prefer to use COD data for input.

6.0 CALCULATION OF BIOCHEMICAL OXYGEN DEMAND FROM EXPERIMENTAL DATA

The procedure for calculating the BOD from test data presented in the 21st edition of *Standard Methods* (APHA et al., 2005) uses the formula for unseeded samples

$$\mathrm{BOD_5,\ mg/L} = \frac{\mathrm{DO_{init}} - \mathrm{DO_{final}}}{F}$$

where

- $\mathrm{DO_{init}}$ = initial DO in BOD bottle immediately after preparation;
- $\mathrm{DO_{final}}$ = final DO in BOD bottle after 5 days of incubation; and
- F = fraction of sample in dilution, expressed as a decimal, for example, 10% = 0.10.

If the sample is seeded, the equation has the form

$$\mathrm{BOD,\ mg/L} = \frac{\mathrm{DO_{init}} - \mathrm{DO_{final}} - R[S_{\mathrm{init}} - S_{\mathrm{final}}]}{F}$$

where

- S_{init} = initial seed control DO before incubation;
- S_{final} = final seed control DO after 5 days of incubation; and
- R = ratio of seed in each sample dilution to seed in seed control.

These equations take no account of the depletion of DO exhibited by the dilution water blank. Instead, if the dilution water blank exhibits greater than a 0.2 mg/L DO depletion, the test is considered invalid and the cause of the excessive blank must be determined and corrected. On the whole, this is excellent quality control, and the authors of this book are in complete agreement that requiring practitioners to pay critical attention to the conditions and reagents used in the procedure is the proper manner in which the test should be conducted.

Notwithstanding our professional insistence on a demonstrated display of high-quality laboratory skills, many technicians who perform the test take exception to the dilution water maximum depletion requirement as an unduly restrictive measure that results in numerous retests. The complaint has also been voiced that ignoring the contribution of the dilution water introduces an error, albeit very small, to the BOD calculation that is abhorrent to purists in the wastewater community.

Alternate calculations of the BOD are available that do account for the dilution water depletion. The first alternate procedure was presented almost 70 years ago (Akerlindh, 1953; Rhame, 1956). It requires that two different dilutions of the waste generate acceptable DO depletions in the test; that is, at least a 2.0 mg/L DO depletion and at least 1.0 mg/L residual DO. A BOD equation is written for each of the dilutions:

$$\text{BOD} = F_1 \, (D_1 - L_B)$$

$$\text{BOD} = F_2 \, (D_2 - L_B)$$

where BOD is the BOD of the sample, F_1 and F_2 are the dilution factors for the two different dilutions, D_1 and D_2 are the respective DO depletions observed in the two different dilutions, and L_B is the BOD of the dilution water. The two equations form a set of simultaneous equations and are combined using standard mathematical techniques to give

$$\text{BOD} = \frac{F_1 F_2 [D_1 - D_2]}{F_2 - F_1}$$

where the order of the assignment of D_1 and D_2 is chosen to give a positive number. This equation factors out the dilution water contribution, which can be calculated if needed using the formula

$$L_B = \frac{F_2 D_2 - F_1 D_1}{F_2 - F_1}$$

The graphical technique (Klein & Gibbs, 1979; Woodring & Clifford, 1988) is an extension of the two-point Akerlindh method. In the graphical method, three or more points are used to develop a regression equation between the amount of sample in the particular bottle and the resulting DO depletion. For example, say that beginning with an initial DO of 9.1, a 5.0-mL sample produces a 5-day DO of 7.1, a 10.0-mL sample gives 4.8, a 15.0-mL sample gives 3.2, and a 20.0-mL sample gives 0.9. These data are graphed with a regression line in Figure 6.7.

The slope and intercept of the regression line, –0.406 and 9.08, respectively, are used in the formula

$$\text{BOD} = -300(\text{slope}) - y\text{-intercept} + S$$

where S is the initial DO of the sample. Entering the appropriate values from this example gives

$$\text{BOD} = -300(-0.406) - 9.08 + 9.10 = 122 \text{ mg/L}$$

Note that a value with slightly lower than minimum residual DO is included in the example and does not affect the result, whether it is included or not.

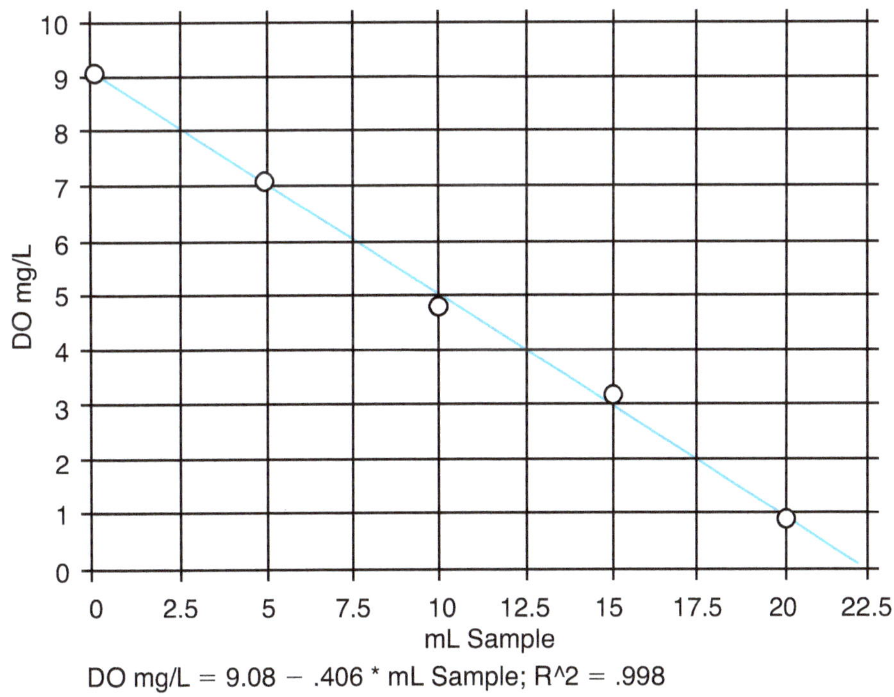

FIGURE 6.7. Plot of dissolved oxygen versus milliliter sample.

Further, seed depletion and dilution water depletion are not explicitly calculated or included, as they are both factored out in the regression procedure.

7.0 CONCLUSION

There are great differences between BOD embodied as a test procedure and BOD as perceived in biological wastewater treatment or water quality. On the one hand, in wastewater treatment and receiving water, there is a steady-state balance in the relative populations of the individual bacterial and protozoan species, and the absolute numbers of these organisms are relatively high. The concentrations of food and essential nutrients are relatively constant, as they are continually added to the system, with the metabolic byproducts being continuously removed. The DO content of the system is relatively constant. The BOD test, on the other hand, is characterized by a minimal concentration of essential nutrients and a minute initial population of bacteria. When commercial seed mixtures are used, there are no predators in the mix and no nitrifying bacteria, although they may be inadvertently added from the diluted sample. The DO concentration begins at a maximum level, then decreases during the test. These differences mean that, as the systems are examined in finer derail, the resulting models tend to diverge and there is less cross-applicability of the treatment models to the BOD test.

Understanding wastewater treatment and river dynamics through models is a vital area of research, as regulatory pressures have stressed the need for higher quality effluents and better water quality predictors. These regulatory pressures translate directly to the availability of funds that researchers can draw on to support their efforts. There has been a noted paucity of money to support research into fundamental studies of the BOD test itself.

Finally, COD and its various associated subfractions have largely supplanted BOD as measurable model inputs. This is especially notable in the literature where, in papers written before 1970, treatment facility efficiencies are predominantly measured in percentage BOD reduction and less commonly as the percentage reduction of COD or TOC. After 1970, efficiencies are increasingly measured as COD reductions, such that it is rare to find efficiency studies reported after 1985 with any mention of BOD at all. This transition can be ascribed to the COD test being much quicker, having less variation in obtained values, and overall being a technically easier testing procedure than BOD.

All of these factors have led to enormous improvements in the precise mathematical descriptions of biological treatment systems and water quality, while at the same time advancements in the rigorous description of the BOD test process have lagged. In fact, since 1970, there have been few developments that contribute significantly to the modeling of the BOD test. The major advance (Braun & Berthouex, 1970; Gaudy, 1972; Jennelle & Gaudy 1970; Simkins & Alexander, 1985) is the application of Monod kinetics (Monod, 1942, 1950) to the BOD test. A book (Gaudy & Gaudy, 1980) examines and summarizes the microbiology and kinetics of BOD, concentrating on the applicability of Monod and Haldane expressions, without actually presenting any equations for BOD tests. Monod kinetics have been applied to the oxygen uptake of starving organisms as they change to a state of food assimilation (Mona et al., 1979). The bacterial growth rate and oxygen-consumption profiles have been related through Monod kinetics (Huang et al., 1985).

A second advance in BOD modeling is the application of stochastic methods to first-order BOD test kinetics (LeDuc et al., 1986). Another advance examines an autocatalytic model of BOD (Maoyu, 1990), whereas the last presents an empirical general BOD exertion equation to model all parts of the BOD curve (Swamee & Ojha, 1991). All of these have been described in detail previously.

Does this mean that the description of the BOD test has reached completion, that no more work remains to be done on it? The answer is obviously no. There is still much more development that can be made in the theoretical description of the test. It is just too complicated a system of interacting forces to be completely characterized by the efforts made to date. Even though researchers seem to have bypassed the BOD test in favor of procedures that are easier to understand and perform, the tremendous number of persons who conduct the BOD test on a daily basis in fulfillment of regulatory obligations should serve as ample justification to continue efforts toward a more complete understanding. Most of the mathematical tools for a complete description seem to be in place; it is just a matter of assembling them.

8.0 REFERENCES

Adams, C. E., Eckenfeldcr, W. W., Jr., & Hovious, J. (1975). A kinetic model for design of completely mixed activated sludge treating variable strength industrial wastewater. *Water Res.*, *9*, 37.

Akerlindh, G. (1953). Computations of BOD for seeded samples of high dilution. *Sew. Ind. Wastes*, *25*, 293.

American Public Health Association, American Water Works Association, & Water Environment Federation (2005). *Standard methods for the examination of water and wastewater* (21st ed.). American Public Health Association.

Andrews, J. F. (1968). A mathematical model for the continuous culture of microorganisms utilizing inhibitory substrates. *Biotechnol. Bioeng.*, *10*, 707.

Apoteker, A., & Thevenot, D. R. (1983). Experimental simulation of biodegradation in rivers: Oxygen, organic matter, and biomass concentration changes. *Water Res.*, *17*, 1267.

Arbabi, M., & Elzinga, J. (1975). A general linear approach to stream water quality modeling. *Water Resour. Res.*, *11*, 191.

Arbabi, M., et al. (1974). The oxygen sag equation: New properties and a linear equation for the critical deficit. *Water Resour. Res.*, *10*, 921.

Argamam, Y., Raize, O., Eckenfelder, W. W., & O'Reilly, A. J. (2000). Applicability of batch test data for industrial wastewater continuous flow process design. *Water Environ. Res.*, *72*, 348.

Arrhenius, S. (1915). *Quantitative laws in biological chemistry.* London.

Aziz, J. A., & Tebbutt, T. H. Y. (1980). Significance of COD, BOD, and TOC correlations in kinetic models of biological oxidation. *Water Res.*, *14*, 319.

Bachanek, S. (1971). Effect of effluent concentration on the kinetics of treatment by the activated sludge process. *Pr. Inst. Gospod. Wodnej.*, 6, 71; *Abstr. Bull. Inst. Paper Chem.*, *43*, 9617 (1973).

Bagchi, D., & Chaudhuri, N. (1970). Simplified graphical evaluation of BOD constants. *J. Water Pollut. Control Fed.*, *42*, 136.

Barker, P. S., & Dold, P. L. (1997a). General model for biological nutrient removal activated-sludge systems: Model application. *Water Environ. Res.*, *69*, 985.

Barker, P. S., & Dold, P. L. (1997b). General model for biological nutrient removal activated-sludge systems: Model presentation. *Water Environ. Res.*, *69*, 969.

Bathala, C. T., et al. (1979). Assimilative capacity of small streams. *J. Environ. Eng.*, *105*, 1049.

Beck, M. B., & Young, P. C. (1975). A dynamic model for DO–BOD relationship in a non-tidal stream. *Water Res.*, *9*, 769.

Beck, M. B., & Young, P. C. (1976). Systematic determination of DO–BOD model structure. *J. Environ. Eng.*, *102*, EE5, 909.

Benedict, A. H. (1977). Discussion of temperature effects on BOD stoichiometry and oxygen uptake rate. *J. Water Pollut. Control Fed.*, *49*, 864.

Berthouex, P. M., & Hunter, W. G. (1971a). Problems associated with planning BOD experiments. *J. Sanit. Eng. Div., Proc. Am. Soc. Civ. Eng.*, *97*, 333.

Berthouex, P. M., & Hunter, W. G. (1971b). Statistical experimental design: BOD tests. *J. Sanit. Eng. Div., Proc. Am. Soc. Civ. Eng.*, *97*, 393.

Bhatla, M. N., & Gaudy, A. F., Jr. (1965a). Role of protozoa in the diphasic exertion of BOD. *J. Sanit. Eng. Div., Proc. Am. Soc. Civ. Eng.*, *91*(SA3), 63.

Bhatla, M. N., & Gaudy, A. F., Jr. (1965b). Sequential substrate removal in a dilute system by heterogeneous microbial populations. *Appl. Microbiol.*, *13*, 345.

Bhatla, M. N., & Gaudy, A. F., Jr. (1965c). Studies on the plateau in oxygen uptake during exertion of BOD by pure cultures of bacteria. *Biotechnol. Bioeng.*, 7, 387.

Bhatla, M. N., & Gaudy, A. F., Jr. (1966). Studies on the causation of phasic oxygen uptake in high-energy systems. *J. Water Pollut. Control Fed.*, *38*, 1441.

Black, W. M., & Phelps, E. B. (1911). The discharge of sewage in New York Harbor. Report to the Board of Estimate and Apportionment of New York City. *Mass. Inst. Tech. Contrib. Sanit. Res. Lab. Sew. Exp. Stn.*, *7*, *65*.

Bonomi, A., & Frederickson, A. G. (1976). Protozoan feeding and bacterial wall growth. *Biotechnol. Bioeng.*, *18*, 239.

Bosko, K. (1965). Self-purification of surface waters polluted with effluents from pulp manufacture. *Paper Celulosa*, *20*(4), 101.

Bouveng, H. O., & Solyom, P. (1973). Long term stability of waste lignins in aquatic systems. *Sven. Pappeerstidn.* (Sweden), *76*, 26.

Boyden, B. H., et al. (1994). Treatment of bleachery effluents from kraft mills pulping mature eucalyptus. *Water Sci. Technol.*, *29*, 274.

Braun, H. B., & Berthouex, P. M. (1970). Analysis of lag-phase BOD curves using the Monod equations. *Water Resour. Res.*, *6*, 838.

Brutsaert, W. F. (1975). Water quality modeling by Monte Carlo simulation. *Water Resour. Bull.*, *11*, 229.

Burges, S. J., & Lettenmaier, D. P. (1975). Probabilistic methods in stream quality management. *Water Resour. Bull.*, *11*, 115.

Burkhead, C. E., & Waddell, S. L. (1971). Composition studies of activated sludges. *Proc. 24th Ind. Waste Conf, Purdue University*, *135*, 576.

Burn, D. H., & McBean, E. A. (1986). Linear stochastic optimization applied to BOD-DO modelling. *Can. J. Civ. Eng.*, *13*, 249.

Busch, A. W. (1958). BOD progression in soluble substrates. *Sew. Ind. Wastes*, *30*, 1336.

Busch, A. W., & Myrick, N. (1960). Food-population equilibria in bench-scale bio-oxidation units. *J. Water Pollut. Control Fed.*, *32*, 949.

Busch, A. W., & Myrick, N. (1961). Aerobic bacterial degradation of glucose. *J. Water Pollut. Control Fed.*, *33*, 897.

Busch, A. W., Myrick, N., McKinney, K. W., & Peterson, J. W. (1960). Bench-scale bio-oxidation studies, batch vs. continuous systems. *Proc. Third Biol. Waste Treatment Conf.*, Manhattan College.

Buswell, A. M. (1955). The BOD test and total load. *Sew. Ind. Wastes*, *27*, 1297.

Buswell, A. M., Mueller, H. F., & Van Meter, I. (1954). Bacteriological explanation of rate of oxygen consumption in the BOD test. *Sew. Ind. Wastes*, *26*, 276.

Butterfield, C. T. (1929a). Studies of natural purification in polluted waters. II. Development of a suitable dilute medium. *Public Health Rep.*, *44*, 2647; *Chemical Abstracts*, *24*, 450 (1930).

Butterfield, C. T. (1929b). Studies of natural purification in polluted waters. III. A note on the relation between food concentration in liquid media and bacterial growth. *Public. Health Rep.*, *44*, 2864.

Butterfield, C. T. (1933). Observations on changes in numbers of bacteria in polluted water. *Sew. Works J.*, *5*, 600.

Butterfield, C. T. (1935). Biological and bacteriological relationships in water purification and sewage treatment. *Kans. Water Sew. Works Assoc.*, *4*, 106; *Chemical Abstracts*, *30*, 7743 (1936).

Butterfield, C. T. (1937a). The oxidation of sewage by bacteria in pure cultures. *J. Bacteriol.*, *34*, 207.

Butterfield, C. T. (1937b). The purification of sewage by bacteria in pure cultures. *J. Bacteriol.*, *33*, 83.

Butterfield, C. T. (1940). Some functions of bacteria in the purification of polluted water. *J. Bacteriol.*, *39*, 527.

Butts, T. A., & Kothandaraman, V. (1970). Fitting first and second order BOD reaction equations to stream data. *Water Sew. Works*, *117*, 276.

Caldwell, D. H., & Langelier, W. F. (1948). Manometric measurement of the biochemical oxygen demand of sewage. *Sew. Works J.*, *20*, 202.

Camp, T. R. (1963). *Water and its impurities*. Reinhold.

Canale, R. P. (1969). Predator–prey relationships in a model for the activated process. *Biotechnol. Bioeng.*, *11*, 887.

Canale, R. P., & Cheng, F. Y. (1974). Oxygen utilization in bacterial–protozoan community. *J. Environ. Eng., 100*, 171.

Castellan, G. W. (1972). *Physical chemistry* (2nd ed.). Addison–Wesley.

Chen, R. L., et al. (1983). Characterization of aerobic chemical processes in reservoirs: problem description and model formulation. Technical Report E-83-16, U.S. Army Waterways Exp. Stn. (Vicksburg, MS).

Churchill, M. A., & Buckingham, R. A. (1956). Statistical method for analysis of stream purification capacity. *Sew. Ind. Wastes., 28*, 517.

Cohen, R. R. H. (1990). BOD and algae: Fractionation of phytoplankton and nonphytoplankton in a large river. *Water Resour. Res., 26*, 671.

Constable, T. W., & McBean, E. A. (1977). Parameter estimation for the first-order BOD equation using nonlinear techniques. *Can. J. Civ. Eng., 4*, 462.

Constable, T. W., & McBean, E. A. (1979). Kalman filter modeling of the speed river ruality. *J. Environ. Eng., 105*, 961.

Cooper, A. B. J. (1986). Developing management guidelines for river NOD. *J. Water Pollut. Control Fed., 58*, 845.

Corman, A., & Pave, A. (1983). On parameter estimation of monod's bacterial growth model from batch culture data. *J. Gen. Appl. Microbiol., 29*, 91.

Custer, S. W., & Krutchkoff, R. C. (1969). Stochastic bodel for BOD and DO in estuaries. *J. Sanit. Eng. Div., Proc. Am. Soc. Civ. Eng., 95*, 865.

Daigger, G. T., & Grady, C. P. L., Jr. (1977). A model for the bio-oxidation process based on product formation concepts. *Water Res., 11*, 1049.

Daigger, G. T., & Grady, C. P. L., Jr. (1982). The dynamics of microbial growth on soluble substrates. *Water Res., 16*, 365.

Damaskos, S. D., & Papadopoulos, A. S. (1983). A general stochastic model for predicting BOD and DO in streams. *Int. J. Environ. Stud., 21*, 229.

Deai, J., Tida, T., Gong, Y., Jianrong, Z., & Yicheng, S. (1991). Factors affecting the relationship between the NBOD values and the amount of nitrogenous pollutants: A field study of the Lee River. *Water Res., 25*, 485.

Dewey, R. J. (1984). Application of stochastic dissolved oxygen model. *J. Environ. Eng., 110*, 412.

DiToro, D. M., O'Connor, D. J., & Thomann, R. V. (1967). Discussion of risk evaluation in sewage treatment plant design. *J. Sanit. Eng. Div., Proc. Am. Soc. Civ. Eng., 93*, SA6, 268.

DiToro, D. M., et al. (1990). Sediment oxygen demand model: Methane and ammonia oxidation. *J. Environ. Eng., 116*, 945.

Dobbins, W. E. (1964). BOD and oxygen relationships in streams. *J. Sanit. Eng. Div., Proc. Am. Soc. Civ. Eng.*, *90*(SA3), 53.

Dochain, D., Vanrolleghem, P. A., & van Daele, M. (1995). Structural identifiability of biokinetic models of activated sludge respiration. *Water Res.*, *29*, 2571.

Droste, R. L., & Sanchez, W. A. (1986). Modeling active mass in aerobic sludge digestions. *Biotechnol. Bioeng.*, *28*, 1699.

Eckenfelder, W. W., Jr. (1989). *Industrial water pollution control.* McGraw-Hill.

Edeline, F., & Lambert, G. (1974). A simple simulation method for river self-purification studies. *Water Res.*, *8*, 297.

Edeline, F., & Lambert, G. (1979). Self-purification of rivers a two-phase model for bacterial biodegradation. *Water Res.*, *13*, 473.

Edwards, V. H. (1970). The effect of high substrate concentration on microbial kinetics. *Biotechnol. Bioeng.*, *12*, 679.

Ekama, G. A., Dold, P. L., & Marais, G. v. R. (1986). Procedures for determining influent COD fractions and the maximum specific growth rate of heterotrophs in activated sludge systems. *Water Sci. Technol.*, *18*(6), 91.

Eliassen, R., Grune, W. N., Jaffe, T., & Diachishin, A. N. (1950). BOD of radioactive sludge. Prog. Rep., U.S. Atomic Energy Commission, NYO 1510 (NYU-6), 4; *Public Health Eng. Abstr.*, *30*(9), 18 (1950).

Ellis, T. G., Barbeau, D.S., Smets, B. F., & Grady, C. P. L., Jr. (1996a). Changes in measured biodegradation kinetics during the long-term operation of completely mixed activated sludge bioreactors. *Water Sci. Technol.*, *34*, 35.

Ellis, T. G., Barbeau, D. S., Smets, B. F., & Grady, C. P. L., Jr. (1996b). Respirometric technique for determination of extant kinetic parameters describing biodegradation. *Water Environ. Res.*, *68*, 917.

Ettinger, M. B. (1958). Universal factors in aerobic biological purification. *Sew. Ind. Wastes*, *30*, 14.

Fair, G. M. (1936). the log difference method of estimating the constants of the first stage of the BOD curve. *Sew. Works J.*, *8*, 432.

Fair, G. M., Moore, E. W., & Thomas, H. A., Jr. (1941). The natural purification of river muds and pollutional sediments. *Sew. Works J.*, *13*, 270.

Firth, B. K. (1984). Status of water quality modeling in the pulp and paper industry. *J. Water Pollut. Control Fed.*, *56*, 1131.

Flegal, T. M., & Schroeder, E. D. (1976). Temperature effects on BOD stoichiometry and oxygen uptake rate. *J. Water Pollut. Control Fed.*, *48*, 2700.

Frisk, T. (1982). An oygen model for Lake Haukivesi. *Hydrobiologia, 86*, 133.

Frisk, T., & Kyla-Harakka, T. (1982). Mathematical methods for predicting DO concentrations and phytoplankton biomass in lakes loaded by waste waters of the pulp and paper industry. *Aqua Fennica, 10*, 21.

Fujie, K., Hu, H.-Y., Lim, B.-R., & Xia, H. (1997). Effect of biosorption on the damping of influent fluctuation in activated sludge aeration tanks. *Water Sci. Technol., 35*(7), 79.

Fullen, D. J., et al. (1982). *Photosynthesis and respiration rates in the Montecello experimental streams: 1976/1977: Diel field data and computed results*. U.S. EPA-600/3-82-065.

Gameson, A. L. H., & Wheatland, A. B. (1958). The ultimate oxygen demand and course of oxidation of sewage effluents. *J. Proc. Inst. Sew. Purif., 106*.

Gannon, J. J. (1966). River and laboratory BOD rate considerations. *J. Sanit. Eng. Div., Proc. Am. Soc. Civ. Eng.*, 92(SA1), 135.

Garrett, M. T. (1951). Kinematics of removal soluble BOD by activated sludge. Unpublished thesis, Massachusetts Institute of Technology.

Garrett, M. T., & Sawyer, C. N. (1952). Kinematics of removal of soluble BOD by activated sludge. *Proc. 7th Ind. Waste Conf., Purdue University, 79*, 51.

Gates, W. E., et al. (1969). The application of bacterial process kinetics in stream simulation and stream analysis. *Water Res., 3*, 663.

Gates, W. E., & Ghosh, S. (1971). Biokinetic evaluation of BOD concepts and data. *J. Sanit. Eng. Div., Proc. Am. Soc. Civ. Eng., 97*, 287.

Gaudy, A. F., Jr. (1972). Biochemical oxygen demand. In *Water Pollution Microbiology*. Wiley & Sons.

Gaudy, A. F., Jr., Bhatla, M. N., Follett, R. H., & Abu-Niaaj, F. (1965). Factors affecting the existence of the plateau during the exertion of BOD. *J. Water Pollut. Control Fed., 37*, 444.

Gaudy, A. F., Jr., & Engelbrecht, R. S. (1960). Basic biochemical considerations during metabolism in growing vs. respiring systems. *Proc. 3rd Conf. Biol. Waste Treatment, Manhattan College*.

Gaudy, A. F., Jr., & Gaudy, E. T. (1980). *Microbiology for environmental scientists and engineers*. McGraw-Hill.

Gaudy, A. F., Jr., Ramanathan, M., & Rao, B. S. (1967). Kinetic behavior of heterogeneous populations in completely mixed reactors. *Biotechnol. Bioeng.*, 9(3), 387.

Gaunt, P., & Abbott, W. E. (1928). Dissolved oxygen absorption time relation of activated sludge effluents. *J. Soc. Chem. Ind.*, *47*, 14.

Gerardin, A. (1875). Alterations de la Seine aux abords de Paris depuis Novembre, 1874, Justua Mai, 1875. *Compt. Rend.*, *80*, 1326.

Goldsmith, H. I., & Watson, K. S. (1980). Use of equivalent BOD as a new waste management. *J. Water Pollut. Control Fed.*, *52*, 372.

Gotaas, H. B. (1949). Effect of sea water on biochemical oxidation of sewage. *Sew. Works J.*, *21*, 818.

Grady, C. P. L., Jr., Smets, B. F., & Barbeau, D. S. (1966). Variability in kinetic parameter estimates: a review of possible causes and a proposed terminology. *Water Res.*, *30*, 742.

Gram, A. L. (1956). Reaction kinetics of aerobic biological processes. *IER. Ser. 90*, Rep. No. 2 (University of California, Berkeley).

Grau, P., et al. (1982). Recommended notation for use in the description of biological wastewater treatment processes. *Water Res.*, *16*, 1501.

Griffiths, P. (1994). Modifications to the IAWPRC Task Group activated sludge model. *Water Res.*, 28, 657.

Grune, W. N. (1950). Calculation and statistical analysis of the BOD velocity constant. *U.S. Atomic Energy Comm.*, unclassified doc. NY00-1501; *Nucl. Sci. Abstr.*, *4*, 395.

Gujer, W., et al. (1995). The activated sludge model no. 2: Biological phosporus removal. *Water Sci. Technol.*, *31*(2), 1.

Gujer, W., Henze, M., Mino, T., & van Loosbrecht, M. (1999). Activated sludge model no. 3. *Water Sci. Technol.*, *39*(1), 183.

Hagawara, K. (1956). Studies on the BOD velocity constants and immediate oxygen demand. *Bull. Inst. Public Health* (Japan), *5*, 15; *Public Health Abstr.*, *37*(3), 24 (1957).

Haldane, J. B. S. (1930). *Enzymes*. Longmans Green, MIT Press.

Hansford, G. S., et al. (1978). A steady state model for the rotating biological disc reactor. *Water Res.*, *12*, 855.

Hao, O. J., & Li, C. T. (1987). Effect of slowly biodegradable organics on kinetic coefficients. *J. Environ. Eng.*, *113*, 205.

Harley, N. H. (1964). A nomograph for the solution of stream pollution equations. *Water Sew. Works*, *111*, 144.

Hartmann, L., & Wilderer, P. (1968). Physical and biological aspects of BOD kinetics. *Vodni Hospod. A or B* (Czech.), 6, 258; *Chemical Abstracts*, *69*, 8391 (1968).

Hatcher, K. J. (Ed). (1986). *Sediment oxygen demand: Processes, modeling, and measurement*. Inst. Natl. Res.

Henze, M., et al. (1982). The use and abuse of notation in biological wastewater treatment. *Water Res.*, *16*, 755.

Henze, M., Grady, C. P. L., Jr., Gujer, W., Marais, G. v. R., & Matsuo, T. (1987). Activated sludge model no. 1. *Sci. Tech. Rep. No. 1*. Int. Assoc. Water Pollut. Res. Control.

Henze, M., Gujer, W., Mino, T., Matsuo, T., Wenzel, M., & Marais, G. v. R. (1994). Activated sludge model no. 2. *Sci. Tech. Rep. No. 3*. Int. Assoc. Water Pollut. Res. Control.

Henze, M., Gujer, W., Mino, T., Matsuo, T., Wentzel, M. C., & Marais, G. v. R. (1995). Wastewater and biomass characterization for the activated sludge model no. 2: Biological phosphorus removal. *Water Sci. Technol.*, *31*, 13.

Henze, M., Gujer, W., Mino, T., Matsuo, T., Marais, G. v. R., & van Loosdrecht, M. (1999). Activated sludge model no. 2D, ASM2D. *Water Sci. Technol.*, *39*(1), 165.

Herman, A. (1956). Graphic determination of BOD. *Bull. Cent. Etude Document Eaux*, *71*, 414; *Public Health Eng. Abstr.*, *37*(1)0, 20 (1957).

Heukelekian, H., & Rand, M. C. (1955). BOD of pure organic compounds. *Sew. Ind. Wastes*, *27*, 1040.

Hewitt, J. P., & Hunter, J. V. (1975). A comparison of the methods used to calculate first order BOD equation constants. *Water Res.*, *9*, 683.

Hewitt, J., et al. (1979). A multiorder approach to BOD kinetics. *Water Res.*, *13*, 325.

Hoover, S. R., Jaswicz, L., & Porges, N. (1952). Biochemical oxidation of dairy wastes. IV. Endogenous respiration and stability of aerated dairy waste sludge. *Sew. Ind. Wastes*, *24*, 1144.

Hoover, S. R., Jaswicz, L., & Porges, N. (1953). An interpretation of BOD test in terms of endogenous respiration of bacteria. *Sew. Ind. Wastes*, *25*, 1163.

Hoover, S. R., & Porges, N. (1951). Treatment of dairy wastes by aeration. II. Continuous aeration studies. *Sew. Ind. Wastes*, *23*, 694.

Huang, J. Y. C., Cheng, M.-D., & Mueller, J. T. (1985). Oxygen uptake rates for determining microbial activity and application. *Water Res.*, *19*, 373.

Ibrahim, G., & Abasaeed, A. E. (1995). Modeling of sequencing batch reactors. *Water Res.*, *29*, 1761.

Ingram, M. (1939a). The endogenous respiration of *Bacillus cereus*. I. Changes in the rate of respiration with the passage of time. *J. Bacteriol.*, *38*, 599.

Ingram, M. (1939b). The endogenous respiration of *Bacillus cereus*. II. The effect of salts on the rate of absorption of oxygen. *J. Bacteriol.*, *38*, 613.

Ingram, M. (1940). The endogenous respiration of *Bacillus cereus*. III. The changes in the rate of respiration causes by sodium chloride in relation to hydrogen-ion concentration. *J. Bacteriol.*, *40*, 683.

James, A. (1978). *Mathematical models in water pollution control*. Wiley & Sons.

Jansa, V., & Akerlindh, G. (1941). A method for computing the oxygen curve in a polluted stream with special reference to the oxygen consumption of sludge desposits. *Sew. Works J.*, *13*, 551.

Jennelle, E. M., & Gaudy, A. F., Jr. (1970). Studies on the kinetics and mechanism of BOD exertion in dilute systems. *Biotechnol. Bioeng.*, *12*, 519.

Kappeler, J., & Gujer, W. (1992). Estimation of kinetic parameters of heterotrophic biomass under aerobic conditions and characterization of wastewater for activated sludge modelling. *Water Sci. Technol.*, *25*(6), 125.

Kawashima, H., & Suzuki, M. (1989). Numerical simulation model for prediction of BOD removal rate in streams. *Water Sci. Technol.*, *21*, 1003.

Klein, R. L., & Gibbs, C. R. (1979). Graphical method for calculating BOD. *J. Water Pollut. Control Fed.*, *51*, 2257.

Koga, S., & Humphrey, A. E. (1967). Study of dynamic behavior of the chemostat system. *Biotechnol. Bioeng.*, *9*, 375.

Koivo, A. J., & Phillips, G. R. (1972). On determination of BOD and parameters in polluted stream models from DO measurements only. *Water Resour. Res.*, *8*, 478.

Koizumi, J., & Aiba, S. (1984). Reassessment of the dynamic K_La method. *Biotechnol. Bioeng.*, *26*, 1131.

Kothandaraman, V. (1970). Probabilistic variations in ultimate first stage BOD. *J. Sanit. Eng. Div., Proc. Am. Soc. Civ. Eng.*, *96*, 27.

Koussis, A. D., et al. (1990). Modeling DO conditions in streams with dispersion. *J. Environ. Eng.*, *116*, 601.

Landine, R. C. (1971). Second order and first order kinetics of BOD data. *Water Sew. Works*, *118*, R-45.

LeDuc, R., Unny, T. E., & McBean, E. A. (1986). Stochastic model of first-order BOD kinetics. *Water Res.*, *20*, 625.

Lee, J. D. (1951). Simplified method for analysis of BOD data: A discussion. *Sew. Ind. Wastes*, *23*, 164.

Lenk, S., et al. (1990). Process analysis and control of biological treatment steps using a plateau-bod measuring instrument. *Wasserwirtsch-wassertech.* (Germany), *40*(1), 4.

Lenov, A. V. (1974). Generalization, typification, and kinetic analysis of biochemical consumption of oxygen curves based on BOD experiments. *Okeanologiia*, *14*, 82.

Lewandowski, Z., et al. (1985). Inhibition coefficient (K_i). determination in activated sludge. *Water Res.*, *19*, 671.

Liang, D. H., et al. (1979). Bipopulation approach to CMAS simulation. *J. Environ. Eng.*, *105*, 905.

Liebman, J. C., & Loucks, D. P. (1966). A note on oxygen sag equations. *J. Water Pollut. Control Fed.*, *38*, 1963.

Lofy, R. J. (1969). Illustrative problems using a nomograph for the Streeter–Phelps equation. *Water Sew. Works*, *116*, 162.

Logan, B. E., Hermanowicz, S. W., & Parker, D. S. (1987a). A fundamental model for trickling filter process design. *J. Water Pollut. Control Fed.*, *59*, 1029.

Logan, B. E., Hermanowicz, S. W., & Parker, D. S. (1987b). Engineering implications of a new trickling filter model. *J. Water Poll. Control Fed.*, *59*, 1017.

Lohani, B. N. (1982). Stochastic water quality index. *Environ. Technol. Lett.*, *3*, 521.

Loucks, D. P. (1967). Risk evaluation in sewage treatment plant design. *J. Sanit. Eng. Div., Proc. Am. Soc. Civ. Eng.*, *93*(SA1), 25.

Loucks, D. P., et al. (1981). *Water resource systems planning and analysis.* Prentice-Hall.

Ludzack, F. J., Krieger, H. L., & Ettinger, M. B. (1964). Interactions of waste feed, activated sludge, and oxygen as traced by radioactive carbon. *J. Water Pollut. Control Fed.*, *36*, 782.

Ludzack, F. J., Moore, W. A., & Ruchhoft, C. C. (1953). A comparative study of derivation procedures for BOD velocity constants. *Sew. Ind. Wastes*, *25*, 875.

Mao, H., & Smith, D. W. (1995). A mechanistic model for assessing biodegradability of complex wastewaters. *Water Res.*, *29*, 1957.

Maoyu, T. (1990). An autocatalytic model for the kinetics of BOD test. *Water Res.*, *24*, 1091.

Marais, G. v. R., & Ekama, G. A. (1976). The activated sludge process part I: Steady state behavior. *Water S.A.*, *2*, 163.

Marais, G. v. R., & Ekama, G. A. (1977). The activated sludge process part II: Dynamic behavior. *Water S.A.*, *3*, 18.

Marsili-Libelli, S. (1980). Reduced-order modeling of the activated sludge process. *Ecol. Modeling*, *9*, 15.

Marske, D. M., & Polkowski, L. B. (1972). Evaluation of methods for estimating BOD parameters. *J. Water Pollut. Control Fed.*, *44*, 1987.

Masutani, G. K., & Stenstrom, M. K. (1991). Dynamic surface tension effects on oxygen transfer. *J. Environ. Eng.*, *117*, 126.

McBride, G. B. (1982). Nomographs for rapid solutions for the Streeter–Phelps equations. *J. Water Pollut. Control Fed.*, *54*, 378.

Michaelis, L., & Menten, M. L. (1913). Die kinetik der ivertinweikung. *Biochem. Z.*, *49*, 333.

Mignone, C. F., & Ertola, R. J. (1984). Measurement of oxygen transfer coefficient under growth conditions by dynamic model moment analysis. *J. Chem. Tech. Biotechnol.*, *34B*, 121.

Milojevic, M. (1967). Contribution of steam-sterilized algae to BOD. *J. Water Pollut. Control Fed.*, *39*, 1201.

Mines, R. O., & Sherrard, J. H. (1987). Biological enhancement of oxygen transfer in the activated sludge process. *J. Water Pollut. Control Fed.*, *59*, 19.

Mona, R., Dunn, I. J., & Bourne, J. R. (1979). Activated-sludge process dynamics with continuous TOC and oxygen uptake measurements. *Biotechnol. Bioeng.*, *21*, 1561.

Monod, J. (1942). *Reserches sur la croissance des culture bacteriennes.* Herman.

Monod, J. (1950). La technique de culture continue, theorie et applications. *Ann. Inst. Pasteur*, *79*, 390.

Moodie, S. P., & Greenfield, P. F. (1979). Substrate removal kinetics in trickling filters. *J. Water Pollut. Control Fed.*, *51*, 1063.

Moore, E. W. (1941). Long time BODs at low temperatures. *Sew. Works J.*, *13*, 561.

Moore, E. W., Thomas, H. A., Jr., & Snow, W. B. (1950). Simplified method for the analysis of BOD data. *Sew. Ind. Wastes*, *22*, 1343.

National Research Council. (1946). Sewage treatment at military installations. II. Characteristics of military sewage. *Sew. Works J.*, *18*, 830.

Navone, R. (1960). A new method for calculating K and L for sewage. *Water Sew. Works*, *107*, 285.

Nisbet, R. M., Cunningham, A., & Gurney, W. S. C. (1983). Endogenous metabolism and the stability of microbial prey–predator systems. *Biotechnol. Bioeng.*, *25*, 301.

Nolasco, D. A., Daigger, G. T., Stafford, D. R., Kaupp, D. M., & Stephenson, J. P. (1998). The use of mathematical modeling and pilot plant testing to develop a new biological phosphorus and nitrogen removal process. *Water Environ. Res., 70*, 1205.

Novak, L., Larrea, L., & Wanner, J. (1995). Mathematical model for soluble carbonaceous substrate biosorption. *Water Sci. Technol., 31*, 2, 67.

Novotny, V., & Krenkel, P. A. (1975). A waste assimilative capacity model for a shallow turbulent stream. *Water Res., 9*, 233.

Nylander, G. (1964). A study of waste waters from forest product industry. 5. A study of the course of biochemical oxidation. *Sven. Papperstidn.* (Sweden), *67*, 562.

Oner, M. D., & Erickson, L. E. (1983). Estimation of true growth and product yields in aerobic cultures. *Biotechnol. Bioeng., 25*, 631.

Ong, S. L. (1983). Least-squares estimation of batch culture kinetic parameters. *Biotechnol. Bioeng., 25*, 2347.

Orford, H. E., & Ingram, W. T. (1953a). Deoxygenation of sewage. I. Critical review of the monomolecular formula. *Sew. Ind. Wastes, 25*, 419.

Orford, H. E., & Ingram, W. T. (1953b). Deoxygenation of sewage. II. The logarithmic formula as applied to sewage. *Sew. Ind. Wastes, 25*, 424.

Orford, H. E., & Ingram, W. T. (1953c). Deoxygenation of sewage. III. The logarithmic formula applied to oxidations other than sewage. *Sew. Ind. Wastes, 25*, 566.

Organization for Economic Cooperation and Development. (1998). *Guidelines for testing of chemicals.* 10th Suppl.

Padgett, W. J., & Rao, A. N. V. (1979). Estimation of BOD and DO probability distribution. *J. Environ. Eng., 105*, 525.

Padgett, W. J., et al. (1977). A stochastic model for BOD and DO in streams where pollutants are discharged over a continuous stretch. *Int. J. Environ. Stud., 11*, 45.

Paolini, A. E. (1979). Development of mathematical models for the treatment of an industrial wastewater by means of biological rotating disc reactors. *Water Res., 13*, 751.

Papadopoulos, A. S., Tiwari, R. C., & Muha, M. J. (1991). Bootstrap procedures for time series analysis of BOD D. *Ecol. Modell., 55*(1), 57.

Parisod, J. P., & Schroeder, E. D. (1978). BOD progression in mixed substrates. *J. Water Pollut. Control Fed., 50*, 1827.

Patry, G. G., & Chapman, D. (1989). *Dynamic modeling and expert systems in wastewater engineering.* Lewis Publishers.

Pawlowsky, U., & Howell, J. A. (1973). Mixed culture biooxidation of phenol. I. Determination of kinetic parameters. *Biotechnol. Bioeng.*, *15*, 889.

Phelps, E. B. (1909). The disinfection of sewage and sewage filter effluents. *U.S. Geol. Surv. Water Supply Paper*, *229*, 74.

Pleisnner, M. (1910). Uber die abhanggigkeit der sauerstoffzehrung naturlicher wasser von der versuchsdauer und der versuchstemperatur. *Arb. Kais. Gesundh.*, *34*, 230.

Porges, N., Jasewicz, L., & Hoover, S. R. (1956). Simple method of estimation of oxygen value of wastes. *Ind. Wastes*, *1*, 95.

Ramanathan, M., & Gaudy, A. F., Jr. (1969). Effect of high substrate concentration and cell feedback on kinetic behavior of heterogeneous populations in completely mixed systems. *Biotechnol. Bioeng.*, *11*, 207.

Ramkrishna, D., Fredrickson, A. G., & Tsuchiya, H. M. (1967). Dynamics of microbial propagation: Models considering inhibition and variable cell composition. *Biotechnol. Bioeng.*, *9*, 129.

Ranjie, H., & Huimin, L. (1987). Modeling of BOD–DO dynamics in an ice-covered river in Northern China. *Water Res.*, *21*, 247.

Rao, B. S., & Gaudy, A. F., Jr. (1966). Effect of sludge concentration on various aspects of biological activity in activated sludge. *J. Water Pollut. Control Fed.*, *38*, 794.

Rathburn, R. E. (1977). Reaeration coefficients of streams: State of the art. *J. Hydraul. Div., Proc. Am. Soc. Civ. Eng.*, *103*, 409.

Reckhow, K. H., & Chapra, S. C. (1983). *Engineering approaches for lake management: Volume 1: Data analysis and empiricle modeling; Volume 2: Mechanistic modeling*. Butterworth.

ReVelle, C. S., Lynn, W. R., & Rivera, M. A. (1965). Bio-oxidation kinetics and a second order equation describing the BOD reaction. *J. Water Pollut. Control Fed.*, *37*, 1679.

Rhame, G. A. (1956). A two point method for estimation of the first stage BOD. *Sew. Ind. Wastes*, *28*, 1087.

Rinaldi, S., & Soncini-Sessa, R. (1978). Sensitivity analysis of Streeter–Phelps models. *Adv. Water Res.*, *1*, 144.

Rinaldi, S., et al. (1979a). *Modeling and control of river quality*. McGraw-Hill.

Rinaldi, S., et al. (1979b). Parameter estimation of Streeter–Phelps model. *J. Environ. Eng.*, *105*, 75.

Ros, M. (1993). *Respirometry of activated sludge*. Technomic.

Rozich, A. F. (1992). Design and operational analysis of activated sludge processes using respirometrically calibrated models. *Water Sci. Technol.*, *26*, 2061.

Rozich, A. F., & Gaudy, A. F., Jr. (1984). Critical point analysis for toxic waste treatment. *J. Environ. Eng.*, *110*, 562.

Rozich, A. F., & Gaudy, A. F., Jr. (1992). *Design and operation of activated sludge processes using respirometry*. Lewis.

Ruchhoft, C. C. (1950). Simplified method for analysis of BOD data: A discussion. *Sew. Ind. Wastes*, *22*, 1353.

Ruchhoft, C. C., Placak, O. R., & Ettinger, M. B. (1948a). Correction of BOD velocity constants for nitrification. *Sew. Works J.*, *20*, 832.

Ruchhoft, C. C., Placak, O. R., & Ettinger, M. B. (1948b). Variations in BOD velocity constants of sewage dilutions. *Ind. Eng. Chem.*, *40*, 1290.

Sampathkumaran, M. A. (1953). The effects of organic loading upon diurnal changes in the dissolved oxygen cycle. Unpublished thesis, Johns Hopkins University.

Schmied, J. (1973). Response of river water to biochemically degradable substances in pulp mill waste waters. *Papiera Celuloza*, *28*, 117; *Chemical Abstracts*, *79*, 45430 (1973).

Servizi, J. A., & Bogan, R. H. (1964). Thermodynamic aspects of biological oxidation and synthesis. *J. Water Pollut. Control Fed.*, *36*, 607.

Shasty, J. S., et al. (1973). Nonlinear parameter estimation in water quality modeling. *J. Environ. Eng.*, *99*, 315.

Sheehy, J. P. (1960). Rapid method for solving first-order equations. *J. Water Pollut. Control Fed.*, *32*, 646.

Shen, H. W. (Ed.). (1979). *Modeling of rivers*. Wiley & Sons.

Simkins, S., & Alexander, M. (1985). Nonlinear estimation of the parameters of monod kinetics that best describe mineralization of several substrate concentrations by dissimilar bacterial densities. *Appl. Environ. Microbiol.*, *50*, 816.

Simkins, S., et al. (1986). Two approaches to modeling kinetics of biodegradation by growing cells and application of a two-compartment model for mineralization kinetics in sewage. *Appl. Environ. Microbiol.*, *51*, 1153.

Simpson, J. R. (1968). A second-order equation, not based on ultimate oxygen demand, describing the first phase of oxygen uptake in BOD reactions. *Water Pollut. Control*, *67*, 433.

Sletten, O. (1966). Determining BOD curve parameters. *Water Sew. Works*, *113*, 133.

Smith, R.-K. (1999). *Handbook of environmental analysis* (4th ed.). Genium Publishing.

Snodgrass, W. J., & Ng, P. S. (1985a). Biochemical models for the hypolimnetic oxygen depletion in lakes impacted by wastewater discharges. 1. Alternative models. *Arch. Hydrobiol. Suppl.*, *72*(1), 89.

Snodgrass, W. J., & Ng, P. S. (1985b). Biochemical models for the hypolimnetic oxygen depletion in lakes impacted by wastewater discharges. 2. Phytoplankton biomass model. *Arch Hydrobiol. Suppl.*, *72*(2), 220.

Sollfrank, U., & Gujer, W. (1990). Simultaneous determination of oxygen uptake rate and oxygen transfer coefficient in activated sludge systems by an on-line method. *Water Res.*, *24*, 725.

Sollfrank, U., & Gujer, W. (1991). Characteristics of domestic wastewater for mathematical modelling of the activated sludge process. *Water Sci. Technol.*, *23*, 1057.

Sornberger, G. C., & Keshavan, K. (1973). Simulation of dissolved oxygen profile. *J. Environ. Eng.*, *99*, 479.

Spitta, O., & Weldert, R. (1906). Indicatoren fur die beurteilung biologisch gereinigter abwasser. *Mitt. Koenig. Pruf. Wasserversorg*, *6*, 160.

Stones, T. (1963). Kinetics of the biochemical oxidation of settled domestic sewage. *J. Proc., Inst. Sew. Purif.*, *3*, 285: *Chemical Abstracts*, *59*, 11092 (1963).

Stones, T. (1979). Kinetics of biochemical oxidation of sewage. *Effluent Water Treat. J.*, *19*, 28.

Stones, T. (1981). A resume of the kinetics of the BOD test. *Water Pollut. Control*, *80*, 513.

Stones, T. (1984). The decreasing rate of biochemical oxidation of sewage. *Water Pollut. Control*, *83*, 435.

Stover, E. L., & Kincannon, D. F. (1982). Rotating biological contactor scale-up and design. *Water Eng. Manage.*, *129*, Ref. Handbook 4, 49.

Streeter, H. W. (1925). The rate of atmospheric reaeration of sewage polluted streams. *Proc. Am. Soc. Civ. Eng.*, *51*, 1829.

Streeter, H. W. (1930). Comparative rates of stream purification under natural and controlled conditions. *Ind. Eng. Chem.*, *22*, 1343.

Streeter, H. W. (1934). A formulation of bacterial changes occurring in polluted water. *Sew. Works J.*, *6*, 208.

Streeter, H. W., & Phelps, E. B. (1925). A study of the pollution and natural purification of the Ohio River. III. Factors concerned in the phenomena of oxidation and reaeration. *U.S. Public Health Sew. Bull.*, *146*.

Stumm, W., & Morgan, J. J. (1996). *Aquatic chemistry* (3rd ed.). Wiley & Sons.

Swamee, P. K., & Ojha, C. S. P. (1991). Modeling of BOD exertion curve. *Water Res.*, *25*, 901.

Sykes, R. M. (1976). Microbial product formation and variable yield. *J. Water Pollut. Control Fed.*, *48*, 2046.

Tabak, H. H., Gao, C., Desai, S., & Govind, R. (1992). Development of predictive structure-biodegradation relationship models with the use of respirometrically generated biokinetic data. *Water Sci. Technol.*, *26*, 763.

Thayer, R. P., & Krutchoff, R. G. (1967). Stochastic model for BOD and DO in streams. *J. Sanit. Eng. Div., Proc. Am. Soc. Civ. Eng.*, *93*, 59.

Theriault, E. J. (1925). The determination of dissolved oxygen by the Winkler method. *U.S. Public Health Serv. Bull.*, *151*.

Theriault, E. J. (1927). The Oxygen Demand of Polluted Waters. I. A Critical Review. II. The Rate of Deoxygenation. *U.S. Public Health Serv. Bull.*, *173*.

Therien, N., Le Clave, P., & Jones, P. (1984). A respirometric study of the influence of aliphatic alcohols on activated sludges. *Water Res.*, *18*, 905.

Thomas, H. A., Jr. (1937). The slope method of evaluating the constants of the first stage BOD curve. *Sew. Works J.*, *9*, 424.

Thomas, H. A., Jr. (1940). Analysis of the BOD curve. *Sew. Works J.*, *12*, 504.

Thomas, H. A., Jr. (1950). Graphical determination of the BOD constants. *Water Sew. Works*, *97*, 123.

Thyssen, N., & Erlandsen, M. (1987). Reareation of oxygen in shallow, macrophyte rich streams. II. Relationship between the reareation rate coefficient and hydraulic properties. *Int. Rev. Ges. Hydrobiol.*, *72*, 575.

Tsivoglou, E. C. (1967). *Tracer measurement of stream reaeration.* Fed. Water Pollut. Control Admin., U.S. Dept. Interior.

Tsivoglou, E. C., Cohen, J. B., Shearer, S. D., & Godsil, P. J. (1968). Tracer measurements of stream reaeration. II. Field studies. *J. Water Pollut. Control Fed.*, *40*, 285.

Tyteca, D. (1981). Nonlinear programming model of wastewater treatment plant. *J. Environ. Eng.*, *107*, 747.

Uhlmann, D. (1979). BOD removal rates of waste stabilization ponds as a function of loading, retention time, temperature, and hydraulic flow pattern. *Water Res.*, *13*, 193.

U.S. Public Health Service. (1958). Oxygen relationships in streams. *Robert A. Taft Eng. Cent., Tech. Rep.*, W58-2.

Van Beneden, G. (1956). Role of seeding in the measurement of BOD and the influence of carbohydrates and associated flora on the values of BOD. *Bull. Cent. Belge Etude et Document Eaux*, *70*, 363.

van Loosdrecht, M. C. M., & Henze, M. (1999). Maintenance, endogeneous respiration, lysis, decay, and predation. *Water Sci. Technol.*, *39*(1), 107.

Vanrolleghem, P., et al. (1995). Practical identifiability of a biokinetic model of activated sludge respiration. *Water Res.*, *29*, 2561.

Varma, M. M., & Lindley, H. H. (1966). BOD of pure carbohydrates and its relation to the chemical structure. *Water Sew. Works*, *113*, 102.

Walker, R. R., & Snodgrass, W. J. (1986). Model for sediment oxygen demand in lakes. *J. Environ. Eng.*, *112*, 25.

Wang, M. H., Kao, J.-F., Wang, L. K., Wen, C.-G., & Vielkind, D. (1979). Computer-aided stream pollution control and management, part I. *J. Environ. Manage.*, *9*, 165.

Wang, M. H., Wang, L. K., Kao, J.-F., & Wen, C.-G. (1979). Computer-aided stream pollution control and management, part II. *J. Environ. Manage.*, *9*, 185.

Watkin, A. T., & Eckenfelder, W. W. (1984). Development of pollutant specific models for toxic organic compounds in the activated sludge process. *Water Sci. Technol.*, *17*, 279.

Watkin, A. T., & Eckenfelder, W. W. (1985). A new design procedure for activated sludge based on active mass. *Environ. Technol. Lett.*, *6*, 421.

Wen, C.-G., et al. (1980). Modeling of stream water quality by a new moment method: theoretical development. *J. Environ. Manage.*, *10*, 1.

Wen, C.-G., & Kao, J. F. (1981). Mathematical modeling of stream water quality by a new moment method: Field investigation of tidal river. *J. Environ. Manage.*, *12*, 127.

Wezernak, C. T., & Gannon, J.J. (1967). Oxygen–nitrogen relationships in autotrophic nitrification. *Appl. Microbiol.*, *15*, 1211.

Whipple, W., Jr., & Hunter, J. V. (1981). Desk top methodology for modeling BOD in streams. *Water Resour. Bull.*, *17*, 678; Natl. Tech. Inf. Serv., PB-80-176142.

Wiegand, J. H. (1954). An improved form of monomolecular formula. *Sew. Ind. Wastes*, *26*, 160.

Willis, R., et al. (1976). Transient water quality modeling in streams. *Water Resour. Bull.*, *12*, 157.

Woodring, S. L., & Clifford, D. A. (1988). A comparison of the graphical and standard methods for the determination of BOD. *J. Water Pollut. Control Fed.*, *60*, 537.

Woodward, R. L. (1953). Deoxygenation of sewage: A discussion. *Sew. Ind. Wastes*, *25*, 918.

Worley, J. L, Burgess, F. J., & Towne, W. W. (1965). Identification of low-flow augmentation requirements for water quality control by computer techniques. *J. Water Pollut. Control Fed.*, *37*, 659.

Wu, Y. C., et al. (1980). Modeling of rotating biological contactor systems. *Biotechnol. Bioeng.*, 22, 2055.

Young, J. C., & Clark, J. W. (1965). Second order equation for BOD. *J. Sanit. Eng. Div., Proc. Am. Soc. Civ. Eng.*, *91*(SA1), 43.

Zanoni, A. E. (1967). Waste water deoxygenation at different temperatures. *Water Res.*, *1*, 543.

Zison, S. W., et al. (1978). *Rates, constants, and kinetic formulations in surface water quality modeling.* U.S. EPA-600/3-78-105.

7

Biochemical Oxygen Demand Method and Practice

No other subject will cause a better argument than biochemical oxygen demand (BOD). Why? Because this determination is apparently subject to more variations and discrepancies due to, as yet, unknown causes than any other test used in the sewage works laboratory (Eldridge, 1933).

The current procedures for determining (BOD) in water are the result of more than 100 years of study, development, tinkering, and tweaking. Most of this experimentation was driven by work aimed at stabilizing sanitary wastes and finding quantifiable ways of measuring the rate and degree of stabilization. There are a number of published optimized BOD procedures that relate to specific waste streams (Batta & Leclerc, 1939; Chaudhuri et al., 1992; Grune & Luckens, 1949; Hammer, 1971; Okura, 1956; Riehl, 1948; Rogers, 1949; Tom, 1951; Wells, 1971). The procedure that is most frequently employed in treatment facility and commercial environmental laboratories, and is most frequently cited by regulatory agencies as the approved method for BOD, is that presented in *Standard Methods* as Method 5210 B (APHA et al., 1998). Today, successful use of this method requires that the procedural steps contained within the method be understood and followed. Also, if one wants to avoid a great deal of frustration with this test, it is important to realize that this is a bioassay, not an analytical chemistry test.

As stated in the introduction to Method 5210 in the 20th edition of *Standard Methods* (APHA et al., 1998), the BOD test is an empirical determination that uses standardized procedures and analytical chemistry tools to measure the desired endpoint of a biological process. The primary "instruments" for carrying out this assay, however, are living organisms, whose number, type, and ability to utilize the aggregate organic matter in a sample as an energy and food source can vary widely. It will behoove the analyst to understand this basic point and keep it in mind as the test is learned and applied in the laboratory. Test results are method dependent, and a great deal of research and experience have gone into standardizing the steps in the method to limit the variability that will result from using living organisms as a basis for the analytical test system. The most precise and accurate measurement of sample volumes and dissolved oxygen (DO) will not save or improve a test in which the analyst used inadequate technique in preparing seed, dilution water, or nutrient buffer, or failed to correctly control sample pH and incubation temperatures. Likewise, failure to remove residual chlorine or failure to seed dechlorinated samples with a viable population of bacteria will yield poor results, no matter how well the analytical chemistry techniques are applied.

As presented in Chapter 2 relating the history of the BOD test, those who first attempted to determine the condition of a water were unaware of the role of bacteria in the oxygen status of the stream. They were aware, however, that low DO was an indicator of poor water quality. Once the realization took hold that microbes were involved, the focus of the test became less on the method for determining DO and more on how to normalize the test for quantitative comparisons. Selection of the proper test temperature, optimization of nutrients and dilution water, and standardization of test duration all represent important facets in the development of the BOD test.

The 21st edition of *Standard Methods* (APHA et al., 2005) offered three distinctly different BOD procedures: a 5-day test (BOD_5, 5210 B), a 30- to 90-day test (ultimate BOD, 5210 C), and a respirometric method (5210 D), which allows monitoring of oxygen uptake on a continuous basis for the desired test duration. Each of these methods has different applications and purposes; this chapter will focus on the practice of the 5-day dilution bottle BOD, with some mention of the ultimate BOD procedure. Respirometric techniques are covered in Chapter 8.

1.0 A 5-DAY TEST

The historical overview in this book summarizes many of the popular stories that have been told over the years to explain why the most common BOD test duration is 5 days. The reader should draw from that review this basic point: short of running an ultimate BOD test, the 5-day test (BOD_5) is widely accepted to achieve the best balance between presenting a reproducible and representative relationship to total demand and the amount of time spent in performing the test. Widely accepted does not mean that 5 days is the only accepted time period to the exclusion of all other intermediate time periods for estimating total demand, such as the 7-day standard used in Scandinavian countries (Broch-due et al., 1994; Krogh, 1970; Ropponen, 1979; Seppovaara & Starck, 1977; Strom & Solyom, 1978; Sundberg, 1975). Nonetheless, 5 days is the accepted incubation period in the United States, dating from the technique published as the *Uniform Procedure for Biochemical Oxygen Demand* by the Federation of Sewage Works Associations (FSWA, 1932), which was itself based on other work (Anonymous, 1932a, 1932b; Childs, 1925; Mohlman et al., 1928; Theriault, 1927). This technique forms the basis for the test and has been closely followed in *Standard Methods* ever since. To be sure, other methods (for example, the ultimate BOD or respirometric methods) are used for a variety of purposes; however, after the 10th edition of *Standard Methods* (APHA et al., 1955), use of slightly shorter or longer incubation periods disappeared from approved usage. Since 1955, the BOD_5 has taken on an identity of its own, and has become a fixed endpoint frequently endowed with "bright line" regulatory significance, while the surrogate origins have been largely forgotten.

So, what does this mean to the analyst? It means run the test for 5 days. (Refer to the Questions and Answers section at the end of this chapter for an idea about the types of questions typically posed by analysts regarding the 5-day incubation period.)

2.0 KINETICS

An astute analyst can challenge the foundation for the BOD_5 test along the lines of "What is the relationship between the BOD_5 and total demand, and what does all of that old research data tell us about the uncertainty in such an estimate?" For our inquisitive analyst, the practical answer may well be "Never mind! Just run the test for 5 days like the method says." However unpalatable that answer may be to most individuals, finding a real answer can involve a great deal of tedium and frustration in the laboratory or the library. The test is widely used as an estimate of wastewater strength, waste loading, and treatment facility efficiencies (Ruchhoft, 1938) and for complying with various regulatory operating permits. Fortunately, for our inquisitive analyst, the 5-day paradigm has grown up in an atmosphere of an internal consistency in experience derived primarily from domestic waste applications. It is frequently observed in domestic waste that, when a comparison is made between long-term definitive oxygen use determinations and a BOD_5 test, a fairly consistent relationship is found.

However, there may be a practical reason to understand something of the kinetics involved in bacterial utilization of oxygen as it occurs in the

BOD test to lay the groundwork for an improved understanding of how the test operates. And so this chapter will introduce some important elements within the context of running the test. A more detailed treatment of theory is provided elsewhere in this volume.

The test is intended to measure the molecular oxygen (O_2) utilized during the biological oxidation of organic matter (carbon) to produce carbon dioxide and water. This process is what traditionally has been viewed as waste stabilization. The utilization of oxygen occurs via biochemical reactions involving microbial enzyme systems that have evolved to process nutrients for life support and propagation of the organisms. In our test, we seldom encounter either pure cultures of a single organism or strictly carbonaceous substrates. Bacteria can use many reduced forms of different elements to successfully derive the energy needed to support life. In wastewater systems, nitrifying bacteria and nitrogen species are usually present to some degree. In fact, in treated effluents, they may be prevalent and can exert a demand over and above that of carbon.

Phelps (1909) is attributed with the first attempt to develop a mathematical expression for the rate of loss of oxygen in wastewater stabilization. This is basically a first-order rate equation predicated on the assumption that the rate or velocity of bacterial reaction with substrate, C, varied directly with the concentration of the substrate supplied, $[Ct]$, times some constant, k. In simplest form, the first-order rate equation is of the form

$$-\mathrm{d}\,C/\mathrm{d}t = k[Ct]$$

where the differential d $C/\mathrm{d}t$ is the velocity of the reaction. This relationship essentially requires that the rate-determining factor is the amount of organic carbon [C] in the sample at any given time, t. Phelps went on to determine rate constants in wastewater, finally deriving an empirical expression, called the *monomolecular equation* for the time to stabilize a waste. However, lacking a means to accurately determine [C], Phelps had to substitute oxygen demand at any time, t, in his model. It has since become common to derive waste stabilization rates in terms of demand rather than [C] (Ruchhoft, 1938).

In 1927, Theriault published an extensive experimental verification of the first, or carbonaceous, stage of deoxygenation of polluted waters using Ohio River water. His results showed conformance with Phelps's monomolecular expression derived for waste stabilization of raw wastewater. However, in water resource recovery facility (WRRF) effluents, this simple first-order relationship did not always hold, probably due to rapid onset of nitrification. The concept of a two-phase process of oxygen utilization, carbonaceous and nitrogenous, became widely accepted in this time period. Much research was conducted over the next couple of decades to find numeric relationships between oxygen-consumption rates, waste stabilization times, and the BOD test.

In 1953, it was found that a logarithmic equation predicts the rate of deoxygenation better than the monomolecular formula, which had been studied and applied for nearly 50 years (Orford & Ingram, 1953a, 1953b, and 1953c). In terms of the BOD_5 test, plotting BOD results on semilog paper versus time yields a straight-line relationship that can be used to derive equations to predict the oxidation time of domestic wastewater, industrial wastes, and polluted streams. In the 1960s, the use of bacterial growth expressions based on the Monad equation for uninhibited growth and the Haldane equation for inhibitory substrates was introduced. These expressions have become popular as researchers attempt to more closely model the minutiae of the biological processes involved in oxygen utilization and waste treatment that the first-order equation tends to gloss over. As described in the chapter on BOD theory, these developments represent fundamentally different mathematical relationships than the monomolecular equation. Despite these advances, the main point to understand about these differing approaches is that each is only an approximation of a true rate relationship exhibited in any given waste treatment system. The effects of organism distribution and acclimation, protozoan contribution to demand, and the onset of nitrification can easily affect the empirical oxygen demand profiles and confound attempts to derive tidy rate equations. Because of these difficulties, the first-order monomolecular equation is typically used as a starting place for estimating rate functions in waste treatment (El-Baroudi & Moawad, 1967).

The carbonaceous BOD (CBOD) test can be divided into two reaction periods, one of rapid growth and another of slower endogenous respiration. Empirically, the rapid-growth phase involves assimilation of available nutrients by the biomass within about 24 hours. Such observations fit the organic carbon decay curves observed in a wide variety of waste and water samples, and are also consistent with first-order enzyme kinetics in biochemical reactions and exponential growth rates for bacterial populations. The implication of this seems to be that the population of organisms must quickly attain a level at which only minor variations occur for the reaction rate to adhere empirically to first-order or pseudo-first-order mechanics. Quite likely, then, this may be why so many researchers over the years have reported erratic BOD results for tests of short duration (e.g., 1 to 2 days) (O'Brien & Clark, 1962). Low populations of viable organisms in the sample, inadequate seed, or toxic chemicals cause a lag in this rapid-growth phase and even more erratic results.

One of the predictions of the first-order equation is that the half-life, $t(1/2)$, or the reaction time for oxidation of 50% of the substrate, can be easily estimated for a waste stream. The integrated form of the first-order equation yields

$$t(1/2) = 0.693/k$$

Thus, if the value of k for a waste is determined empirically by measuring the oxygen demand at various time intervals, then the time necessary to achieve successive half-life (i.e., 50%, 75%, etc.) reductions of the initial substrate concentration can be estimated for a waste stream. Because the value of k remains the same for a given reaction after first-order kinetics, regardless of the substrate concentration, a decay curve such as that in Figure 7.1 provides a generalized picture of what is happening to organic matter inside a BOD bottle.

Another consequence of assuming a first-order rate relationship is that it is possible to calculate the percentage of ultimate BOD exerted in 5 days by integrating the basic first-order rate equation. This yields the expression

$$\text{BOD}_5 = \text{ultimate BOD}(1 - 10^{-5k})$$

which can be further generalized for any time, t, instead of 5 days. Again, k must be experimentally determined for a given waste stream.

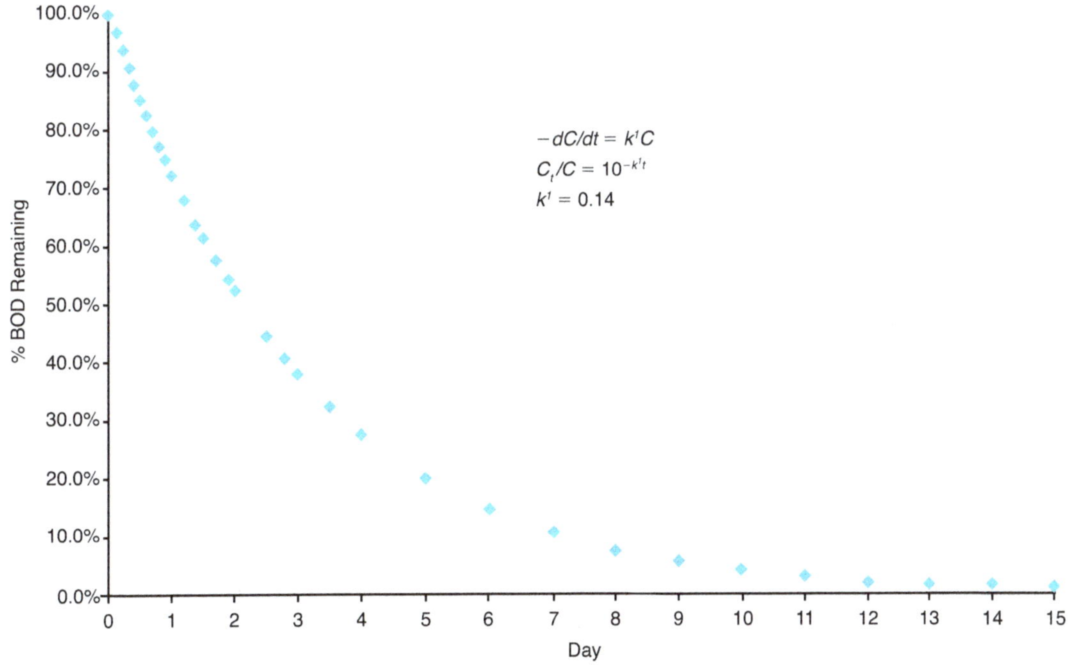

FIGURE 7.1. Percent organic carbon remaining as BOD.

Because oxygen is consumed in direct proportion to the organic carbon oxidized, the oxygen-depletion curve in the BOD test should follow a similar but inverse parabolic curve to Figure 7.1. As it turns out, the degree to which this holds is dependent on the amount of reduced nitrogen present as well as the population of nitrifying organisms.

3.0 CARBONACEOUS VERSUS NITROGENOUS BIOCHEMICAL OXYGEN DEMAND

The oxidation of organic nitrogen and ammonia to nitrite and nitrate is termed nitrification (Schloessing & Muntz, 1879). Originally regarded as a strictly inorganic process (Hazen, 1891; Munro, 1886), nitrification is known to be biologically mediated and has long been viewed as the "second stage" of oxygen demand in the stabilization of wastes (Cooper, 1921; Gowda, 1924; Heukelekian, 1942; Mohlman, 1938; Purvis & Coleman, 1907; Purvis et al., 1911; Scheringa, 1920; Theriault & McNamee, 1934). Some amount of nitrification begins to take place early in the BOD test, and the degree to which the oxygen consumed by nitrification represents a significant interference to the carbonaceous demand depends on the nature of the sample and the organisms present (Washington et al., 1983). Some of the problems that researchers have had in getting observed demand data to fit the simple monomolecular equation over the past century are rooted in some variable degree to nitrification in the waste streams they chose to study. Nitrification complicates the calculation and interpretation of velocity coefficients unless specific steps are taken to isolate the nitrogenous from carbonaceous processes (Ruchhoft et al., 1948). The initial rate of increase of nitrifying organism biomass, although slower than the growth of carbonaceous organisms, has a significant effect on the rate expression for nitrogenous demand (Buswell et al., 1950a, 1950b, 1954; Zanoni, 1969). In compliance monitoring BOD tests, disregarding the contribution of nitrification to total oxygen demand can result in significant excursions from permit limits (Hall & Foxen, 1983).

Traditionally, the BOD test has been used for two primary purposes. The first is to assess loading and removal efficiencies in WRRFs. The second is to assess loading on receiving waters to control oxygen depletion therein. The importance of nitrogen interference in the BOD_5 test should be understood in these different applications.

Sawyer's interpretation of the vast amount of BOD research conducted before 1946 was that oxygen demand during the first 10 days is largely carbonaceous (Sawyer & Bradney, 1946). Only after the carbonaceous demand is nearly stabilized do the conditions begin to favor rapid growth of nitrifying populations, with a subsequent inflection in the BOD curve. In this interpretation, then, BOD_5 is measuring only carbonaceous demand. For determining waste-loading parameters in treatment facility operation applications, this is probably a satisfactory concept. However, in a number of other scenarios, failure to recognize and account for nitrogenous interference can lead to some rather significant misinterpretations.

Problems associated with running BOD_5 on WRRFs operating in a partial nitrification mode have been described (Sawyer & Bradney, 1946). In these types of effluents, the population of nitrifying organisms is already quite high, nitrogenous demand competes with carbon oxidation, and seemingly poor BOD removals are calculated for the treatment facility as a result of the nitrogen interference in the test.

Nitrogen interference can also cause significantly higher BOD_5 results in the more highly diluted test bottles. This has been misinterpreted as toxicity in the dilution water (Sawyer & Bradney, 1946). In this example, researchers had been investigating erratic BOD results from daily samples in aeration units within the treatment facility. Eventually, through much study, it was determined that the amount of nitrification occurring in the aeration units was the cause of both problems.

Traditionally, BOD_5 has been used to evaluate the effects of effluent loading on receiving waters. Whether the nitrogenous demand, as part of the BOD test, should be included in these estimates is debatable. It can be argued that only carbonaceous BOD should be considered, and that the nitrogen loading should be evaluated through determination of specific nitrogen species as part of an overall nutrient loading profile. This perspective carries with it the implicit assumption that carbon oxidation is the only process that irreversibly removes oxygen from the system. The ultimate product of carbon oxidation—carbon dioxide—renders the oxygen unavailable for other purposes. Nitrogen oxidation converts part of the DO into nitrate and nitrite ions. These anions remain in the water body as potential sources of oxygen and delay anaerobic conditions from developing. The counterargument here is that anaerobic conditions, as such, are not the point; rather, concern needs to be expressed over maintenance of a minimum DO level in the receiving water. Because the BOD test is not intended to evaluate nitrogen loading, and because the onset of nitrification in ambient waters is dependent on a number of interrelated parameters, it becomes apparent that the effects of nitrogen loading on a water body need to be studied independently of the BOD test. Absent such evaluations, the BOD_5 is a rather insufficient tool to evaluate receiving water loading, regardless of whether nitrification inhibition is used.

Use of seed material from nitrifying treatment facilities provides a potential source of "spreading" the nitrification problem to other samples of wastes that might be subjected to BOD testing, especially if these wastes are rich in organic nitrogen. Commercially available seed is intentionally devoid of organisms that contribute to nitrification, and its use can help suppress cross-contamination between samples. However, unless the sample is sterilized before the BOD test, nitrifying organisms may well be present even when a commercial seed is used.

Clearly, a means of controlling nitrogen-demand interference in the BOD test is needed. Fortunately, this need, too, has been studied thoroughly for more than a century and has yielded rather simple chemical solutions.

4.0 NITRIFICATION INHIBITION

Nitrification is a two-step process itself, primarily involving two groups of organisms, *Nitrosomonas* and *Nitrobacter*. *Nitrosomonas* obtain the needed energy for growth from the oxidation of ammonia to nitrite; *Nitrobacter* oxidize nitrite ion to nitrate ion. Other organisms that have been found to contribute to nitrification include *Nitrosococcus* and *Nitrosolobus*, which can convert ammonia to nitrite, and *Nitrospina*, *Nitrococcus*, and *Nitrospira*, which convert nitrite to nitrate (Gaudy & Gaudy, 1980; Ingraham & Ingraham, 1995; Tortora et al., 1998; Weinberger, 1949). These organisms generally use carbon dioxide, bicarbonate, or both as the sole source of carbon for cell growth. Both nitrification reactions utilize molecular oxygen. Both groups of organisms metabolize some of the available nitrogen into cellular amino acids, proteins, enzymes, and other materials, so direct stoichiometric calculation of nitrogenous oxygen demand from determination of nitrogen species is not comprehensive. Empirical relationships have been proposed for estimating nitrogenous oxygen demand from nitrogen species (Wezernak & Gannon, 1968) but are probably not universally applicable to all sample types.

Some attempts have been made to separate nitrogenous demand from carbonaceous demand using physical separation of the responsible organisms (Washington et al., 1983) or measurement apparatus manipulations (Symons & Labonte, 1965). On the whole, these approaches are not very successful and can needlessly complicate determinations.

Fortunately, nitrifying organisms are sensitive to physical and chemical conditions and are generally quite a bit more susceptible to the effects of toxic additives than carbonaceous organisms. Brief pasteurization is effective for the control of nitrification in the BOD test (Sawyer & Bradney, 1946). The 10th edition of *Standard Methods* (APHA et al., 1955) contains the recommendation to use pasteurization. Pasteurization works, but has some drawbacks that ultimately led to a search for other solutions. "Flash" pasteurization involves heating samples to 80 °C, followed by rapid cooling to

20 °C. Pasteurization, even in the flash mode, does affect the heterotrophic organisms to some extent (Gaffney & Heukelekian, 1958). Thus, samples are afterwards necessarily seeded with viable organisms from an appropriate source. Temperature and time control are critical, as are the source and amount of seed added. In some laboratories, the use of heating blocks and temperature-control baths is effective for processing large numbers of samples, and use of seeded dilution water provides a reasonable level of controlling seed addition variability. Other laboratories are able to control temperatures better at 60 °C, and inhibition of nitrifiers by extended exposure to 60 °C is just as effective as flash pasteurization at 80 °C. One effect resulting from using pasteurization processes is that resident organisms responsible for carbonaceous demand may also be inhibited, and the CBOD results obtained may not be representative of the actual amount of CBOD that is characteristic of the receiving stream.

In a quest for simpler approaches, various researchers examined chemical additives: methylene blue (Abbott, 1948a, 1948b), sulfuric acid, and ammonium chloride have all been reported as inhibitors to nitrification in the BOD test. Although use of methylene blue as a nitrification inhibitor appeared in the 10th edition of *Standard Methods*, it was dropped from the 11th edition (APHA et al., 1960). Head-to-head comparison between pasteurization and methylene blue in a partially nitrified activated sludge facility effluent shows methylene blue to be almost totally ineffective (see Case Study No. 1 later in this chapter). Increased amounts of methylene blue are also found to inhibit carbonaceous demand (Young, 1973). A comparison of nitrification inhibition using methylene blue, ammonium chloride, or sulfuric acid is available (Nikitine, 1970).

A rather straightforward method of inhibiting nitrifiers is to add sulfuric acid to drop the pH to 2, then neutralize the sample to pH 7 with sodium hydroxide after 15 minutes (Hurwitz et al., 1947). The neutralized samples are then reseeded and analyzed normally. Control of the acidification and neutralization steps can be problematic in poorly buffered water but, in general, this is an easy procedure to use.

Relying on the decades-old knowledge that chlorine addition can lower the oxygen demand necessary to stabilize wastewater, a chlorine addition step to control nitrification has been investigated (Sawyer & Bradney, 1946). In comparative tests with pasteurization, chlorination can control as much as 50 mg/L or more of nitrogenous demand. However, it requires that chlorine be carefully removed with sodium sulfite before setting up the BOD test, and that seed must be added. Excess sulfite ion can interfere with the BOD test, and any chlorine not neutralized is toxic to the seed.

Excess ammonium ion added to the BOD test to a level of approximately 0.1 M can effectively control nitrification (Siddiqui et al., 1967).

However, this method is pH sensitive, being ineffective at pH levels greater than 8. In fact, failure to control pH lower than 8 with strong buffer solutions can allow nitrification to proceed quite well, and the added ammonia under these conditions serves as an additional source of nitrification interference.

Dissatisfaction with all of these alternatives led to continued searches for more effective nitrification inhibitors. Thiourea was found to specifically inhibit *Nitrosomonas* oxidation of ammonia (Painter, 1963). Two other chemical additives have been found, 2-chloro-6-(trichloromethyl)pyridine (TCMP), and allylthiourea (ATU), for the control of nitrification (Young, 1973). Allylthiourea functions by selectively inhibiting the conversion of ammonia to nitrite by *Nitrosomonas*, similar to the action of thiourea (Surmacz-Gorska et al., 1995). As ammonia is generally many more times abundant than nitrite in water, the inhibition of ammonia oxidation serves to restrict the entire process. Selective inhibition of the conversion of nitrite to nitrate by *Nitrobacter* can be achieved with sodium chlorate ($NaClO_3$) (Surmacz-Gorska et al., 1995). Both organisms are inhibited by TCMP.

Allylthiourea is effective for nitrification at the 5-mg/L level (Raff, 1985). In both BOD_5 and 10-day respirometric tests, TCMP often demonstrates effective nitrification control without affecting carbonaceous BOD when added at 10 to 20 mg/L; however, for complete suppression of nitrification, a maximum of 50 mg/L may be needed (Raff, 1985). On the other hand, allylthiourea allows *breakthrough* of nitrification after 10 to 15 days of testing with all dosages tested. It is unclear whether the allylthiourea is degraded during the course of the test or if the nitrifying organisms acclimate to the toxic effects of the chemical. The TCMP has limited solubility in water so that addition of precise amounts to BOD tests is difficult to control. However, the availability of a commercial product in which the TCMP is coated onto a carrier salt makes uniform addition a trivial step and the solubility problems moot (Young, 1973). Although it is concluded that both allylthiourea and TCMP are effective for the control of nitrification in the BOD_5 test, allylthiourea is not recommended for BOD tests that extend beyond 5 days.

It was several years before nitrification inhibitors showed up in *Standard Methods*. Even though mention was made of nitrification inhibition in the 14th edition (APHA et al., 1975), specific methods for control had been dropped from the 12th and 13th editions. Not until the 15th edition (APHA et al., 1980) was TCMP included as a standardized step. However, many laboratories began to use TCMP as a replacement for pasteurization or chemical addition many years before publication of the 15th edition. A direct comparison of pasteurization and TCMP on 35 representative samples of activated sludge secondary effluent, river water, and estuarine water indicated there is no statistically significant differences between the two methods of nitrification inhibition (see Case Study No. 2 later in this chapter for details of these findings). Allylthiourea is not included in past editions of *Standard Methods* because of concerns that its biodegradability makes it a potential contributor to CBOD and because it interferes with the Winkler DO determination. Use of allylthiourea, however, is approved in the European environmental community. In addition, allylthiourea is approved for nitrification inhibition by the Standard Methods Task Group for the 21st edition (APHA et al., 2005) after additional study and recommendations from the Task Group.

Traditionally, it has been believed that raw and primary wastewater contain too few viable nitrifying organisms, and that nitrification is therefore not a concern in the BOD_5 test. Based on interpretations of Method 5210, however, the U.S. Environmental Protection Agency and some states began to require inhibition of nitrification on raw and primary influents. This practice results in underestimation of the strength of these wastes and, perhaps, underestimation of treatment facility design or removal efficiency (Albertson, 1995). In addition, there are assertions that TCMP is actually toxic to the organisms responsible for carbonaceous demand (Albertson, 1993, 1995). The potential for toxic inhibition from TCMP has been examined in detail, and no evidence for toxicity has been found (Young et al., 1997). This finding is consistent with several earlier evaluations (Slayton & Travato, 1979; Young, 1973). The results of this extensive study did show that nitrification takes place in primary wastewater to a greater extent than previously believed. The effects of a little nitrification are amplified when multiplied by the large dilution factors normally used for high-strength wastewaters. The conclusion is that unexpectedly low CBOD results, compared with those obtained from the uninhibited BOD_5 on raw and primary wastewater (Albertson, 1995; Dague, 1981), are probably due to inhibition of previously unsuspected nitrification rather than any toxicity to the carbonaceous organisms arising from use of TCMP (see Case Study No. 3 later in this chapter for a summary of these arguments).

Nitrification control can become an important issue when it comes to understanding the test and interpreting the results. For many applications, it is probably necessary to perform side-by-side determinations of BOD and CBOD. The difference between these two determinations is regarded as the nitrogen biological oxygen demand, and has been interpreted as a measure of the nitrogen-containing pollution in the water. This interpretation may not be as straightforward as it may seem on the surface (Deai et al., 1991), as there are many synthetic organic nitrogen compounds that are resistant to biological rectification, for instance, pyridine bases (Ettinger et al., 1954). In other instances, it is not too difficult to estimate the extent that nitrification is proceeding by analyzing total nitrate and nitrite before and after incubation.

5.0 THE 5-DAY BIOCHEMICAL OXYGEN DEMAND AS A DILUTION TEST

For most wastewaters, the BOD exceeds the amount of DO available in the test bottle from air saturation alone. It is thus necessary to dilute these samples to a strength that is commensurate with the available oxygen, which at saturation is approximately 9 mg/L at one atmosphere pressure and room temperature. There must be more than enough oxygen to meet the demand in each bottle to get a valid result. The dilution method is based on the belief that the rate of biochemical oxidation of organic carbon is directly proportional to the amount of available unoxidized material remaining at any time, as discussed in the first section of this chapter. Belief in this state of affairs in the early days of waste stabilization study was probably one of the most important aspects in development of a stable, reliable test method for measuring BOD.

Initially, the "excess oxygen test," as the dilution method was called, had strong competition from other methods that originally appeared in the first and second editions of *Standard Methods* (APHA, 1905, 1912). Some of these methods, such as the putrescibility test and oxygen consumed test, relied on chemical indicators like methylene blue or potassium permanganate as surrogates of biological activity rather than attempting to measure biological oxygen consumption directly. Early experiments with shaking, bubbling, use of open containers, or pressurization to speed up oxidation process were largely ineffective (Drown, 1892; O'Brien & Clark, 1962). The Royal Commission on Sewage Disposal recommended the excess oxygen method in 1908 (Adeny, 1908a, 1908b, 1908c, 1908d, 1928).

The third edition of *Standard Methods* (APHA, 1917) was the first edition to contain a test for BOD. An improved Winkler method for determining DO had generally been accepted by then, and several studies suggested that, within reasonable bounds, the rate of oxygen uptake was not significantly governed by the amount of DO present (discussed in O'Brien & Clark, 1962). However, the dilution method was slow to catch on until several independent researchers found a link between the quality of the dilution water used and the reliability of the test results. It was pointed out that the oxygen-demand formula in the third edition of *Standard Methods* was useless if no oxygen was available or if the dilution water had a significant demand of its own (Frank & Rhynus, 1919).

Numerous efforts to examine dilution water variables were made (Butterfield, 1929; Eldridge & Mallman, 1931; Garner, 1922; Heukelekian & Chamberlain, 1931; Holderby & Lea, 1934; Theriault & Hommon, 1918; Theriault et al., 1931). It was recommended that dilution water be stored until it had no more demand, and that control blanks be run as well to correct these problems (Theriault & Hammon, 1918). If attention were paid to the dilution water issues, then the oxygen demand was held to be independent of the DO in the incubation bottle, provided that the depletion was maintained between 20% and 90% of the initial DO.

Subsequent editions of *Standard Methods* contained virtually no change in the demand method as it appeared in the third edition. Publication of *Uniform Procedure for Biochemical Oxygen Demand* (FSWA Committee Report, 1932), which was based on several earlier published procedures (Anonymous, 1932a, 1932b; Childs, 1925; Mohlman et al., 1928; Theriault, 1927, 1931), proved to be a defining document for the dilution test. It stated that the oxygen demand of the dilution water for the 5-day test should be less than 0.2 ppm; dilution water should be brought to saturation by storage at 20 °C; the water should be free from interfering substances such as iron or copper salts, chlorine, chloramine, and nitrite; and the pH and mineral content should be favorable to biological growth. Much of this instructional information is extant today in the BOD_5 method.

The latter recommendation about the mineral salts itself grew out of a great deal of study. It had become clear by the time that the FSWA Committee Report was published that distilled water, when used by itself for diluting wastewater and industrial wastes, was not conducive to obtaining reliable BOD results. There was a plethora of papers that examined the effects of added phosphate, bicarbonate, sulfate, and other nutrients and minerals to the dilution water. It was demonstrated that the plain sodium bicarbonate–fortified dilution water in vogue at this time was inadequate for BOD_5, in that the BOD results were only about one-eighth of that produced when dilution water is supplemented with trace amounts of minerals that also included calcium, magnesium, phosphate, and sulfate (Lea, 1941; Lea & Nichols, 1936, 1937). This work also suggested, for the first time, that trace amounts of ammonia were required to round out the nutrients. The addition of 2.5 ppm of ammonium sulfate was a departure from earlier recommendations (Theriault, 1927) and apparently did not gain immediate acceptance. However, in 1941, the *Standard Methods* Committee (Ruchhoft, 1941) described the landmark study of more than 1100 samples that established the buffer components and nutrient mix that is still employed. In this recommendation, the dilution water concentrations listed in Table 7.1 were established. Some discussion was presented about preparing this as two stock solutions, one containing phosphate buffer salts and the other a

TABLE 7.1. Nutrient mineral salts used in the BOD_5 test.

Nutrient	Concentration, mg/L
$FeCl_3$	0.06
$(NH_4)_2SO_4$	1.87
$MgSO_4$	14.7
$CaCl_2$	30.0
$KNaHPO_4 + KH_2PO_4$	47.5

mixture of the iron, magnesium, and calcium salts. However, because the ferric chloride did not store well, a preference was given to using four separate solutions, with the ammonium sulfate added to the phosphate buffer. This preference has been retained for several decades in *Standard Methods*.

6.0 PAY ATTENTION TO THE DILUTION WATER

The modern-day BOD test, with its nutrient and buffer solutions, is the result of a studied effort to optimize the balance of salts necessary for optimum performance. Use of a phosphate buffer, and added magnesium sulfate, calcium chloride, ammonium chloride, and ferric chloride are all essential to running the test in a uniform manner on a wide range of samples. These will ensure that the pH and nutrient balance remains within a range suitable for bacterial growth.

Dilution water should be prepared so that the analyst always has dilution water of proper quality available for testing BOD. The initial choices of water to use for dilution water preparation include tap water, deionized water, distilled or higher-grade reagent water, or ambient receiving waters. However, the source must be free of toxic contaminants that might inhibit test organisms (e.g., chlorine, copper, zinc) and should be free from measurable, degradable organic compounds. Innumerable problems occur in the BOD test when the initial source of dilution water is contaminated, and the analyst will save a significant amount of frustration and grief by paying attention to this point at the outset. With the ready availability of high-quality bottled water or laboratory water purification systems, it makes little sense to routinely use some of the sources listed in *Standard Methods*. Chlorinated feed water, stills, deionizing resin beds, water piping, and contaminated water storage containers are common culprits in poor quality water. The surveillance must not stop here. Once dilution water is prepared, hidden contamination possibilities abound for the analyst. Remembering that we have created a nutrient-rich water suitable for microbial growth and that we may introduce seed organisms to the dilution water, the conditions are right for bacterial growth in dilution water containers and the tubing used to deliver the water to sample bottles (Lloyd, 1937).

Dilution water blanks provide a check on the quality of unseeded dilution water. A bottle of unseeded dilution water should be incubated with each batch of samples tested. The DO uptake of the blank water should not be greater than 0.1 to 0.2 mg/L, based on the difference between the initial and final DO readings. A DO depletion of less than 0.1 mg/L is preferred. Dilution water with a demand greater than 0.2 mg/L is unsuitable and should be discarded.

This point has resulted in one of the most frequently asked questions from analysts: "Why can't I just subtract the blank value from the final DO readings?" The typical answer is shown in the Questions and Answers section at the end of this chapter. Resist the temptation to use poor-quality water, as it will ultimately cause chronic problems.

Dilution water may be stored after preparation, provided that dilution water blank criteria are met. Although a maximum of 24 hours is recommended in Method 5210 B, longer storage is acceptable if the blank criteria are met. Experience has shown that the quality of some ambient waters may be improved by storage, but most other water types are more likely to deteriorate. The method requires that dilution water be at 20 °C before use, with an allowable variation of 17 to 23 °C, to stabilize DO concentrations within a narrow range. It, therefore, makes sense to store dilution water if at all possible at 20 °C to keep it ready to use with a minimum of extra logistical steps.

If the dilution water is not near oxygen saturation when prepared, it should be aerated with a source of clean, filtered air. The use of low-grade compressed air is not recommended due to the frequent hydrocarbon contamination that is found in compressed air cylinders. Air compressors, including the small air pumps used for aquarium aeration, are also subject to hydrocarbon contamination. The preferred technique is to shake vigorously a partially filled container of dilution water and then allow it to settle. Large containers of dilution water can be vigorously stirred with either magnetic or overhead stirrers; however, these techniques require considerably more time to achieve saturation than simple shaking. The ambient air in the laboratory can be a source of contamination and the use of an isolated room with its own supply of conditioned, filtered air can often be beneficial for dilution water preparation.

Stored water should be kept in containers with cotton plugs that allow air saturation to be maintained, but prevent outside dust and microbial contamination. Nitrification inhibitor should be added to stored dilution water to prevent oxidation of the trace amount of ammonia added as a nutrient for the carbonaceous bacteria during the test. This is very important to avoid low bias in the final BOD, which would result from nutrient-deprived seed organisms. This requirement for stored water presents a problem for the analyst not intent on using a nitrification inhibitor on all the samples to be tested and may dictate use of a higher quality water and a protected storage area at the outset to avoid storage issues.

7.0 SEEDING IS IMPORTANT

The addition of viable organisms in sufficient quantity to carry out the test is called seeding. It may be obvious at this point that samples such as raw and primary wastewater and certain polluted ambient waters may not need to have seed added. However, in WRRFs where the effluents have been highly treated, in chlorinated or pasteurized samples, and in industrial wastes that could be devoid of viable organisms due to toxic levels of metals or other inhibiting chemicals, addition of seed is essential.

Early seeding practices included addition of various amounts of wastewater, lake, or river water. As described in the FWSA Committee Report (1932), 1 mL of wastewater or 10 to 20 mL of river water per liter of dilution water were typical. It was pointed out that the type of seeding material used, the method of application, and the quantity dosed had not yet been sufficiently standardized at the time of the FWSA Committee Report but was dependent on available material and the judgment of the analyst (Eldridge, 1933). In a series of experiments, it was demonstrated that, by using different amounts of several different wastewater sources, the BOD would increase in proportion to the amount of wastewater seeded into the sample. It was also demonstrated with standards of sugars that a significant lag in oxygen-consumption rate occurred if insufficient seed was added, the net result being low and erratic BOD_5 results. This time lag could be as much as 2 days with the lowest amount of seed added.

The current suggestions in *Standard Methods* for choosing a seed source include settled wastewater, undisinfected effluent from a biological treatment facility, or mixed liquor from an activated sludge treatment facility. Seed from a biological treatment process is usually preferred because the oxygen uptake from degradation of organic matter should be minimized, although a nitrification inhibitor will likely be needed for samples with reduced nitrogen compounds. Standard seed is available from several commercial sources (Paixao et al., 2000; Young, 1996), which is certified to be free of nitrifying organisms and may offer a consistent quality of seed. However, the analyst must be cautious with such proprietary seed material and pay attention to the glucose-glutamic acid (GGA) check standard results to ensure that sufficient seed amounts are being added from batch to batch of the commercial material.

Some sample sources, especially industrial discharges, may contain organic compounds that are not readily degraded by the organisms found in the recommended seed sources (Gellman & Heukelekian, 1955; Lee & Oswald, 1959; Razzell, 1972). Under these conditions, *Standard Methods* instructs that an acclimated seed is required. Experiments using commercial seed in these situations reveal that good precision (reproducibility) is generally obtained; however, the results are almost always biased low compared with results obtained from an acclimated seed (Fitzmaurice & Gray, 1989). Appropriate seed organisms may be obtained from a biological treatment facility that treats these types of wastes or from a stream segment that receives the discharge from such a system.

Once the effort is made to acclimate a seed, thought should be given to preserving the seed for future use. Simply freezing the seed suspension has been used many times (Hargrave & Tyler, 1965; Lee & Oswald, 1958) and generally is successful. Another technique is freeze-drying, which allows the preserved seed to be stored at room temperature.

Steps to acclimate seed material are given in Method 5210 B for instances when an ambient source is not available. These are somewhat more labor intensive initially, but if particular wastes are tested for BOD on a regular basis, the investment in seed development is worth the effort to the analyst in terms of reliable results. Development of an acclimated seed from a soil suspension of organisms, from gradual acclimation of waste-fed aerated sludge in a beaker, or a commercial seed product should all produce viable seed for running BOD on difficult wastes. The proof, however, is up to the analyst. One way to check the acclimation would be to run successive BOD tests as the seed development progressed and look for an increase in BOD. Another would be to start with determination of the BOD/chemical oxygen demand (COD) for the waste and look for an increase in this ratio over time as the seed became more active with respect to the intractable organics in the sample. For acclimated seed, there is no assurance that the GGA check standard will yield results in the prescribed range. For this reason, in the 15th edition of *Standard Methods* (APHA et al., 1980), the Joint Task Group on BOD recommended (Young et al., 1981) that the analyst consider using a major identifiable compound present in the waste to measure effectiveness of the acclimated seed. Obviously, this requires that some knowledge of the waste is available and that a database of typical BOD response be obtained for the compound used. In normal laboratory operations, these methods should be viewed as special cases for applications in which treatment system development is needed or the same waste stream is tested routinely (e.g., daily). See Question and Answer No. 10 later in this chapter for more discussion on this topic.

8.0 SEED CONTROL MEASURES

Standard Methods section 5210 B provides procedures to allow the analyst to determine the amount of seed that should be added to samples. The goal is to add enough seed to ensure utilization of the carbon in the sample bottle without a significant lag in oxygen uptake so that carbonaceous demand is met and stabilized in 5 days. The analyst is instructed to use a GGA mixture to ensure that sufficient seed strength exists for all but acclimated seed, where other identifiable organics may be more suitable, as discussed in the previous section.

Start by running a BOD_5 on the candidate seed material. This should be a normal dilution test, possibly with some extra dilutions added if the source is not familiar. The highest strength dilution should give at least a 50% DO depletion. All results used for determining seed DO uptake must meet the criteria of at least a 2.0-mg/L uptake, and at least a 1.0-mg/L residual in the respective bottles. One of two methods may be used to determine the seed DO uptake, either a graphical or averaging method.

9.0 THE 5-DAY BIOCHEMICAL OXYGEN DEMAND OF SEED

The amount of seed added to each sample should contribute a 0.6- to 1.0-mg/L DO depletion in the sample or give a GGA result in the range of 168 to 218 mg/L BOD. Initially, the analyst should focus on the 0.6- to 1.0-mg/L DO depletion approach as a starting place to establish seed volumes and only modify the amount if there are apparent problems with obtaining correct GGA values. If different potential sources of seed are being evaluated, it would be advisable to use the one for which the 0.6- to 1.0-mg/L criterion gave the correct GGA response, unless acclimation issues were overriding, as discussed above.

Using the BOD results from each dilution of the seed material run for the seed control, plot on square-ruled x–y graph paper the DO depletion in mg/L versus milliliter of seed in each bottle. This should be a straight line, whose slope indicates the DO depletion per milliliter of seed. The intercept of the DO axis represents the DO depletion of the dilution water, which should be less than 0.1 mg/L.

A second technique described in *Standard Methods* is the averaging method. This is somewhat simpler, but requires that a dilution water blank be run separately to have this information and it will not be readily apparent that a linear relationship exists. For each dilution that met the above criteria for minimum depletion and residual DO in the seed control BOD test, divide the DO depletion (mg/L) by the volume of seed that gives results falling in the allowed deletion range to provide a good average. Ideally, at least three seed control bottles will be within the allowed depletion range to provide a good average. Average the acceptable results to obtain a value for the DO uptake per milliliter of seed. From this average value, the analyst can easily determine the amount of this seed material that must be added to each sample to be in the 0.6- to 1.0-mg/L DO uptake range as specified. The 21st edition version of 5210 B (APHA et al., 2005) refers to this as the ratio method. The average value of the DO uptake per milliliter of seed represents the change, or of DO per milliliter of seed in each bottle. This factor is referred to as S_0 in the calculation section of this chapter.

Seed can either be added directly to each sample bottle or to the dilution water. Method 5210 gives explicit direction on techniques for adding seed to dilution water, or to volumetric containers used for dilution or directly into BOD bottles. Seeding of individual dilution containers or individual bottles avoids any problems that might be encountered from a declining amount of seed with increasing dilution (sample strength) in cases where wide dilution ranges are used. However, within the bounds of the procedures, the analyst can choose which of these seeding techniques is more suitable for routine use. Seed should be added after the dilution container is one-half to two-thirds full to allow sufficient mixing. Nitrification inhibitor, if used, should be added at this stage as well.

The amount of DO uptake due to seed is factored into the calculations using the seed control value initially determined. This is accomplished through use of the ratio, f, of amount of seed in diluted samples to amount of seed in the seed control. See either the Calculations section later in this chapter or a current edition of *Standard Methods* for the correct equations for use of either seeded dilution water or direct seeding of sample bottles.

10.0 GLUCOSE–GLUTAMIC ACID CHECK STANDARD

Let it be stated uncategorically at the beginning that the use of check standards in the BOD test is not to establish accuracy or precision in the traditional interpretation of these terms as used in analytical chemistry. The intended underlying use of check standards is to detect toxins in the dilution water, detect deficiencies in the phosphorous-to-nitrogen ratio in the nutrient additives, and to determine the viability of the seed organisms (Sawyer et al., 1950). Let us repeat that it is not to measure accuracy in the test.

Many different chemicals have been used to evaluate BOD performance, most being various carbohydrates. Early efforts used mixtures of lactose and maltose to evaluate seed performance (Eldridge, 1933). Glucose is a very popular substrate to use in BOD tests; however, it does present problems. Pure glucose solutions are not stable due to the conversion to ethyl alcohol and acetic acid by yeasts as the DO level drops in the storage bottle. Glucose is somewhat erratic in BOD yield in the 5-day test due to these fermentative pathways and, therefore, not completely suitable as a check standard. In a comparison of possible standards to use in the test, glucose metabolism is noted to be stabilized when nitrogen-containing chemicals such as amino acids are present. A mixture of glutamic acid and glucose exhibits the needed stability and has been used as the standard ever since (Sawyer et al., 1950).

Glucose–glutamic acid is readily biodegradable but not completely converted by microorganisms to carbon dioxide, ammonia, and water in analogy with a combustion reaction. The difference between the theoretical (320 mg/L) and observed (250 to 285 mg/L) ultimate BOD values of a 300-mg/L GGA solution is due to partial conversion of the carbon to refractory cellular components that are undigestable by other bacteria during the death and reuse cycling that occurs during the extended incubation (Sawyer et al., 1994).

In the 12th edition of *Standard Methods* (APHA et al., 1965), a GGA mixture is recommended as a check standard for BOD_5. The underlying rationale for the GGA mix is that it gives an empirically derived rate constant similar to that of domestic wastewater, it is a stable mixture, and it provides a consistent 5-day utilization value. The BOD_5 GGA check standard mixture has a 300-mg/L concentration with equal weights of glutamic acid and glucose. A series of laboratory studies (U.S. EPA, 1986), involving more than 100 laboratories using many different dilution waters and seed sources provides the foundation for the standard BOD_5 mean value of 198 mg/L, with a permissible range of 167.5 to 228.5 mg/L for the GGA solution.

The GGA value is affected by the number of viable organisms added in the seed. This leads to the recommendation to adjust the amount of seed needed to achieve this range. The amount of added seed may need to be adjusted up or down to generate values within the acceptance range of the standard.

The BOD_5 procedure specifies using a 2.00% dilution of the GGA standard, which makes the concentration of GGA in the BOD bottle 6.00 mg/L. This GGA concentration requires a 5.08- to 6.92-mg/L oxygen demand to be within the check standard criteria. Editions of *Standard Methods* before the 20th (APHA et al., 1998) are perhaps unclear on this point. The availability of commercial GGA mixtures at different concentrations than 300 mg/L has caused some confusion in various laboratories regarding how much should be added.

It is important to realize that the GGA check standard is not intended to be an accuracy check of the method. Rather, GGA serves as a positive check on the bioassay organisms to ensure that they are viable in the proper range of activity and that the test conditions (dilution water quality, temperature, test setup, etc.) are acceptable. There is not a recommended way to determine the accuracy of the BOD_5 test per se. Inability to provide either an adequate biomass in terms of numbers of viable organisms or maintenance of consistent test conditions will result in certain bias and increased variability. Due to the great amount of variability that is inherent in a bioassay, the GGA result is intended as an indicator of the possible need to adjust the test procedure. This is fundamentally different from the traditional interpretation of quality control for a chemically based test, in which results outside the acceptance range indicate a lack of exactness in the application of the procedure. Due primarily to the ever-changing state of the seed, a GGA test is required with each batch of samples.

11.0 SAMPLING

The BOD test is performed on a sample that is collected from any of a number of locations, such as final effluent points, oxidation ditches, activated sludge basins, and facility influents. The success of the analysis hinges to a great extent on how the sample is collected, preserved, and transported to the laboratory. Sampling considerations have been discussed in detail (Smith, 1999a).

The starting point for a sampling event always has to be the laboratory, in that there must be coordination between the laboratory and the samplers so that delay between the moment of sampling and the initiation of analysis is minimized. This is particularly important when commercial laboratories are involved, as they tend to set up BOD tests only on certain days of the week, for instance, Wednesday and Friday. Samples collected on a Monday and sent to the laboratory could well sit around until Wednesday, thus exceeding maximum holding times and invalidating the sample result.

In addition to closely scheduling the sampling, the laboratory has to be involved in the selection and preparation of the sample containers. Sample containers should be constructed of glass. There are data supporting the often-held belief that plastic containers should be avoided (Theobald, 1959) for two reasons. The first reason is the possible leaching of biodegradable substances from the container by the sample, thus increasing the BOD result for the sample. The second reason is the absorption of organic materials from the sample into the walls of the container, thereby decreasing the BOD of the sample. These two effects are not equal in magnitude, and it is not correct to assume that they cancel each other out.

The cleanliness of the sample container is a key area of concern. The common practice of simply rinsing the container out with sample before filling the container cannot be defended as an appropriate technique to use in lieu of regular direct determination of the cleanliness of the containers through the analysis of container blanks. If sample containers are reused, an unsuitable practice to begin with, the containers must be cleaned between each use. Appropriate cleaning techniques for sample containers are the same as those employed for the BOD dilution bottles, that is, washing with detergent, rinsing with reagent water, acid, and reagent water, then air drying. The cleaning procedure must be checked periodically by determining the BOD of a cleaned container blank. The introduction of BOD-inhibitory substances to the containers by the cleaning process needs to also be checked periodically by performing GGA determinations in dilution water stored in representative cleaned containers.

There is no recommended preservation in *Standard Methods* for BOD samples other than cooling to 4 °C. The question of preservation has been examined several times, and even the simple expedient of cooling has been found to contribute to nonrepresentative results (Carpenter & Berger, 1978; Funk, 1977). The addition of borax for preservation of dairy effluent samples has been examined (Jasewicz et al., 1954); however, it is not recommended as a general technique. Chemical means of dechlorination are discussed in *Standard Methods* 5210 B; however, they should be performed in the laboratory rather than in the field.

12.0 THE PROCEDURE

So, by now, our intrepid analyst is ready to run samples. Seed source has been characterized and seed prepared, dilution water readied, performance checks with GGA demonstrated, glassware cleaned, samples collected, and all reagents prepared according to the latest edition of *Standard Methods*. What could go wrong? What have we overlooked? Well, at this point, it would be useful for the analyst to know as much as possible about the samples to be analyzed. Within a WRRF, or perhaps in a commercial laboratory serving a small variety of clients, this goal might be reasonably easy to reach. However, in large laboratories that provide analytical services for a wide range of samples or clients, the reality is that little may be known about the samples. For such cases, it will prove prudent, if not mandatory, to perform a few tests on the samples for which no background information is available.

The minimum of tests needed include determinations of pH and residual chlorine. *Standard Method* 5210 B gives specific instructions for pH and residual chlorine testing. Additional information concerning specific oxidants or reductants may be necessary for certain industrial waste streams, as well as the presence of potentially toxic metal or organic components in the source. Of course, the classical interference from nitrification must be considered, and a decision made on whether it is appropriate to add an inhibitor. This latter decision may be driven by nontechnical parameters, such as regulatory requirements or client specification. Characterization of the sample's DO concentration, temperature, and an approximation of the BOD itself will facilitate selecting dilutions efficiently. Our analyst at this point should have sufficient information to perform modifications to the test procedure, if needed.

Samples found to be outside the pH range of 6.0 to 8.5 need to be neutralized to a range of 6.5 to 7.5 (see Questions and Answers section later in this chapter), and the neutralization step should not dilute the sample by more than approximately 0.5%. Although the dilution water is buffered, it is important that the highest sample strength dilution not overpower the buffer capacity of the added dilution water. All pH-adjusted samples must be seeded.

All samples that exhibit residual oxidants must be neutralized to prevent toxicity to the test organisms. Chlorine left over from disinfection of the wastewater is one of the most common oxidants encountered; however, there are others that can occasionally be present in the sample. These alternate oxidants may or may not display the toxicity of chlorine; however, all are treated as if they are toxic. This step requires that a residual chlorine test be run on a portion of the sample initially. The test chosen may be selective for residual chlorine, such as a N,N-diethyl phenylene diamine-based technique (*Standard Methods* 4500-Cl F or G); however, a more general test for residual oxidants such as an iodometric titration (*Standard Methods* 4500-Cl B or C) will also provide sufficient information. If letting the sample sit for 1 to 2 hours does not cause the apparent oxidant to dissipate, the analyst must use chemical methods to remove residual oxidant. This requires determination of the exact amount of dechlorinating agent to be used from the residual oxidant test. Method 5210 B gives a "cookbook" method for the analyst to do this, using sodium sulfite as the dechlorinating reagent. Although other neutralization steps may be used in specific situations, this simple procedure is the method of choice. As sulfite ion will itself exert an oxygen demand, the analyst must be careful not to "overdose" the sample during dechlorination. All chlorinated samples must be seeded after neutralization, and this is a prudent course to follow for any sample exhibiting residual oxidant.

Usually, samples will be close enough to DO saturation to not need any aeration. A slight supersaturation or undersaturation will normally be resolved during addition of dilution water and temperature stabilization (FWSA Committee Report, 1932). The physical nature of some waste streams may render them greatly supersaturated at the time of sampling. Suspect this problem in very cold samples, in samples from algae oxidation ponds, and in samples subjected to ozone or hydrogen peroxide treatment. If the DO has not equilibrated at test time by simple temperature control, the initial DO of the dilutions will be too high, and there is a risk of oxygen outgassing after the test is set up and placed in the incubator. This loss would be counted as demand and lead to erroneous results. Shaking or aeration with clean nitrogen gas or air may be necessary at this point.

Samples that have been collected or transported on ice, cold-weather ambient water samples, and some process waters may be unusually cold when delivered to the laboratory and will not reach 20 °C in a timely manner. Likewise, some arid region waters, treatment facility effluents, cooling tower, or other process streams may be significantly warmer than the necessary 20 °C test temperatures. To minimize problems with initial DO in setting up the samples, normalize initial starting temperatures for all dilutions and prevent adverse conditions for incubator temperature control, 5210 B recommends that samples be adjusted to 19 to 21 °C before dilution. This is most conveniently accomplished by use of a temperature-controlled water bath and allows the analyst to take care of other matters while samples are approaching the appropriate temperature.

Most samples will require that a number of dilutions be prepared. Two points are important here: the approximate BOD of the sample and the allowable depletion range of the test. Dating back to the FWSA Committee Report (1932), it has been known that BOD values based on oxygen depletion of less than 2 mg/L are not reliable. By the time seed correction has been applied and dilution factors included, the statistical error propagation can be large. It has also been learned and relearned that BOD response is not very significant at DO levels in the range of 0.5 to 1.0 mg/L. Therefore, it is important that at least 1 mg/L DO remain at the end of the test to ensure carbonaceous demand is met. For nitrified effluents, it has been reported that at least 2 mg/L residual is desirable (FWSA Committee Report, 1932), although this is not a recommendation that has been retained in *Standard Methods*.

Given an initial DO of 9 mg/L, a seed DO uptake in the 0.6- to 1.0-mg/L range, and either the 2.0- or 1.0-mg/L recommended DO residual, it is necessary to have dilutions that will deplete between 2 and 7 mg/L of

DO in each bottle. From these numbers, it is easy to calculate the approximate BOD range that any given dilution scheme can cover. For example, a 100% sample (i.e., no dilution) will cover the range of < 2 to 7 mg/L; a 10% sample, 20 to 70 mg/L; a 1% sample, 200 to 700 mg/L; and a 0.1% sample, 2000 to 7000 mg/L. The analyst can interpolate any number of convenient percentages or dilution factors from this relationship and put them in a table for easy reference.

Other techniques for estimating the appropriate dilution are available. The analyst can determine the COD of the sample and then use the following formula:

$$\text{Sample, mL} = \frac{1800}{\% \times \text{COD}}$$

where % is the on-site determined relation between BOD and COD, to arrive at the approximate amount of sample that needs to be added to the dilution bottle (Brown et al., 1973). Methods using nomographs have been published (Horne, 1954; Wolf, 1950) for determination of needed dilutions based on an estimated BOD of the sample. Single-dilution techniques may be employed to estimate the BOD, then the estimated value used to establish the needed multiple dilutions (Orford et al., 1953). A short-term, 1- or 2-day BOD can be used to arrive at an estimate (Orford & Matusky, 1959). A quick estimation of the BOD can be accomplished by respirometry (Gellman & Heulekekian, 1951).

For samples with well-known characteristics, the analyst may be allowed a convention that only requires a single dilution. Under such circumstances, it is typical to run these in duplicate or triplicate. For totally unknown samples, three to five dilutions (*Standard Methods* recommends five) are typically set up to cover between two and three orders of magnitude in possible BOD levels, in the hope of getting two or three valid dilutions from which BOD may be calculated.

When some knowledge of approximate BOD strength exists, perhaps through past experience or by estimation from COD data, a useful strategy is to select a dilution factor that brackets the approximate range. Then apply a factor of two on either side of the primary factor for the other two dilutions. For example, if the BOD of an industrial sample is expected to be approximately 500 mg/L, a 1.0% sample mixture would encompass a 200 to 700 mg/L range. Selection of 0.5% and 2.0% sample dilutions would then provide a test range from 100 to 1400 mg/L BOD. Preparing two additional dilutions for a total unknown BOD, say 0.25% and 5.0%, would extend the useable range to 40 to 2800 mg/L. *Standard Methods* gives suggestions on the approximate ranges to bracket for a number of sample types, and the analyst can calculate the most useful dilution series within each range.

If nitrification inhibition is determined to be necessary, follow the recommendations in the current edition of *Standard Methods* or the instructions provided by the supplier of any proprietary inhibitor. Pure TCMP is not readily water soluble, so addition of the powdered reagent is not very effective. Commercial and "homemade" reagents made by coating TCMP onto a soluble saltlike crystalline sodium sulfate decahydrate can be easy to use, but attention must be paid to getting the proper dosage. Allylthiourea is more water soluble and should not present this type of problem for the analyst. However, the analyst needs to be aware of the potential limitations of allylthiourea as discussed above.

Standard Methods does not, at this writing, recommend that all nitrification-inhibited samples be seeded. However, as discussed above, there have been enough reports in the literature regarding problems with nitrification in wastewater and treated effluents (Albertson, 1995; Young, 1973; Young et al., 1997) that the analyst should rightly be wary. A significant portion of the viable bacterial population in such samples may be composed of nitrifiers. Adding an inhibitor could leave such samples short on viable organisms in the BOD test. Therefore, it is suggested here that the analyst give serious consideration to seeding samples that have been treated with a nitrification inhibitor. Refer to Case Study No. 3 later in this chapter for more details on this subject.

13.0 PREPARING DILUTIONS

At this point, our analyst is more intrepid than inquisitive, and more than ready to dispense samples. Except now, a decision has to be made whether to prepare dilutions in graduated cylinders, volumetric flasks, or the BOD bottle. As a practical guide, the analyst should use the best grade of glassware commensurate with the nature of the sample and apply the best technique possible. The current procedure requires using Class A volumetric glassware, that is, volumetric flasks, graduated cylinders, transfer pipets, and measuring (Mohr) pipets (Smith, 1995), for each step. Note that *Standard Methods* specifies glassware, not stamped plastic laboratory ware, which is unavailable in Class A standards. It is strongly suggested that calibrated BOD bottles be used and that the calibration be consistent with Class A volumetric standards (Smith, 1995).

There may be nothing sillier than watching an analyst new to the BOD laboratory trying to pipet 5.00 mL of raw wastewater or mixed liquor through a narrow-tip Class A volumetric or Mohr pipet that is normally used for chemical reagents. While the narrow-tip pipet may work well for tertiary effluent or clean river water, use measuring pipets with wide-bore tips for wastewater and other heterogeneous samples. Wide-bore measuring pipets, even when constructed of glass, are not calibrated to the Class A standards. As most samples do contain some particulates that will clog the commercially available Class A volumetric or Mohr pipets, the analyst needs to verify the calibration of the wide-bore pipets that will be used (Smith, 1995). After an analyst has gone to the trouble of calibrating their wide-bore pipets and BOD bottles, these should be clearly marked and segregated from all other glassware used in the laboratory. The increased reproducibility that comes from use of calibrated glassware is well worth the initial investment of time.

Match the size of the dilution vessel to the sample as well. Do not attempt to pipet 0.50 mL from a 25.0-mL total capacity measuring pipet. When working with very small volumes of sample to prepare high dilution levels, good analytical technique argues that it is more reproducible to serially dilute larger and more representative samples in Class A graduated cylinders or Class A volumetric flasks than it is to rely on smaller volumetric pipets. As a guide, when diluting samples greater than 100:1, prepare the initial dilution of the sample in a larger flask or cylinder, then make final dilutions into a calibrated BOD bottle. This will minimize error propagation from use of very small quantities of heterogeneous samples. Ultimately, the analyst must bear responsibility for the techniques used, so our advice is to pay attention to these issues, check out the performance, and be consistent in application of techniques to specific sample types. There are some other advantages to preparing dilutions in separate containers, especially where wide ranges in dilution factors may be used for seeded samples. Seed added directly to the container (BOD bottle, graduated cylinder, or volumetric flask) instead of to the dilution water will prevent any effect that might occur from a declining seed ratio with declining amount of dilution water.

One technique that has proven very useful in preparing dilutions of a sample that is repetitively analyzed involves a 1- or 2-L separatory funnel (Swango & Lamb, 1950). The separatory funnel is calibrated to Class A standards by gravimetric determinations (Smith, 1995, 1999a). Sufficient sample to prepare all of the dilutions is added to the funnel by weight or volume, then dilution water is added to a calibrated mark. The first set of BOD bottles is filled with part of the contents of the funnel, then additional dilution water is added to another calibrated mark to make the second dilution. The second bottle set is filled and the process continued by adding more dilution water and filling bottles until all of the dilutions have been prepared. The advantage to this process is that sample is measured only one time and in an amount much greater than required in any single bottle. This greatly improves the measurement accuracy and substantially improves precision.

14.0 INITIAL DISSOLVED OXYGEN MEASUREMENT

Over the years, various publications have dealt with the issue of immediate dissolved oxygen demand (IDOD) and the possible effects it might have on determining both the initial DO and final BOD in sample dilutions (O'Brien & Clark, 1962; Sawyer et al., 1994; and Young et al., 1981). Before advent of the DO probe, the chemical procedures for determining initial DO in the sample were fairly time consuming and sometimes prone to matrix interference. Practitioners therefore sought shortcuts to avoid measuring the initial DO in every dilution. As a consequence, especially in high-strength wastewater, biological process wastes, or chemical wastes with rapidly changing chemical demand, there was concern about the loss of DO in the 10 to 15 minutes that might elapse between setting up dilutions and measuring the initial DO. The FWSA Committee Report (1932) suggested measuring DO in both the sample before setup, as well as in the dilution water, and then determining the initial DO arithmetically. Then, if the computed value were different from the initial DO, a sample was determined to have IDOD, and analysts were instructed to use the computed value for initial DO. When several dilutions were set up, it was allowable to just test the initial DO in one of the dilutions and calculate the value for the other dilutions as long as there was no significant IDOD. Subsequent practices were recommended, based on the dilution factor (Sawyer et al., 1994), in which sample dilutions of less than 20% sample were aerated to saturation then assumed to be at the same DO as the dilution water. In this scenario, DO still had to be measured directly on the full-strength waste for dilutions greater than 20%.

By the time the 14th edition of *Standard Methods* was published (APHA et al., 1975), the instructions for dealing with IDOD had become rather convoluted. Furthermore, the introduction of DO probes revolutionized the way BOD samples were set up and analyzed. The 14th edition instructions for dilution setup and calculation allowed BOD to be calculated using either the initial DO at time zero, or the DO 15 minutes after the dilutions were prepared. One of these alternatives made IDOD part of the BOD, whereas the other did not. The 15th edition of *Standard Methods* (APHA et al., 1980) dropped reference to the IDOD for a number of reasons, including its arbitrary definition and application, limited use, and questionable accuracy in most cases (Young et al., 1981). It cited a lack of any direct evidence that the

15-minute timeframe stated in the 14th edition was sufficient to ensure that high concentrations of chemical interference such as ferrous salts, sulfite, sulfide, or others were completely oxidized. No separate measurement of IDOD has been included in *Standard Methods* since the 14th edition; however, the current method instructs that the initial DO measurement should be taken immediately after preparation of each dilution for samples that are suspected of having any materials with a rapid demand. Any demand that might occur within the first few minutes is thus included in the BOD result. For samples without such concerns, the method only requires that DO be measured within the first 30 minutes. However, with the use of the DO probe, there is really no reason to delay measuring the initial DO.

15.0 INCUBATION TEMPERATURE

Samples should be incubated at 20 °C for 5 days with an acceptable range of 19.0 to 21.0 °C. Temperature control within the incubator is critical. Experience has shown that the BOD can change approximately 5% for every 1 °C change in incubation temperature in BOD_5 (FWSA Committee Report, 1932). In laboratories processing many samples per day, a BOD incubator may be opened and closed several times. Although one should expect that there would be a small recovery period each time the box is opened and closed, it is important that the samples going into the box be uniformly close to the controlled temperature range. This is especially true when large incubators are used and when many samples are taken in and out daily. In general, the air temperature inside the incubator responds to equilibration faster than the filled BOD bottles. For this reason, as well as for initial control of oxygen saturation during setup, *Standard Methods* has specified fairly narrow tolerances on the temperature of the dilution water and samples.

16.0 SEALING SAMPLE BOTTLES

The contemporary BOD_5 procedures call for sealing the BOD bottles. This practice goes back to the 1920s and 1930s (FWSA Committee Report, 1932). The purpose is to prevent the tapered glass stoppers from drying out and allowing oxygen exchange with the atmosphere inside the incubator. Early attempts to keep the incubator humidified to prevent drying were successful, but clumsy. Equally clumsy was the technique of keeping the BOD bottles completely immersed in a water bath. The most common means of providing a seal now are either water seals or plastic collar caps. Some analysts choose to use both. Currently, *Standard Methods* recommends the water seal technique, which requires use of specially designed BOD bottles with a flared mouth. These are routine laboratory supplies from many scientific equipment vendors.

17.0 CALCULATIONS

Following the 5-day incubation period, the final DO is to be determined in each sample dilution bottle. For each dilution that met the 2.0-mg/L initial DO depletion and 1.0-mg/L residual DO requirements, the analyst should calculate the BOD result. A description of the mathematics of the BOD test is available (Gifft, 1943). The contemporary methods of calculation are relatively streamlined, compared with early versions of the method (FWSA Committee Report, 1932). The analyst may want to streamline routine BOD calculations further by setting up a computer spreadsheet that can act as an electronic bench sheet for tracking sample information as well.

Two equations are used for calculation of BOD: one for unseeded samples, the other for seeded. If the sample is not seeded, the form of the equation is

$$\mathrm{BOD_5,\ mg/L} = \frac{\mathrm{DO_{init}} - \mathrm{DO_{final}}}{F}$$

where

- $\mathrm{DO_{init}}$ = initial DO in BOD bottle immediately after preparation;
- $\mathrm{DO_{final}}$ = final DO in BOD bottle after 5 days of incubation; and
- F = fraction of sample in dilution, expressed as a decimal, for example, 10% = 0.10.

If the sample is seeded, the equation has the form

$$\mathrm{BOD_5,\ mg/L} = \frac{\mathrm{DO_{init}} - \mathrm{DO_{final}} - R[S_{\mathrm{init}} - S_{\mathrm{final}}]}{F}$$

where

- S_{init} = initial seed control DO before incubation;
- S_{final} = final seed control DO after 5 days of incubation; and
- R = ratio of seed in each sample dilution to seed in seed control.

The ratio R can be calculated either in terms of percent for seeded dilution water:

$$R = \frac{\text{Seed in sample dilution, \%}}{\text{Seed in seed control, \%}}$$

or volume of seed added to BOD bottles

$$R = \frac{\text{Seed in sample dilution, mL}}{\text{Seed in seed control, mL}}$$

Obviously, then, the analyst must keep accurate track of all of this information from the outset. This latter equation can cause some difficulty if multiple dilutions are used in the seed controls set with a batch of samples. The analyst may not know which seed control dilution was associated with each individual BOD dilution bottle. The 21st edition version of 5210 B (APHA et al., 2005) provides a solution:

$$\text{BOD, mg/L} = \frac{\text{DO}_{\text{init}} - \text{DO}_{\textit{final}} - S_o \times V_s}{F}$$

where

- V_s = seed in each respective BOD bottle, mL, and
- S_o = oxygen uptake of seed in respective bottle.

Note that S_o is a Δ DO per milliliter seed added per bottle

The oxygen depletion of dilution water blanks is not to be subtracted, as discussed in the dilution water section of this chapter.

Currently, *Standard Methods* does not contain any statistical recommendations on how to deal with the results from the individual dilutions of a given sample, except to average them. Under normal circumstances, the analyst should simply average the results from all dilutions that meet the minimum depletion and residual DO stipulations. However, if the differences are greater than 5% to 10%, the analyst should take a close look at the data. If the discrepancies are between dilutions at opposite ends of the dilution range, the dilutions giving the most oxygen depletion would ordinarily be considered most reliable because any errors contributed by the blank DO or other sources would be minimized. Of course, this statement is predicated on preparing samples that require great dilution as stock solutions in larger containers, as recommended in the procedure. Some narrative regarding toxicity, in particular sample dilutions, is presented, but no guidance is provided in *Standard Methods*. The reader can refer to the Questions and Answers section at the end of this chapter for more information on this special case of discrepancies in dilution results.

There are a number of alternate techniques for calculation of BOD results, none of which is in compliance with the formulas described in *Standard Methods*. Graphical methods have been proposed (Herman, 1956). A graphical method has been described that corrects for seed variation and dilution water blanks greater than 0.2 mg/L (Woodring & Clifford, 1988). Another formula was devised for use with seeded high-strength samples (Akerlindh, 1953). These are described in detail in the chapter on BOD theory. Still another paper described a BOD calculator (Bruner, 1943). Simplified techniques for analysis of BOD data have been described (Moore et al., 1950; Ruchhoft, 1950). As a summary of these alternate calculations, it bears repeating that they are not in compliance with the formulas presented in *Standard Methods* 5210 B, although in certain situations they may prove useful.

18.0 QUALITY CONTROL

The BOD procedure contains a number of parameters that should be tracked for quality control purposes. Seed control data, dilution water blanks, and GGA check standard results give the analyst valuable information regarding the routine status of major elements of the test. Further, the analyst must ensure that data are within specifications and look for trends that might suggest an imminent problem. Incubator temperature and water bath temperatures should be checked at least daily to ensure that tolerances are met. Dates on buffer and nutrient solutions, as well as observations like pH, discoloration, and turbidity can provide useful information if problems arise. If commercial products such as seed or nitrification inhibitors are used, then lot numbers, dates, and data relevant to their usage should be cataloged for easy reference.

The DO membrane electrode should be used in accordance with *Standard Methods* 4500-O G (APHA et al., 1998). If a titrimetric method is used, follow *Standard Methods* 4500-O C, the azide modification of the Winkler method. Some states require the electrode to be calibrated against an independent method on a periodic basis; method 4500-O C is the appropriate procedure. Follow all quality control guidelines given for these methods and retain the data for easy reference.

The use of spreadsheets for data entry and calculation can provide rapid calculations without the attendant error propagation associated with manual data transposition and hand calculations. These tools also provide convenient storage and retrieval capabilities for all of the important data associated with the test. The analyst should verify that the calculation routines in spreadsheets have been validated against hand calculations before implementation in routine analysis. Also, the analyst should provide data backup protocols to avoid record losses in case of equipment malfunction.

New analysts should be thoroughly trained; in fact, "indoctrinated" is probably a better term for the training needed in the BOD test. The test is so different from most classical chemical test methods that a new analyst needs to understand how critical the techniques are to the outcome of the test. Use of in-house written standard operating procedures to augment *Standard Methods*, additional resources such as may be found in this book, and the benefit of tutoring from an experienced BOD analyst can all help ensure that test performance is satisfactory from the outset. Analysts should ensure that they can prepare acceptable quality dilution water, seed material, and GGA reference data before running samples. Many regulatory laboratory certification programs require the new analyst perform an initial demonstration of ability with GGA samples before performing analysis of any samples that are reported for permit compliance purposes.

Glucose–glutamic acid is used for ensuring that seed strength and viability are adequate and that the dilution water and techniques are in control. The current acceptable range for a 300-mg/L GGA mix is 167.5 to 228.5 mg/L for any given run. This is based on long-term comparisons from more than 100 laboratories. However, the analyst can gain much more quality control information from the ongoing GGA results obtained from single laboratory evaluations. This is typically done with control charts (*Standard Methods*, section 1020 [APHA et al., 1998]; see also Smith, 1999a, 1999b), which provide objective trends and warning limits to the astute analyst. Such information can allow problems to be headed off before they occur. Also, because single-laboratory performance on GGA should be less variable than the multiple-laboratory database, the analyst is assured that the single-laboratory data are well within the range specified in *Standard Methods*.

Some laboratory quality assurance programs require that 5% to 10% of all BOD samples be duplicated for control chart purposes. An issue that frequently arises is how this should be done. Some analysts just assume that, because they are already running several dilutions, this will suffice. It will not. A true replicate sample is one that has been prepared, from start to finish, at the identical dilutions, with the same technique, in the same analytical batch. From these data, the analyst can establish warning and control limits for test precision, independent of the dilution factor. The within-sample precision of different dilutions is an entirely different matter, as discussed earlier in this chapter.

The literature has suggested that use of a nitrification inhibitor will yield a lower GGA check than the range prescribed in *Standard Methods*. Long-term data accumulated by the U.S. EPA (1986) provides additional leverage to the assumption that this is "normal." Nitrification-inhibited BOD_5 results ($CBOD_5$) of the 300-mg/L GGA solution are reported to have a mean value of 164 mg/L, with an acceptance range of 133.3 to 194.7 mg/L (Smith, 1999a, 1999b). This range is used in several environmental laboratory certification programs (those of South Carolina and Florida can be cited as examples) as quality control standards for $CBOD_5$. Study of this issue has concluded that these observations are a result of inhibiting nitrifiers in the seed rather than any fundamental suppression of heterotrophic bacterial metabolism of GGA (Young et al., 1997). The solution is to use a higher seed amount until the GGA for the inhibited procedure is in the control range for uninhibited BOD_5, thereby ensuring that there is a sufficient population of viable carbonaceous organisms.

With careful attention to detail and strict adherence to quality control indicators, it is possible to obtain very reproducible results with the BOD_5 test (National Council for Air and Stream Improvement [NCASI], 1985). A study involving 10 laboratories demonstrated that total error in the procedure could be reduced to less than 1 mg/L (Analytical Quality Control Committee, 1985).

19.0 LIMIT AND RANGE OF BIOCHEMICAL OXYGEN DEMAND RESULTS

Finally, we come to the issue of detection level. This chapter will avoid the conventional and controversial calculation of a precision-based method detection limit (MDL), following U.S. EPA protocols (Smith, 1999a; U.S. EPA, 40 CFR 136, Appendix B). The reader will undoubtedly pardon the authors for not encouraging a process that requires setting up a minimum of 35 dilutions for the 7 replicates it takes for a single iteration of the MDL process. Such a process with the BOD test is tedious and next to meaningless in terms of the test itself. Instead, we will cut to the salient issues for establishing a detection level for this test. A detection level of 2 mg/L is fixed by the requirement for a minimum DO depletion of 2.0 mg/L. Therefore, per the method, an undiluted sample must have a DO depletion of 2.0 mg/L to be valid. Period. Detection values reported less than 2 mg/L are not possible. The sensitivity of the DO analytical method, its precision, and any interference near the sensitivity limitations of the DO test itself are not factors in this determination of detection level.

The working range of the test is dependent on two factors: (1) the dilution factor(s) of the sample, and (2) the difference between the maximum initial DO (i.e., 9 mg/L), the minimum depletion, and the minimum DO residual of 1.0 mg/L specified in the test. Therefore, the allowable DO difference, 6 (from 9 − 2 − 1), divided by the lowest and highest dilution factors for a given sample, define the working range for that sample.

20.0 ULTIMATE BIOCHEMICAL OXYGEN DEMAND

Ultimate, or total, BOD is measured by extending the incubation time beyond 5 days (Ademoronti, 1987; Stamer et al., 1979; Walden et al., 1972; Whittemore & Hovis, 1982). Incubation for 20 days is commonly believed to be sufficient; however, biodegradation studies for determination of ready biodegradability normally are performed over a 28-day incubation period. The ultimate BOD procedure described in the 20th edition of *Standard Methods*, 5210 C (APHA et al., 1998), recommends incubating the samples from 30 to 60 days or longer.

The procedure describes preparing either one large container (2 L or larger) or at least six replicates of small containers (300-mL BOD bottles) of a single dilution of the sample. Nutrients and minerals may be added to the sample, depending on the intent of the test. A dilution water blank may be used as a control. The initial DO is measured. During the incubation period, the oxygen uptake is determined by measuring the DO every 2 to 5 days. A common practice is to add marbles to the container to replace sample displaced by the insertion of the DO probe. Method 5210 C recommends maintaining a reservoir of extra diluted sample and replacing lost sample from the test container with sample from the reservoir.

If the oxygen level approaches 2 mg/L, the sample is reaerated and the test continued. Reaeration is accomplished in the large single container by shaking the sample in the presence of air. If replicate small containers are used, the replicates are combined, the combined sample reaerated by shaking, then the small containers refilled (Elmore, 1955).

The end of the test is determined by graphing the accumulated oxygen uptake against time. When the oxygen-uptake rate approaches zero, the test is completed.

The calculation of the ultimate BOD is similar to that used for the 5-day procedure; however, the effect of the dilution water blank can be factored into the result. *Standard Methods* 5210 C should be consulted for more details on the technique.

21.0 AUTOMATION OF THE BIOCHEMICAL OXYGEN DEMAND TEST

Within the last 20 years, there have been great strides made by scientific equipment suppliers toward development of computer-controlled instruments that automate various analytical procedures. The BOD test has been subject to these efforts, and several automated and semiautomated instruments are available (Mellbin, 1990; Michalik et al., 1991).

The use of these instruments is not specifically discussed in *Standard Methods* 5210 B; however, their use is not precluded, either. As a general statement, it is up to the user to determine whether the procedure exhibited by any particular automated instrument is in compliance with the method presented in *Standard Methods* 5210 B. The limited experience of the authors with these types of instruments is that they do not comply with *Standard Methods* 5210 B, particularly in the areas of calculation. Two specific areas of concern are the inclusion of the dilution water blank result in the calculation and the way the seed depletion factor is treated. It is to be hoped that manufacturers of automated BOD instruments will, in the future, strive to comply with the regulatory agency–approved methodology, as this is the standard by which their customers are judged.

22.0 CONCLUSION

The BOD test is a bioassay rather than a chemical analysis. Being a bioassay, it contains many more factors that affect variability and error in the final result. The BOD test is extraordinarily sensitive to conditions in and around the BOD bottle, and is very unforgiving of sloppy technique. The test can produce results that are reproducible; however, the reproducibility is directly related to the analyst's level of attention to detail in exactly following the standard procedure.

The BOD test is not going to go away, despite all the controversy and complaining that surround it. Analysts need to recognize this fact and strive to master all the minutiae that accompany correct performance of BOD measurements.

23.0 CASE STUDIES

23.1 Case Study No. 1: Comparison of Nitrification-Inhibition Method

23.1.1 Summary

Direct comparisons of pasteurization, methylene blue, acidification, and chlorine addition to control nitrification in samples from a partially nitrified activated sludge showed that pasteurization was the most effective and consistent of these methods.

23.1.2 Background

This project began as a result of chronically high BOD results in the effluent of a new activated sludge treatment facility. Based on the idea that the COD and suspended solids data would predict a much lower set of BOD results, the laboratory staff proceeded on the assumption that nitrification was the source of the problem. This work, done circa 1964, had to cope with a lack of firm direction in the 9th to 11th editions of *Standard Methods* (APHA & AWWA, 1946; APHA et al., 1955; APHA et al., 1960) regarding nitrification. The ninth edition made no mention of nitrification in the BOD test, whereas the 10th edition recommended methylene blue. However, the 11th edition had dropped methylene blue, with no other option provided. Initial work with methylene blue in this case study yielded virtually no improvement in the data.

The facility's secondary effluent was showing BOD values in the 20- to 40-mg/L range, whereas the seeded chlorinated/dechlorinated effluent was in the range of 4 to 7 mg/L. Based on this knowledge and literature reports, staff evaluated chlorination, acidification, and pasteurization for inhibition of nitrification.

Table 7.2 shows, first, just how much nitrification can occur in a partially nitrified effluent in a 5-day BOD, and second, the effects of several inhibition techniques. Acidification was not proposed because of concern that the pH adjustment would be difficult in low-alkalinity waters, and pH meters tended to be on the sluggish side in terms of response.

Treatment was deemed to be unnecessary for chlorine-disinfected effluents, but pasteurization was chosen as the most controllable procedure for all other samples. Pasteurization as practiced at this utility for nearly a decade thereafter consisted of rapid heating to 60 °C, followed by cooling to room temperature before setting up the BOD test (Stewart, 1964).

TABLE 7.2. Five-day BOD (mg/L) by various nitrification control methods.

Composite	Untreated	Acidified	Pasteurized	Chlorinated
Tertiary 1	20	5	2	7
Tertiary 2	25	4	5	7
Tertiary 3	27	5	5	7
Tertiary 4	27	4	5	8
Tertiary 5	36	8	6	<2
Tertiary 6	21	6	5	2

23.2 Case Study No. 2: Comparison of 2-chloro-6-(trichloromethyl) Pyridine

23.2.1 Summary

Thirty-five samples from activated sludge treatment facilities, river, and estuary locations were compared using both inhibition methods. No statistical differences were found for BOD_5 ranging from less than 2 to 57 mg/L. Subsequently, TCMP replaced pasteurization as the method of choice.

23.2.2 Background

Before acceptance of TCMP as a nitrification inhibitor in the 14th edition of *Standard Methods* BOD procedure (APHA et al., 1975), a number of literature publications appeared, as did a commercial preparation of the material. Laboratory staff, faced with an increasing sample load for BOD testing, needed comparative data between pasteurization and TCMP before recommending a change.

All chlorinated effluent samples were dechlorinated and seeded. Pasteurized samples were seeded after pasteurization. All samples treated by pasteurization or TCMP were seeded with the same material. Pasteurization yielded slightly increased results (2.5 mg/L average) than TCMP 12 times, decreased results (1.6 mg/L less) 11 times, and the equivalent value 12 times. The standard deviation for these comparisons was 1 mg/L.

In addition, experiments were run to determine if TCMP was toxic to the seed material used. Dosage up to 10 times the recommended amount of TCMP gave no difference in BOD values for two different treatment facilities compared with one-half of the recommended amount. The conclusion from this study was that TCMP was a more cost-effective method of inhibiting nitrification, and was not toxic to carbonaceous organisms in the seed. This utility thus adopted TCMP for all routine BOD tests requiring inhibition (Jenkins, 1975).

23.3 Case Study No. 3: The Debate Over 2-chloro-6-(trichloromethyl) Pyridine Toxicity in Carbonaceous Biochemical Oxygen Demand

23.3.1 Summary

Detailed studies were conducted to examine possible inhibition of CBOD determinations in raw and primary wastewater and GGA check standards. There was no evidence of toxic response resulting from treatment with various formulations of TCMP in either the 5-day or respirometric tests. When TCMP-treated samples gave lower BOD results than untreated samples, the differences were attributable entirely to nitrification. It was concluded that insufficient seeding was the cause of much of the trouble asserted in various literature reports.

23.3.2 Background

In the early 1990s, many states began requiring CBOD tests instead of total BOD_5 for analysis of raw and primary wastewater strength. This seems to have been primarily required where nitrification inhibition was required for the treated effluent of any given facility. The basis or reasoning for this requirement apparently was that one needed to compare apples with apples when assessing BOD removals. Additionally, there was some understanding that nitrification could actually occur in the BOD test of raw or primary wastewater, so inhibition of nitrification seemed warranted. A number of public utilities began to note that the actual strength of their facility influents was being underestimated. Furthermore, it was asserted (Albertson, 1993, 1995) that use of the CBOD test data was resulting in a 20% to 40% underdesign of WRRFs as well as violations of the 85% BOD removal rule in secondary treatment facilities. The TCMP was implicated as the culprit interfering with the bacteriological oxidation of organic matter in these wastes. It was assumed that TCMP was causing widespread toxicity in the BOD test.

A further complication arising from use of TCMP arose within the U.S. EPA quality control check standard database for GGA, in that the "acceptance" limits stated for the DMR-QA Study 17 for CBOD at 133.3 to 194.7 instead of the 167.5 to 228.5 stated by the U.S. EPA and *Standard Methods* for uninhibited BOD_5. The inference drawn by many was that TCMP inhibited the carbonaceous organisms as well as nitrifiers. However, this assumption was contrary to published data on TCMP and allylthiourea (Young, 1973), contrary to data published by the European testing community (Denmark Water Quality Institute, 1992), and certainly at odds with data in the hands of the BOD Joint Task Group members of *Standard Methods*.

In response to these assertions, a series of studies was conducted to investigate the effect of the nitrification inhibitor TCMP on carbonaceous BOD. This examination included seed with nitrifiers, a commercial seed material without nitrifiers, different inhibitor formulations (TCMP and allylthiourea), GGA standard results, raw and primary effluent results from different treatment facilities in different states, and industrial wastewater. Both BOD_5 and respirometric oxygen demand tests were run, and nitrogen species and COD were tested. Dilution and respirometric BOD tests were performed according to *Standard Methods* (APHA et al., 1995).

The TCMP has been shown to block the conversion of ammonia to hydroxylamine, which is an essential step in the oxidation of ammonia to nitrite (Goring, 1962). However, this step is not a pathway in the oxidation of organic carbon, so TCMP should be quite specific to inhibiting nitrifiers unless it was toxic by some other mechanism to the carbonaceous organisms. The literature is quite consistent that such toxic effect is nonexistent (Manoharan, 1994; Slayton & Trovato, 1979; Young, 1973; Young et al., 1981).

The results of the detailed study indicated that TCMP was not toxic to carbonaceous organisms in the dilution BOD test in the dose regime recommended in the method. Respirometric oxygen-uptake experiments showed no significant difference between TCMP-treated samples and controls, except where analysis of nitrogen species confirmed that nitrification had occurred in the untreated controls. The study concluded that low GGA results or lower-than-expected CBOD results reported in the literature are most likely due to improper test setup and insufficient seed amounts, rather than toxicity to carbonaceous bacteria in the test (Reilly & Young, 1995; Young, 1996; Young et al., 1997).

23.4 Case Study No. 4: The Research Study Gone Bad–Check Your Blanks!

The following correspondence was received by the authors: "I just completed a project where I ran over 250 samples of primary effluent through the BOD test. In evaluating the data, I find that the dilution water blanks were in the range of 0.3 to 0.6 mg/L. I also found that for more than half of my samples, the final DO was less than 1.0 mg/L. Is there a statistical way to correct these data?"

First, it is difficult to understand why this writer failed to recognize that he or she had both excessive blank DO depletion and excessive sample DO depletion in such a large percentage of samples until more than 250 tests were run. The writer also neglected to say whether serial dilutions had been run. Samples analyzed with poor dilution water are generally considered unreliable, as are samples with low residual DO. It is impossible to evaluate the magnitude of the problem from the information provided. However, it may be possible to drop the low DO residual dilutions from the dataset and use the satisfactory dilutions (if they exist) to calculate the BOD from what remains. As a group, these sample BOD values could be statistically compared with the acceptable set of samples using simple null hypothesis statistics to see if they were comparable. One could also compare the total dataset with just the acceptable data to see if conclusions are altered by using the poor data. Depending on the quality of data needed for acceptable confidence in the writer's study conclusions, the excessive depletion may not damage the overall outcome of the writer's study. Under no circumstances, however, should these data be considered of publishable quality or suitable for regulatory use.

As far as understanding the consequences of excessive blank DO depletion, the analyst could compare the contribution of the BOD contamination in the blank to the sample BOD. It would then be possible to quantify the damage and provide an objective basis for deciding whether the suspect data could be resurrected for purposes of the study. However, one can only hope that this writer was not involved in some regulatory activity or working on an academic thesis, because the problem data in no way meet test requirements as described in *Standard Methods* (Baird, 1992).

23.5 Case Study No. 5: The Trouble With Glucose–Glutamic Acid

The following letter was received by the authors: "I am having trouble using our settled raw sewage for a seed source, as the GGA results are too low. Until about two months ago, the GGA checks were fine. Since then, we have tried many things to correct the problem. First, we tried purchasing distilled water from a scientific supplies source, but the checks were still low. Next, we tried adding inhibitor to the check standard and seed, since one of our industrial discharges contains high ammonia, nitrate, nitrite, and organic nitrogen. The checks were still low. We have also tried final effluent and mixed liquor as seed material, with the same result. I tried a new supply of GGA from a commercial source, with no improvement. We have had the facility influent and effluent checked for heavy metals, but nothing seems out of line with those results. I have tried one commercial source of seed, and am looking for another commercial source. Do you have any other suggestions?"

This writer has already done a great deal of work trying to track down what initially appeared to be a seed problem. However, at this point, we might be tempted to determine that this is not a seed problem. Several sources of seed are giving the same low results. A new source of reagent water was tried in hopes that a dilution water toxicity problem would become apparent, but to no avail. Missing from this case history are (1) a statement of how low the results are; (2) assurance that the seed control determination is being followed; (3) how much seed is being added to the samples; (4) whether the writer had tried increasing the amount of seed to see if the GGA results would increase with increasing amounts of seed; (5) a check of the incubator temperature control; and (6) calibration checks of the DO probe against the Winkler titration. If increasing amounts of seed are used, one must be sure to correct for the extra BOD being added from the seed itself. And finally, it would be wise to add inhibitor to the seed control. This latter may be the most important hint to follow, as it will ensure that the seed control response is due to carbonaceous organisms alone, without contribution from nitrifiers. If nitrifiers are present in large numbers in the settled raw wastewater or other seed sources then, effectively, an insufficient amount of carbonaceous seed would end up being added to samples getting inhibition or to samples (and GGA!) that have no ammonia.

Troubleshooting this problem would also require that the nutrient-buffer solution was correct. The ferric chloride solution should be examined to ensure that it is fresh. A solution that has been sitting around turns brown from precipitated ferric oxide (rust). The phosphate buffer should be checked to ensure that the correct salts have been mixed in the proper proportion. There are many names for the phosphate salts. For instance, sodium hydrogen phosphate and sodium phosphate dibasic are the same material. Sodium dihydrogen phosphate and sodium phosphate monobasic are the same salt. But the monobasic and the dibasic salts are different. Further, most of the phosphate salts are available in both anhydrous and hydrated forms. Use of an anhydrous salt when a hydrated salt is needed will add too much of the salt and the resulting buffer pH will be off. The ammonium chloride solution should be checked for both pH and ammonia content. For the latter, the analyst mentioned adding inhibitor to the seeded dilution water, which would prevent nitrogenous organisms from oxidizing the ammonia, without improvement, so it seems unlikely that loss of nitrogen from the nutrient mix would be a problem.

The analyst should also check the validity of the GGA solution being purchased. The best way to do that is to prepare the GGA according to *Standard Methods* from purchased glucose and glutamic acid. If the starting concentration is low then, of course, one can expect lower final BOD. Other things to check in this category include making sure that D-glutamic acid is not chosen as the source.

As a last point, we should find out something about how the glassware is being cleaned between samples. If the glassware is just being rinsed and reused for the same sample types each time, it would be prudent to acid-wash with sulfuric acid, rinse with reagent water, and reset the GGA, although a thorough scrubbing with a good laboratory detergent followed by copious reagent water rinse to remove all traces of the detergent each time the bottles are used will prevent these problems (Mascarenhas & Klein, 1958). Occasionally, glassware handling has proven to be a problem in causing low results, presumably because of residual scum buildup on the glass surfaces of the BOD bottles or dilution water container. Usually, though, this leads to increased rather than decreased results (Baird, 2000).

24.0 QUESTIONS AND ANSWERS

24.1 Question and Answer No. 1

Question: Older editions of *Standard Methods* referred to the possibility of using alternative incubation periods. Does *Standard Methods* still stand by this? When and why was this provision removed? It would be a lot more convenient to take some tests off after 4 or 6 days. What has to be done to get recognition or approval for alternate incubation periods?

Answer: There are both regulatory and technical aspects to these questions. One must go back to the 10th edition (APHA et al., 1955) to find any mention of conversion from different incubation periods to a 5-day result. This edition stated, "Conversion of the data from one incubation period to another, or from one temperature to another, may be approximated only after special studies have been made." Although laboratories are known to have followed this practice, it takes a great deal of effort to characterize and tabulate results for different waste streams at, for example, 4, 5, and 6 days. Furthermore, this does not represent a standardized approach to conversion. There may be analysts who still practice this type of modification for their own particular purposes, but compliance with regulatory limits is probably not one of those purposes. The language was changed in subsequent editions to resemble its current wording because it was recognized that the rate constant for oxygen uptake (usually considered to be a logarithmic function) can vary radically for different waste streams. Therefore, no single conversion factor could be considered universal. No provision is made in the current edition, nor is any intended, to allow alternative incubation periods in 5210 B for the 5-day BOD. Provisions do exist for respirometric procedures (5210 D), which allow one to follow the course of the BOD over specified times using a standardized approach. Interestingly, the 10th edition followed a similar logic for industrial waste samples, recommending respirometric BOD methods.

Standard Methods is an evolving work, and the *Standard Methods* committee strives to present the best available technique for each of the methods. As time has passed, more is discovered about each method. In some instances, procedural mistakes and incorrect science have been presented in past editions, then corrected in later editions. And sometimes, procedures have been dropped from *Standard Methods* because better techniques have become available. The most current edition then is the edition that should be consulted for the most correct procedure.

24.2 Question and Answer No. 2

Question: Why is ammonium chloride no longer included in the method for inhibiting nitrification in the BOD test? There are several references in the literature recommending this, and it is safer and more cost effective than the TCMP.

Answer: Ammonium chloride lost support as a nitrification inhibitor because its effectiveness was deemed to be pH dependent and, consequently, erratic. The preferred inhibitor has been TCMP since the 15th edition (APHA et al., 1980).

24.3 Question and Answer No. 3

Question: We test wastewater with 4 mg/L of ammonia–nitrogen. Because NH_3 is toxic, will this amount of ammonia inhibit CBOD values?

Answer: The simple answer is "No!" Historically, the literature demonstrates that sufficient nutrients are needed in samples to get consistent BOD results. The buffer solution in 5210 B.3 was developed based on this early research and contains ammonium ion. Furthermore, experience has shown no evidence of toxicity in the BOD test when used routinely on secondary effluents containing upwards of 50 mg/L of NH_3–N.

24.4 Question and Answer No. 4

Question: Does the presence of ammonia in the buffer (1.7 g of NH_4Cl/L of buffer) increase the chances of nitrogen contributing to the BOD results?

Answer: No, the amount of ammonium ion added would only be approximately 1 mg/L of BOD if it were all oxidized to nitrate ion. This is an insignificant amount. In most cases, this should not contribute to DO depletion in the 5-day test anyway, as the N is incorporated to cell growth.

24.5 Question and Answer No. 5

Question: We understand that allylthiourea may be an appropriate inhibitor of nitrogenous demand when determining CBOD. Our laboratory is getting satisfactory results with allylthiourea. What is the current status of allylthiourea in *Standard Methods*?

Answer: As discussed above, allylthiourea has pros and cons. It has been in use by the European testing community for some time. The *Standard Methods* Joint Task Group for BOD took a very cautious approach to accepting allylthiourea as an acceptable nitrification inhibitor. With publication of the 21st edition (APHA et al., 2005), however, allylthiourea will be included in Method 5210, with some caveats. The material is itself biodegradable under some conditions and may, therefore, contribute to the determined BOD.

24.6 Question and Answer No. 6

Question: Section 4.h calls for a dilution water blank, but the calculation does not contain a term for the blank. Why?

Answer: The dilution water blank is just a check on the dilution water quality; as stated, it should exhibit a BOD less than 0.1 mg/L. It should be discarded if the demand is greater than 0.2 mg/L. Technically, sample data obtained when the dilution water exceeds 0.2 mg/L are invalid (although perhaps not unusable for some purposes). In this case, the analyst should search for the cause of the high oxygen demand in the dilution water and correct the problem. Common sources of high depletion include use of poor quality deionized water, use of plastic or rubber tubing, and contaminated glassware.

24.7 Question and Answer No. 7

Question: Method 5210 B, Section 4.b, "Dilution Water Check," contains a statement that dilution water not stored for quality improvement should have a DO uptake of 0.05 to 0.1 mg/L; this seems erroneous and should really be 0.6 to 1.0 mg/L, as stated in 4.d. l.

Answer: The instructions are not erroneous; if dilution water is stored, the method recommends adding a small amount of seed (sufficient to contribute not more than 0.05 to 0.1 mg/L to the depletion) and then assessing the dilution water quality as in section 4.g. The range of DO depletion after 5 days should then be no more than 0.2 mg/L and preferably not more than 0.1 mg/L, as stated. This is the normal depletion allowance for clean water, as stated in Section 4.h. The 0.6 to 1.0 mg/L depletion referred to in the question is the recommended range of seed contribution to DO uptake in a BOD test (Section 4.d.2). (*Note*: the BOD Joint Task Group for the

20th edition [APHA et al., 1998] revised the recommendations on dilution water storage and deleted the narrative regarding addition of seed to stored water for quality improvement purposes. Although steps recommended in the 19th and earlier editions may work for improving dilution water quality, the Task Group unanimously voted to simplify this by making stored water quality a performance criterion in which dilution water blanks must be less than 2.0 mg/L. The change came about as a result of persistent confusion about dilution water check and dilution water blank. With the availability of ultrapure water systems, deionization systems, and bottled water as possible sources of good quality water, there was little perceived need for storage as a quality improvement step.)

24.8 Question and Answer No. 8

Question: If my dilution water check is 0.6 mg/L instead of the 0.2 mg/L recommended, can I just compensate for this in the seed blank correction if the seed control is within the bounds stated in the method?

Answer: If the oxygen depletion for the dilution water is 0.6 mg/L and the seed adds another 0.6 to 1.0 mg/L of demand, the depletion in the seed control could be as high as 1.6 mg/L. If, however, the seed only adds 0.2 mg/L to the seed control measurement, then the seed control criterion could be met; however, there probably would not be enough seed to provide a good BOD measurement. This could quite likely be reflected in poor results in the GGA check.

24.9 Question and Answer No. 9

Question: Section 4.d specifies the use of seed for industrial wastewaters and indicates the use of a nitrification inhibitor if the seed source is effluent from a biological treatment process. If the inhibitor is added in such a test, would the results he reported as BOD or as CBOD? If it is reported as CBOD, is there any variation in the test method that could be used to report total BOD when such seed is added to industrial samples?

Answer: There is no approved test variation that would allow defining this as total BOD. If it is known that there is no nitrogen in the industrial wastewater, then nitrification would not be much of an issue. If nothing is known about nitrogen content, a commercial seed product might be considered. However, Section 4.d recommends developing an adaptive seed if there is a great likelihood that the sample contains organic compounds that might be refractory to the organisms in a seed derived from typical settled domestic waste. It would be reasonable to suspect a commercial seed of lacking adaptive properties under these conditions as well, so some experimentation would be necessary at the outset.

24.10 Question and Answer No. 10

Question: It appears that use of an adaptive seed is the most proper way of running the BOD test because it will ensure the presence of necessary microorganisms. Yet, section 4.d.1 does not seem clear on this point in discussing seed source. Can you clarify?

Answer: Section 4.d.1 begins by stating, "It is necessary to have present a population of microorganisms capable of oxidizing the biodegradable organic matter in the sample." This, in effect, recommends the use of an adapted seed for any sample. For most cases, the mixed liquor source is satisfactory as described. The section describes how an acclimated seed may be obtained or developed if not directly available.

It has become fairly routine practice in the municipal wastewater treatment field to use mixed liquor supernatant from a domestic WRRF as seed source for testing BOD, not only for the treatment facility samples but also for industrial effluent samples tributary to the facility, regardless of whether it has been demonstrated that the seed was adequately adapted to the waste stream. Long practice has shown this to be a viable and preferred method of determining BOD loading to the treatment facility, absent toxic effects from the waste stream being tested. However, because it is impossible to cover all possible waste types and BOD scenarios, assessment of BOD of industrial effluents that directly discharge to a receiving water may need acclimated seed. Other scenarios should be evaluated on a case-by-case basis. This could obviously present some logistical problems for laboratories that may have a large number of clients needing BOD tests for discharge permits. Sample holding times and irregular client schedules preclude using the acclimation steps described in Method 5210 B. In these cases, absent specific knowledge of the waste stream, the preferred seed is the domestic waste material, as described.

24.11 Question and Answer No. 11

Question: On numerous occasions, laboratories have seemingly misinterpreted the paragraph on seed control uptake and run seed control so that only a 0.6 to 1.0 mg/L depletion is achieved. We have interpreted that the seed control sample should show a DO depletion of approximately 50% of initial DO, a depletion of at least 2.0 mg/L, and a residual DO of at least 1.0 mg/L. Then, the seeded dilution water added to the samples should contribute approximately 0.6 to 1.0 mg/L of the demand. Which is correct? (Note that this question comes from a regulatory agency.)

Answer: This individual is correct. The seed control steps are performed to aid selection of the proper amount of seed to add to samples. Using a consistent source of seed over time, the goal is to add enough seed to utilize the carbon in all samples tested. The GGA is used as an ongoing check. Paragraph 4.d.2 in the 20th edition (APHA et al., 1998) describes running a BOD on the seed material (seed check). This test should contain several dilutions, with the highest strength aliquot giving at least a 50% depletion. All other dilutions used in this plotting method should have at least a 2 mg/L depletion and a minimum of 1 mg/L residual DO after 5 days. Dilutions that do not meet these criteria should not be used to determine seed strength. The volume of seed determined from this process for addition to samples or GGA should correspond to the amount calculated to contribute in the 0.6 to 1.0 mg/L DO depletion range.

24.12 Question and Answer No. 12

Question: Does Section 4.c restrict the use of a GGA standard to only those samples not requiring an adaptive seed?

Answer: No, use of GGA is permitted in all cases. The method also allows using alternative (known) organic compounds as standards if the seed is acclimated to such a known compound and does not use the GGA standard readily. Caution has to be exercised here, in that use of an alternative compound requires that the quality controls be established for the new substrate. The control limits for the GGA will not be valid for a different material.

24.13 Question and Answer No. 13

Question: *Standard Methods* says to use a 2% solution of GGA. I cannot seem to get a 2% solution of glutamic acid to stay dissolved. What am I doing wrong?

Answer: The method states that a 2% dilution of the GGA solution should be used. The solution itself is prepared from a mixture of 150 mg glucose plus 150 mg glutamic acid/L.

24.14 Question and Answer No. 14

Question: *Standard Methods* says to dry reagent grade glutamic acid before preparing the GGA solution. However, there are two forms of glutamic acid, D-glutamic and L-glutamic acid. Which should be used?

Answer: L-glutamic acid is the generic and naturally occurring form. It is obtained commercially from microbial culture and is biologically active. Chemically synthesized glutamic acid is a racemic mixture of the D- and L-forms. D-glutamic acid has some properties that differ from the L-form, but whether seed material utilization is one of these properties has not apparently been reported in the literature. It is recommended that the generic L-form be used.

24.15 Question and Answer No. 15

Question: I cannot understand the BOD test method seed control section. It says that the DO uptake attributable to the seed added to each bottle should be between 0.6 and 1.0 mg/L, but the amount of seed added should be adjusted from this range to get the GGA result in the range of 167.5 to 228.5 mg/L. Does this really mean that the amount of seed added is dependent on the GGA value? I have data that show that the DO uptake of the seed is 1.2 mg/L, but the GGA check is 202 mg/L. Is this acceptable? If so, it seems as if we do not really care what is the seed uptake.

Answer: The seed control section (4.d.2) is designed to guide the analyst in selecting the proper amount of seed for each sample. Clearly, one does not select, condition, and characterize seed just before setting up a test, as the steps described in 4.d require acclimation time, BOD testing, and possibly other lengthy measures. The above question, and others like it, seems predicated on a desire to take shortcuts through the method. This is a normal tendency for analysts continuously running BOD tests who have become familiar with the nature of the seed material, dilution water, and sample characteristics. Be careful! Take the time necessary to perform the test correctly.

The method clearly states to use the amount of seed calculated to give a 0.6- to 1.0-mg/L uptake or adjust this amount to ensure that the GGA is in range. In either case, we do care what the value is because the value is subtracted from the total DO uptake in a sample. Sure, we want enough seed to give acceptable GGA check results. But the more DO uptake from the seed, the closer this value gets to the minimum DO uptake required in a sample, meaning that more uncertainty is going to result in low BOD samples. The 1.2 mg/L cited in this case does not invalidate the test, but the analyst should probably start thinking about another seed preparation if this seed uptake value starts to climb and the GGA stays in range.

24.16 Question and Answer No. 16

Question: How should significant figures be applied or interpreted in the method itself? For instance, a 0- to 2-mg/L range is specified for depletion in the dilution water blank. If an analyst gets a 2.4-mg/L blank, can that be considered within the range by rounding off to 2? Or, if a sample has a pH of 7.8, should the analyst still adjust to 7.5?

Answer: This question has arisen more and more in recent years, probably as a result of more laboratory certification programs and certification inspections. The BOD Joint Task Group has attempted to clarify these details, or at least narrow the interpretive range, for the 21st edition of *Standard Methods* (APHA et al., 2005). Depletion range for blanks should be 0 to 2.0 mg/L, so 2.4 would not be in range. Beginning in the 20th edition (APHA et al., 1998), the pH range was specified as follows, "samples containing caustic alkalinity ($pH > 8.5$) or acidity ($pH < 6.0$) should be neutralized to pH 6.5 to 7.5." In this case, the range and the decimal place were not changed but, rather, the meaning of caustic was specified. So, to answer the question about a sample at pH 7.8, no, it does not need to be neutralized. But, if the sample had been at 8.8, it would have to be neutralized to a value within the specified range. Neutralization to 7.8 would not be satisfactory to conform to the range specified for adjustment. It is important that (1) the sample be in a pH range to which the seed organisms are acclimated, (2) the samples requiring adjustment are adjusted to the same range, and (3) the pH buffer in the dilution water not be affected by the lowest sample dilution.

24.17 Question and Answer No. 17

Question: Method 5210 B specifies the use of sodium sulfite for dechlorinating samples. Is this correct? All other dechlorinating procedures in *Standard Methods* use sodium thiosulfate (e.g., Part 9000: Microbiological Methods). If this is intended, why is a reagent blank for sulfite not included in the method? Does sulfite exhibit an oxygen demand?

Answer: Yes, sodium sulfite is the intended reagent. And, yes, sulfite can indeed interfere with the BOD test if present. There is a stoichiometric reaction between sulfite ion and oxygen to yield sulfate ion. The procedure is pretty clear in the dechlorination instructions, however, that just enough sulfite should be added to titrate the residual chlorine. The method also instructs to recheck for chlorine 10 to 20 minutes after sulfite is added. This should be enough time for small amounts of sulfite to oxidize to sulfate as well, if the analyst mistakenly added a little too much. Lastly, because the dechlorination step is performed on undiluted sample, subsequent sample dilutions and sample handling will further minimize any such mistakes. There is no way to run a "sulfite blank" on a sample that would just be the sample after dechlorination. As for why a different reagent, like thiosulfate or ascorbic acid, is not used, they also have potential problems. Excess ascorbic acid could add oxygen demand, the same problem as cited in the correspondence above. Thiosulfate disproportionates to give sulfur and sulfate in the reaction with chlorine. Sulfur can be oxidized in the test or stimulate growth of other types of organisms that themselves may exert some demand. Stoichiometrically, it also takes more thiosulfate than sulfite to effect dechlorination and would possibly exacerbate the latter problem.

Sulfite present in pulp mill effluents has traditionally presented problems in the BOD test aside from interferences with DO determination (e.g., Watts & Adams, 1993). One approach to removing sulfite interference is to chemically oxidize it to sulfate by adding no more than 0.05 to 0.10 mg/L of cobalt in the form of cobalt(II) chloride, then aerating the sample before preparing the BOD dilutions. This is further discussed in the end of Chapter 4 on DO.

24.18 Question and Answer No. 18

Question: I suspect that I am getting an interference in the chlorine residual determination as written in the BOD method. Is it okay to use a different method to determine chlorine residual, such as another of the 4500-Cl methods?

Answer: Clearly, if you have an interference to the iodide test, then use of a different procedure, such as the amperometric method, would be advisable. You would need to ensure that the interference was not a problem with the alternative test, and determination of the amount of sulfite needed for dechlorinating the sample must be based on the stoichiometry of the reaction in the alternate procedure.

24.19 Question and Answer No. 19

Question: The method does not provide ways of removing other oxidizing agents. Why? For example, our effluent has 100 to 150 mg/L of sulfite ion from a chemical treatment process. This is an oxidizing waste stream, and sulfite interferes with the test.

Answer: Whether sulfite is a strict interference in this situation (i.e., a chemical reductant), or will exert an actual BOD, may be a moot point to argue. It is likely that the oxygen demand exerted on a receiving water by such a high-strength chemical waste would be devastating in either case. A BOD test is probably not definitive in such a scenario but, if deemed important, would require modification to remove this reducing species before testing.

Clearly, chlorine residual must be removed because it is toxic to the seed organisms but, obviously, not all samples are chlorinated. Should all sample streams be tested for the presence of other oxidants or reductants? Probably not, at least not in the name of toxicity. Knowledge of the sample process can direct such needs. Toxic effects from unknown toxicants will probably show up in the dilution test as anomalous depletions in the dilution series. Some test users have written about a variety of potential interferences, such as sulfite and magnesium hydroxide. The latter may interfere with the chlorine residual test, while others simply exert an oxygen demand that may be chemical, biochemical, or both. Reduced nitrogen species are considered by the method to be interference, at least in terms of CBOD (thus, the recommended use of nitrification inhibitors).

24.20 Question and Answer No. 20

Question: I want to know the reason for requiring that samples be brought to 19 to 21 °C before analysis. This narrow range is difficult to achieve without a refrigerated bath, and laboratory auditors are requiring this range now before dilution.

Answer: This step is aimed at minimizing variance in the test; the microbes responsible for driving the BOD are temperature sensitive, as is the solubility of molecular oxygen. So, not only do we want to get DO near saturation in a quick and uniform manner, the start condition for the bacteria must be the same in all dilutions. Typically, samples are delivered to the laboratory at 4 °C. Failure to warm samples before dilution, seeding, and incubation provides opportunity for wide variance in starting conditions, and some potentially serious discrepancies in kinetic rates. Samples that are too cold reach equilibrium at varied and unspecified times in the incubator. Incubator temperature control may also be compromised for lengthy periods if large numbers of samples are introduced cold. Conversely, samples that are overly warm may lose BOD during initial setup before initial DO measurements are made. This would be a particular problem when large batches of samples are being set up.

The current method allows a 1 °C window, whereas earlier editions did not specify a tolerance. Task Group members agreed that this specified tolerance both improves test consistency and is reasonably achieved with typical water baths in widespread use. In older editions, without a specified tolerance, auditors would have been free to require exactly 20 °C. Note that dilution water is to be at 20 °C, with a tolerance of 17 to 23 °C, before use to stabilize DO in the dilution water. Sample temperatures tolerances are set a little tighter because dilution occurs just as initial DO is measured and incubation begins. So the sample itself is not likely to affect DO saturation at this point if the temperature is closer to 20 °C than the dilution water.

24.21 Question and Answer No. 21

Question: Please clarify the holding time allowed for BOD samples. Our regional laboratory certification authority insists that all samples should be finished with setup and in the incubator within 6 hours and cites *Standard Methods*. However, U.S. EPA has promulgated 48 hours, and *Standard Methods* seems to indicate that up to 24 hours is allowed.

Answer: *Standard Methods* recommends that tests be started within 2 hours of collection for samples analyzed at a laboratory near the point of collection. If the test is started more than 2 hours after collection, it should be chilled to at least 4 °C, and analysis should begin within 6 hours of collection. 5210 B.1.b further distinguishes between grab and composite samples. Grab samples should not be held longer than 24 hours before commencing analysis, to allow transport to remote laboratories; composite samples should represent no longer than 24-hour compositing time and should not be held longer than 24 hours after final collection, again, allowing for transport. Studies of sample storage temperature and duration before BOD analysis indicate that the shorter the storage time and more immediate the analysis after sample collection, the better (Carpenter & Berger, 1978; Funk, 1977). It is thus recommended that every effort be made to begin testing within 6 hours, even allowing for transport. However, nowhere is it the stated intention in *Standard Methods* to complete test setup within 6 hours: the procedure clearly states, "Begin analysis within 6 hours."

It is likely that the U.S. EPA-specified 48 hours was derived from the maximum of 24 hours composite time plus 24 hours holding, but this is not clear from published literature. Young et al. (1981) can be consulted for more discussion on this topic.

24.22 Question and Answer No. 22

Question: Why does *Standard Methods* state, "Do not test chlorinated/dechlorinated samples without seeding the dilution water"? I have tested WRRF effluents with high total suspended solids and fecal coliform counts, which even after chlorination/dechlorination give BOD results with 6- to 7-mg/L oxygen depletion at very low dilutions (i.e., 5-mL sample + 295-mL dilution water). Is it really necessary to seed when there is consistent evidence that there is a viable microorganism population even after chlorination?

Answer: The objective of seeding samples that have been chlorinated is to ensure the presence of a viable microbial culture acclimated to the waste being tested. Because there is no way to a priori verify that a chlorinated sample has retained a viable population, the procedure calls for seeding this type of sample. The purpose of the chlorination step in the treatment process was to disinfect, regardless of whether the step was effective. One must assume that some kill was effected and that seed is required because failure to seed chlorinated samples usually yields erratic results in the different dilutions used in the test.

The questioner did not mention what the typical BOD level was in the samples, or whether comparative data had been developed between the chlorinated samples with and without seeding. Both the level and consistency of the BOD results between the two treatments at all dilutions would likely have revealed some effects from chlorination.

24.23 Question and Answer No. 23

Question: We prepare dilutions of 500 mL in a graduated cylinder and the sample is then decanted into incubation bottles, as outlined in Section 4.f. Is the size of the incubation bottle critical? We would prefer to use smaller bottles than the specified 250- to 300-mL size.

Answer: Smaller bottles are satisfactory, provided that there is enough room for a DO probe to be seated inside the bottle (assuming that this is the method being used to determine DO). There has been some concern voiced over the effect of the internal surface area-to-volume ratio on BOD tests, the idea being that increased surface area can provide more anchorage for sessile organisms with subsequent creation of floc mats and greater chance for surface absorption of organics in the sample. An optimum ratio of 0.7 cm^2/mL has been proposed (Kuftarkova, 1972); however, a 500-mL reagent bottle is needed to meet this ratio. The fact that the dilutions were prepared in larger vessels diminishes concern about measuring inappropriately small sample aliquots into the smaller bottle.

24.24 Question and Answer No. 24

Question: How much leeway is given for the time to read the BOD samples on day 5?

Answer: There is nothing definitive on this issue. Other methods that require reading a bacterial metabolism endpoint may specify within 2 hours (e.g., fecal coliform), so some analysts have adopted this as a practical solution. This amount of leeway is reasonable from another standpoint, in that the

BOD is nearing an asymptotic rate after 5 days. Thus, with low carbon and lowered DO, there is not likely to be much depletion occurring at the end of the test if it was seeded properly in the first place.

24.25 Question and Answer No. 25

Question: Why does the BOD method require that at least 1.0 mg/L DO remain in each dilution for valid results? My DO meter can read to much lower values.

Answer: The 1-mg/L residual was established to ensure that there is enough oxygen in the sample at the end of the test to satisfy the needs of the microorganisms in the BOD bottle. If the DO gets much less than 1 mg/L, the BOD may not represent a true 5-day BOD as the kinetics start to slow down.

24.26 Question and Answer No. 26

Question: Should the DO uptake of the blank dilution water be included in the calculations in Section 5?

Answer: The *Standard Methods* Joint Task Group for BOD has long been firmly against allowing correction for the DO uptake in the blank dilution water. It is the standing belief that to allow this correction would only invite analysts to ignore the dilution water quality criteria, and thereby undermine a basic cornerstone of reliable BOD testing. The preferred method of dealing with blank DO uptake is to seed all samples, and thus remove a source of < 0.2-mg/L DO depletion criterion.

24.27 Question and Answer No. 27

Question: What is the proper way of dealing with toxicity in samples? The method is not clear on this topic. Also, how does one accept or reject data from toxic dilutions? Should a toxic outlier be averaged into the results?

Answer: It is widely held that increases in calculated BOD with increasing dilution indicates toxicity. This may or may not be the case, and there is considerable disagreement on this point. In a classical toxicological sense, such an apparent dose-dependent relationship in DO uptake would seem to make this a simple deduction. However, if one considers the possibility that the increasing BOD with increasing dilution could be related kinetically to a food-to-microorganism ratio in the BOD bottles, or confounded by oxidation of reduced forms of nitrogen compounds in waters with low carbonaceous BOD, the simple deduction looks less clear. In highly oxidized treated effluents, toxicity is probably never the root cause of observing an increased BOD with increasing dilution (absent undetected residual chlorine). These highly efficient processes could not successfully treat toxic influent to yield the typical BOD values in the 8- to 15-mg/L range and still have toxicity in the BOD test. A safe practice is to suspect nutrient effects or systematic analyst error in low-BOD, highly treated municipal effluents; suspect toxicity in unknown industrial effluents with low dilution factors and/or an apparent outlier in the highest strength dilution (less diluted sample) compared with the other two dilutions. The latter case could suggest a "threshold" of effect, which is another classical tenet of toxicology.

A reseeding technique has been proposed for detecting toxic interferences in the BOD test (Bordon & Woodcock, 1949).

At this time, there is not a satisfactory discussion of criteria for acceptance or rejection of outliers in the method. To some degree, the analyst may have to rely on judgment, but a better way would be to use some of the statistical tests based on variance or develop control charts based on historical (typical) data. In any case, the apparent toxic dilution should not be averaged into the final results.

25.0 REFERENCES

Abbott, W. E. (1948a). Suppression of nitrification and denitrification during the determination of biochemical oxygen demand. *J. Soc. Chem. Ind.*, *67*, 373; *Chemical Abstracts.*, *43*, 35441 (1949).

Abbott, W. E. (1948b). The bacteriostatic effect of methylene blue on the BOD test. *Water Sew. Works*, *95*, 424; *Public Health Eng. Abstr.*, *30*(3), 21 (1950).

Ademoronti, C. M. A. (1987). Oxygen utilization and biochemical stabilization rate constants for a brewery effluent. *Int. J. Environ. Stud.*, *29*, 145.

Adeny, W. E. (1908a). Nature and rate of the chemical changes which occur in polluted fresh and salt water. Addenum: Capacities of fresh and of sea water to bear pollution within limits of fouling. *Royal Commission Fifth Rep.*, App. I(1), 433.

Adeny, W. E. (1908b). Reprint of papers by Adeny, 1895–1897. *Royal Commission Fifth Rep.*, App. VII(5), 49.

Adeny, W. E. (1908c). Studies in the chemical analysis of fresh and salt waters. I. Applications of the aeration method of analysis to the study of river waters. *Royal Commission Fifth Rep.*, App. VII, 95.

Adeny, W. E. (1908d). Testimony by Adeny. *Royal Commission Fifth Rep.*, App. I, 665.

Adeny, W. E. (1928). *The principles and practice of the dilution method of sewage disposal.* Cambridge University Press.

Albertson, O. E. (1993). The $CBOD_5$ test: More trouble than its worth. *Water Environ. Technol.*, *5*(11), 16.

Albertson, O. E. (1995). Is $CBOD_5$ test viable for raw and settled wastewater? *J. Environ. Eng. (Reston, Va.)*, *121*, 515.

Akerlindh, G. (1953). Computations of BOD for seeded samples of high dilution. *Sew. Ind. Wastes*, *25*, 293.

American Public Health Association. (1905). Standard Methods for the Examination of Water and Sewage. *J. Infect. Dis.* Suppl. No. 1.

American Public Health Association. (1912). *Standard methods for the examination of water and sewage* (2nd ed.).

American Public Health Association. (1917). *Standard methods for the examination of water and sewage* (3rd ed.).

American Public Health Association & American Water Works Association. (1946). *Standard methods for the examination of water and sewage* (9th ed.). American Public Health Association.

American Public Health Association, American Water Works Association, & Federation of Sewage and Industrial Wastes Associations. (1955). *Standard methods for the examination of water and wastewater* (10th ed.). American Public Health Association.

American Public Health Association, American Water Works Association, & Water Pollution Control Federation. (1960). *Standard methods for the examination of water and wastewater* (11th ed.). American Public Health Association.

American Public Health Association, American Water Works Association, & Water Pollution Control Federation. (1965). *Standard methods for the examination of water and wastewater* (12th ed.). American Public Health Association.

American Public Health Association, American Water Works Association, & Water Pollution Control Federation. (1975). *Standard methods for the examination of water and wastewater* (14th ed.). American Public Health Association.

American Public Health Association, American Water Works Association, & Water Pollution Control Federation. (1980). *Standard methods for the examination of water and wastewater* (15th ed.). American Public Health Association.

American Public Health Association, American Water Works Association, & Water Environment Federation. (1995). *Standard methods for the examination of water and wastewater* (19th ed.). American Public Health Association.

American Public Health Association, American Water Works Association, & Water Environment Federation. (1998). *Standard methods for the examination of water and wastewater* (20th ed.). American Public Health Association.

American Public Health Association, American Water Works Association, & Water Environment Federation. (2005). *Standard Methods for the Examination of Water and Wastewater* (21st ed.). American Public Health Association.

Analytical Quality Control (Harmonised Monitoring) Committee. (1985). Accuracy of determination of BOD in river waters: Analytical quality control in the harmonised monitoring scheme. *Analyst*, *110*, 97.

Anonymous. (1932a). Procedure recommended for the biochemical oxygen demand test. *Sixth Annu. Ohio Conf. Sew. Treatment Rep.*, *61*; *Chemical Abstracts*, *28*, 41443 (1934).

Anonymous. (1932b). Tentative procedure for biological oxygen demand on chlorinated sewage. *Sixth Annu. Ohio Conf. Sew. Treatment Rep.*, *64*; *Chemical Abstracts*, *28*, 4144 (1934).

Baird, R. B., Los Angeles County Sewerage Authority, Los Angeles, California. (1992). Personal communication.

Baird, R. B., Los Angeles County Sewerage Authority, Los Angeles, California. (2000). Personal communication.

Batta, G., & Leclerc, E. (1939). The organic matter content and the biological oxygen demand of polluted waters. Measurement of the biological oxygen demand. *Ind. Chim. Belge.*, *10*, 85; *Chemical Abstracts*, *33*, 4718 (1939).

Bordon, G. C., & Woodcock, I. (1949). Reseeding BOD bottles: A means of detecting toxic interferences in the BOD test. *Tappi J.*, *32*, 516.

Broch-due, A., Anderson, R., & Kristoffersen, O. (1994). Pilot plant experience with an aerobic moving bed biofilm reactor for treatment of NSSC wastewater. *Water Sci. Technol.*, *29*(5–6), 283.

Brown, J. A., Paul, J., Rice, L. C., & Lines, J. M. (1973). BOD: Determining the necessary dilution technique. *Water Sew. Works*, *120*(5), 105.

Bruner, P. L. (1943). A BOD calculator. *Sew. Works J.*, 15, 86.

Buswell, A. M., Shiota, T., Lawrence, N., & Van Meter, I. (1954). Kinetic studies on the growth of *Nitrosomonas* as related to the nitrification phase of the BOD test. *Appl. Microbiol.*, *2*, 21.

Buswell, A. M., Van Meter, I., & Gerke, J. R. (1950a). Study of the nitrification phase of the BOD test. *Sew. Ind. Wastes*, *22*, 208.

Buswell, A. M., Van Meter, I., & Gerke, J. R. (1950b). Study of the nitrification phase of the BOD Test. II. Determination of an incubation vessel for multiple determinations. *Sew. Ind. Wastes*, *22*, 1297.

Butterfield, C. T. (1929). Studies of natural purification in polluted waters. II. Development of a suitable dilute medium. *Public Health Rep.*, *44*, 2647; *Chemical Abstracts*, *24*, 450 (1930).

Carpenter, W. L., & Berger, H. F. (1978). Investigation of factors affecting BOD measurement and experience with the 1-day BOD test. *Natl. Council Air Stream Improvement Tech. Bull.*, *305*.

Chaudhuri, N., Tyagi, P. C., Niyogi, N., Thergaonkar, V. P., & Hanna, P. K. (1992). BOD test for tropical countries. *J. Environ. Eng.*, *118*, 298.

Childs, J. A. (1925). The biochemical oxygen demand test. *Public Works*, *56*, 218.

Cooper, A. E. (1921). Denitrification as a measure of sewage purification. *Biochem. J.*, *15*, 513.

Dague, R. E. (1981). Inhibition of nitrogeneous BOD and treatment plant performance evaluation. *J. Water Pollut. Control Fed.*, *53*, 1738.

Deai, J., Tida, T., Gong, Y., Jianrong, Z., and Yicheng, S. (1991). Factors affecting the relationship between the NBOD values and the amount of nitrogenous pollutants: A field study of the Lee River. *Water Res.*, *25*, 485.

Denmark Water Quality Institute (1992). *Interlaboratory Comparison*. Hoershoelm.

Drown, T. M. (1892). The effect of aeration of natural waters. *23rd Annual Report, Mass. State Board Health*.

El-Baroudi, H. M., & Moawad, R. S. (1967). Rate of BOD reduction by oxidation ponds. *J. Water Pollut. Control Fed.*, *39*, 1626.

Eldridge, E. F. (1933). More about the BOD determination. *Sew. Works J.*, *5*, 788.

Eldridge, E. F., & Mallman, W. L. (1931). The influence of diluting waters on the biochemical oxygen demand. *Michigan State. Coll. Eng. Exp. Stn. Bull.*, *39*, 3.

Elmore, H. L. (1955). Determination of BOD by a reaeration technique. *Sew. Ind. Wastes*, *27*, 993.

Ettinger, M. B., Liska, R. J., & Kroner, R. C. (1954). Persistance of pyridine bases in polluted water. *Ind. Eng. Chem.*, *46*, 791.

Federation of Sewage Works Associations Committee Report: Joint Biochemical Oxygen Demand Committee of the Great Lakes and Ohio River Boards of Public Health Engineers. (1932). Progress report: Uniform procedure for biochemical oxygen demand. *Sew. Works J.*, 4, 738.

Fitzmaurice, G. D., & Gray, N. F. (1989). Evaluation of manufactured inocula for use in the BOD test. *Water Res.*, *23*, 655.

Frank, L. C., & Rhynus, C. P. (1919). The treatment of sewage from single houses and small communities. *Public Health Bull.*, *101*, 28.

Funk, W. (1977). Preservation of samples for later analysis in a laboratory. *Vom wasser* (Germany), *48*, 75.

Gaffney, P. E., & Heukelekian, H. (1958). Oxygen demand measurement errors in pure organic compounds; nitrification studies. *Sew. Ind. Wastes*, *30*, 503.

Garner, J. H. (1922). Notes on sewage works analysis and standards of purification. *Surveyor*, *61*, 393.

Gaudy, A. F., Jr., & Gaudy, E. T. (1980). *Microbiology for environmental scientists and engineers*. McGraw-Hill.

Gellman, I., & Heukelekian, H. (1951). Studies of biochemical oxidation by direct methods. I. Direct method for determining BOD. *Sew. Ind. Wastes*, *23*, 1267.

Gellman, I., & Heukelekian, H. (1955). Studies of biochemical oxidation by direct methods. V. Effect of various seed materials on rates of oxidation of industrial wastes and organic compounds. *Sew. Ind. Wastes*, *27*, 793.

Gifft, H. M. (1943). The mathematics of BOD determinations. *Water Works Sew.*, *90*(8), 313.

Goring, C. I. A. (1962). Control of nitrification by 2-chloro-6-trichloromethyl pyridine. *Soil Science*, *93*, 211.

Gowda, R. N. (1924). Nitrification and nitrifying organisms. *J. Bacteriol.*, *9*, 251.

Grune, W. N., & Luckens, M. M. (1949). BOD and standardization of laboratory procedures. *U.S. Atomic Energy Comm. Prog. Rep.* NYOO-1500 (NYU-3), 3.

Hall, J. C., & Foxen, R. J. (1983). Nitrification in BOD_5 test increases POTW noncompliance. *J. Water Pollut. Control Fed.*, *55*, 1461.

Hammer, M. J. (1971). BOD determination for industrial waste waters. *Water Sew. Works*, *118*, R-104.

Hargrave, E. C., & Tyler, L. P. (1965). Preserving sewage seed for BOD analysis. *Water Sew. Works*, *112*, 181.

Hazen, A. (1891). On the Determination of Nitrates in Water. *Chem. News*, *64*, 162.

Herman, A. (1956). Graphic determination of BOD. *Bull. Cent. Etude Document Eaux*, *71*, 414; *Public Health Eng. Abstr.*, *37*(10), 20 (1957).

Heukelekian, H. (1942). The influence of nitrifying flora, oxygen, and ammonia supply on the nitrification of sewage. *Sew. Works J.*, *14*, 964.

Heukelekian, H., & Chamberlain, N. S. (1931). The effect of the dilution water on the BOD determination. *Sew. Works J.*, *3*, 187.

Holderby, J. M., & Lea, W. L. (1934). The relation of the pH of dilution water to the BOD of various wastes. *Sew. Works J.*, *6*, 734.

Horne, R. H. (1954). BOD tests made easy by dilution graphs. *Water Eng.*, *25*, 71; *Public Health Eng. Abstr.*, *34*(5), 15 (1954).

Hurwitz, E., Barnett, G. R., Beaudoin, R. E., & Kramer, H. (1947). Nitrification and BOD. *Sew. Works J.*, *19*, 995.

Ingraham, J. L., & Ingraham, C. A. (1995). *Introduction to microbiology*. Wadsworth Publishing.

Jasewicz, L., Porges, N., & Hoover, S. R. (1954). Borax as a preservative of dairy waste for the BOD test. *Proc. 8th Ind. Waste Conf., Purdue University*, 387.

Jenkins, R. L., (1975). Los Angeles County Sanitation Districts, internal untitled memo.

Krogh, O. (1970). Analysis of organic materials in slaughterhouse sewage water. *Vatten*, *26*(1), 9; *Chemical Abstracts*, *76*, 62953 (1972).

Kuftarkova, E. A. (1972). Determination of BOD in sea water polluted with industrial and every day wastes. *Biol. Morya*, *27*, 162; *Chemical Abstracts*, *77*, 156130 (1972).

Lea, W. L. (1941). The role of nitrogen and phosphorus in BOD dilution water. *University of Wisconsin Symposium Hydrobiol.*, *71*; *Chemical Abstracts*, *36*, 2062 (1942).

Lea, W. L., & Nichols, M. S. (1936). Influence of substrate on biochemical oxygen demand. *Sew. Works J.*, *8*, 435.

Lea, W. L., & Nichols, M. S. (1937). Influence of phosphorus and nitrogen on BOD. *Sew. Works J.*, *9*, 34.

Lee, E. W., & Oswald, W. J. (1958). Preserved inoculum for BOD determinations. *University of California Inst. Eng. Res.*

Lee, E. W., & Oswald, W. J. (1959). Effects of seeding on BOD. *Water Waste Treatment J.*, *7*, 400.

Lloyd, B. (1937). Bacteria in stored water. *J. Roy. Tech. Coll., Glasgow*, *4*, 173.

Manoharan, R. (1994). *Investigation of total BOD_5 versus carbonaceous BOD_5 as a monitoring parameter for sewage works performance and compliance*. Canadian. Ministry Environ. Eng.

Mascarenhas, J. P., & Klein, G. (1958). Evaluation of BOD bottle cleaning techniques. *Sew. Ind. Wastes*, *30*, 976.

Mellbin, G. (1990). Application of laboratory robots in industrial automation. *Anal. Chim. Acta* (Netherlands), *238*(1), 71.

Michalik, W. A., et al. (1991). Automation of the BOD test for wastewater. *Adv. Lab. Autom. Rob.*, *7*, 325.

Mohlman, F. W. (1938). Nitrification. *Sew. Works J.*, *10*, 792.

Mohlman, F. W., Edwards, G. P., & Swoppe, G. (1928). Technique and significance of the BOD determination. *Ind. Eng. Chem.*, *20*, 242.

Moore, E. W., Thomas, H. A., Jr., & Snow, W. B. (1950). Simplified method for the analysis of BOD data. *Sew. Ind. Wastes*, *22*, 1343.

Munro, J. H. M. (1886). The formation and destruction of nitrates and nitrites in artificial solutions and in river and well waters. *J. Chem. Soc.*, *49*, 632.

National Council for Air and Stream Improvement. (1985). A review of the separation of carbonaceous and nitrogenous BOD in long-term BOD measurements. *Tech. Bull. No. 461*.

Nikitine, N. S. (1970). Determinations of 5-day BOD of purified wastewaters. *Bu. Prom.*, *5*, 19; *Chemical Abstracts*, *73*(10), 202 (1970).

O'Brien, W. J., & Clark, J. W. (1962). The historical development of the biochemical oxygen demand test. *Bull. No. 20, Eng. Exp. Stn., New Mexico State University*.

Okura, T. (1956). Determination of BOD. *Bunseki Kagaku, Shimpo Sosetsu* (Japan), *5*, 169; *Chemical Abstracts*, *51*, 9983 (1957).

Orford, H. E., & Ingram, W. T. (1953a). Deoxygenation of sewage. I. Critical review of the monomolecular formula. *Sew. Ind. Wastes*, *25*, 419.

Orford, H. E., & Ingram, W. T. (1953b). Deoxygenation of sewage. II. The logarithmic formula as applied to sewage. *Sew. Ind. Wastes*, *25*, 424.

Orford, H. E., & Ingram, W. T. (1953c). Deoxygenation of sewage. III. The logarithmic formula applied to oxidations other than sewage. *Sew. Ind. Wastes*, *25*, 566.

Orford, H. E., & Matusky, F. E. (1959). Comparisons of short term with 5-day BOD determinations of raw sewage. *Sew. Ind. Wastes*, *31*, 259.

Orford, H. E., Rand, M. C., & Gellman, I. (1953). A single dilution technique for BOD studies. *Sew. Ind. Wastes*, *25*, 284.

Painter, H. A. (1963). The use of the wide-bore dropping mercury electrode for the determination of rates of oxygen uptake and of oxidation of ammonia by microorganisms. *J. Appl. Bacteriol.*, *26*, 471.

Paixao, S. M., Baeta-Hall, L., & Anselmo, A. M. (2000). Evaluation of two commercial microbial inocula as seed in a 5-day BOD test. *Water Environ. Res.*, *72*, 282.

Phelps, E. B. (1909). The disinfection of sewage and sewage filter effluents. *U.S. Geol. Surv. Water Supply Paper*, *229*, 74.

Purvis, J. E., & Coleman, C. J. (1907). The decomposition and nitrification of sewage in alkaline solution and in distilled water. *Proc. Camb. Phil. Soc.*, *14*, 354.

Purvis, J. E., McHattie, A. C. N., & Fisher, R. H. J. (1911). the non-nitrification of sewage in sea water. *J. Royal Sanit. Inst.*, *32*, 442.

Raff, J. (1985). Nitrification suppression by allylthiourea and 2-chloro-6-(trichloromethyl)pyridine for determining carbon-BOD. *Ber. wasserguetwirtsch. gesundheitsingenieurwes. Tech. Univ. Munchen*, *60*, 177; *Chemical Abstracts*, *103*, 109657 (1985).

Razzell, W. E. (1972). Pittfalls and procedures in BOD tests. *Pulp Paper Mag. Can.*, *73*(7), 86.

Reilly, K. A., & Young, J. C. (1995). *Research report: The effect of trichloromethyl pyridine on carbonaceous biochemical oxygen demand in wastewater.* Pennsylvania State University.

Riehl, M. L. (1948). BOD dilution technique. *Sew. Works J.*, *20*, 219.

Rogers, P. (1949). Technique of the BOD determination. *Assoc. Pulp Paper Ind.*, *32*, 481.

Ropponen, M. (1979). Is closing the white water system an optimal method for minimizing 7-day BOD discharges from a newsprint paper mill? *Paperi Puu*, *61*(4a), 233.

Ruchhoft, C. C. (1938). The BOD test and oxidation in sewage treatment. *Sew. Works J.*, *10*, 264.

Ruchhoft, C. C. (1941). Report on the cooperative study of dilution waters made for the standard methods committee of the Federation of the Sewage Works Association. *Sew. Works J.*, *13*, 669.

Ruchhoft, C. C. (1950). Simplified method for analysis of BOD data: A discussion. *Sew. Ind. Wastes*, *22*, 1353.

Ruchhoft, C. C., Placak, O. R., & Ettinger, M. B. (1948). Variations in BOD velocity constants of sewage dilutions. *Ind. Eng. Chem.*, *40*, 1290.

Sawyer, C. N., & Bradney, L. (1946). Modernization of the BOD test for determining the efficiency of sewage treatment processes. *Sew. Works J.*, *18*, 1113.

Sawyer, C. N., Callejas, P., Moore, M., & Tom, A. Q. Y. (1950). Primary standards for BOD work. *Sew. Ind. Wastes*, *22*, 26.

Sawyer, C. N., McCarty, P. L., & Parkin, G. F. (1994). *Chemistry for environmental engineering* (4th ed.). McGraw-Hill.

Scheringa, K. (1920). Denitrification by bacteria and its relation to water analysis. *Central Lab. Public Health Serv. Rep.*, 165; *Chemical Abstracts*, *16*, 1289 (1922).

Schloessing, T., & Muntz, A. (1879). Recherches sur la nitrification. *Compt. Rend.*, *89*, 1074.

Seppovaara, O., & Starck, B. (1977). Receiving water as seeding material for determination of 7-day BOD in pulp and paper mill effluent. *Kemia Kemi*, *4*(5), 193.

Siddiqui, R. H., Speece, R. E., Englebrecht, R. S., & Schmidt, J. W. (1967). Elimination of nitrification in the BOD determination with 0.10 ammonia nitrogen. *J. Water Pollut. Control Fed.*, *39*, 579.

Slayton, J. L., & Travato, E. R. (1979). Simplified NOD determination. *Proc. 34th Ind. Waste Conf., Purdue University.*

Smith, R.-K. (1995). *Water and wastewater laboratory techniques*. Water Environment Federation.

Smith, R.-K. (1999a). *Handbook of environmental analysis* (4th ed.). Genium Publishing.

Smith, R.-K. (1999b). *Lectures on wastewater analysis and interpretation.* Genium Publishing.

Stamer, J. K., et al. (1979). Methods of ultimate CBOD determination. *J. Water Pollut. Control Fed.*, *51*, 918.

Stewart, R. M. (1964). Personal communication. Los Angeles County Sanitation District.

Strom, A., & Solyom, P. (1978). Origin and composition of effluent in a swedish liquid milk dairy. *Nordisk Mejeriindustri*, *5*(2), 41.

Sundberg, T. (1975). Two-stage biological treatment for high 7-day BOD reduction efficiencies. *Pulp Paper Int.*, *17*(12), 58.

Surmacz-Gorska, J., et al. (1995). Nitrification process control in activated sludge using oxygen uptake rate measurements. *Environ. Technol.*, *16*, 569.

Swango, W. H., & Lamb, C. B. (1950). Dilution technique for BOD test. *Sew. Ind. Wastes*, *22*, 939.

Symons, J. M., & Labonte, R. (1965). A procedure for continuous nitrification corrections during warburg respirometer studies. *J. Water Pollut. Control Fed.*, *37*, 842.

Theobald, L. S. (1959). A possible source of error arising in analysis from the use of polyethylene bottles. *Analyst*, *84*, 570.

Theriault, E. J. (1927). The oxygen demand of polluted waters. I. A critical review. II. The rate of deoxygenation. *U.S. Public Health Serv. Bull.*, *173*.

Theriault, E. J. (1931). Detailed Instructions for the performance of the dissolved oxygen and BOD tests. *Public Health Rep., Suppl.*, *46*, 30.

Theriault, E. J., & Homman, H. B. (1918). Studies on the treatment and disposal of industrial wastes. II. The determination of the biochemical oxygen demand of industrial wastes and sewage. *U.S. Public Health Serv. Bull.*, *97*.

Theriault, E. J., & McNamee, P. D. (1934). Experimental studies of natural purification in polluted waters. IX. Nitrification in sewage mixtures. *Public Health Rep.*, *49*, 1169.

Theriault, E. J., McNamee, P. D., & Butterfield, C. T. (1931). Experimental studies of natural purification in polluted water. V. Selection of dilution water for use in oxygen demand tests. *Public Health Rep.*, *46*, 1084.

Tom, A. Q. Y. (1951). Investigations on improving the BOD test. Unpublished thesis, Massachusetts Institute of Technology.

Tortora, G. J., Funke, B. R., & Case, C. L. (1998). *Microbiology, An introduction* (6th ed.), Addison–Wesley.

U.S. Environmental Protection Agency. (1986). *Method-by-method statistics from water pollution (WP). Laboratory performance evaluation studies.* Natl. Qual. Assurance Branch, Environ. Monit. Syst. Lab.

Walden, C. C., et al. (1972). Microbiologial treatment parameters of bleached kraft pulp mill effluents. *Pulp Paper Mag. Can.*, *73*(6), 49.

Washington, B., et al. (1983). Exertion of 5-day nitrogenous oxygen demand in nitrifying wastewaters. *J. Water Pollut. Control Fed.*, *55*, 1196.

Watts, R. J., & Adams, V. D. (1993). Elimination of sulfur (IV). Interference in five-day BOD determinations. *Analyst*, *118*, 1515.

Weinberger, L. W. (1949). Nitrogen metabolism in the activated sludge process. Unpublished thesis, Massachusetts Institute of Technology.

Wells, W. N. (1971). Minor modifications of the BOD test. *Water Sew. Works*, *118*, 207.

Wezernak, C. T., & Gannon, J. J. (1968). Evaluation of nitrification in streams. *J. Sanit. Eng. Div., Proc. Am. Soc. Civ. Eng.*, *94*, 883.

Whittemore, R. C., & Hovis, J. S. (1982). A review of ultimate BOD estimation and its kinetic formulation for pulp and paper mill effluents. *Natl. Council Air Stream Improvement Tech. Bull.*, *382*.

Wolf, J. R. (1950). Nomograph for determining proper dilution in the BOD test. *Water Sew. Works*, *97*, 122.

Woodring, S. L., & Clifford, D. A. (1988). A comparison of the graphical and standard methods for the determination of BOD. *J. Water Pollut. Control Fed.*, *60*, 537.

Young, J. C. (1973). Chemical methods for nitrification control. *J. Water Pollut. Control Fed.*, *45*, 639.

Young, J. C. (1996). Fundamentals of respirometry: Instrument types and basis of operation. *Proc. 51st Ind. Waste Conf., Purdue University*, 441.

Young, J. C., McDermott, G. N., & Jenkins, D. (1981). Alterations in the BOD procedure for the 15th edition of *Standard methods for the examination of water and wastewater. J. Water Pollut. Control Fed.*, *53*, 1253.

Young, J. C., Reilly, K. A., & Baumann, E. R. (1997). Effect of tetrachloromethyl pyridine on carbonaceous BOD. *Proc. 70th Annual Water Environ. Fed. Tech. Exposition Conf.*

Zanoni, A. E. (1969). Secondary effluent deoxygenation at different temperatures. *J. Water Pollut. Control Fed.*, *41*, 640.

8

Respirometric Methods

There has been a realization that a deeper understanding of the processes involved in treatment of wastewater and industrial wastes can be gained by methods, which unlike the standard 5-day biochemical oxygen demand (BOD) test, do not give a single numerical answer, but rather a continuous picture of the whole (biological) process (Jenkins, 1960).

Waste stabilization by microorganisms under aerobic conditions is the oldest and most easily implemented treatment process that is practiced. The term "aerobic" means that the stabilization is conducted in the presence of oxygen. As a first approximation, the microorganisms responsible for aerobic waste rectification require oxygen as the terminal electron acceptor in the metabolism of the waste. Biochemical oxygen demand is fundamentally nothing more than measurement of the utilization of oxygen by microorganisms as they eat and, as such, represents merely a limited application in the overall study of microorganism respiration (Longmuir, 1954).

Microorganisms are not the only organisms that consume oxygen during metabolism, and many of the techniques that have been used in the study of microbial respiration were first developed by medical researchers for the study of tissue respiration. Beginning in the 1890s, these tissue tools were adapted for the study of waste stabilization. Efforts to stabilize wastewater, optimize aerobic treatment processes, estimate wastewater strength as part of facility design, and evaluate industrial waste treatment potential have extensively relied on use of oxygen-consumption measurements. Other studies requiring oxygen-utilization measurements have been focused on assessment of stream effects from discharge of treated wastes, measurement of toxic effects to activated sludge, studies of biodegradation and acclimation rates, chronic effects of metals and other pollutants, and numerous other examples.

Biochemical oxygen demand, then, is not a single test or test result. Rather, it is a general phenomenon, and the widely practiced dilution bottle test is only one of many ways that are available for its measurement. It is perhaps unfortunate that the dilution bottle test is so well known, as this has placed the unreasonable burden on researchers to correlate results from alternate techniques for respiration measurement to those obtained from use of the standard 5-day BOD (BOD_5).

This chapter summaries respirometric measurement techniques and principles. Numerous reviews and books can be referenced for a more complete discussion of the use of respirometry for BOD determinations and other purposes (Dixon, 1943; King & Dutka, 1986; Montgomery, 1967; Ros, 1993; Steinecke, 1976; Tool, 1967; Umbriet et al., 1964; Vasel et al., 1989).

1.0 MEASUREMENT OF RESPIRATION

The term "respirometry" alludes, in a general way, to more or less continuous measurements of the oxygen consumed by organisms. Lest we get lulled into a single mode of thinking about respirometery, organisms that consume oxygen are not the only living things in a treatment facility that can assist in pollutant mineralization. Plants also contribute, and they produce oxygen. Respirometry studies have been performed many times on plants (Dor & Levy, 1987), and although the remainder of this chapter discusses consumption of oxygen, its production can be monitored using very similar techniques.

For the oxygen demand in wastewater and related samples, tests are usually conducted on samples open to an air headspace in some type of closed test system or reactor. That said, the earliest forerunners of this test consisted of open beakers of wastewater, from which aliquots of sample are removed for dissolved oxygen (DO) measurement (Clark, 1896, 1901, 1903). Nevertheless, the significant advances in respirometric technique came about through use of closed systems. These systems rely on some form of calibrated, liquid-filled tube (the manometer) connected to a reaction vessel in which even slight changes in internal pressure can be read by changes in the height of the liquid column. Consumption of gaseous oxygen during bacterial respiration is monitored over time as a change in pressure with these devices, thereby allowing calculation of an oxygen-uptake rate curve for whatever experimental conditions are desired. Addition of more oxygen to the system is possible via use of valves if longer run times are desirable.

The first use of a constant-pressure manometric device in wastewater studies was in 1890 (Adeny, 1897, 1908, 1928; Beach et al., 1995). Adeny's original constant-pressure apparatus was an adaptation of the Haldane device and suffered from leaks as well as the need for constant agitation of a cumbersome apparatus. The Haldane apparatus was a volumetric device developed to measure small quantities of oxygen in blood and oxygen binding by hemoglobin (Barcroft & Haldane, 1902). The equipment, however, was quite clumsy to use and required a great deal of operator expertise. It did not gain any favor with early practitioners in the wastewater field. In terms of BOD testing, most attention in the early 1900s instead focused on the development of the dosed-system dilution method (Adeny, 1908), which evolved into the current BOD_5 procedure. Whereas the latter techniques require that the waste strength be diluted so that there always remains an excess of DO in the test bottle, most respirometric methods relied on some method of oxygen replenishment during the course of the test. It took a number of innovations and advancement in thought and technique, however, for these so-called direct oxygen-utilization tests to evolve into functional procedures.

Respirometric measurements can be performed in several different ways. One is similar in format to the dilution bottle BOD test: a sample is aerated, sealed in a container, then incubated for a period of time. The DO level in the bottle is monitored continuously with a DO probe or periodically by withdrawing aliquots and measuring the DO (Keyes & Gillespie, 1913). The sample can be reaerated as the incubation proceeds to extend the monitoring beyond that allowed by exhaustion of the limited amount of DO at saturation. Reaeration is a tedious process. The results of the DO measurements are plotted against time on graph paper to give a respiration curve.

Most other forms of respirometry measurement expose the liquid sample to an atmosphere that continuously replenishes the oxygen in the liquid as it is used and permits longer-term measurements. The maintenance of a relatively high concentration of oxygen, or oxygen tension, in the sample has a positive effect on the utilization rate of the organisms being examined. Studies on fungi, bacteria, and other microorganisms generally support the idea that up to the oxygen saturation level, higher oxygen tensions relate to higher oxygen utilization rates (Callow, 1924; Georgi & Wilson, 1933; Hansen, 1945; Harris & Gainey, 1944; Nickerson & Chadwick, 1946; Nunheimer & Fabian, 1942; Schlayer, 1936; Sevag & Maiweg, 1934; Sevag & Shelburne, 1942; Shoup, 1929; Stern and Frazier, 1941).

The movement of oxygen from an atmosphere to a liquid is called transfer or reaeration. The basic physical law describing the oxygen concentration relationship between an atmosphere and solution in contact with each other is Henry's law:

$$p_j = k_j x_j$$

where p_j is the partial pressure of the substance j in the gas phase, x_j is the concentration of substance j in the liquid phase, and k_j is a constant called the Henry's law coefficient. The coefficient has a temperature dependence where, for oxygen, greater amounts of the gas are soluble in water as the temperature is lowered. Other gases can exhibit different behavior, and the equation for temperature effects on the Henry's law coefficient is different for each individual gas.

Henry's law is formulated under the assumption that the gas of interest forms an ideal solution with the solvent; in other words, there is no chemical interaction between the water and oxygen. For the most part, this is true in the pure oxygen–water system; however, there are situations in which other materials in water can affect the apparent concentration of oxygen. One example is the presence of particulates, where the oxygen can be involved in a surface absorption phenomenon (Sundararajan & Lu-Kwang, 1995). There are also observable differences in the equilibrium concentrations of oxygen in fresh and saltwater. Even considering the category of freshwater, there are differences between influent, effluent, and receiving stream waters and between hard and soft water (Elmore & Hayes, 1960, 1961). Equations have been derived that estimate Henry's law constant for these different situations (Benson & Krause, 1980).

In most water-related research, K_La is used as the oxygen mass-transfer coefficient, and there is a considerable volume of published literature concerning its determination. The active mass transfer of oxygen that takes place when bubblers are used in samples or aeration basins is of fundamental importance in controlling aerobic digestions (Eckenfelder, 1959; Ippen & Caver, 1954; Ippen et al., 1951; Ippen et al., 1952; Kite and Garrett, 1983), as is the passive reaeration that occurs in rivers and streams under quiescent conditions (Issacs &Gaudy, 1968). Static measurements using DO-sensing galvanic cells are used in the laboratory (Casey & O'Connor, 1978) for transfer rate determinations. A significant finding is that the actual velocity of oxygen molecules as they pass through the air–water interface is no different that the rate of diffusion of oxygen through the bulk water, indicating that surface tension is not a factor in oxygen transfer rates. An important factor in oxygen mass transfer is the surface area of water directly in contact with the air, which can be indirectly affected by surface tension.

Oxygen mass-transfer rates, Henry's law concentrations, and respiration rates are separate but linked concepts. Transfer rates describe the motion of individual molecules from air to water, and is a kinetic concept. Henry's law is a statement of the relative proportions of oxygen molecules in the water and air at equilibrium, but says nothing about how fast the molecules move from one phase to the other or how quickly the equilibrium is achieved.

Equilibrium states are dynamic situations; they are not static. An equilibrium expression such as Henry's law does not describe the movement of an individual molecule and, in fact, some molecules are moving from the air to water and some from the water to the air at all times, even at equilibrium. At equilibrium, a predictable portion of the molecules are in the water and a predictable portion are in the air, and the number of molecules at any instant moving from one phase to the other is balanced by the number moving in the opposite direction.

One way to understand an equilibrium is to picture a situation that contains four molecules. At equilibrium, assume that there will be two molecules in the water and two in the air. All four molecules are moving at every moment, and the direction of movement is completely random in the three dimensions around them. At any particular instant of time, either of the molecules in the water can move from the water to the air, giving three molecules in the air and one in the water. All of the molecules are still moving randomly, but now there is greater probability that one of the three molecules in the air will move to the water rather than the one molecule in the water moving to join the others in the air. Over the course of time, the average state will be two molecules in the water and two in the air.

The assumption of an equal concentration of oxygen in the two phases in this simplified air–water model ignores any ability of water to stabilize the DO molecules. In fact, water is very inefficient in dissolving oxygen, as evidenced by the 9-mg/L saturation level. However, Henry's law accounts for the relative stabilities of oxygen in water and air by the constant k_j and, when this is factored into the model, a statistical distribution quite accurately describes the situation.

As organisms respire, they consume oxygen in the water and upset the equilibrium amounts of oxygen in air and water. Oxygen will move in and out of the water at the transfer rate until a statistical balance between the air and water concentrations is reestablished. Respiration is ultimately limited by the transfer rate rather than the equilibrium concentrations, as respiration rates faster than the transfer rate will serve to exhaust the oxygen from the water and then bacterial respiration will slow down.

Respirometric methods have been used to directly estimate K_La (Howell et al., 1984; Koizumi & Aiba, 1984). Online respirometry techniques are particularly invaluable for these determinations when conditions are changing, as can frequently occur in influents to treatment facilities (Carlsson et al., 1994; Randall et al., 1991; Sollfrank & Gujer, 1990). Knowledge of maximum reaeration transfer rates means that the time needed for aerobic digestion can be adjusted to provide an optimum level of oxygen within the basin. Alternate methods for determination of transfer rates, as discussed in the chapter on theory, include the use of a surrogate gas such as krypton (Hovis, 1984); however, these techniques are generally time consuming and do not lend themselves to immediate response and control. Comprehensive reviews of the literature on transfer rate determination are available (Bennett, 1980; Gulliver, 1991; Rathburn, 1977).

Oxygen-transfer rates and Henry's law distributions affect the measurement of apparent respiration rates of biological samples (Mines & Sherrard, 1987). Particularly in devices that have small-bore tubing and use atmospheric oxygen, local depletions of oxygen can be created by rapidly respiring samples, which is only slowly replenished by diffusion from the oxygen reservoir into the air directly in contact with the sample. The lower oxygen partial pressure at the water surface skews the Henry's law relative balance between water and air and further restricts the respiration rate (Li & Zhang, 1996). Using the mathematical derivative of oxygen uptake against time in respirometry studies has been proposed as one way to correct for this diffusion controlled limitation (Chandran et al., 1996). Another approach is to flood the entire respirometer with 100% oxygen to prevent gaseous diffusion effects (Nakouzi et al., 1996). There can be localized areas in the liquid within the vessel that are depleted in oxygen, a problem especially noted in large (> 20 L) culture vessels (Elsworth et al., 1957; Hixon & Gaden, 1950). The common solution is to stir or vigorously agitate the sample. This provides better mixing within the sample, eliminating local depletion zones and increasing the surface area, providing more contact of the liquid sample with the atmosphere to improve oxygen transfer. Agitation is a standard technique used even in small-volume samples (Hartmann, 1967). One method for estimation of K_La alters the stirring speed of the sample while keeping mechanical reaeration constant and tracks the time required for establishment of a new steady-state value for DO (Mignone & Ertola, 1984). The time lag of the DO probe response to changing DO levels has been identified as a factor that must be taken into account when this technique is used for estimation of K_La and other purposes (Philichi & Stenstrom, 1989; Spanjers & Olsson, 1992).

Most respirometer designs provide for reaeration by direct contact with an atmosphere, although a respirometer that separates the sample from the atmosphere with an oxygen-permeable membrane has been described (Garg & Mathur, 1997a, 1997b). The uptake of oxygen by the organisms in the culture can then be followed by either monitoring oxygen depletion in the liquid phase of the culture or oxygen depletion in the atmosphere.

In-culture monitoring of oxygen uptake is of necessity limited to a very short timeframe and is complicated by the competing addition of oxygen via atmospheric transfer. Despite these limitations, it is a very useful technique for evaluation of the rapid biological utilization of substrate and is a superb technique for determining the acclimation of a culture to a substrate (Burkhead & McKinney, 1969; Ros & Dular, 1990; Ros et al., 1988a, 1988b; Ros et al., 1990; Vernimen et al., 1967). The device consists of a stirred culture vessel containing a DO probe. It is not necessary to have the device enclosed, and vessels open to the air are popular (Ros, 1993). Open-air systems are particularly useful when the respirometer is designed as part of a continuous-flow system rather than a batch analysis (Clarke et al., 1978; Lukasse et al., 1997; Lukasse et al., 2000; Mann, 1972; Spanjers et al., 1993). The rate of endogenous respiration is determined and a visual plot prepared of the DO against time. The line describing the rate of endogenous respiration also includes the contribution from atmospheric oxygen transfer. In a perfectly balanced culture, the rate of endogenous respiration is exactly equal to the rate of reaeration, and no overall depletion is noted. Injection of a small amount of readily degradable substrate causes an immediate drop in the DO level in the culture. After the substrate is consumed, the culture returns to the endogenous oxygen depletion rate. The time required for the return is related to the kinetics of the ability of the culture to degrade the substrate (Cech et al., 1985). Figure 8.1 is a graph of the DO behavior under these conditions.

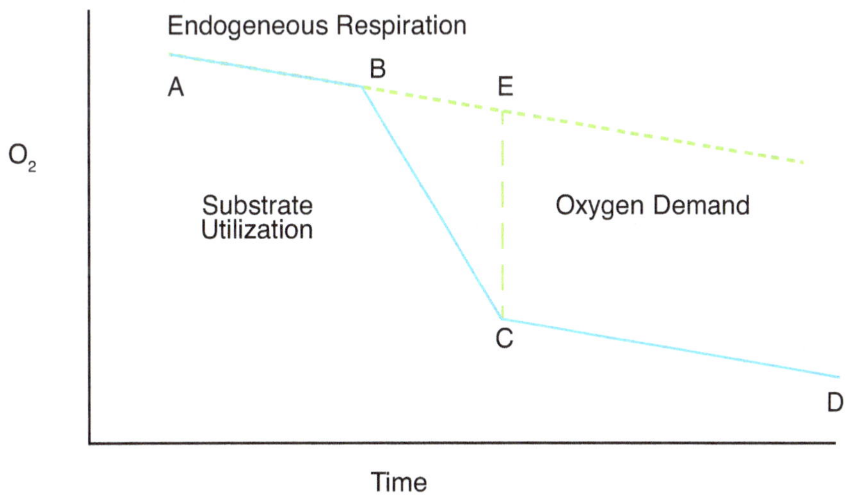

FIGURE 8.1. Idealized plot of DO levels in an open air respirometer for a culture utilizing a readily biodegradable substrate.

With reference to Figure 8.1, the rate of endogenous respiration is determined as the line AB, then substrate is added at point B. Substrate utilization is completed at point C, as indicated by the rate of oxygen utilization returning to the same slope (AB) as endogenous respiration. The oxygen demand is determined by extending the original endogenous respiration rate (AB) through point E and then dropping a vertical (line EC) to line CD.

If respiration is monitored by measuring the amount or depletion rate of oxygen in the atmosphere, a closed system is required. The system can be designed to operate in either a constant-volume or constant-pressure mode.

In a constant-volume mode, the partial pressure of oxygen in the closed atmosphere will decrease. One consequence of the decreasing partial pressure of oxygen is an effect on the Henry's law concentration of oxygen in the liquid culture. Decreasing partial pressure means decreasing concentration of oxygen in the liquid. This adds complications to the determination of the oxygen-utilization rate by the organisms and limits the length of time available for conducting the experiment. Further, as the DO level in the liquid drops, the oxygen tension will be dropping, changing the efficiency of oxygen respiration by the organisms. Nonetheless, use of respirometers of this design has appeared in the literature. A pressure transducer or other pressure-monitoring device is used in the system to provide a metric that is indirectly related to the use of oxygen (Burchard, 1966; Genthe et al., 1972; King & Palmer, 1989). A modification of this basic system eliminates the problems caused by decreasing oxygen partial pressure. The signal from a pressure transducer is used as feedback in a control loop that provides continuous replenishment of the oxygen in the atmosphere as it is used. The source of the oxygen can be a compressed cylinder; however, a more suitable system contains an electrolysis cell where oxygen is made from water as needed. This constant-volume device is called the electrolytic respirometer. The metric is the amount of current used in the electrolysis cell to maintain the partial pressure of oxygen in the system as the organisms respire. Electrolytic respirometers, such as the Sapromat AC, are commercially available and enjoy widespread use (Steinecke, 1967, 1968, 1976).

The constant-pressure mode of operation is embodied in the Haldane, Sierp, Barcroft, and Warburg respirometers. The atmosphere in the chamber holding the test solution is connected to either the outside atmosphere or to a pressure standard via a liquid-filled volumetric measuring tube (the manometer). As oxygen is used in these closed-atmosphere devices, the external pressure pushes on the liquid in the measuring tube and the volume occupied by the internal atmosphere decreases. The decrease in volume is quantitatively determined by a graduated scale on the measuring tube.

A technical key to successful use of the manometer-based respirometers is the liquid in the measuring tube. It must not interact with the gases being measured and, further, must exhibit essentially no thermal expansion or contraction. Two fluid formulations have been commonly used in manometers (Caldwell & Langelier, 1948; Umbriet et al., 1964). The use of Brodie solution has generally been replaced by Krebs solution.

Brodie solution:

- Add 23 g sodium chloride;
- Add 5 g sodium choleate;
- Dilute to 500 mL water, density 1.033; and
- Add 200 mg/L Evans Blue or acid fuchsin for blue or red color.

Krebs solution:

- Add 44 g anhydrous sodium bromide;
- Add 0.3 g Triton X-100;
- Dilute to 1000 mL water, density 1.033 at 20 °C; and
- Add 300 mg/L Evans Blue or acid fuchsin for blue or red color.

Manometers measure the change in volume of the total atmosphere within the closed system of the respirometer. Early work on bacterial metabolism showed that the only gases involved in aerobic bacterial metabolism are oxygen and carbon dioxide, oxygen being utilized and carbon dioxide being produced. The interference caused by carbon dioxide with the volumetric approach has generally been circumvented by removal of the carbon dioxide with an alkaline absorbant such as the bases barium hydroxide or sodium hydroxide. Carbon dioxide reacts with the base forming a carbonate. Some researchers have intentionally monitored metabolic carbon dioxide evolution by titration, conductimetric, or gravimetric analysis of the absorbant (Govind et al., 1997; Porges et al., 1952).

2.0 OPERATION OF MANOMETRIC AND ELECTROLYTIC RESPIROMETERS

Several of the early respirometric devices came to be widely used, including the Sierp apparatus, the Barcroft differential manometer, and the Warburg manometer. These are illustrated in Figures 8.2, 8.3, and 8.4.

The Sierp apparatus (Figure 8.2) is a constant-pressure type of system consisting of a glass bottle reactor that holds the sample, topped by a standard taper fitting through which is mounted a calibrated gas bottle or buret. The two main components are connected by a small-diameter diffusion tube.

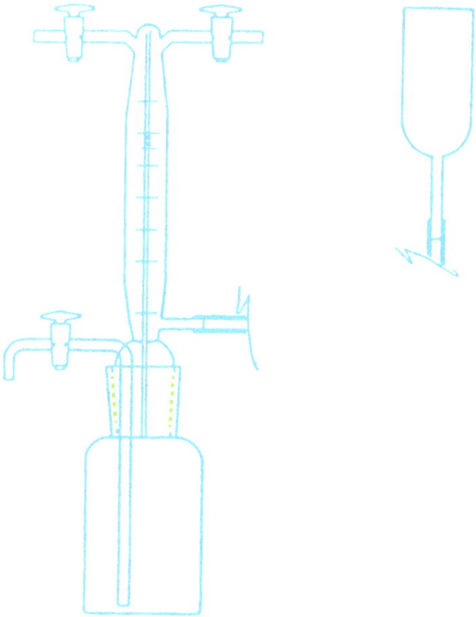

FIGURE 8.2. Sierp apparatus.

A side tube from the reactor is fitted with a stopcock for removal of excess liquid when oxygen is added. The top of the gas bottle has two stopcocks for adding oxygen when needed. A leveling bulb is connected to the bottom of the gas bottle for adjusting liquid levels inside the apparatus. Before use, the volume of the entire apparatus must be determined. To do this, the reactor and gas buret is filled with water, ensuring that all air is excluded from the device. The manometer is then disassembled, and the water is poured into a graduated cylinder, giving the apparatus volume. The reaction vessel is filled with sample, which is forced up the diffusion tube nearly to the top using the leveling bulb. The buret is purged with oxygen for a couple of minutes, then the exit stopcock on top of the buret is closed. The side tube stopcock is then opened and a portion of sample forced out by the pressurized oxygen flow, displacing the sample in the reactor until about 100 mL is collected in a graduate. The volume is recorded. The side tube is closed, and oxygen is allowed to fill the gas buret. As oxygen is consumed, the liquid level in the diffusion tube extending into the buret drops in proportion to the volume of oxygen consumed. The system has to be shaken during the course of the experiment, and the temperature and barometric pressure recorded. Early designs of the system required that carbon dioxide corrections be made in stoichiometric ratio to the oxygen consumed, based on independent determination of the carbon dioxide production. Even when carbon dioxide absorption tubes are added to the device, they are inefficient, and absorption corrections must be determined and applied to the data (Falk & Rudolfs, 1947).

The Sierp manometer was evaluated and modified for use to estimate BOD (Gerloff, 1942) (see Figure 8.2). Despite the reported success of the measurement technique, the apparatus did not gain a huge following of users. The main problem with the original Sierp system is that it tended to leak air during shaking because of the cork or rubber stoppers used to connect the various parts of the system. Leaks are a fatal flaw in respirometric equipment. The proposed modification introduced automated shaking, an improvement over the manual shake, but the shaking apparatus was awkward. And it still did not correct the leakage problem. Leaks remained a problem until the apparatus was redesigned with glass joints replacing the stoppers used to assemble the various parts of the manometer. Agitation of the all-glass respirometer is accomplished by shaking the entire assembly on a mechanical shaker table (Falk & Rudolfs, 1947).

The Barcroft differential manometer is a constant-volume type system (Figure 8.3) and is considerably less clumsy to use. It consists of a reaction flask and a control flask on opposite ends of a U-shaped capillary manometer. Each end of the manometer is fitted with a stopcock to equalize pressure

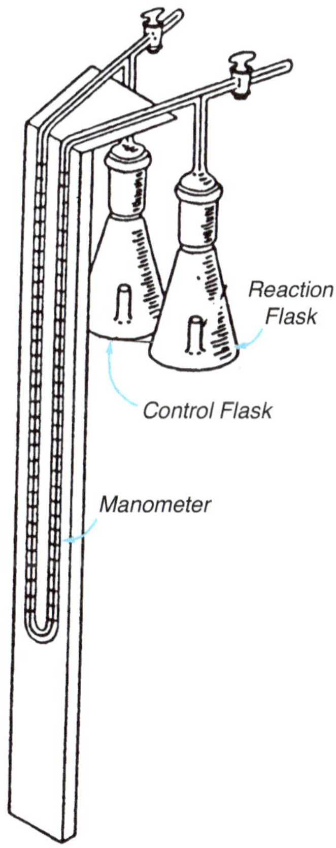

FIGURE 8.3. Barcroft differential manometer.

in each flask at the beginning of the test. The flasks are fitted with a center well to contain alkali solution for absorption of carbon dioxide. In use, one flask contains a small measured amount of sample, and the other a measured amount of clean water. After initial shaking and temperature equilibration in a water bath, the stopcocks are closed, and shaking at constant temperature is continued. As respiration progresses in the sample, the manometer fluid can be read directly at any desired time interval. Readings are not adversely affected by barometric pressure or slight temperature changes because of the self-compensating differential setup, thus the system is considered quite precise. In addition, it does not rely on replenishment with pure oxygen, but simply can be reset with air by opening the stopcocks. An evaluation of the repeatability of measurements from Barcroft differential manometers showed that it is indeed a reliable instrument (Corbet & Wooldridge, 1936).

The Warburg respirometer (Figure 8.4) was originally developed to study tissue respiration (Warburg, 1923, 1926). It is similar in design to the Barcroft

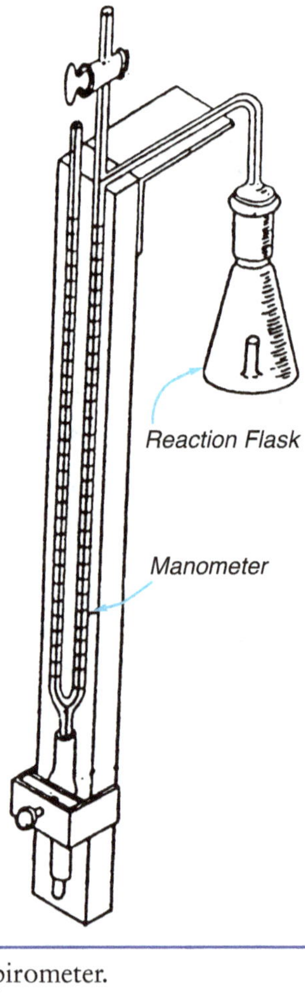

FIGURE 8.4. Warburg respirometer.

differential manometer, but somewhat simpler in that only one respiration flask is used on the U-tube manometer, the other end being open to the atmosphere. A separate control respirometer is used for each series of respirometric measurements. In wastewater applications, an angled side arm can be fitted to the side of the reaction flask for addition of feed or test substrate at any desired time. Compensation for barometric and temperature irregularities are performed with the control flask and manometer. Manometric readings are collected at specified times. The Warburg respirometer is fairly inexpensive, and researchers normally attach them in sets of up to 25 to large circular supports, the whole support being agitated by a mechanical rocker assembly during use.

The first edition of a book presenting standardized Warburg respirometry techniques for the study of tissue respiration appeared in 1946 (Umbreit et al., 1964). It documented analytical methods for the commercially available equipment and was the standard technical reference for many practitioners in the wastewater field. The fourth edition came out almost 40 years ago, but this book is still the prime reference for respirometry techniques.

Most respirometric devices use relatively small samples, frequently in the 10- to 100-mL range. However, some researchers find it desirable to use systems that accommodate larger samples. For example, a 600-mL capacity respirometric vessel has been described (Murray & Coaker, 1956). A 2 L BOD reaction vessel for determining BOD in wastewater and industrial waste was developed (Clark, 1959, 1960). These large-capacity systems are generally designed as electrolytic respirometers and are operated as automated continuous BOD reactors that can be run for long periods of time. Further work demonstrated that the 2-L system can be used for evaluating wastes with 1000- to 5000-mg/L demands without dilution. With such large systems, it is necessary to ensure complete mixing, and a simple mechanical mixing device that can accomplish this goal is described. The 5-day direct uptake BOD is demonstrated to be within 5% of the theoretical carbonaceous BOD. The onset of nitrification in domestic wastewater was also studied.

The electrolytic respirometer underwent several modifications, and a commercial version was marketed as the Sapromat AC, developed in 1967. An improved version that is quite compact is described (Young et al., 1965; Young & Baumann, 1976a, 1976b). This improvement greatly simplified the device, relying to a large extent on stock glassware with some custom modification. The system consists of a 1-L reagent bottle with standard taper ground glass neck and magnetic stir bar. The carbon dioxide absorber is a modified standard taper adapter, and the electrolysis cell is housed in a glassblower's creation fitted inside another ground glass adapter, all of which is mounted atop the reagent bottle in a compact apparatus. The cell houses platinum wire switching electrodes, hydrogen electrode, oxygen electrode, and electrolyte for generation of oxygen on demand. This equipment is compact enough to fit several units inside a normal BOD incubator. These systems provide continuous readout of oxygen uptake and have proven themselves to be very precise compared with the dilution method (Young et al., 1965). This equipment has been used to examine the factors affecting oxygen-uptake measurements in electrolytic respirometers and demonstrates the utility of these instruments in wastewater resource recovery facility (WRRF) control (Young & Baumann, 1976a, 1976b). These publications marked a turning point in respirometric methods and pointed the direction away from the classical volumetric displacement and manometric methods based on the Sierp and Warburg apparatus.

The main limitations on operation of the electrolytic respirometer are mixing rates and the concentration or strength of the test material, both of which can be overcome by simply providing variable stirring and an oxygen-enriched gas phase with the electrolysis cell design. Decisions on whether to use full-strength or diluted waste samples depend on whether the investigator wishes to evaluate treatment facility kinetics or try to mimic the BOD_5 dilution test.

Respirometric equipment has been tested for use in routine monitoring by laboratory and operations staff at nine WRRFs (Young & Baumann, 1976b). With the electrolytic respirometer, it is possible to run tests for days or weeks without interruption, on relatively large sample volumes, and with very good precision. The need for nitrification control was demonstrated and, under those conditions, it was possible to get good correlation with the dilution test at 3-, 4-, and 5-day incubation periods for some sludges. However, for other treatment facilities, the equivalent BOD (EBOD, as it was termed) at 2.5 days most closely resembled the BOD_5. Industrial wastewaters from different locations were evaluated within this study for acclimation and degradability assessment. The study of acclimation rates for toxic materials has become a standard application of electrolytic respirometers (Madden & Tittlebaum, 1984).

3.0 COMPARISON OF RESPIROMETERS

In 1948, a review of available manometric systems for evaluating wastewater appeared (Caldwell & Langelier, 1948). It presented detailed comparisons of the two most widely used and reliable respirometers available, the Barcroft differential manometer and the Warburg respirometer. It included data from extensive tests using the modified Sierp system. The constant-pressure Sierp volumetric system was reported to be less accurate than either Warburg or Barcroft systems. The Warburg respirometer was recommended as being the most suitable for routine use, as it is somewhat simpler to use. This publication is a bit of a landmark, in that it develops the theory and practice of using the Warburg apparatus for the analysis of BOD and provides suggested modifications of commercially available equipment for manometric determinations. Even though the Warburg had been in use for tissue respiration studies for 20 years, this review presented an approach for its use in wastewater studies. It described preferred ratios of sample volume to flask size, an optimum method for removal of carbon dioxide, and introduced the use of buffered samples. Another innovation was evaluation of test precision as a quality control measure. Mathematical descriptions of the rate of oxygen diffusion into the liquid phase, both with and without mixing or shaking, were included. In the experimental section, the rate constants for oxygen uptake by activated sludge obtained using manometric methods were described as considerably higher than those obtained from the dilution method. The 24-hour respirometer-derived BOD results were found to be approximately 75% of the dilution bottle BOD_5 values.

The oxygen-uptake rate of activated sludge under a variety of conditions using both the Warburg and Barcroft manometers has been studied (Dawson & Jenkins, 1949). This work started in England during World War II then, after being interrupted for several years, was completed near the end of the 1940s. Their efforts in some ways paralleled the work of Caldwell and Langelier (1948). Lacking commercial-grade equipment, they constructed their own mechanical shaker and constant-temperature system for the manometers. In two papers, the use of manometers was demonstrated by producing a wealth of information on the behavior of activated sludge. The effects of photosynthesis, homogenization, and various inert substrates on oxygen uptake of activated sludge were evaluated, as were the effects of metal cations and anions, biocidal organics, sugars, proteins, amino acids, peptides, carbohydrates, acids, and alcohols (Dawson & Jenkins, 1949 and 1950).

The International Water Association (IWA) maintained a website at the time the first edition of this book was published listing more than two dozen commercial respirometric systems, most of which can be connected to a personal computer (PC) for data logging, calculations, and modeling (http://www.spb.wagoningen-ur.nl/mt/IAWQSTRPROJECT/). The IWA Task Group on Respirometry also published a technical report summarizing the principles of use of respirometry in control of activated sludge (Spanjers et al., 1996). These resources should be consulted for a detailed overview of technology and equipment. It is salient to point out here that the prime motivation for IWA is the strong interest within the European community to understand the limitations of activated sludge in treating toxic organic compounds (OECD, 1998) and to model activated sludge treatment systems for optimization of facility performance. This focused interest on biodegradability and operations control has steadily increased since the early publications of the 1960s and 1970s on these topics, and was assisted greatly by the continued development and commercialization of automated respirometric equipment.

4.0 A STANDARD RESPIROMETRIC METHOD

The purpose of a standardized respirometric method is not to provide an alternative to the BOD_5 method that will mimic the BOD_5 results, although many have tried to do just this, as illustrated in Table 8.1. The techniques are generally far too tedious to be of any practical application in that regard, and short-term determination of fewer than 5 days rarely provides a good comparison with BOD_5 or ultimate BOD results. However, these methods do provide a means of following the oxygen-consumption pattern when BOD tests do not proceed normally, when toxicity may be involved, or when bacterial acclimation to a waste might be necessary. For many practitioners, respirometric methods have offered the opportunity to simulate treatment facility conditions in ways that simply can not be achieved in a BOD bottle. The effects from various chemicals, waste types, or treatment conditions can be observed with more or less continuous oxygen demand observations, and meaningful characterizations of these are often available in fewer than 5 days.

A standardized method for determining respirometric BOD, using both the Sierp and Warburg respirometers, was published in *Standard Methods* format as an appendix to a paper (Gellman & Heukelekian, 1951, 1953). A modification to the Sierp apparatus ensured that all carbon dioxide was removed from the system, thus solving the problems that plagued earlier work. This problem had already been solved in the Warburg instrument, where it was routine practice to use a highly alkaline medium placed in a small well inside the reaction vessel to absorb the carbon dioxide. For the Sierp apparatus, an additional vial was added to the lower connecting tube of the system to improve gas-phase contact with the alkaline solution. The necessity of continuous agitation or mixing in respirometric tests was also demonstrated. In a companion paper, work on the oxidation of wastewater and industrial wastes using respirometer techniques examined the effect of factors such as pH, nutrient addition, substrate concentration, effects of seed source and volume, and effect of adapted seed (Heukelekian & Gellman, 1951).

It was these developments with the Sierp, Barcroft, and Warburg equipment in the late 1940s and into the 1950s that set the stage for a *Standard Methods* section on respirometric BOD. In 1955, the 10th edition of

TABLE 8.1. Respirometric determinations of BOD_5.

Reference	Comment
Symons & Buswell, 1929	Sierp respirometer and dilution BOD method compared
Sierp & Fransemeier, 1931	Sierp respirometer used to determine BOD
Wooldridge & Standfast, 1936	Barcroft manometer used to estimate BOD
Heukelekian, 1947	Respirometric determination of BOD
Gellman, 1950	Respirometric determination of BOD
Heukelekian & Gellman, 1951	Warburg determination of BOD
Ludwig et al., 1951	BOD determined with manometric respirometer
Hoover & Porges, 1952	Related microbial synthesis to respirometric oxygen consumption
Jaeger & Niemitz, 1952	BOD determined with Warburg respirometer
Berger, 1955	BOD calculation from respirometric measurements
Elmore, 1955	BOD determined with reaeration respirometer
Heukelekian & Gellman, 1955	Warburg determination of BOD
Murray, 1955	Barcroft manometer used for BOD determinations
Wheatland & Lloyd, 1955	Respirometric determination of BOD
Mendia, 1956	Respirometric determination of BOD
Murray & Coaker, 1956	Modified Barcroft manometer used in BOD determinations
Dillingham et al., 1958	Manometric respirometer for determination of BOD
Donlon, 1958	Rapid respirometric method for BOD
Papel et al., 1958	Volumetric respirometer determination of BOD
Clark, 1959	Electrolytic respirometer determination of BOD
Snaddon & Harkness, 1959	Manometric determination of BOD
Clark, 1960	Electrolytic respirometer determination of BOD
Dillingham & Jose, 1960	Manometric respirometer for determination of BOD
Clark, 1961a, 1961b	Electrolytic respirometer determination of BOD
Busch et al., 1962	Short-term BOD by respirometry
Dozank, 1965	Warburg estimation of BOD
Knorr, 1965	BOD determination with continuous electrolytic respirometer
Zielinski et al., 1965	Manometric respirometer for BOD measurements
Burchard, 1966	Barometric respirometer use related to BOD
Bolling, 1967	Conversion of Warburg and Sapromat data to BOD results
Steinecke, 1967	Sapromat AC respirometer introduced for BOD measurement
Vernimem et al., 1967	Short-term BOD measured with respirometery

(continued)

TABLE 8.1. Respirometric determinations of BOD_5 (*continued*).

Reference	Comment
Fuhs, 1968	Factors affecting BOD determination by manometers
Ingols, 1968	Respirometric method for short-term BOD determination
Steinecke, 1968	Use of Sapromat AC respirometer to measure BOD
Bridie, 1969	BOD determined from electrolytic respirometer
Buczowska & Jarnuszkiewicz, 1969	First-stage BOD from respirometer oxygen-uptake rate
Gomolka, 1970	Respirometer apparatus for determination of BOD and carbon dioxide
Montgomery & Gardiner, 1971	Bacterial inoculums for use in respirometric BOD determinations
Farkas & Literathy, 1972	Device to measure integrated respiration to determine BOD
Fleps, 1973	Constant-pressure manometer used to determine BOD
Arthur, 1974	Manometric determination of BOD
Nay et al., 1974	Respirometry used as a dynamic BOD test
Verstraete et al., 1974	BOD measurements made with Sapromat manometric respirometer
Fleps, 1975	Automated continuous-flow respirometer for BOD determination
Kuz'min et al., 1975	Automated electrometric respirometer used to determine BOD
Young & Baumann, 1976a, 1976b	Electrolytic respirometer used to determine BOD
Hung & Eckenfelder, 1976	Warburg and Marais respirometric results correlated to BOD
Peberdy, 1976	Commercial manometric respirometer used to measure BOD
Tebbutt & Berkum, 1976	Large-volume respirometer for BOD determination
Golovatyi et al., 1977	Online respirometer oxygen-uptake rate calibrated to BOD
Atolarczyk, 1978	Oxygen-generating electrolytic cell feeds sealed reaction chamber
Gantner, 1979	BOD determined with electrolytic respirometer
Bauer et al., 1981	40-minute determination of BOD degradability of pollutants
Burchard, 1981	Construction of an online BOD respirometer
Farkas, 1981	BOD measured w ith integrated respiration rate device

(*continued*)

TABLE 8.1. Respirometric determinations of BOD_5 (*continued*).

Reference	Comment
Heddle & Tavener, 1981	Carbonaceous BOD and BOD results from a respirometer compared
Riegler, 1984	Continuous respirometric measurement of BOD
Kulikov & Vasil'eva, 1985	Automated respirometer for determination of BOD
Koehne et al., 1986	Dilution and respirometer determined BOD compared
Suschka & Ferreira, 1986	Use of DO probe in respirometric BOD measurements
Schreiber & Neumainer, 1987	Electrolytic respirometer used to determine BOD
Cailas & Gehr, 1989a, 1989b	Respirometer oxygen-uptake rate data and determination of BOD
Allot, 1990	3-minute online bioreactor respirometer related to BOD
Lenk et al., 1990	Online respirometer determines plateau BOD values
Le Gall, 1991	Review of short-term BOD measurements with respirometers
Spanjers & Klapwijk, 1991	Short-term BOD measured by respirometry
Londong, 1992	BOD in 3 minutes with a respirometer
Spanjers, Olsson, & Klapwijk, 1994; Spanjers, Temmink, & Klapwijk, 1994	Short-term BOD measured by online respirometry
Watts & Garber, 1995	Online respirometry used to estimated BOD
Kong , Aerewijck, & Werstraete, 1996; Kong, Vanrolleghem et al., 1996	Respirometric biosensor used to determine BOD
Lukasse et al., 1997	Continuous-feed respirometer used for BOD determinations
Lukasse et al., 1997; Lukasse et al., 2000	Continuous-flow respirometer used for BOD determinations

Standard Methods contained the first tentative method for manometric BOD procedures. This procedure was written to allow use of either the Warburg or Sierp apparatus and was included in the industrial waste section of the book. The BOD_5 method also referred the analyst to use of respirometric methods for specific waste streams or study of alternate incubation periods. A number of features characteristic of the 5-day dilution bottle method were included: dilution water nutrients, use of glucose–glutamic acid (GGA) check standards, pH adjustment, and a check on the phosphorus-to-nitrogen ratio for the seed organisms. The method was provided as an adjunct method to the BOD_5 test to facilitate treatability studies for industrial waste or high-strength wastes in activated sludge. Unlike the dilution method, it allowed measurement of oxygen-uptake characteristics at concentrations normally encountered in waste treatment processes and facilitated determination of oxidation rate constants, effects of pH and nutrient changes, and observations of the effects of adaptive seed. The method was designed to be up to a 5-day test, with manometer readings to be taken at whatever interval the operator desired. Sample volumes up to 50 mL in the Warburg and 100 mL in the Sierp systems could be accommodated with the right equipment, although in the commercial Warburg equipment, samples less than 10 mL were more common. Using an oxygen atmosphere, the Sierp procedure was capable of measuring BOD concentrations up to 10 000 mg/L, the Warburg up to 2300 mg/L. Many of the conditions set forth were from the earlier publication (Gellman & Huekelekian, 1951). The 11th edition of *Standard Methods* was published in 1960, with no change in the tentative respirometric method from the 10th edition. By the 12th edition (APHA et al., 1965), however, respirometric methods disappeared from the publication, presumably because of a lack of point-to-point correlation with BOD_5 and perhaps a perceived lack of need for a standardized procedure for what was viewed by the Joint Task Group as primarily a research tool. The method does not seem to have disappeared from *Standard Methods* because of a lack of interest by wastewater treatment operators in the field, as interest in use of the method for a variety of purposes was becoming more popular, evidenced by the increasing number of articles in the trade journals and presentations at conferences.

Since that time, at least three organizations have presented modern attempts at standardizing respirometric methods: the Standard Methods Committee in the 19th edition of *Standard Methods* (APHA et al., 1995), the International Water Association (IWA) (Spanjers et al., 1996), and American Society of Testing Materials (ASTM, 1994). It seems obvious that professional organizations have become convinced of the value of these methods in the water and wastewater field and have seen the need for providing common frameworks of usage. The reader has already been referred to the IWA publications for a comprehensive overview of contemporary equipment and methodology (Spanjers et al., 1996). The ASTM publications and *Standard Methods* may be reviewed for more “cookbook” procedural templates. We will here review the *Standard Methods* respirometric procedure.

Between the 11th edition and 19th edition of *Standard Methods*, no methods were offered for respirometric BOD. However, in the late 1980s, the BOD Joint Task Group began working on a draft procedure to reintroduce respirometry to the BOD section of the book. Production of an approved method took many iterations and attempts to satisfy different users to develop a more universal framework accommodating various equipment and application variables. Method 5210 D finally appeared in the 19th edition of *Standard Methods* (APHA et al., 1995).

Method 5210 D is written as a basic guideline. As such, it recognizes that there are inherent features of the available test equipment that are different, and that there is a wide range of possible applications, seed cultures, and sample types or mixtures that may be tested. As with most other types of analytical instrumentation, analysts have to rely on the manufacturers' instructions and manuals for operating the equipment chosen for a specific application. It is important that available equipment be reviewed with knowledge of the intended application before acquisition. Keep in mind that the wide range of design features in commercial equipment result from inventiveness driven by the perceived differences in applications. An analyst with a need for online respirometry may find that a set of small electrolytic respirometers designed for batch testing in the laboratory will present serious adaptive challenges. A device like a Warburg respirometer is suitable only for offline measurements.

The introduction to 5210 D classifies the available respirometers into four main types, which should now be recognizable to the reader of this chapter even though we have not chosen to separate them thusly: (1) manometric, (2) volumetric, (3) electrolytic, and (4) direct oxygen input. All modern commercial respirometers can be interfaced to a PC or microprocessor for data accumulation, and this is an essential feature for most uses. While some of the classical equipment like Warburg manometers are certainly serviceable, they are much more labor intensive than PC systems in terms of data acquisition, storage, calculations, and display. So, any intended use of respirometric methods should prompt consideration of modern data-processing equipment.

The 5210 D procedure assumes the existence of some common features among the acceptable equipment, including (1) a means of mixing the sample during the course of the test, (2) the ability to quantitatively remove carbon dioxide on a continuous basis using some type of scrubber system, (3) effective temperature control or at least amenability to external temperature control, and (4) accurate metering and measurement systems. Use of appropriate control mechanisms minimizes interference to data acquisition and interpretation from carbon dioxide or ambient temperature and barometric pressure changes.

Although the purpose of the method is not to reproduce the BOD_5, the presentation recognizes that it is possible, in some cases, to establish a good correlation between respirometric and 5-day dilution tests for specific samples. The test duration described, then, is 5 days. The procedure is primarily intended to take advantage of respirometry's advantages in terms of continuously tracking oxygen consumption and thus providing valuable information about biodegradation, treatability, toxicity, and acclimation. Continuous or periodic oxygen readings can be taken; data sampling at 15-minute to 6-hour intervals are mentioned specifically as possible choices. The utility in determining long-term, ultimate BOD is also noted, although that is a separate procedure described in Method 5210 C.

The BOD Joint Task Group made a deliberate effort to have Method 5210 D be consistent with protocols established by the European Organisation for Economic Cooperation and Development (OECD, 1998), which require oxygen-uptake methods. Some of these features include the strength of the nutrient buffer and use of trace element and nutrient augmentation solutions, use of nitrification inhibitor (see Chapter 7), and use of GGA check standard. Dilution water sources should be free of extraneous organic matter, chlorine, chloramine, and heavy metals. The pH should not affect the buffer capacity of the nutrient mix added for testing.

The nutrient and mineral buffer prescription is somewhat different from the dilution BOD_5 method, based on the findings that organism nutrient requirements are different in direct oxygen-consumption tests than they are in the dilution test (OECD, 1998). Two different approaches are provided as options. The method instructs the addition of sufficient ammonia nitrogen to provide a ratio of chemical oxygen demand (COD):nitrogen:phosphorus equal to 100:5:1, or in terms of total organic carbon (TOC), 30:5:1. Whether the trace mineral solutions are added depends on knowledge of whether these constituents will be available at sufficient concentration in the diluted sample. If phosphate buffer is used, use 1 mL of the standardized phosphate buffer per 50 mg/L COD in the diluted sample. As stated, this will normally be sufficient to maintain the pH in the range of 6.8 to 7.2 for the duration of the test. As an option, the analyst can substitute the OECD nutrient mix as specified in the method.

Many of the prescriptive and recommended sample-handling steps in the respirometric method are the same as the *Standard Methods* procedures for the BOD_5 dilution method. Collection and storage conditions for grab and composite samples are equivalent. While analysis of fresh sample is always recommended, samples should be iced if held more than 2 hours; testing should normally begin within 6 hours of collection but, if not possible, then at most within 24 hours of the end of the collection period. Because of the critical effect of temperature on both oxygen-consumption rate and gas-phase measurement, the sample, seed, and controls need to be brought to within 1 °C of test temperature conditions before initiation of test steps like dilution or mixing. Chlorinated samples should be dechlorinated before mixing with seed. Either aerate the sample, let it stand in ambient light for 1 to 2 hours, or dechlorinate stoichiometrically with sulfite as described in BOD_5 Method 5210 B.

Samples need to be well mixed, and if large settleable particles exist, the sample can be blended. Mechanically homogenized samples need to be subsampled quickly and transferred to the test containers before they resettle. Volumetric measurement of mechanically blended samples may be problematic if entrained air bubbles cannot be dispersed. Also, pay attention to the initial DO with these samples as well, because they are apt to be supersaturated with oxygen. These samples may have to be brought to within normal saturation conditions before testing by bubbling clean air or nitrogen gas through the sample. However, this may not be necessary with all test conditions or equipment. For example, respirometers that use pure oxygen would not require any concern about supersaturation of the homogenized sample.

The volume of sample selected for the test depends on a number of considerations, including the volume of the apparatus, the expected oxygen demand, and the method of supplying air or oxygen in the test device itself. For low-strength wastes, use the largest volumes possible with the equipment to improve reliability of the measurements. If studying treatability of an industrial waste, neutralize it to pH 7 before introduction to the seed activated sludge. If toxicity is suspected, it will likely be necessary to evaluate different dilutions, perhaps with and without mixing with settled wastewater to understand and select a routine test protocol. Method 5210 D is not written in a form that encourages an experimental design for simulating an activated sludge treatment system. However, one of the main advantages of using respirometric methods is to be able to test oxygen uptake on full-strength wastes, as they might be diluted within the treatment system itself. Therefore, on a case-by-case basis, the analyst will have to decide how much the dilution and seeding recommendations outlined in 5210 D can be stretched and still be considered within the boundaries of using a standardized method.

Method 5210 D is written as if the primary interest in running seeded tests is to follow oxygen uptake using seed-to-sample ratios consistent with the BOD_5. In this regard, selection of seed source, seed acclimation, and seed control steps are similar to those prescribed in the BOD_5 method. The seed volume must be checked with the GGA check standard. However, the seed-corrected oxygen uptake after 5 days of incubation is generally 25% to 30% larger than for the dilution method BOD_5. The current control range for GGA in the 5-day respirometric procedure is 230 to 290 mg/L. Standard values outside of this range need to be rechecked, and different volumes of seed or perhaps different sources of seed should be considered. Respirometric methods offer the added dimension of observing changes in the oxygen-uptake rate. If insufficient seed is added to the respirometer in the GGA check, an initial lag in demand rate may occur while the seed biomass increases. Try to avoid using seed volumes that demonstrate long lag times (keep fewer than 6 hours), but also avoid seed volumes that cause the initial GGA-uptake rate to greatly exceed the initial rate in the seeded samples (absent toxicity in the sample, of course).

When nitrification control is desirable, the method currently recommends adding 10 mg of 2-chloro-6-(trichloromethyl)pyridine (TCMP) per liter of sample. The reader should review the discussion of nitrification and nitrification inhibition in the BOD_5 chapter if unsure about the need or desire for control of nitrifying organisms during the conduct of the test. If nitrification begins to occur, the result may be apparent in the oxygen-uptake curve developed during the course of the test. The analyst will then have to make a decision with regard to further experimentation with that particular sample source, as to whether nitrification control is necessary. While the updated version of BOD Method 5210 B includes the option of using allythiourea (ATU), the current version of 5210 D does not contain that choice. The analyst will, therefore, have to decide on a case-by-case basis whether substitution of ATU for TCMP is an appropriate deviation from the standardized framework of 5210 D.

Calculation of the results obtained from the various equipment is, again, not a straightforward element of the procedure, as written in 5210 D. Each different type of instrument has its own recommendations and description for converting instrument output to oxygen uptake. It is, again, the analyst's decision on what to do with that information. Because the method is written in terms of BOD, one can simply represent the results in terms of total oxygen demand (mg/L BOD). Or, it is possible to numerically or graphically represent an oxygen uptake rate (OUR). Personal computer software packages often provide several options, so that the advantages of having continuous or nearly continuous data are not wasted for want of facile and capable data handling. However, the analyst must remember to correct for any seed and dilution contributions to the BOD or OUR determinations. This requires that a seed control is run concurrently with the test and that accurate trackings of seed volumes and undiluted sample volumes are performed. Method 5210 D provides an adequate description of the formula used for these corrections. Compensation for temperature and barometric conditions are specific for each type of respirometer, and should be completely described for each particular device.

Quality control/quality assurance measures are unique to the respirometric method. Overall method control is performed on triplicate GGA/nutrient standards (TCMP added) that have been seeded at a seed strength to prevent greater than 6-hour lag times in OUR and sufficient to give a GGA result of 230 to 290 mg/L in a 5-day test run. However, this must be a seed-corrected value. Therefore, concurrently run seed blanks as described.

Basically, the seed blanks are 1:1 dilutions of the seed material fortified with the same amount of nutrient buffer and TCMP as in the check standards and samples. The oxygen uptake per milliliter of seed is used to perform the seed correction of GGA and samples. The frequency at which this exercise is carried out should be great enough to establish a normal range of expected results with the seed source(s) typically used. If enough respirometers are available for any given experimental protocol, it is advisable to run a GGA check with each batch of samples.

Each instrument manufacturer should provide instrument sensitivity information. The available equipment is described as being in the range of 0.05 to 1.0 mg of oxygen. It is the analyst's responsibility to ensure that sample volume and dilutions are within the working range specified for the particular equipment in use. The coefficient of variation for standards should be between 3% and 5%, depending on concentration (APHA et al., 1995). Although the accuracy of the method cannot be determined per se, proper usage of the method can be demonstrated using the GGA check standard criteria as a basis. Analysts should develop control charts by running several series of GGA checks. Using a minimum of 25 check results meeting the 230- to 290-mg/L requirement, determine the mean and standard deviation of these results. It is further suggested that a control limit of three standard deviations be established. For single laboratories, it is expected that these control limits will be narrower than the GGA check criterion based on multiple laboratories. However, this provides the analyst a way to identify problems and deal with data trends that may wind up being problems if not resolved.

5.0 APPLICATIONS OF RESPIROMETRY

Respirometry has long been used as a tool to study BOD. A number of homemade respirometric inventions showed up in the literature at the end of the 1930s to further this study. A device called the Nordell Odeometer was used to compare the dilution and direct absorption BOD methods on domestic wastewater (Burtle & Buswell, 1937; Kessler, 1936). With this constant-pressure device, a second rise in the oxygen-demand curve was found and demonstrated to be the onset of nitrification. In these studies, nitrification was demonstrated to begin at approximately 46 hours; both methods gave BOD results within 10% of each other. This work was also important in that it compared four different versions of the COD test with each other and with both BOD methods. They demonstrated that, although there was a loose consistency between the various tests, there was no numeric correlation between the chemical oxidation and biochemical oxidation methods. Other studies on the relation between COD and OUR results found it is normally coincidence rather than any fundamental linkage (Chandra et al., 1987).

A relatively simple mechanized respirometer, called the biometer, was developed to determine oxygen demand of different industrial wastes in 15 minutes, or gauge activated sludge condition through a 90-minute period of respirometric data collection (Bloodgood, 1938). This invention used a 100-mL gas buret and pumped the gas through a caustic solution to scrub the carbon dioxide, then returned the gas to the reactor in a closed system. Oxygen can be renewed from a separate source.

A last homemade apparatus from this era is called the Oxy-Utilometer (Sawyer, 1939). This was also a closed system and used a rotary compressor to circulate the air through alkali to remove carbon dioxide before returning it to the sample in the reactor for mixing. It used volumetric analysis for determination of oxygen consumed. Oxygen is added from a separate reservoir as necessary. The device was highlighted in a series of four papers examining various aspects of activated sludge oxidation (Sawyer, 1939; Sawyer & Nichols, 1939a, 1939b; Sawyer & Rohlich, 1939).

A survey of the literature reveals a wide range of different types of respirometry applications, especially in the study of activated sludge and secondary treatment of a variety of wastewaters where oxygen consumption is widely held to be the primary process variable (Ekama & Marais, 1979; Stenstrom & Andrews, 1979; Takamatsu et al., 1982). The applicability of respirometric methods to solve operational problems was wide, primarily because the controlled conditions of the respirometer gave reliable and reproducible results. As chemical methods for COD and TOC came into popular use through the 1950s to 1970s, manometric methods were used to develop complementary data concerning treatment of emerging pollutants. Eventually, manometric methods became a commonplace tool in the characterization of the behavior of trace pollutants in treatment systems. A number of summaries and books are available, giving a thorough treatment (Arthur, 1984, 1994; Jenkins, 1960; Ros, 1993; Young, 1996). However, here we present only a few examples that are of particular interest as indicators of the direction in which respirometry developed.

Warburg respirometry, using 24-hour incubation periods, provides monitoring and control of treatment facilities receiving industrial wastewater (Jaeger & Niemitz, 1952; Pasveer, 1955). The Warburg apparatus was used to study the effects of radioactivity on activated sludge treatment (Skrindle & Sawyer, 1952a, 1952b).

The dilution BOD method and Warburg method results were compared for activated sludge treatment (Lee & Oswald, 1954). The differences between the 5-day dilution test and the Warburg results were studied, and the practicality of using either method on a routine basis in a treatment facility laboratory examined. The conclusions were that the Warburg gave, on average, approximately 15% higher results, required much higher buffer strength, and was potentially affected to a greater extent by seed volumes. When large numbers of daily samples are being tested, the Warburg is not a practical method.

The Warburg apparatus was used to evaluate the role that different types of organisms played in the BOD test (Lee & Oswald, 1959). The results showed that, under controlled conditions, the reaction rate was a function of the amount and type of organisms present in the seed. A maximum seed dilution of 100:1 was recommended for the Warburg system.

Both Warburg and dilution methods were used to study the degradation of surfactants in activated sludge (Bogan & Sawyer, 1954, 1955). Straight- and branched-chain alkyl benzene sulfonates, alkyl sulfonates, sulfonated amides, and sulfonated fatty acid ester anionics were characterized, and a series of non-ionic alkylphenoxy polyethoxylate synthetic detergents were chosen. It was found that all of these anionic and non-ionic surfactants were biodegradable to some degree, but that there was considerable variability related to structural properties within a compound class. In spite of apparent degradation, most of these compounds were responsible for persistent foaming and interference with oxygen transfer in the activated sludge process (Sawyer et al., 1956).

Treatability and degradation studies on specific compounds or compound classes in activated sludge have become standard fare for respirometric studies with activated sludge (see Table 8.1). Acclimation of activated sludge to high concentrations of aniline as a sole carbon source was found to enhance treatment of aliphatic amines as well as other aromatic amines and alkyl benzenes (Malaney, 1960). This enhanced ability to treat a wide variety of petroleum hydrocarbons raised many questions about endogenous versus exogenous enzyme activity by activated sludge organisms, and established that the lag time or acclimation time of organisms is a critical part of the overall oxidative process.

The Warburg method was used to study the ability of activated sludge to recover from heavy metal–induced toxicity (Malaney, 1960). After short periods of inhibition, the organisms recovered most of their ability to oxidize the wastes. However, the recovery was seldom complete in the study periods used. This work provided a basis for understanding and tracking the chronic effects of heavy metals in activated sludge treatment. Ultimately, it was widely demonstrated that even trace levels of heavy metals could concentrate in activated sludge, eventually causing chronic poor performance or acute upset conditions.

Warburg respirometry and gas chromatography (GC) were used to study the effects of phenol, cresols, and chlorinated phenols on activated sludge (Baird et al., 1974). Toxicity, acclimation, and biodegradation were evaluated under simulated treatment conditions. As little as 1 mg/L phenol was toxic to unacclimated sludge, but acclimated organisms could tolerate up to 100 mg/L. With the exception of ortho-substituted phenolics, all of the activated sludges evaluated could degrade 80% to 100% of the test compounds within 6 hours at 10 mg/L or less. At 100 mg/L, degradation was markedly reduced to 20% or less, without a good correlation with respiration. The lack of correlation between toxicity and removal of test compounds was attributed to production of toxic metabolites of the test compounds.

The biodegradability of 15 carcinogenic organic compounds (aromatic amines) in activated sludge treatment using a combination of Warburg respirometry, GC, and mass spectrometry (MS) was examined (Baird et al., 1977). In general, most of the compounds studied showed some inhibition of respiration at 20 mg/L in unacclimated sludge, and about one-half of the compounds studied showed poor removal characteristics. The work suggested that degradation required that the aromatic ring positions ortho- and para- to the amine group had to be unsubstituted for facile removal by the activated sludge. These positions are susceptible to electrophilic attack and are similar to mechanisms involved in animal metabolism and carcinogenicity. However, the implication that similar stable carcinogenic intermediates might be formed in the treatment process could not be corroborated by mass spectrometry.

In the United States, respirometric methods have been used for essentially the same purposes as set forth by IWA since the 1980s (Tabak et al., 1981; Tabak et al., 1990). These examples underscore the participation of the U.S. EPA and others in the efforts. In the latter report, respirometric biodegradability, biokinetic profiles, inhibitory effects, and acclimation of a wide range of priority pollutants and related target organic compounds are examined. A 12-place commercial electrometric respirometer using sample volumes of 250 mL is described. Adjunct chemical measurements of substrate were performed, including GC and GC/MS, dissolved organic carbon (DOC), COD, and solids, to complement the respirometric data.

Other uses of computerized direct oxygen utilization equipment have been directed at assessing the biodegradation and treatability of a wide range of mixed industrial waste and individual organic compounds in both activated sludge (Magliette et al., 1992) and soil remediation studies (Beach et al., 1995; Graves et al., 1991; Haines et al., 1996). A direct oxygen input system for study of the toxicity of eight- and ten-component organic chemical mixtures to activated sludge organisms was reported (Hall et al., 1996). The effects are found to be simply additive, rather than synergistic. Quantitative structure activity relationships (QSAR) were used to model the toxicity data in a predictive approach to estimate effects from numerous compound types on the organisms.

A multiple-batch reactor respirometric system was used to evaluate the toxicity and inhibition potential of industrial wastewaters (Beach et al., 1995). The systems were commercial electrometric systems capable of either on- or offline batch operation. The online system used a fixed-growth biological filter, and DO was measured in the effluent of the reactor. A graphical interpolation of respirometric curves was presented to establish 50% effective-concentration values for grading waste treatability for batch experiments and measured DO drop from the biological filters for the online system.

A flow-through microcosm was described to evaluate nutrient requirements for sustaining oil spill biodegradation under different tidal flow conditions (Haines et al., 1996). A respirometer recorded oxygen uptake over time within the microcosm. This work was done as part of the cleanup effort for the Exxon *Valdez* disaster and proved useful for timely information on degradability of oil residue, volatile organics, phenolics, and a variety of related hydrocarbon mixtures.

There are scores of similar published examples of the use of modern respirometers and adjunct chemical methods, modeling techniques, and other innovations aimed at addressing similar problems in a wide variety of environmental settings. As with classical manual manometric respirometry, these techniques provide a profile of kinetic data, but with a distinct advantage in speed and labor savings. The computerized methods also make possible continuous, long-term, bench-scale experiments, which are of particular importance in soil and groundwater remediation efforts. The reader is encouraged to refer to any of the cited papers and reviews to find articles pertinent to specific interests and needs.

Not all of the interest in respirometry has been fueled directly by the desire to study the biokinetics of organic pollutant mineralization. There is an obvious desire to simplify the laboratory procedures used in classical WRRF design and operation. And there is a significant body of knowledge regarding the growth kinetics and oxygen-uptake mechanisms of microorganisms in modern treatment. In other words, the field has gained a great deal of knowledge about factors affecting wastewater treatment and purification in the 100 years since Adeny and others pioneered the research. As an example, there have been continued studies of the factors affecting biokinetic constants in activated sludge treating both inhibitory and nontoxic wastes (Gaudy et al., 1990). Use of respirometry allows relatively short-term perturbations and resultant kinetics to be modeled and predicted under a wide variety of conditions.

The use of respirometry has been examined as a treatment system audit and assessment tool, as an alternative to the more traditional and laborious approach of through-facility sampling surveys (Watts & Cavenor-Shaw, 1998). Dissolved-oxygen-type respirometer measurements are linked to a modem to transmit oxygen-uptake rate data to the facility operations center to track effects of diurnal loading, shock loads, storm flow, and mixed liquor return rates.

6.0 SHORT-TERM BIOCHEMICAL OXYGEN DEMAND MEASUREMENTS

Interwoven through the literature of modern electrometric respirometers are the recurring themes of rapid, short-term BOD measurements and the special variation, online BOD measurements. Historically, researchers have sought to reduce the time required to evaluate BOD, from 60 days or more, to 20 to 30 days, to 10 to 12 days, and finally, to the venerable 5-day BOD. Efforts in the dilution test have focused on shortening the incubation time, optimizing nutrients, aeration, and mixing. Others sought nonbiological methods, such as use of chemical oxidation methods, combustion, or UV photolysis (or combinations). Still other approaches seek to use short-term respirometry in various ways to correlate with dilution BOD results or as a separate metric altogether. Of course, the application of online methodology can be as simple as sequential batch DO depletion measurements with a DO probe in a container that is filled and emptied on a cyclic or continuous basis. When applied to activated sludge, such homemade systems resemble Method 2710 B, Oxygen Consumption Rate in *Standard Methods* (APHA et al., 1998). However, of interest to us at this juncture are the more flexible instrumental systems and methods capable of a wider range of applications, including fast turnaround for desired information.

The ideal of having short-term BOD data and development of a method that can replace the 5-day test has been the holy grail of wastewater investigators since the very inception of the BOD_5. Although the BOD_5 has taken on a life of its own, via regulation and convention, there have been many practical-minded investigators in this field who think the test should be relegated to the junk heap of obsolescence. The quest for a replacement method, however, has never quite found the true standard. To be sure, many investigators have found their own replicas to be suitable in narrow circumstances, and these examples dot the literature with perpetual hope.

A brief review of the literature on rapid BOD test methods through 1974 (LeBlanc, 1974) represents the first 90 years of the quest. Discounting the chemical oxidation techniques, most of the attempts to develop rapid alternatives can be placed in either of two categories: those that sought modification of the dilution BOD method, and those that were variations of

DOC tests. Herein lies the fundamental difficulty, in that the two approaches have different kinetic properties. While further attempts to devise either a respirometric procedure or modified dilution method that give nearly universal agreement with each other or with the BOD_5 within some arbitrary short time frame seems ill advised, it is instructive to provide an overview of the attempts thus far.

Variations of the dilution test aimed at providing short-term equivalency to the 5-day BOD include use of elevated temperatures (Orford & Matusky, 1959; Robbins, 1961) and use of specialized seed (Tidwell & Sorrells, 1956; Zehnpfennig & Nichols, 1953). Satisfactory results are found for specific wastes in times ranging from 18 hours to 3 days, and with temperatures up to 37 °C, but the prospects for universal replacement apparently are dim, as each of these findings are unique to the conditions used.

Most manometric techniques are capable of giving short-term oxygen-demand data, but in terms of correlating with the 5-day BOD, the results have a checkered history. Some reported successes are worth noting, however. Results between 18-hour Warburg and BOD_5 with pulp and paper wastes, river water, and GGA were comparable, but only at elevated (37 °C) temperatures (Dillingham & Jose, 1960; Dillingham et al., 1958). A paper describes the development of a commercial (Hach Co.) respirometer, which is compared with operation of Warburg, Sierp, Barcroft, and early electrolytic manometers (Tool, 1967). This instrument is marketed specifically for short-term BOD and is calibrated to read directly in milligrams per liter BOD. Elevated temperatures are used to achieve 2.5-day results, which correlated well (within 5%) with BOD_5.

Using an electrolytic respirometer, a good correlation with BOD_5 is demonstrated in as few as 2.5 days for some sludges (Young & Baumann, 1976b). This endpoint is termed equivalent BOD (EBOD), and the method is touted as very precise. However, for a variety of activated sludges, 3 to 5 days is sometimes required for EBOD to match BOD_5.

A respirometric method for short-term BOD using an automated system has been described (Arthur & Hurstra, 1968). The method gives reasonable agreement between a 4-hour, short-term prediction of the 5-day BOD and the conventional BOD_5. Others have studied this idea of short-term BOD (STBOD) and presented variations. One example is a respirometric technique that gives results in 30 to 60 minutes (Vernimen et al., 1967). This system, as with most similar approaches, requires a correlation be established between the short-term oxygen demand and BOD_5 for a given application before the STBOD is useful. A short-term method for continuous estimation of BOD for influent to an activated sludge treatment facility, which is termed BODST, has been presented (Lukasse et al., 1997, 2000; Spanjers et al., 1993). Operationally, this continuous-flow, online system is claimed to give acceptable agreement with BOD_5, but in real time (Vanrolleghem & Spanjers, 1994). The BODST offered advantages that are impossible with long-term tests in the form of interactive feed control adjustment to the facility and response to oxidative inhibition.

7.0 ONLINE RESPIROMETRY

It has become apparent that respirometric BOD measurements have value in their own right, unrelated to comparison with the BOD_5. Further attempts to standardize a respirometric procedure to mimic a 5-day dilution BOD are probably not warranted. Not only do such efforts wind up being case specific, but they minimize and trivialize the true value of respirometric tools. One of the clear values is its ability to be used online in a full-scale, pilot, or other simulated treatment system.

Several of the electrometric respirometers listed on the IWA website are intended to or can be adapted to online type systems (Spanjers et al., 1996). Basically, there are two types of systems available, batch and continuous flow. Of the batch systems, some measure DO depletion and others use manometric or other pressure-sensing techniques to analyze oxygen consumption or to measure the amount of oxygen needed to maintain a constant system pressure. The batch systems rely on an automated fill and draw cycle for the reactor vessel, and the desired mixture of sample must be metered in. The continuous-flow types of systems can be made to mimic a treatment process. The sample to be tested is pumped continuously through the reactor. Some systems rely on mixing of mixed liquor or other biomass with the sample, or metering the organisms into the reactor, while other designs utilize fixed-growth surfaces. As with the batch systems, respiration can be measured by either DO concentration in the liquid phase or oxygen concentration in the gas phase of the reactor. The gaseous oxygen (manometric) sensing type of systems seem to be used more for research applications, whereas the DO-type systems appear to be more popular for in-facility control.

Because the purpose of online oxygen-uptake measurements is to provide a continuous data record of microbial oxygen demand, the ability to collect the data on a PC is highly desirable (Fujimoto et al., 1981). Once that technological step has been made, operator-friendly software and appropriate algorithms are the next necessities.

Activated sludge treatment facilities typically have a consistent pattern of oxygen uptake, perhaps with a diurnal fluctuation or other pattern reflecting changes in the influent load. While most facilities use DO electrodes in a feedback system to regulate aeration rates within the different stages of aeration, these provide little advance notice of major events or treatability problems that upset process control. Batch online systems provide data that are typically read as change in oxygen uptake per unit of time. The resultant curves have time scales of a few minutes to a few hours, depending on system design and operator usage. Comparison of these utilization rate curves for each reactor fill cycle or to a parallel endogenous respirator provide a patterned monitoring tool characteristic of the treatment process. Continuous-flow systems provide oxygen-uptake rate on a continuous basis, and reactor profiles provide the ability to observe typical oxygen-uptake patterns for a given facility. In systems with multiple reactors, it is possible to utilize endogenous biomass respiration or response to standard reference solutions such as GGA for baseline information.

Influx of toxic materials can cause an acute effect on respiration rate and is reflected as a rapid decrease in oxygen uptake in an online respirometer. Slug loads of high-strength, soluble organic matter have just the opposite effect, causing a rapid rise in demand. At the tail end of a biological treatment process, online respirometers are applicable to monitoring the consistency of effluent quality. In any of these scenarios, advance or real-time knowledge of significant perturbations offer facility operations personnel the opportunity to take immediate remedial action.

A few examples of these applications will provide an avenue for further study by the interested reader. Using two different types of respirometers to evaluate activated sludge respirometry provides an early warning sign for toxicity (Temmink et al., 1993). Because of the way raw feed is introduced to the sludge in the different reactors, each had its own characteristic behavior as a biological early warning device. The first system, called RODTOX, is a batch device, which results in a more sensitive and faster-responding (30 to 120 minutes) system than the facility itself for detecting the presence of toxicants. A flowthrough system, called RA-1000, is designed to mimic the operation of the facility, and it provides influx response time to toxicity in as few as 15 minutes. The flowthrough system is configured to measure both the DO coming into as well as exiting the reactor. The data processor computes the respiration rate across the reactor every minute. In the batch reactor, however, sludge is fed every 30 minutes, and the operators follow substrate oxidation with a DO probe in the reactor. These readings give DO profiles with three important variables to be interpreted: peak height, peak area, and peak slope. The change in these variables define the shape of the DO-uptake peak for every feeding cycle, and comparison of these profiles from cycle to cycle allows interpretation of toxic activity.

Respirometry in the activated sludge process provides insights to the food-to-microorganism ratio (F:M), the influent strength, and the effects of organic shock loading (Abu, 1995; Bjorlenius & Reinius, 1998; Harlan,

1991; Marsch & Drnevich, 1977; Mona et al., 1979; Smith, 1953; Spanjers & Vanrolleghem, 1995; Vargas-Lopez et al., 1989; Wong & Smith, 1992). The term "carbon tension" was proposed to indicate the relationship between respirometry and available food (Schroeder & Irvine, 1979); however, the term "readily biodegradable COD" has become more widely accepted (Witteborg et al., 1996; Xu & Hasselblad, 1996). Based on the respirometrically derived F:M, aeration and cell recycle rates can be adjusted to provide a maximum operating efficiency (Haas, 1979). The major effect of shock loading is a decrease or increase in the numbers of viable bacteria as indicated by respirometry (Blok, 1974, 1976).

The ability of an online respirometer to identify toxic inputs to an activated sludge treatment facility can be significant (Bess & Conway, 1966; Cotter, 1974; Geenans & Thoeye, 1998). Using the RODTOX online system, it is possible to isolate toxicity sources in a large urban upstream wastewater system tributary to the facility. It is observed that low levels of respiratory inhibition, approximately 10% in the respirometers, do not necessarily manifest in the treatment facility condition, whereas 43% inhibition results in an acute or a total washout toxic response of the sludge. In one example, monitoring respirometric data over long periods (10 days) is useful for determining that a virtually continuous influx of toxic material is occurring in the facility. These observations demonstrate the necessity of understanding the relationship between respirometer response and the operational condition of the facility itself.

Use of online respirometry shows that it is a viable alternative to conventional process monitoring for continuously or periodically evaluating process efficiency (Watts & Cavenor-Shaw, 1998). Two case studies at different treatment facilities are described. The monitoring system requires DO sensors in each aeration unit; one respirometer and one optical mixed liquor analyzer per unit; one respirometer in the return sludge line and another in the primary effluent; and flow-levelers/meters at the influent and effluent points of the clarifiers, return sludge, and feed lines. These units are linked together through a multichannel data logger and phone modem to a central computer and data system. Although complex, these systems allow remote unattended operation and data review in almost any timeframe (Aidun & Smith, 1989).

8.0 MODELING FACILITY OPERATIONS

Most of the models developed for process control using respirometric and adjunct data are based on the knowledge that net microbial growth rates can be described as a function of the concentration of the feed (substrate), and that the oxygen-uptake rate is a function of substrate and microorganism concentration. It has been demonstrated that respirometric measurements can be used to calculate biomass growth or substrate utilization curves, which can then be used to derive rate expressions for these variables (Benefield et al., 1979). The respirometric data (i.e., oxygen-uptake measurements) cannot be used directly as a substitute for growth or utilization rates, but must be converted to these via parametric relationships. The art is in accurately deriving these relationships for populating models. Respirometry has been proven to be an essential tool in making decisions about which growth mode, be it Monod, double-Monod, Haldane, or exponential, is most appropriate for individual parameter functions in models (Dochain et al., 1995; Vanrolleghem et al., 1995). Respirometry has also proven itself as the quickest technique for determining parameter interdependencies such as temperature and endogenous OUR (Giona & Annesini, 1979; Giroux et al., 1996). A review of the uses of respirometry for determination of kinetic coefficients in models is available (Rozich, 1992).

A number of examples exist in the literature in which investigators use respirometry for model calibration of various aspects of treatment facility operation. A summary of the uses of respirometry in modeling is found in the discussion on fractionation in the COD chapter in this book (Chapter 12). Respirometric protocols are used to calibrate full-scale activated sludge treatment facility performance to models (Colvin et al., 1988). These methods obviate much of the labor-intensive data collection efforts required by previous methods for predicting facility operational performance and provide predictive values that compare well with actual facility measurements. A set of general guidelines for implementing biokinetic methods and data in normal facility operations is described, and it was suggested that this process could be flexible enough to model treatment of intermittent toxic and hazardous wastes that might be in the influent.

A model-based characterization of an industrial WRRF has been described (Coen et al., 1998). Batch respirometry coupled with tracers provided an understanding of the treatment facility itself and, ultimately, the use of models for facility control. This is a relevant example that shows that respirometry can be used at an industrial facility for process and effluent control.

A multireactor batch system is used to profile oxygen-uptake data for various activated-sludge facility biomass challenged with different industrial wastes to assess impacts on the operation of the facilities during different time periods (Palmateer & Thomas, 1996). These biokinetic data establish normal and abnormal growth and respirometric patterns to be able to control facility operations. Essential to the approach is use of the respirometer to establish biomass growth rates at different feed levels following Monad kinetics. Then, in routine operation, when a feed source containing toxic components is received at the facility, the change in growth is observed and quantified. Using this approach, even chronic effects on activated sludge treatment can be seen. The results provide the impetus to require offending industrial facilities to meet respirometric criteria for sewered waste streams.

As mentioned, the plethora of automated, continuous respirometers that generate real-time data for computer capture potentially can provide a great deal of flexibility, and the research literature is replete with examples of this equipment performing more than simply monitor process control. In addition to QSAR and other models cited for degradability of specific organic compounds (Hall et al., 1996), a great deal of effort has gone into modeling the activated sludge process using both off- and online electrometric systems.

The use of dynamic models has become more widespread, especially in the European community (Vanrolleghem et al., 1999). The ASM1 model developed by the International Association on Water Quality (IAWQ) is described as the most popular and is generally considered state-of-the art by practitioners engaged in simulating wastewater treatment systems (Henze et al., 1987; Vanrolleghem et al., 1999). Selected examples of the use of batch and continuous-flow systems in the laboratory to simulate online respirometry are described, along with how such data are used to frontload mathematical models such as the ASM1 model (Vanrolleghem et al., 1999). These systems have been used to study rapidly and slowly biodegradable substrates (Degremont Societe Generale, 1975; Henze, 1992); nitrification (Brouwer et al., 1998; Spanjers & Vanrolleghem, 1995); heterotrophic and autotrophic biomass growth; hydrolysis constants; and various feed, respirometric, and growth variables in mixed pilot- and facility-scale activated sludge systems. These reports emphasize the need to carefully select the design and application of respirometric studies if the resultant models are to accurately predict full-scale treatment facility behavior. The type of reactor setup, feeding regimes, collection of supporting physical and chemical data, and assumptions made in interpreting respirometric data that are needed to track facility operations parameters are detailed in references presented in Chapter 6 of this book.

An extensively instrumented activated-sludge facility and four online respirometers to provide data input via PC acquisition software to model the operations using the IAWQ model have been reported (Stokes et al., 1993). This system provides data that are asserted to provide an accurate simulation of the facility operation. The in-facility instrumentation must be carefully chosen to provide accurate data from the facility, data that are absolutely essential to model calibration. The oxygen and ammonia probes in the mixed-liquor respirometers are noted to need recalibrating every 12 hours, a step performed by the automated respirometer itself. A 1-day simulation can be performed in 15 seconds.

However, as a comparison, simple models based on aggregate parameters such as BOD are not able to simulate correct oxygen demand patterns in the aeration tanks if the influent contains several organic fractions. In particular, when easily oxidized organics, slowly degraded organics, and organics that required acclimation are all present, complications in the model occur. The frequency of this situation supports the need to use more complicated models and to have information about the degradation rates of the various fractions of organics to adequately predict performance and efficiently operate a treatment facility.

The ASM1 model has been modified, adding phosphorus-removal predictions, and published as the ASM2 (Gujer et al., 1995). Others have evaluated the model for a variety of treatment systems (Cinar et al., 1998). The models were calibrated in one treatment facility, then applied to three other facilities as demonstrations. Removal efficiencies for nitrogen and phosphorus are predicted to within 7% of operational results for only one of the facilities. The authors were also able to utilize the model to recommend modifications to the one facility for improved phosphorus removal. The authors point out that the model does not work well with oxidation ditch treatment systems and is very sensitive to DO measurements.

A modified respirometer is used in a continuous-flow mode to study the rate constant for oxidizing slowly degradable organic material (Vanrolleghem & Spanjers, 1998). In normal simulated treatment studies, the respirometer feeds the sludge with fresh material, and so the resulting net rate parameters are dominated by rapid oxygen demand of the fresh wastewater. In this modified system, the investigators maintained an aerobic and anaerobic reactor, in which activated sludge can be circulated continuously from one to the other. In principle, respiration is due to two processes, growth and decay. Using DO decay profiles in both reactors resulting from the cycling events between the two reactors produces data that the investigators can interpret graphically or use to develop model rate parameters for any given feed introduced to the dual-reactor system. The outcome is an empirical approach to estimating rate constants for both rapidly and slowly degraded organic matter in loading and operation of full-scale, activated-sludge treatment systems.

Although there are many respirometry-based, activated-sludge treatment control strategies in the literature, there are far fewer successful, practical demonstrations or simulation-based evaluations reported (Spanjers et al., 1998). The state of the art provides insufficient justification for the development of standard how-to procedures for such control strategies in full scale. Simulation studies should be based on stringent, rigorous methodologies and performance standards if a benchmark paradigm is ever to be realized for respirometric-based control strategies. This paper is essentially a literature synthesis of a benchmarking strategy that proposes tools, simulation model, process model, influent, and parametric assumptions that need standardization. The role of online respirometers within the activated-sludge facility itself was limited to specific assumptions, including one key assumption that the data were perfect. The need for determining the variability in respirometric measurements must be emphasized to ensure that such variability does not taint demonstration of model veracity.

Since the completion of the original version of this work, several reports have appeared that describe successful use of online respirometry for process control of WRRFs. Zermeno et al. (2002) and Iranpour & Miguel (2008) are real-time comparisons of several designs: microbes in a biomembrane (Japan design, Nissin Electric, BOD-2000), the fluidized bed of bacteria carriers (German–USA design, ISCO-STIP, BIOX-1010), and cascade activated sludge bioreactors (USA design, LAR Biomonitor). The BIOX-1010 stable population of microbes is maintained under controlled conditions by using an immobilized biofilm on a multitude of small, hollow, cylindrical plastic carriers. The quantity of biomass is thus fixed by the surface area to which it adheres. The study found the BIOX-1010 to be the more reliable. A study (Young et al., 2003) of a variety of possible methods for short-term BOD suggests that a high-rate respirometric method used in a process control context provides the most consistent BOD-like data.

9.0 CONCLUSION

While the BOD_5 chapter makes it apparent that the 5-day dilution test will continue to be used well into the 21st century, primarily because of regulatory requirements, future understanding of BOD processes would appear to rest with respirometric methods. The evolution of direct oxygen methods and the concurrent increase in knowledge regarding activated sludge treatment facility operations, waste treatability, and receiving water behavior suggests that these are the techniques that researchers and practitioners will seek out to solve their problems. The advent of electrometric methods, PCs, and data-handling software would seem to be the foundation for development of better modeling approaches and feedback-based treatment facility operations. Will these be facile tools for the operator, or will they be primarily the hardware and software of applied research? Only time will tell.

10.0 REFERENCES

Abu, Z. H. (1995). Oxygen uptake rate as an extended aeration process control parameter. *J. Environ. Sci. Health*, *A30*, 951.

Adeny, W. E. (1897). The course and nature of fermentative changes in natural and polluted waters and in artificial solutions as indicated by the composition of the dissolved gases. Parts I–IV. *Trans. Royal Dublin Soc.*, 1895.

Adeny, W. E. (1908). Nature and rate of the chemical changes which occur in polluted fresh and salt water. Addendum: Capacities of fresh and of sea water to bear pollution within limits of fouling. *Royal Commission Fifth Rep., App. I*, *1*, 433.

Adeny, W. E. (1928). *The principles and practice of the dilution method of sewage disposal.* Cambridge University Press.

Aidun, B., & Smith, D. W. (1989). Design and evaluation of an automated oxygen uptake rate measurement system. *Can. J. Civ. Eng. 15*, 1015.

Allot, K. (1990). Sizing up the strength of organic waste. *Process Eng.*, *71*, 46.

American Public Health Association, American Water Works Association, & Federation of Sewage and Industrial Wastes Associations. (1955). *Standard methods for the examination of water and wastewater* (10th ed.). American Public Health Association.

American Public Health Association, American Water Works Association, & Water Pollution Control Federation. (1960). *Standard methods for the examination of water and wastewater* (11th ed.). American Public Health Association.

American Public Health Association, American Water Works Association, & Water Pollution Control Federation. (1965). *Standard methods for the examination of water and wastewater* (12th ed.). American Public Health Association.

American Public Health Association, American Water Works Association, & Water Environment Federation. (1995). *Standard methods for the examination of water and wastewater* (19th ed.). American Public Health Association.

American Public Health Association, American Water Works Association, & Water Environment Federation. (1998). *Standard methods for the examination of water and wastewater* (20th ed.). American Public Health Association.

American Society for Testing and Materials, Subcommittee E47.06. (1994). *Standard test method for determining biodegradability of organic chemicals in semi-continuous activated sludge (ACAS) 11*(5), E1625–94.

Arthur, R. M. (1974). Let's upgrade the BOD test. *Water Sew. Works*, *121*, 100.

Arthur, R. M. (1984). Twenty years of respirometry. *Proc. 39th Ind. Waste Conf., Purdue University.* Butterworth Publishers, 861.

Arthur, R. M. (1994). Thirty years of respirometry. *Proc. 49th Ind. Waste Conf., Purdue University*. Lewis Publishers, 169.

Arthur, R. M., & Hurstra, W. N. (1968). Short term BOD using the automatic respirometer. *Proc. 23rd Ind. Waste Conf., Purdue University.*

Atolarczyk, L. G. (1978). BOD measuring device. U.S. Patent 4105523.

Baird, R. B., Carmona, L., & Jenkins, R. L. (1977). Behavior of benzidine and other aromatic amines in aerobic wastewater treatment. *J. Water Pollut. Control Fed.*, *49*, 1609.

Baird, R. B., Kuo, C. L., Shapiro, J. S., & Yanko, W. A. (1974). The fate of phenolics in wastewater: Determination by direct injection GLC and Warburg respirometry. *Arch. Environ. Contam. Toxicol.*, *2*, 165.

Barcroft, J., & Haldane, J. B. S. (1902). A method of estimating the oxygen and carbonic acid in small quantities of blood. *J. Physiol.*, *28*, 232.

Bauer, N. J., et al. (1981). A simple, rapid bioassay for detecting effects of pollutants on bacteria. *Bull. Environ. Contam. Toxicol.*, *27*, 577.

Beach, M. S., Beach, J. S., Jr., & Cadena, F. (1995). Respirometric methods for rapid toxicity testing/inhibition assessment of industrial wastewater. *Proc. 50th Purdue Ind. Waste Conf.*, 619.

Benefield, L. D., Lawrence, D., & Randall, C. (1979). Effect of activated sludge viability on biokenetic coefficient evaluation. *J. Water Pollut. Control Fed.*, *51*, 187.

Bennett, G. F. (1980). Oxygen mass transfer rates, mechanisms, and applications in biological wastewater treatment. *Crit. Rev. Environ. Control*, *9*, 301.

Benson, B. B., & Krause, D. (1980). The concentration and isotopic fractionation of gases dissolved in freshwater in equilibrium with the atmosphere. 1. Oxygen. *Limnol. Oceanogr.*, *25*, 662.

Berger, H. F. (1955). BOD from oxygen consumed determination. *Tappi J.*, *38*, 534; *Public Health Eng. Abstr.*, *36*(7), 18 (1956).

Bess, F. D., & Conway, R. A. (1966). Aerated stabilization of synthetic organic chemical wastes. *J. Water Pollut. Control Fed.*, *38*, 939.

Bjorlenius, B., & Reinius, L. G. (1998). Use of on-line data to evaluate the activity in the biological stage at a wastewater treatment plant. *Water Sci. Technol.*, *37*(9), 33.

Blok, J. (1974). Respirometric measurements on activated sludge. *Water Res.*, *8*, 11.

Blok, J. (1976). Measurements of the viable biomass concentration in activated sludge by respirometric techniques. *Water Res.*, *10*, 919.

Bloodgood, E. D. (1938). Biologic oxidation. *Sew. Works J.*, *10*, 927.

Bogan, R. H., & Sawyer, C. N. (1954). Biochemical degradation of synthetic detergents. I. Preliminary studies. *Sew. Ind. Wastes*, *26*, 1069.

Bogan, R. H., & Sawyer, C. N. (1955). Biochemical degradation of synthetic detergents. II. Studies on the relation between chemical structure and biochemical oxidation. *Sew. Ind. Wastes*, *27*, 917.

Bolling, A. (1967). Automatic BOD measurement. *Papiermacher.*, *17*, 1, 5.

Bridie, A. L. (1969). Determination of BOD with continuous recording of oxygen uptake. *Water Res.*, *3*, 157.

Brouwer, H., Bloemen, M., Klapwijk, B., & Spanjers, H. (1998). Feedforward control of nitrification by manipulating the aerobic volume in activated sludge plants. *Water Sci. Technol.*, *38*, 245.

Buczowska, Z., & Jarnuszkiewicz, I. (1969). Evaluation of the short-term BOD test on the basis of comparative investigations of sewage. *Bull. Inst. Mar. Med. Gdansk*, *20*(1–2), 29; *Chemical Abstracts*, *71*, 24, 287 (1969).

Burchard, C. H. (1966). Barometric vibration method for BOD. *Gas-u. Wasserfach, Wasser Abwasser* (Germany), *107*(2), 42; *Chemical Abstracts*, *64*, 7862 (1966).

Burchard, C. H. (1981). Study of the predictive value of variations in the curve of oxygen concentration in activated sludge basins. Part 3: Construction and testing of a BOD monitor. *Stuttg. Ber. Siedlungswasserwirtsch*, *71*, 84; *Chemical Abstracts*, *95*, 174882 (1981).

Burkhead, C. E., & McKinney, R. E. (1969). Energy concepts of aerobic microbial metabolism. *J. Sanit. Eng. Div., Proc. Am. Soc. Civ. Eng.*, *95*(SA2), 253.

Burtle, J., & Buswell, A. M. (1937). Oxygen demand studies. *Sew. Works J.*, *9*, 224.

Busch, A. W.; Grady, L., Jr., Rao, T. S., & Swilley, E. L. (1962). Short-term total oxygen demand test. *J. Water Pollut. Control Fed.*, *34*, 354.

Cailas, M. D., & Gehr, R. (1989a). The oxygen uptake rate approach for analyzing respirometric BOD data: I. Method development. *Water Res.*, *23*, 985.

Cailas, M. D., & Gehr, R. (1989b). The oxygen uptake rate approach for analyzing BOD data: II. Application and evaluation. *Water Res.*, *23*, 993.

Caldwell, D. H., & Langelier, W. F. (1948). Manometric measurement of the biochemical oxygen demand of sewage. *Sew. Works J.*, *20*, 202.

Callow, A. B. (1924). The oxygen uptake of bacteria. *Biochem. J.*, *18*, 507.

Carlsson, B., et al. (1994). On-line estimation of the respiration rate and the oxygen transfer rate at Kungsangen WTP in Uppsala. *Water Sci. Technol.*, *30*, 273.

Casey, T. J., & O'Connor, P. E. (1978). New light on mechanism of air–water oxygen exchange in quiescent systems. *Water Res.*, *12*, 649.

Cech, J. S., Chudoba, J., & Grau, P. (1985). Determination of kinetic constants of activated sludge microorganisms. *Water Sci. Technol.*, *17*, 259.

Chandra, S., Mines, R. O., & Sherrard, J. H. (1987). Evaluation of oxygen uptake rate as an activated sludge process control parameter. *Toxic Hazard. Wastes*, *19*, 51; *J. Water Pollut. Control Fed.*, *59*, 1009.

Chandran, K., Accashain, J. V., & Smets, B. F. (1996). Discussion of oxygen transfer limitation of a respirometer. *Water Environ. Res.*, *68*, 1084.

Cinar, O., Daigger, G. T., & Graef, S. P. (1998). Evaluation of IAWQ activated sludge model No. 2 using steady state data from four full-scale wastewater treatment plants. *Water Environ. Res.*, *70*, 1216.

Clark, H. W. (1896). Experiments upon the purification of sewage and water at the Lawrence Experiment Station. *Rep. Mass. State Board Health, 1895–1896*, 427.

Clark, H. W. (1901). *Mass. State Board Health Rep., 1900–1901*, 373.

Clark, H. W. (1903). *Charles River Dam Comm. Rep., 1903.*

Clark, J. W. (1959). New method for biochemical oxygen demand. *New Mexico State Univ. Eng. Exp. Stn. Bull., No. 11.*

Clark, J. W. (1960). Continuous recording BOD determination. *Water Sew. Works*, *107*, 140.

Clark, J. W. (1961a). Biochemical oxygen demand apparatus. *New Mexico State Univ. Eng. Exp. Stn. Bull., No. 18.*

Clark, J. W. (1961b). BOD's for plant operation. *Water Sew. Works*, *108*, 246.

Clarke, A. N., Eckenfelder, W. W., & McMullen, E. D. (1978). Development of a continuous respirometer. *Water Res.*, *12*, 799.

Coen, F., Petersen, B., Vanrolleghem, P., Vanderhagen, B., & Henze, M. (1998). Model-based characterization of hydraulic, kinetic, and influent properties of an industrial WWTP. *Water Sci. Technol.*, *37*, 12, 317.

Colvin, R. J., Rozich, A. F., Gaudy, A. F., Jr., & Martin, J. (1988). Application of a process model calibrated with respirometry to predict full-scale activated sludge performance. *Proc. 45th Annu. Ind. Waste Conf., Purdue University.* Lewis Publishers, 501.

Corbet, A. S., & Wooldridge, W. R. (1936). An investigation of the accuracy of the Barcroft differential manometer in respiration studies. *Biochem. J.*, *29*, 561.

Cotter, J. E. (1974). Best effluent the goal. *Water Wastes Eng.*, *11*, 2, 46.

Dawson, P. S. S., & Jenkins, S. H. (1949). The oxygen requirements of activated sludge determined by manometric methods. *Sew. Works J.*, *21*, 643.

Dawson, P. S. S., & Jenkins, S. H. (1950). The oxygen requirements of activated sludge determined by manometric methods. II. Chemical factors affecting oxygen uptake. *Sew. Ind. Wastes*, *22*, 490.

Degremont Societe Generale d'Epuration et d'Assainissement. (1975). Continuous determination of variations of biodegradable pollution in water. French Patent 2269717; *Chemical Abstracts*, *85*, 67790 (1976).

Dillingham, E. O., & Jose, A. G. (1960). The correlation and precision of manometric and five-day BOD in the range of one to twenty parts per million. *Tappi J.*, *43*, 626.

Dillingham, E. O., Knuth, D. T., & Wessman, C. E. (1958). A manometric method for the rapid practical determination of BOD. *Tappi J.*, *41*, 321.

Dixon, M. (1943). *Manometric methods*. Macmillan Publishing.

Dochain, D., Vanrolleghem, P. A., & van Daele, M. (1995). Structural identifiability of biokinetic models of activated sludge respiration. *Water Res.*, *29*, 2571.

Donlon, J. (1958). Rapid method for estimating BOD values of wastes. *Sanitalk*, *6*(5), 22; *Public Health Eng. Abstr.*, *38*(10), 30 (1958).

Dor, I., & Levy, I. (1987). Single-sample technique for measuring oxygen production and consumption in macrophytic algae. *Aquat. Bot.*, *27*, 323.

Dozank, W. (1965). Influence of nonionic detergent (alphenol 710) on BOD process occurring in sewage. *Rocz. Panstw.*, *16*, 363; *Chemical Abstracts*, *64*, 1805 (1966).

Eckenfelder, W. W. (1959). Absorption of oxygen from air bubbles in water. *J. Sanit. Eng. Div., Proc. Am. Soc. Civ. Eng.*, *35*(SA4), I, 89.

Ekama, G. A., & Marais, G. v. R. (1979). Dynamic behavior of the activated sludge process. *J. Water Pollut. Control Fed.*, *51*, 534.

Elmore, H. A., & Hayes, T. W. (1960). Solubility of atmospheric oxygen in water. *J. Sanit. Eng. Div., Proc. Am. Soc. Civ. Eng.*, *86*(SA4), I, 41.

Elmore, H. A., & Hayes, T. W. (1961). Solubility of atmospheric oxygen in water. Closure by Comm. Sanit. Eng. Res. Sanit. Eng. Div. *Proc. Am. Soc. Civ. Eng.*, *87*(SA5), I, 73.

Elmore, H. L. (1955). Determination of BOD by a Reaeration Technique. *Sew. Ind. Wastes*, *27*, 993.

Elsworth, R., Williams, V., & Smith, R. H. (1957). A systematic assessment of dissolved oxygen supply in a 20-litre culture vessel. *J. Appl. Chem.*, 7(5), 261.

Falk, L. L., & Rudolfs, W. (1947). Factors affecting the direct oxygen utilization method. *Sew. Works J.*, *19*, 1000.

Farkas, P. A. (1981). The use of respirography in biological treatment plant control. *Water Sci. Technol.*, *13*, 125.

Farkas, P. A., & Literathy, P. (1972). Apparatus for continuous or intermittent automatic determination of the BOD and (or) the rate of consumption of oxygen of overground waters and sewage. *Hung. Teljes* (Patent), 3817; *Chemical Abstracts*, *77*, 66034 (1972).

Fleps, W. (1973). Automatic recording respirometer for determination of the BOD. *Hung. Sci. Instrum.*, *27*, 9; *Chemical Abstracts*, *81*, 96259 (1974).

Fleps, W. (1975). An automatic, continuous flow respirometer, its description and use. *Prog. Water Technol.*, 7(2), 1.

Fuhs, G. W. Z. (1968). Some factors affecting the BOD as determined in manometric or manostatic devices. *Wasser Abwasser Forsch.*, *5*, 161.

Fujimoto, E., et al. (1981). Automatic measurement device of the respiration rate and experimental investigation on the constant by using the device for activated sludge process. *Water Sci. Technol.*, *13*, 193.

Gantner, H. (1979). BOD_5 determination by direct coulometric determination of the oxygen consumed. *Fresenius' Z. Anal. Chem.*, *299*, 42.

Garg, R. P., & Mathur, R. P. (1997a). Development of a membrane respirometer. *Water Res.*, *31*, 23.

Garg, R. P., & Mathur, R. P. (1997b). Development of a polyethylene film bag as a respirometer. *Water Res.*, *31*, 262.

Gaudy, A. F., Jr., Eckambaram, A., Rozich, A. F., & Colvin, R. J. (1990). Comparison of respirometric methods for determination of biokinetic constants for toxic and non-toxic wastes. *Proc. 44th Ind. Waste Conf., Purdue University,* 393.

Geenans, D., & Thoeye, C. (1998). The use of an on-line respirometer for the screening of toxicity in the Antwerp Wastewater Treatment Catchment Area. *Water Sci. Technol.*, 37(12), 55.

Gellman, I. (1950). The direct oxygen utilization method and its application to the study of oxidation characteristics of wastes. Unpublished thesis, Rutgers University.

Gellman, I., & Heukelekian, H. (1951). Studies of biochemical oxidation by direct methods. I. Direct method for determining BOD. *Sew. Ind. Wastes*, *23*, 1267.

Gellman, I., & Heukelekian, H. (1953). Studies of biochemical oxidation by direct methods. III. Oxidation and purification of industrial wastes by activated sludge. *Sew. Ind. Wastes*, *25*, 1196.

Genthe, W. K., et al. (1972). Go on-line; it's vital for automation. *Water Wastes Eng.*, *9*(7), 42.

Georgi, C. E., & Wilson, P. W. (1933). The influence of the tension of oxygen on the respiration of *Rhizobia*. *Arch. Microbiol.*, *4*, 543.

Gerloff, W. (1942). The development of measuring devices for the operation of sewage treatment plants. *Kleine Mitt. Mitglied. Ver. Wasser-, Boden-, u. Lufthyg.*, *18*, 278; *Chemical Abstracts*, *39*, 4177 (1945).

Giona, A. R., & Annesini, M. C. (1979). Oxygen uptake in the activated sludge process. *J. Water Pollut. Control Fed.*, *51*, 1009.

Giroux, E. Y., Spanjers, H., Patry, G. G., & Takacs, I. (1996). Dynamic modelling for operational design of a respirometer. *Water Sci. Technol.*, *33*(1), 297.

Golovatyi, E. I., et al. (1977). Automatic monitoring of the BOD of city wastewater. *Vodosnabzh. Sanit. Tekh.* (Russia), 6, 13.

Gomolka, E. (1970). Respiratory apparatus for simultaneous determination of BOD and carbon dioxide. *Gaz, Woda Tech. Sanit.* (Poland), *44*(2), 56; *Chemical Abstracts*, *73*(4), 249 (1970).

Govind, R., Gao, C., Lai, L., & Tabak, H. H. (1997). Continuous, automated and simultaneous measurement of oxygen uptake and carbon dioxide evolution in biological systems. *Water Environ. Res.*, *69*, 73.

Graves, D., Lang, C., & Leavitt, M. (1991). Respirometric analysis of the biodegradation of organic contaminants in soil and water. *Appl. Biochem. Biotechnol.*, *28*, 29.

Gujer, W., et al. (1995). The activated sludge model no. 2: Biological phosporus removal. *Water Sci. Technol.*, *31*(2), 1.

Gulliver, J. S. (1991). Introduction to air–water mass transfer. In S. C. Wilhelms & J. S. Gulliver (Eds.), *Air–water mass transfer*. Am. Soc. Civ. Eng.

Haas, C. N. (1979). Oxygen uptake rate as an activated sludge control parameter. *J. Water Pollut. Control Fed.*, *51*, 938.

Haines, J. R., Holder, E. L., & Venosa, A. D. (1996). Respirometry as a tool to assess effects of environmental variables in bioremediation research and field testing. *Proc. 51st Ind. Waste Conf., Purdue University*. Ann Arbor Press, 463.

Hall, E., Sun, B., Prakash, J., & Nirmalakhandan, N. (1996). Toxicity of organic chemicals and their mixtures to activated sludge microorganisms. *J. Environ. Eng.*, *122*, 424.

Hansen, P. A. (1945). The respiration of bacteria. *J. Bacteriol.*, *49*, 110.

Harlan, W. (1991). Continuous and versatile process control. *Oper. Forum*, *8*(6), 32.

Harris, J. O., & Gainey, P. L. (1944). Respiration of resting *Azotobacter* cells as affected by the respiratory menstrum. *J. Bacteriol.*, *48*, 689.

Hartmann, L. (1967). Influence of turbulence on the activity of bacterial slimes. *J. Water Pollut. Control Fed.*, *39*, 958.

Heddle, J. F., & Tavener, A. (1981). Evaluation of the total carbonaceous BOD method, using a compensated recording respirometer. *Appl. Environ. Microbiol.*, *41*, 580.

Henze, M. (1992). Characterization of wastewater for modelling of activated sludge process. *Water Sci. Technol.*, *25*(6), 1.

Henze, M., Grady, C. P. L., Jr., Gujer, W., Marais, G. v. R., & Matsuo, T. (1987). Activated sludge model no. 1. *Sci. Tech. Rep. No. 1, Int. Assoc. Water Pollut. Res. Control.*

Heukelekian, H. (1947). Use of the direct method of oxygen utilization in waste treatment studies. *Sew. Works J.*, *19*, 875.

Heukelekian, H., & Gellman, I. (1951). Studies of biochemical oxidation by direct methods. II. Effect of certain environmental factors on the biochemical oxidation of water. *Sew. Ind. Wastes*, *23*, 1546.

Heukelekian, H., & Gellman, I. (1955). Studies of biochemical oxidation by direct methods. IV. Effect of toxic metal ions on oxidation. *Sew. Ind. Wastes*, *27*, 70.

Hixon, A. W., & Gaden, E. L., Jr. (1950). Oxygen transfer in submerged fermentation. *Ind. Eng. Chem.*, *42*, 1792.

Hoover, S. R., & Porges, N. (1952). Assimilation of dairy wastes by activated sludge. II. The equation of synthesis and the rate of oxygen utilization. *Sew. Ind. Wastes*, *24*, 306.

Hovis, J. S. (1984). The development and application of a method for determining gas transfer in full-scale biological treatment systems. *Natl. Council Air Stream Improvement Tech. Bull.*, *446*.

Howell, J. A., et al. (1984). On-line measurement of respiration and mass transfer rates in an activated sludge aeration tank. *J. Water Pollut. Control Fed.*, *56*, 319.

Hung, Y., & Eckenfelder, W. W. (1976). Comparison of Warburg and Marais methods of determining oxygen uptake rate constants. *Water Pollut. Control*, *114*(2), 6.

Ingols, R. S. (1968). Short term BOD. *Water Sew. Works*, *115*, 258.

Ippen, A. T., & Carver, C. E., Jr. (1954). Basic factors of oxygen transfer in aeration systems. *Sew. Ind. Wastes*, *26*, 813.

Ippen, A. T., Campbell, L. G., & Carver, C. E., Jr. (1952). The determination of oxygen absorption in aeration processes. *Mass. Inst. Tech. Hydrodynamics Lab. Tech. Rep.*, *7*.

Ippen, A. T., Yoseph, R. S., & Posthill, B. N. (1951). The continuous measurement of oxygen concentration during aeration process. U.S. Public Health Proj. No. RG-863, Massachusetts Institute of Technology.

Iranpour, R., & Miguel, Z. (2008). Online biochemical oxygen demand monitoring for wastewater process control: Full scale studies at Los Angeles Glendale Wastewater Treatment Plant. *Water Environ. Res.*, *80*(4), 298–307.

Issacs, W. P., & Gaudy, A. F., Jr. (1968). Atmospheric oxygen in a simulated stream. *J. Sanit. Eng. Div., Proc. Am. Soc. Civ. Eng.*, *94*(SA2), 319.

Jaeger, K., & Niemitz, W. (1952). BOD determination of sewage by the direct method using a Warburg apparatus. *Staedtehygiene* (Germany), *9*, 1; *Sew. Ind. Wastes*, *25*, 631 (1953).

Jenkins, D. (1960). The use of manometric methods in the study of sewage and trade wastes. In P. C. G. Issac (Ed.), *Wastewater treatment*. Pergamon Press.

Kessler, L. H. (1936). The Odeometer: Its place in the control of activated sludge plants. *Water Works Sewerage*, *83*, 13.

Keyes, F. G., & Gillespie, L. J. (1913). The absorption of oxygen by growing cultures of *B. coli* and *B. welchii*. *J. Biol. Chem.*, *13*, 305.

King, E. F., & Dutka, B. J. (1986). Respirometric techniques. In G. Bitton & B. J. Dutka (Eds.), *Toxicity testing using microorganisms, Vol 1*. CRC Press.

King, L. R., & Palmer, H. J. (1989). Measurement of bacterial oxygen uptake kinetics by a modified respirometric technique. *Biotechnol. Bioeng.*, *33*, 932.

Kite, O. A., & Garrett, M. E. (1983). Oxygen transfer and its measurement. *Water Pollut. Control*, *82*(1), 21.

Knorr, F. (1965). Examination of brewery effluent. *Brauwissenschaft* (Germany), *18*, 191; *Chemical Abstracts*, *63*, 17677 (1965).

Koehne, M., Sierpmann, F. W., & Heeses, D. (1986). A comparison of the BOD_5 and the continuous short-time BOD (BOD-M3). *Korresp. Abwass.* (Germany), *33*, 787.

Koizumi, J., & Aiba, S. (1984). Reassessment of the dynamic K_La method. *Biotechnol. Bioeng.*, *26*, 1131.

Kong, Z., Aerewijck, M., & Werstraete, W. (1996). On-line BOD measurement and toxicity control of wastewater with a respirographic biosensor [letter]. *Environ. Technol.*, *17*, 399.

Kong, Z., Vanrolleghem, P., Willems, P., & Verstraete, W. (1996). Simultaneous determination of inhibition kinetics of carbon oxidation and nitrification with a respirometer. *Water Res.*, *30*, 825.

Kulikov, A. I., & Vasil'eva, A. N. (1985). Automated respirometers: process monitoring. *Vodosnabzh. Sanit. Tekh* (Russia), *4*, 7; *Chemical Abstracts*, *103*, 146853 (1985).

Kuz'min, A. A., et al. (1975). Automatic device for determining the BOD of waste and natural waters. *Otkrytiya Isobret'. Prom. Obraztsy, Tovarnye Znaki*, *52*, 59; *Chemical Abstracts*, *84*, 140518 (1976).

LeBlanc, P. J. (1974). A review of the BOD_5 test. *Pulp Paper Mag. Can.*, *75*(3), 79.

Lee, E. W., & Oswald, W. J. (1954). Comparative studies of the dilution and Warburg methods for determining BOD. *Sew. Ind. Wastes*, *26*, 1097.

Lee, E. W., & Oswald, W. J. (1959). Effects of seeding on BOD. *Water Waste Treatment J.*, *7*, 400.

Le Gall, D. (1991). Continuous short-time BOD measurement. *Eau. Ind. Nuisances*, *148*, 55.

Lenk, S., et al. (1990). Process analysis and control of biological treatment steps using a plateau-BOD measuring instrument. *Wasserwirtsch-Wassertech.* (Germany), *40*(1), 4.

Li, K. Y., & Zhang, Y. B. (1996). Oxygen transfer limitation in a respirometer. *Water Environ. Res.*, *68*, 36.

Londong, J. (1992). Strategies for optimizing nitrate reduction with primary denitrification. *Water Sci. Technol.*, *26*, 1087.

Longmuir, I. S. (1954). Respiration rates of bacteria as a function of oxygen consumption. *Biochem. J.*, *57*, 81.

Ludwig, H. F., Oswald, W. J., & Gotaas, H. B. (1951). Manometric technique for measurement of BOD. *Univ. Calif. San. Eng. Lab. Inst. Eng. Res.*, 442.

Lukasse, L., Keesman, K., Spanjers, H., & Bloemen, M. (2000). Diagnosis and identification of respirometer dynamics and sludge kinetics in continuous-flow respirometers. *Water Environ. Res.*, *72*, 340.

Lukasse, L. J. S., Keesman, K. J., & Van Straten, G. (1997). Estimation of BOD_{st}, respiration rate, and kinetics of activated sludge. *Water Res.*, *31*, 2278.

Madden, M., & Tittlebaum, M. (1984). Oxygen uptake rates associated with biological treatment of pentachlorophenol wastewater. *J. Environ. Sci. Health A*, *19*, 321.

Magliette, R. J., McKinney, D., & Venkataramani, E. S. (1992). Computerized respirometric method for determining biodegradation parameters of organic chemicals. *Proc. 46th Ind. Waste Conf., Purdue Univ.* Lewis Publishers.

Malaney, G. W. (1960). Oxidation abilities of aniline-acclimated activated sludge. *J. Water Pollut. Control Fed.*, *32*, 1300.

Mann, T. (1972). Continuous measurement of respiratory activity of activated sludge. *Chem. Ingr. Tech.*, *44*, 996; *Chemical Abstracts*, *77*, 117898 (1972).

Marsch, L. C., & Drnevich, R. F. (1977). Autothermal aerobic digestion. *J. Water Pollut. Control Fed.*, *49*, 296.

Mendia, L. (1956). Respirometric determination of the BOD of sewage. *Ing. Sanit.*, *4*, 159; *Public Health Eng. Abstr.*, *36*(8), 22.

Mignone, C. F., & Ertola, R. J. (1984). Measurement of oxygen transfer coefficient under growth conditions by dynamic model moment analysis. *J. Chem. Tech. Biotechnol.*, *34B*, 121.

Mines, R. O., & Sherrard, J. H. (1987). Biological enhancement of oxygen transfer in the activated sludge process. *J. Water Pollut. Control Fed.*, *59*, 19.

Mona, R., Dunn, I. J., & Bourne, J. R. (1979). Activated-sludge process dynamics with continuous TOC and oxygen uptake measurements. *Biotechnol. Bioeng.*, *21*, 1561.

Montgomery, H. A. C. (1967). Determination of BOD by respirometric methods. *Water Res.*, *1*, 631.

Montgomery, H. A. C., & Gardiner, D. K. (1971). Experience with a bacterial inoculum for use in respirometric tests for oxygen demand. *Water Res.*, *5*, 147.

Murray, K. A. (1955). Use of the Barcroft respirometer for assessing the oxygen uptake of sewage liquors and sewage effluents. *S. Afri. Ind. Chem.*, *9*, 198.

Murray, K. A., & Coaker, A. W. M. (1956). a new manometric apparatus and its use in sewage fermentation studies. *J. S. Afri. Chem. Inst.*, *9*, 71.

Nakouzi, S. R., Salmeen, I. T., & Kim, B. R. (1996). Avoiding conditions where electrolytic respirometers fail to function as oxygen chemostats. *Water Res.*, *30*, 1333.

Nay, M. W., Jr., et al. (1974). biological treatability of trinitrotoluene manufacturing wastewater. *J. Water Pollut. Control Fed.*, *46*, 485.

Nickerson, W. J., & Chadwick, J. B. (1946). On the respiration of the *Dematophytes*. *Arch. Biochem.*, *10*, 81.

Nunheimer, T. D., & Fabian, F. W. (1942). Respiratory studies of the micrococci. *J. Bacteriol.*, *44*, 215.

Orford, H. E., & Matusky, F. E. (1959). Comparisons of short term with 5-day BOD determinations of raw sewage. *Sew. Ind. Wastes*, *31*, 259.

Organisation for Economic Cooperation and Development. (1998). *Guidelines for testing of chemicals.* 10th Suppl.

Palmateer, G. A., & Thomas, P. C. (1996). Industrial waste impact assessment on domestic activated sludge treatments plants using repirometry. *Proc. 51st Ind. Waste Conf., Purdue Univ.* Lewis Publishers, 453.

Papel, F., Hunken, K. H., & Steinnecke, H. (1958). Simplified method for the volumetric determination of the bod of water and sewage. *Gas.-Wasserfach. Wasser Abwasser* (Germany), *99*, 897; *Chemical Abstracts* *53*, 1602 (1959).

Pasveer, A. (1955). Research on activated sludge. V. Rate of biochemical oxidation. *Sew. Ind. Wastes*, *27*, 783.

Peberdy, M. F. (1976). A Simple system for the measurement of BOD of an effluent sample. *N. Z. J. Dairy Sci. Technol.*, *11*, 129.

Philichi, T. L., & Stenstrom, M. K. (1989). Effects of dissolved oxygen probe lag on oxygen transfer parameter estimation. *J. Water Pollut. Control Fed.*, *61*, 183.

Porges, N., Jasewicz, L., & Hoover, S. R. (1952). Measurement of carbon dioxide evolution from activated sludge. *Sew. Ind. Wastes*, *24*, 1091.

Randall, E. W., et al. (1991). An instrument for the direct determination of the oxygen transfer rate. *Water S.A.*, *17*(1), 11.

Rathburn, R. E. (1977). Reaeration coefficients of streams: State of the art. *J. Hydraul. Div., Proc. Am. Soc. Civ. Eng.*, *103*, 409.

Riegler, G. (1984). Continuous short-term BOD measurement: A new method with many promising application possibilities. *Korresp. Abwass.* (Germany), *31*, 369; *Chemical Abstracts*, *101*, 176911 (1985).

Robbins, M. H. (1961). Short-term substitute for the BOD test. *Natl. Council Air Stream Improvement Tech. Bull.*, *146*.

Ros, M. (1993). *Respirometry of activated sludge.* Technomic.

Ros, M., & Dular, M. (1990). Determination of adaption of the activated sludge for industrial wastewater. *Z. Wasser Adwasser Forsch.* (Germany), *23*, 189.

Ros, M., Dular, M., & Farkas, P. A. (1988a). An improved method using respirography for the design of activated sludge aeration systems. *Water Res.*, *22*, 1483.

Ros, M., Dular, M., & Farkas, P. A. (1988b). Measurement of respiration of activated sludge. *Water Res.*, *22*, 1405.

Ros, M., Zagorc-Koncan, J., Dular, M., & Pompe, M. (1990). Respirometric determination of wastewater biodegradation in biological treatment plants. *Z. Wasser Abwasser Forsch.* (Germany), *23*, 236.

Rozich, A. F. (1992). Design and operational analysis of activated sludge processes using respirometrically calibrated models. *Water Sci. Technol.*, *26*, 2061.

Sawyer, C. N. (1939). Activated sludge oxidations. III. Effects prolonging rapid oxygen uptake rates. *Sew. Works J.*, *11*, 595.

Sawyer, C. N., Bogan, R. H., & Simpson, J. R. (1956). Biochemical behavior of synthetic detergents. *Ind. Eng. Chem.*, *48*, 236.

Sawyer, C. N., & Nichols, M. S. (1939a). Activated sludge oxidations. I. Effect of sludge concentration and temperature upon oxygen utilization. *Sew. Works J.*, *11*, 51.

Sawyer, C. N., & Nichols, M. S. (1939b). Activated sludge oxidations. II. A comparison of oxygen utilization by activated sludges from four wisconsin communities. *Sew. Works J.*, *11*, 462.

Sawyer, C. N., & Rohlich, G. A. (1939). Activated sludge oxidations. IV. The influence of temperature upon the rate of oxygen utilization by activated sludges. *Sew. Works J.*, *11*, 946.

Schlayer, C. (1936). The influence of oxygen tension on the respiration of pneumococci (Type I). *J Bacteriol.*, *31*, 181.

Schreiber, J. D., & Neumainer, E. E. (1987). BOD for agricultural runoff. *J. Environ. Qual.*, *16*, 6.

Schroeder, E. D., & Irvine, R. L. (1979). Parameter choice for process control. *J. Environ. Eng.*, *105*, 217.

Sevag, M. G., & Maiweg, L. (1934). The respiration mechanism of pneumococcus. III. *J. Exp. Med.*, *60*, 95.

Sevag, M. G., & Shelburne, M. (1942). The respiration of *Streptococcus pyogenes*. I. Optimal conditions of respiration. *J. Bacteriol.*, *43*, 411.

Shoup, C. S. (1929). The respiration of luminous bacteria and the effect of oxygen tension upon oxygen consumption. *J. Gen. Physiol.*, *13*, 27.

Sierp, F., & Fransemeier, F. (1931). Progress in the knowledge of BOD. *Tech. Gemeindeblatt*, *34*, 233, 250, 260, 271, 289, 301, 313; *Chem Abstr.*, *26*, 1368 (1932).

Skrindle, R., & Sawyer, C. N. (1952a). Application of the Warburg respirometer to industrial waste analysis, special studies on radioactive wastes. *Proc. 7th Ind. Waste Conf., Purdue Univ.*, 481.

Skrindle, R., & Sawyer, C. N. (1952b). Effect of beta radiation upon biochemical oxidation in polluted water. U.S. Atomic Energy Comm., NWO-4438 (Sept.); *Public Health Eng. Abstr.*, *33*(10), 26 (1953); *Nucl. Sci. Abstr.*, 7, 404.

Smith, D. B. (1953). Measurements of respiratory activity of activated sludge. *Sew. Ind. Wastes*, *25*, 767.

Snaddon, X. Y. M., & Harkness, N. (1959). A manometric apparatus for the measurement of oxygen uptake by activated sludges, sewage effluents, and trade wastes. *Water Waste Treatment J.*, 7, 250.

Sollfrank, U., & Gujer, W. (1990). Simultaneous determination of oxygen uptake rate and oxygen transfer coefficient in activated sludge systems by an on-line method. *Water Res.*, *24*, 725.

Spanjers, H., & Klapwijk, A. (1991). Continuous estimation of short term oxygen demand from respiration measurements. *Water Sci. Technol.*, *24*, 29.

Spanjers, H., & Olsson, G. (1992). Modelling of the dissolved oxygen probe response in the improvement of a continuous respiration meter. *Water Res.*, *26*, 945.

Spanjers, H., Olsson, G., & Klapwijk, A. (1993). Determining influent short term BOD by combined respirometry and estimation. *Water Sci. Technol.*, *28*(11–12), 401.

Spanjers, H., Olsson, G., & Klapwijk, A. (1994). Determining short-term BOD and respiration rate in an aeration tank using respirometry and estimation. *Water Res.*, *28*, 1571.

Spanjers, H., Temmink, H., & Klapwijk, A. (1994). Monitoring of the maximum respiration rate. *Water Sci. Technol.*, *30*, 285.

Spanjers, H., & Vanrolleghem, P. (1995). Respirometry as a tool for rapid characterization of wastewater and activated sludge. *Water Sci. Technol.*, *31*, 105.

Spanjers, H., Vanrolleghem, P., Nguyen, K., Vanhooren, H., & Patry, G. (1998). Towards a simulation-benchmark for evaluating respirometry-control strategies. *Water Sci. Technol.*, *37*(12), 219.

Spanjers, H., Vanrolleghem, P., Olsson, G., & Dold, P. (1996). Respirometry in control of the activated sludge process. *Water Sci. Technol.*, *34*(3–4), 117.

Steinecke, H. (1967). Significance of determination of the BOD. *Papier*, *21*(3), 109; *Chemical Abstracts*, *66*, 118715 (1967).

Steinecke, H. (1968). Automatic recording and measuring instruments for the determination of BOD in samples of waste water. *Stuttg. Ber. Siedl Wasserw.*, *34*, 158.

Steinecke, H. (1976). Direct determination of the BOD. *Gas-Wasserfach, Wasser Abwasser* (Germany), *117*, 454.

Stenstrom, M. K., & Andrews, J. F. (1979). Real-time control of activated sludge process. *J. Environ. Eng.*, *105*, 245.

Stern, R. M., & Frazier, W. C. (1941). Physiological characteristics of lactic acid bacteria near the maximum growth temperature. II. Studies on respiration. *J. Bacteriol.*, *42*, 501.

Stokes, L., Takacs, I., Watson, B., & Watts, J. B. (1993). Dynamic modeling of an ASP sewage works: A case study. *Water Sci. Technol.*, *28*(11–12), 151.

Sundararajan, A., & Lu-Kwang, J. (1995). Biological oxygen transfer enhancement in wastewater treatment systems. *Water Environ. Res.*, *67*, 848.

Suschka, J., & Ferreira, E. (1986). Activated sludge respirometric measurements. *Water Res.*, *20*, 137.

Symons, G. E., & Buswell, A. M. (1929). Comparison of the dilution and absorption methods for the determination of BOD. *Ind. Eng. Chem. Anal. Educ.*, *1*, 214.

Tabak, H. H., Desai, S., & Govind, R. (1990). Determination of biodegradability kinetics of RCRA compounds using respirometry for structure-activity relationships. *Proc. 44th Ind. Waste Conf., Purdue Univ.* Lewis Publishers.

Tabak, H. H., Quave, S. A., Mashni, C. I., & Barth, E. F. (1981). Biodegradability studies with organic priority pollutants. *J. Water Pollut. Control Fed.*, *35*, 1503.

Takamatsu, T., et al. (1982). On-line monitoring and control of an activated sludge process for wastewater using MMOUR. *Eur. J. Appl. Microbiol. Biotechnol.*, *14*, 187.

Tebbutt, T. H. Y., & Berkum, M. (1976). Respirometer determination of BOD. *Water Res.*, *10*, 613.

Temmink, H., Vanrolleghem, P., Klapwijk, A., & Verstraete, W. (1993). Biological early warning systems for toxicity based activated sludge respirometry. *Water Sci. Technol.*, *28*, 415.

Tidwell, W. L., & Sorrels, J. H. (1956). The relative stability test. *Sewage Ind. Wastes*, *28*, 136.

Tool, H. R. (1967). Manometric measurement of the BOD. *Water Sew. Works*, *114*, 211.

Umbriet, W. W., Burris, R. H., & Stauffer, J. F. (1964). *Manometric techniques* (4th ed.). Burgess Publishing.

Vanrolleghem, P., & Spanjers, H. (1994). Comparison of two respirometric principles for the determination of short-term BOD. *Proc. 49th Ind. Waste Conf., Purdue Univ.* Lewis Publishers, 177.

Vanrolleghem, P., & Spanjers, H. (1998). A hybrid method of more reliable assessment of activated sludge model parameters. *Water Sci. Technol., 37*(12), 237.

Vanrolleghem, P., et al. (1995). Practical identifiability of a biokinetic model of activated sludge respiration. *Water Res., 29*, 2561.

Vanrolleghem, P. A., Spanjers, H., Petersen, B., Ginestet, P., & Takacs, I. (1999). Estimating (combinations of). ASM1 parameters and components by respirometry. *Water Sci. Technol., 39*(1), 195.

Vargas-Lopez, C. E., et al. (1989). Discussion of: Evaluation of oxygen uptake rate as an activated sludge process control parameter. *J. Water Pollut. Control Fed., 61*, 99.

Vasel, J. L., Warnier, P., Juspin, H., & Schrobiltgen, P. (1989). *Comparison of different respirometric methods to estimate rapidly BOD*. Foundation Universitaire Luxembourgeoise.

Vernimen, A. P., Henken, E. R., & Lamb, J. C., III. (1967). A short-term BOD test. *J. Water Pollut. Control Fed., 39*, 1006.

Verstraete, W., et al. (1974). Three-step measurement by the sapromat to evaluate the BOD_5, the mineral imbalance, and the toxicity of water samples. *Water Res., 8*, 1077.

Warburg, O. (1923). Experiments on surviving carcinoma tissue. Methods. *Biochem. Zeit. 142*, 317.

Warburg, O. (1926). *Uber den stoffwechsel der turmoren*. Springer.

Watts, J. B., & Cavenor-Shaw, A. R. (1998). Process audit and asset assessment using on-line instrumentations. *Water Sci. Technol., 37*(12), 55.

Watts, J. B., & Garber, W. F. (1995). Online respirometry: A powerful tool for activated sludge plant operating and design. *Pap. Technol., 36*(9), 25.

Wheatland, A. B., & Lloyd, R. (1955). A respirometer for the study of the oxygen demand of polluted water and sewage. *Sew. Ind. Wastes, 27*, 1111.

Witteborg, A., van der Last, A., Hamming, R., & Hemmers, I. (1996). Respirometry for determination of the influent S_s-concentration. *Water Sci. Technol., 33*(1), 311.

Wong, W. K., & Smith, D. W. (1992). Evaluation of an on-line specific oxygen uptake rate monitoring apparatus at the gold bar wastewater treatment plant. *Environ. Technol.*, *13*, 45.

Wooldridge, W. R., & Standfast, A. F. B. (1936). Certain factors that influence the rate of activated sludge and sewage oxidations. *Biochem. J.*, *30*, 156.

Xu, S., & Hasselblad, S. (1996). A simple biological method to estimate the readily biodegradable organic matter in wastewater. *Water Res.*, *30*, 1023.

Young, J. C. (1996). Fundamentals of respirometry: instrument types and basis of operation. *Proc. 51st Ind. Waste Conf., Purdue Univ.* Ann Arbor Press., 441.

Young, J. C., & Baumann, E. R. (1976a). The electrolytic respirometer I. Factors affecting oxygen uptake measurements. *Water Res.*, *10*, 1031.

Young, J. C., & Baumann, E. R. (1976b). The electrolytic respirometer II. Use in water pollution control plant laboratories. *Water Res.*, *10*, 1141.

Young, J. C., Chaing, C. F., Kim, I. S. (2003). Evaluation of short-term BOD methods. *Proceedings of the Water Environment Federation*. WEF.

Young, J. C., Gaines, W., & Clark, J. W. (1965). An improved apparatus for biochemical demand. *Anal. Chem.*, *37*, 784.

Zehnpfennig, R., & Nichols, M. S. (1953). Inoculation studies in BOD determination on sewage and pulp mill waste. *Sew. Ind. Wastes*, *25*, 61.

Zermeno, D., Bianchi, D., Miller J., Flaig, K. W., & Iranpour R. (2002). Rapid BOD measurement technology to prevent industrial waste impact on municipal wastewater plants. *Proceedings of the Water Environment Federation*. WEF.

Zielinski, J., Leszczynski, C., & Skwara, Z. (1965). Determination of the interrelation between the manometrically determined and the BOD after 5 days for waste water from pulp mills. *Zesz. Nauk. Politech. Slaska., Inz. Sanit.* (Poland), *8*, 187; *Chemical Abstracts*, *67*, 36222 (1967).

9

Alternate Measures of Biochemical Oxygen Demand

The major alternates to the dilution bottle biochemical oxygen demand (BOD) technique are the use of respirometers, chemical oxygen demand (COD) and related techniques, and total organic carbon (TOC). These are described in the appropriate chapters. There are still more methods that have been developed, including the use of gas chromatographs for detection of oxygen depletion; headspace BOD (HBOD); the use of alternate sources of oxygen such as hydrogen peroxide, nitrate, and chlorate; microcalorimetry; the biomass activity; and some other miscellaneous methods. These could be viewed as intellectual curiosities that really have little possibility of being developed into practical techniques with widespread application, although they are discussed below. However, there is one direction of current research in BOD measurement that has tremendous potential for practical application, and that is the development of microbial bioelectrodes and other biosensors (Lynggaard-Jensen & Harremoes, 1996).

1.0 GAS CHROMATOGRAPHY AND HEADSPACE BIOCHEMICAL OXYGEN DEMAND TECHNIQUES

Gas chromatography is an instrumental technique for measuring volatile and semi-volatile materials. The related instrumental techniques of liquid chromatography, ion chromatography, and capillary electrophoresis can be used for measuring soluble nonvolatile materials. Although these instruments can be adapted for determination of the food content of the waste, and some applications are discussed in the chapters on COD and TOC, they are generally not used as a surrogate for BOD in this sense. Rather, gas chromatography, using a molecular sieve column, helium carrier gas, and thermal conductivity detector, has been most commonly used to determine the amount of oxygen initially present or remaining after biological treatment of samples (Andrews, 1968; Bilstad et al., 1978; Logan & Kohler, 2001; Logan & Patniak, 1997; Logan & Wagenseller, 1993).

Gas chromatography is used as the determinative method in the HBOD technique (Logan & Patniak, 1997). The HBOD differs from the standard dilution bottle BOD in that the amount of dissolved oxygen (DO) in the sample is continuously replenished from atmospheric oxygen in a sealed reaction tube. This keeps the DO level higher during the incubation period than the dilution bottle BOD and allows faster bacterial utilization of the substrate. A 3-day incubation period in the HBOD procedure is equivalent to a 5-day standard BOD incubation. Another advantage is that the total amount of oxygen in the HBOD is substantially greater than in the closed bottle test, and fewer dilutions need to be prepared. Table 9.1 presents ranges of BOD that can be determined in 28-mL closed tubes by simply varying the amount of tested sample.

The HBOD test has a significantly different calculation than is used in the standard BOD procedure. The initial and final concentrations of oxygen

TABLE 9.1. Range of measurable headspace BOD as a function of headspace and liquid volumes for a DO change of > 1 mg/L and residual DO > 2 mg/L for 28-mL sealed tubes (from Logan & Patniak, 1997).

Headspace volume, mL	Sample volume, mL	HBOD range, mL
5	23	7–50
8	20	12–86
10	18	17–117
15	13	39–236
18	10	51–364
20	8	71–503

in the headspace is the metric for the test, and factors such as barometric pressure, water vapor pressure, and relative humidity on the day the test is initiated must be included. The formula, without any correction for BOD exertion by the seed, is

$$\text{HBOD} = (P_0 - 0.01 p_{0,w} r_0)\left[1 - \frac{\mathrm{A}_N}{\mathrm{A}_{0,n}}\right] \times \left[\frac{107.2}{T_0 + 273.15}\left\{\frac{V_T}{V_1} - 1\right\} + \frac{DO}{760 - P_{0,w}}\right]$$

where P_0 is the initial barometric pressure, $p_{0,w}$ is the initial water vapor pressure, r_0 is the initial relative humidity, A_n is the chromatographic area of the oxygen peak on day n, $\mathrm{A}_{0,n}$ is the chromatographic area of the oxygen peak in a control tube on day n, T_0 is the initial temperature, V_T is the total volume of the tube used for the test, V_1 is the sample volume, and DO is the initial water-saturated DO level at 760 mm Hg pressure and T_0. A seed correction can be included by separate determination of the HBOD of the seed, then factoring out the contribution of the seed from the calculated sample HBOD, taking into consideration the fraction of seed added to the sample.

2.0 ALTERNATE OXYGEN SOURCES

Problems with BOD measurements based on DO consumption have been recognized since the early 1900s. These problems center upon the limited amount of oxygen that is in water, which generally requires that the sample be diluted. Efforts to replace molecular oxygen with other oxygen sources have met with mixed results. Compounds that have been evaluated include nitrate (Lederer, 1913, 1914a, 1914b, 1915a, 1915b), chlorate (Bryan & Rohlich, 1954; Patniak, 1996), and hydrogen peroxide (Chin & Hicks, 1970).

The sodium nitrate method was introduced to the third edition of *Standard Methods* (APHA, 1917) but later dropped because of the lack of correlation with dilution bottle BOD results (Theriault, 1927). Later studies demonstrated that nitrate is very inefficient as a replacement oxygen source (Porges et al., 1952; Vlekke et al., 1988).

The same cannot be said of the use of chlorate as an alternate oxygen source (Bryan & Rohlich, 1954). Chlorate is reduced to chloride by the action of microorganisms present in wastewater; however, tests with pure seed cultures on sterilized wastewater demonstrate that *Escherichia coli, Aerobacter aerogenes, Serratia marcescens, Proteus vulgaris, Bacillus macerans*, and *Pseudomonas fluorescens* are not the causitive agents. Further, pasteurization at 85 °C for 10 minutes of an acclimated seed did not destroy the ability of the seed to utilize chlorate. Earlier work indicated that fungus species of *Penicillum, Aspergillus*, and *Fusarium* may be capable of utilizing chlorate (Aslander, 1928), which may explain the pasteurization results. Further, the common bacterial disinfectants chlorite and hypochlorite, which are nominally intermediates in the reduction of chlorate to chloride, are less effective on fungi.

A chlorate-based BOD procedure has been demonstrated to give 10-day results that are linearly related to 5-day DO-based BOD results up to 250 mg/L BOD, with a slope of approximately 0.71 (Bryan & Rohlich, 1954). Comparative standard deviations on repeated testing of synthetic wastewater for the two types of BOD are 5.35 mg/L, with a mean of 32.2 mg/L for the chlorate procedure, and 9.08 mg/L, with a mean of 45.5 mg/L for the regular procedure, indicating that the chlorate procedure has substantially increased degree of reproducibility. The use of a chlorate-acclimated seed allows the test to be shortened to a 5-day procedure and the seed can tolerate initial chlorate concentrations up to 1000 mg/L. These data further suggest that fungi are the responsible organisms (Cooke et al., 1956).

The chlorate oxygen demand of the test is calculated by multiplication of the difference between the initial and final chloride concentrations by a factor of 0.82. The chloride concentrations can be determined by the Mohr titration (silver nitrate with a dichromate indicator) or other more environmentally friendly procedures such as ion chromatography or capillary ion electrophoresis (Smith, 1999a, 1999b).

Hydrogen peroxide has been tested as an alternate oxygen source for BOD (Chin & Hicks, 1970; Sekoulov & Heinrich, 1981). Advantages to hydrogen peroxide include stable stock solutions for dosing with precisely known initial concentrations of oxygen and numerous simple methods based on either direct combination or enzyme chemistry for the determination of residuals. Peroxidase-based analytical procedures are widely used. A photometric method determines low concentrations of hydrogen peroxide from peroxidase catalyzed oxidation of *N, N*-diethyl phenylenediamine (Bader et al., 1988). Other peroxidase-catalyzed reactions have been used to form crystal violet (Mottola et al., 1970) and malachite green (Zepp et al., 1988) from the leuco forms of the dyes. The oxidized dyes are measured colorimetrically. Fluorescent compounds that are measured fluorometrically are formed from the direct reaction of hydrogen peroxide with *p*-hydroxyphenylacetic acid (Guilbaut et al., 1968; Hwang & Dasgupta, 1986; Kok et al., 1986; Kunen et al., 1983). Determination of hydrogen peroxide is possible through the destruction of fluorescent compounds such as scopoletin (Andreae, 1955; Perschke & Broda, 1961; van Baalen & Marler, 1966). Chemiluminescent methods have been described for the direct action of hydrogen peroxide on

3-amino-phthalhydrazide, also known as luminol (Kok et al., 1978), and bis(2,4,6-trichlorophenyl) oxalate (Klockow et al., 1986). A very sensitive version of the luminol reaction is catalyzed by hemin (Feigl & Anger, 1972).

Hydrogen peroxide is added to the sample in the form of a 3% aqueous solution, with dosings possible up to 430 mg/L without significant impairment of the microorganisms. The extended range of the hydrogen peroxide version of the BOD test is very attractive, as fewer dilutions need to be prepared. The downside is that the BOD with hydrogen peroxide as an oxygen source is slower, with 6 to 7 days of incubation needed to approximate results from the oxygen-based 5-day BOD (BOD_5) (Chin & Hicks, 1970).

3.0 MICROCALORIMETRY AND BIOMASS ACTIVITY

Calorimetry is the study of heat transfer in chemical reactions. The metabolic reactions of microorganisms generate heat just like other exothermic chemical reactions, with the amount of heat given off by the reaction being proportional to the concentration of the reactants. Tiny temperature fluctuations are easily measured with sensitive microscale thermocouples (Smith, 1995) and a twin-cell stop-flow microcalorimeter has been examined for use as a surrogate for BOD determinations. The heat generated by the interaction of microorganisms with food has been found to be linearly proportional to BOD measurements (Beaubien & Jolicoeur, 1986). The ease of the measurement and low maintenance requirements of the instrument have led to the suggestion to use microcalorimetry as a process-monitoring tool in treatment facilities.

The measurement of the heat generated by the metabolism of food is an indirect measure of the biomass activity. Several direct measures have been found. The microbial decolorization of dyes such as methylene blue and indigo carmine is the basis for the relative stability test, which is in essence an activity measure (Jorgensen, 1984). Another metric of potential biomass activity is the determination of various biochemicals used in or created by metabolism. Oxygen uptake of course is a fundamental metric for biomass activity in aerobic systems, and is discussed in detail throughout this book (Huang et al., 1985). The determination of adenosine triphosphate, the major energy storage compound of microbes, is a probe of the viable biomass and has been related to BOD in wastewater effluents and sediments (Kang et al., 1984; Lee et al., 1971a, 1971b; Levin, 1972; Nelson & Lawrence, 1980; Patterson et al., 1970). Dehydrogenase activity is easily measured through the decoloration of triphenyl tetrazolium chloride (TTC). The TTC test has been found to correlate with oxygen-uptake rates and BOD (Klapwijk et al., 1974; Legeron & Ben Aim, 1976; Lelonkiewicz, 1972).

Protease activity has been examined as a viable biomass probe (Lenhard, 1967) and is claimed to be more indicative than dehydrogenase activity. Cellular ribose has been proposed as a surrogate for biomass activity and is found to be linear with oxygen-uptake rates (Banks et al., 1976). Enzyme activities in batch reactors have been used as predictors of biodegradation of pollutant compounds (Sayler et al., 1984). The production of microbial fatty acids has been examined as a metric for bacterial activity and measurement of bacterial growth kinetics (Werker & Hall, 2000). The addition of hydrogen peroxide to the sample, followed by determination of the catalase activity through monitoring the increase in oxygen in the solution, gives a measure of the biomass activity (Guway et al., 1998).

One operational definition of microbial activity that is intimately related to the observed oxygen demand of sludge has been reported. Activity is defined as the minimum amount of sludge necessary to reduce in 21 hours the soluble COD of a wastewater by one-half of the value equivalent to the BOD_5 (Peters et al., 1975).

4.0 MISCELLANEOUS TECHNIQUES

The amount of carbohydrate in water is often related to the BOD_5, particularly when carbohydrates are the major food source in the wastewater. The anthrone reaction for quantitative determination of carbohydrate was introduced in the 1960s (Gaudy & Gaudy, 1980). Anthrone is mixed with the sample and heated at 95 °C for 2 to 3 minutes, then the colored product is determined spectrophotometrically. Results from this test have been found on several occasions to be strongly correlated with BOD_5 (Garceau & Lo, 1978).

The relation of UV spectrophotometry to TOC and dissolved organic matter is discussed in the TOC chapter. Correlation of BOD_5 to UV absorption at 254 nm (Markva, 1983) and 280 nm (Brookman, 1997) for selected effluents has been studied. The fluorescence at 650 nm has been highly correlated ($r^2 = 0.91$) with BOD_5 for another effluent, and is useful for online monitoring (Comber et al., 1996).

Total dissolved solids have been related to BOD_5 for selected effluents, with the metric being the conductivity of the sample (MacLeod, 1974). The capacitance of a wastewater stream has been correlated with BOD (Formaro & Trasatti, 1968). Other reports discuss the relationship between oxidation-reduction potential (ORP) and BOD (Coin et al., 1952; Hewitt, 1936). While ORP readings are indicative of the concentration of DO and can be used as surrogates for oxygen-uptake rates (Baumberger, 1939; Horton, 1954; Koch

& Oldham, 1985; Moore & Ruchhoft, 1943; Rohlich, 1943), any relation to BOD is coincidental.

As a general statement, there is a lack of any fundamental correlation to BOD using any of these techniques, although they may be applicable to selected effluents as a surrogate parameter. The primary reason for any correlation that is achieved is that the characteristics of the wastewater are very consistent in both a qualitative and a quantitative sense, and a coincidental relationship is found.

5.0 MICROBIAL BIOELECTRODES AND OTHER BIOSENSORS

The first published report of a bioelectrode used for estimation of BOD was in 1977 (Karube et al., 1977). This report describes two different designs of measuring systems for BOD. The first is a fuel cell with an electrode constructed from *Clostridium butyricum* immobilized in a polyacryamide matrix on a nylon mesh support in one side of the cell and a platinum electrode in the other. An anion-permeable membrane is used to separate the two sides of the cell. There has been little work reported since then that pursues this design.

In contrast to the lack of success of the first design, the second design has been tremendously popular to researchers and is simple enough for ready construction and use by any treatment facility laboratory technician (Li & Chu, 1991; Tan et al., 1992; Tanaka et al., 1994). It is, in essence, a miniaturized open-chamber respirometer with continuous reaeration, and many of the properties of this style of respirometer are exhibited by the BOD biosensor. Oxygen and substrate in the sample diffuse first through a biofilter and then any excess oxygen over that unused by the organisms in consumption of the substrate diffuses through the Teflon membrane into the body of the electrode for measurement. The DO determined by the electrode is indirectly related to the amount of readily biodegradable substrate in the sample.

In brief (An et al., 1998), a representative microbial population from the treatment facility is cultured in the laboratory. Microorganisms are isolated from the culture by centrifugation, then resuspended in phosphate buffer. The concentration of organisms in the suspension can be determined by plate count, and the volume of the seed suspension can be adjusted to give the same concentration, ideally around 4.0×10^6 organisms/mL, for consistent results, although other researchers have used prepared membranes with up to 10^9 organisms/mL. Aliquots (1 mL) of the suspension are transferred to

0.45-mm or smaller-pore-sized acetylcellulose (An et al., 1998), acrylamide (Karube et al., 1977), collagen (Karube et al., 1977), or polycarbonate (Qian & Tan, 1998) membrane filters approximately the same diameter (5 mm) as the inside of the endcap of the DO electrode that will be used. Suction is used to remove the water, leaving an even dispersion of the organisms over the surface of the filter. A second membrane filter is laid over the one containing the microorganisms, and the edges of the two are sealed together with a biologically inert adhesive. Silicone-based glues are not totally inert, as they offgas acetic acid during curing, but may be satisfactory when used in minimal amounts with adequate ventilation for the 24 hours needed for complete curing. A medical adhesive that does not contain germicides may also be satisfactory. The prepared biofilters can be stored at 4 °C for up to 1 year.

Before use, the biofilter is activated by suspension in a phosphate-buffered glucose–glutamic acid (GGA) solution for 24 hours, then placed on top of the Teflon membrane that seals the end of the DO electrode. The retaining O-ring holds the Teflon membrane in place, and the endcap seals the biofilter against the Teflon with the biofilter outermost. The bioelectrode can also be activated after assembly, but in a production environment, it saves time to activate several biofilters before assembly, in case one fails to give appropriate response. A schematic of a BOD bioelectrode is illustrated in Figure 9.1.

In operation, the modified DO electrode is placed in aerated buffer and the endogenous respiration state of the organisms in the biofilter measured as the difference between the DO level in the buffer, as measured by an

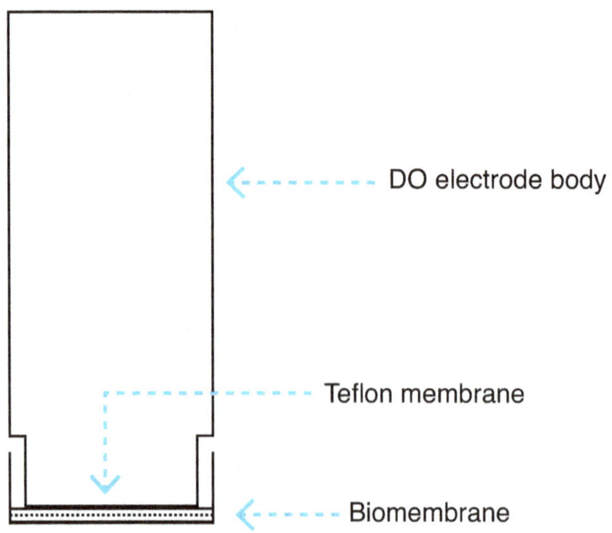

FIGURE 9.1. A schematic of a BOD bioelectrode.

unmodified probe, and the value determined by the bioprobe. The output current in microamps can also be used as the metric. The bioprobe is then transferred to a stirred portion of calibration solution, made up from GGA. Once the bioprobe achieves a steady-state reading in the calibration solution, which can take up to 6 minutes, the electrode output is recorded. The electrode response is plotted against the BOD of GGA solutions and a calibration curve is prepared. The current decrease or DO decrease compared with the endogenous respiration state of the bioelectrode is linearly related to the BOD of the sample.

A study of the characteristics of a bioelectrode prepared from a mixed-organism activated sludge population (An et al., 1998) indicates a highly correlated (r = 0.997) linear response to GGA calibration from 3.3 to 32.8 mg/L BOD. Another preparation using a pure culture of *Trichosporon cutaneum* on a microelectrode gave linear response over a 0.2- to 18-mg/L BOD range (Yang et al., 1996). The range of the bioelectrode response depends on the number of metabolically active organisms in the biofilter. This number is affected by the relative rates of cell death in the endogenous respiration mode and cell reproduction in the active measurement mode. Heavy use of the electrode can lead to increasing the population in the biofilter and an increased level of response, while only occasional use leads to decreasing response. Thus, the bioelectrode requires at least daily calibration with verification of response to the lowest and highest calibration levels, and frequent calibration checks, such as one for every 10 to 20 samples, are prudent. When the death and reproduction rates are approximately in balance, the expected lifetime for a bioelectrode operated daily at 30 °C and stored at 4 °C overnight is at least 48 days, although some researchers consider the electrodes to be disposable (Yang et al., 1996).

Living mixed populations of activated sludge or effluent organisms give the most representative results, particularly when the effluent requires acclimation (An et al., 1998; Karube et al., 1977; Kong et al., 1994; Strand & Carlson, 1984; Vanrolleghem et al., 1994); however, long-term data comparability is at the mercy of the consistency of the treatment facility microorganism population (Hikuma et al., 1979). Pure cultures of organisms such as the yeasts *Trichosporon cutaneum* (Hikuma et al., 1979; Hikuma et al., 1980; Hikuma et al., 1981; Hikuma et al., 1992; Marty et al., 1997; Preininger et al., 1994; Riedel et al., 1988; Riedel et al., 1990; Tanaka et al., 1994; Yang et al., 1996), *Torulopsis candida* (Sangeetha et al., 1996), *Arxula adeninivorans* (Lehmann et al., 1999), *Hansenula anomala* (Kulys & Kasziauskienne, 1980; Li & Chu, 1991), and various bacteria such as *Vibrio fischeri* (formerly *Photobacterium phosphoreum*) (Hyun et al., 1993), *Bacillus subtilis* (Riedel et al., 1988), *Pseudomonas* sp. (Li & Chu, 1991), and *B. licheniformis* (Tan et al., 1992) have been tested. The bioelectrode prepared from *V. fischeri*, also known as *P. phosphoreum* (Hyun et al., 1993), is noteworthy as one of the responses to the presence of assimilable food in this species is the generation of light (Harvey, 1928), which is recorded using a photodiode rather than a DO sensor. Results from use of bioelectrodes prepared from a mixed population of *Bacilli* have been reported (Li & Tan, 1994a, 1994b; Li et al., 1994; Tan et al., 1992). There has been only one report of biosensors prepared from commercial BOD seed, and this report compared results from living and thermally killed preparations (Tan & Wu, 1999). Desirable characteristics for bioelectrodes to become commercially viable have been discussed (Rawson et al., 1989), and several commercially available yeast-based biosensors have been tested (Iranpour et al., 1997; Marty et al., 1997). Somewhat erratic results were obtained during field use, although good results were obtained in a laboratory environment. Other work has concentrated on optimization of field-usable biosensors (Yamane et al., 1991a, 1991b).

Bioelectrodes have been prepared from dead cells rather than living (Qian & Tan, 1998, 1999; Tan & Qian, 1997; Tan & Wu, 1999). Dead-cell sensors offer the advantage of long-term stability and consistency of response, along with lower maintenance and activation requirements. The idea behind the dead-cell preparation is to incapacitate (kill) the growth and reproductive characteristics of the bacteria without denaturing the enzyme systems that are responsible for the initial steps of food acquisition. Heating vacuum-dried cells of *B. substilis* at 280 °C for 2.5 minutes in an oven has been demonstrated to accomplish this (Tan & Qian, 1996). The dead cells are then suspended in phosphate buffer and filtered through a polycarbonate membrane to prepare the biomembrane with a 5.4×10^9 cell concentration.

Examination of the characteristics of the dead-cell bioelectrode reveals it to exhibit a linear relationship to the BOD_5 of GGA and numerous other soluble, readily biodegradable standards (Qian & Tan, 1998). The linear response range of the dead-cell bioelectrode is up to 70 mg/L BOD, although the strength of the signal is approximately one-half of that exhibited by a living cell preparation. A better correlation is obtained with the 8-day BOD values of GGA standards. The response from the bioelectrode remains constant for periods of several months with simple storage in phosphate buffer at room temperature between uses. Repeated testing of a GGA standard exhibiting 50.0 mg/L 8-day BOD gave an average value of 51.8 mg/L, with a standard deviation of 1.23 over a 134-day period. Even starch, which requires extracellular hydrolysis before transport through the cell wall by living cells, elicits response from the dead cells, albeit up to 20 to 30 hours are required for development of maximal response rather than the 15 minutes needed for easily assimilable substrates. The authors did demonstrate that starch diffuses very slowly through the polycarbonate membrane and requires approximately 25 hours to reach equilibrium. They suggest that the diffusion rate is the key factor in the slow response.

The effect of metal ions on the response of dead-cell bioelectrodes has been exhaustively studied (Qian & Tan, 1999) and is approximately the same as that exhibited by living-cell electrodes. An increase in response by tin(II), manganese(II), and zinc(II) is noted. A decrease in response is observed because of the copper(II), and the sensor is completely deactivated by silver(I) and mercury(II). No effect is noted by aluminum(III), cadmium(II), cobalt(II), chromium(III), iron(II and III), and lead(II). Various successful protective measures that have been reported are the uses of ammonium chloride–fortified buffer solution and coating the bioelectrode with a vinylpyridine membrane to alleviate the silver and mercury effect (Li & Tan, 1994a, 1994b; Qian & Tan, 1999). The addition of complexing agents such as ethylenediamine tetraacetic acid and sodium diethyl dithiocarbamate for silver and mercury interference is observed to be, for the most part, minimally effectual.

One design of a BOD biosensor does not use a DO electrode as the base. Instead, an oxygen-sensitive part of the probe is constructed from a layer of tris (4,7-diphenyl-1, 10-phenanthroline) ruthenium(III) perchlorate that is excited by a 480-nm source, and the emitted red light is collected at 610 nm with a photodiode. The presence of oxygen reversibly quenches the fluorescence of this material. The fluorescent sensor is covered with a transparent layer of the yeast *T. cutaneum* immobilized in a polyvinyl alcohol matrix. In the presence of food, the amount of oxygen permeating the membrane decreases, which results in an increase in the fluorescence emmitted from the ruthenium salt. The sensor takes from 3 to 10 minutes to develop maximum response from a sample. The response is linearly correlated with BOD_5 up to 110 mg/L BOD (Preininger et al., 1994).

A variation on the microbe-based biosensor for direct BOD determination uses an electronic nose. The nose is a commercially available device (Model D Neotronics NOSE, Neotronics ScieAntific, Poole, Dorset, UK) that contains 12 nonspecific polypyrrole-conducting polymer sensors. The mixed output from the nose is not correlated to BOD_5 in the general sense; however, very good correlations can be developed for specific effluents (Fenner & Stuetz, 1999; Stuetz et al., 1999a, 1999b).

Pretreatment of the sample (Chee et al., 2001) with a 6-watt black light tube and TiO2 semiconductor gives optimum response with a BOD biosensor. The pretreatment can be as short as 4 minutes for optimum response.

An amperometric BOD biosensor has been reported by Purnima et al. (2008). Analysis times of 90 minutes gave results comparable to standard BOD_5 procedures. A mixed-cell BOD biosensor has been reported by Ampai et al. (2008). Response time of 10 to15 minutes is reported.

6.0 CONCLUSION

Although many different approaches to determination of BOD have been examined over many years, the only one that shows great promise for future commercialization and use is the bioelectrode. Future developments of these devices are eagerly awaited.

7.0 REFERENCES

American Public Health Association. (1917). *Standard methods for the examination of water and sewage* (3rd ed).

Ampai, K., Proespichaya., K., Punnee, A., Bo, M., Panote., T. (2008). Microbial BOD sensor for monitoring treatment of wastewater from a rubber latex industry. *Enzyme Microb. Technol.*, *42*(6), 483–91.

An, L., Niu, H., & Zeng, H. (1998). A new biosensor for rapid oxygen demand measurement. *Water Environ. Res.*, *70*, 1070.

Andreae, W. A. (1955). A sensitive method for the estimation of hydrogen peroxide in biological materials. *Nature*, *175*, 859.

Andrews, J. F. (1968b). Chromatographic analysis of gaseous products and reactants in biological processes. *Water Sew. Works*, *115*, 54.

Aslander, A. (1928). Experiments on the eradication of Canada thistle, *Ceirsium arvense*, with chlorates and other herbicides. *J. Agric. Res.*, *36*, 915.

Bader, H., Sturzenegger, V., & Hoigne, J. (1988). Photometric method for the determination of low concentrations of hydrogen peroxide by the peroxidase catalyzed oxidation of DPD. *Water Res.*, *22*, 1109.

Banks, C. J., et al. (1976). Biological and physical characterization of activated sludge: A comparative experimental study at ten treatment plants. *Water Pollut. Control*, *75*, 492.

Baumberger, J. P. (1939). Relation between oxidation–reduction potential and oxygen consumption rate of yeast cell suspensions. *Proc. Cold Spring Harbor Symposia Quantitative Biol.*, *7*, 195.

Beaubien, A., & Jolicoeur, C. (1986). Application of flow microcalorimetry to process control in biological treatment of industrial wastewater. *J. Water Pollut. Control Fed.*, *57*, 95.

Bilstad, T. B., et al. (1978). Simultaneous in situ determination of dissolved nitrogen and oxygen by gas chromatography. *Prog. Water Technol.*, *10*, 519.

Brookman, S. K. E. (1997). Estimation of BOD in slurry and sediments using ultra-violet spectrophotometry. *Water Res.*, *31*, 372.

Bryan, E. H., & Rohlich, G. A. (1954). Biological reduction of sodium chlorate as applied to measurement of sewage BOD. *Sew. Ind. Wastes*, *26*, 1315.

Chee, G.-J., Nomura, Y., Ikebukuro, K., & Karube, I. (2001). Biosensor for the evaluation of biochemical oxygen demand using photocatalytic pretreatment. *Sens Actuators, B Chem*, *80*(1), 15.

Chin, C., & Hicks, M. G. (1970). Hydrogen peroxide studies of oxygen demand. *J. Water Pollut. Control Fed.*, *42*, 1327.

Coin, L., Mathis, M., & Bonneson, H. (1952). Relation between BOD and oxidation reduction potential. *Bull. Cent. Belge Etude Document Eaux*, *17*, 191; *Chemical Abstracts*, *47*, 12712 (1953).

Comber, S. D. W., Gardner, M. F., & Gunn, A. M. (1996). Measurement of absorbance and fluorescence as potential alternatives to BOD. *Environ. Technol. (Lett.)*, *17*, 771.

Cooke, W. B., Moore, W. A., & Kabler, P. (1956). Comparative satisfaction of BOD by fungi in pure culture and in competition with domestic sewage seed. *Sew. Ind. Wastes*, *28*, 1075.

Feigl, F., & Anger, V. (1972). *Spot tests in inorganic analysis* (6th ed.). Elsevier Science.

Fenner, R. A., & Stuetz, R. M. (1999). The application of electronic nose technology to environmental monitoring of water and wastewater treatment activities. *Water Environ. Res.*, *71*, 282.

Formaro, L., & Trasatti, S. (1968). Capacitance measurements on platinum electrodes for the estimation of organic impurities in water. *Anal. Chem.*, *40*, 1060.

Garceau, J. J., & Lo, S. N. (1978). A rapid estimation of BOD_5 of spent sulfite liquor. *Tappi J.*, *61*(12), 85.

Gaudy, A. F., Jr., & Gaudy, E. T. (1980). *Microbiology for environmental scientists and engineers*. McGraw-Hill.

Guilbaut, G. G., Brignac, P. J., Jr., & Juneau, M. (1968). New substrates for the fluorometric determination of oxidative enzymes. *Anal. Chem.*, *40*, 1256.

Guway, A. J., Buckland, H., Hawkes, F. R., & Hawkes, D. L. (1998). Active biomass in activated sludge: Comparison of respirometry with catalase activity measured using an on-line monitor. *Water Res.*, *32*, 3705.

Harvey, E. N. (1928). The oxygen consumption of luminous bacteria. *J. Gen. Physiol.*, *11*, 469.

Hewitt, L. F. (1936). *Oxidation–reduction potentials in bacteriology and biochemistry* (4th ed.). P. S. King & Son, Ltd.

Hikuma, M., Matsuoka, H., Kawari, M., & Karube, I. (1992). Use of microbial electrodes for observation of microbiological nitrogen elimination process. *Biotechnol. Bioeng.*, *40*, 1, 130.

Hikuma, M., Obana, H., Yasuda, T., Karube, I., & Suzuki, S. (1980). Amperometric determination of total assimilable sugars in fermentation broths with use of immobilized whole cells. *Enzyme Microbiol. Technol.*, *2*, 234.

Hikuma, M., Suzuki, H., Yasuda, A., Karube, I., & Suzuki, S. (1979). Amperometric estimation of BOD by using living immobilized yeasts. *Eur. J. Appl. Microbiol. Biotechnol.*, *8*, 289.

Hikuma, M., Yasuda, T., Karube, I., & Suzuki, S. (1981). Applications of microbial sensors to the fermentation process. *Ann. New York Acad. Sci.*, *369*, 307.

Horton, J. P. (1954). Measurement of O_2 uptake by oxidation–reduction potential. *Instr. Automation*, *27*, 1312.

Huang, J. Y. C., Cheng, M.-D., & Mueller, J. T. (1985). Oxygen uptake rates for determining microbial activity and application. *Water Res.*, *19*, 373.

Hwang, H., & Dasgupta, P. K. (1986). Fluorometric flow injection determination of aqueous peroxides at nanomolar level using membrane reactors. *Anal. Chem.*, 58, 1521.

Hyun, C. K., Tamiya, E., Takeuchi, T., & Karube, I. (1993). A novel BOD sensor based on bacterial luminescence. *Biotechnol. Bioeng.*, *41*, 1107.

Iranpour, R., Straub, B., & Jugo, T. (1997). Real time BOD monitoring for wastewater process control. *J. Environ. Eng.*, *123*, 154.

Jorgensen, K. P. (1984). Determination of the enzyme activity of activated sludge by methylene blue reduction. *J. Water Pollut. Control Fed.*, *56*, 89.

Kang, S. J., et al. (1984). ATP as a measure of active biomass concentration and inhibition in biological wastewater treatment process. *Proc. 38th Ind. Waste Conf., Purdue University.* Ann Arbor Sci. Publ., 751.

Karube, I., Matsunage, T., Mitsuda, S., & Susuki, S. (1977). Microbial electrode BOD sensors. *Biotechnol. Bioeng.*, *19*, 1535.

Klapwijk, A., et al. (1974). A modified procedure for the TTC-dehydrogenase test in activated sludge. *Water Res.*, *8*, 121.

Klockow, D., Jacob, P., & Becher, K. H. (1986). *Zur rolle des wasserstoffperoxids in der atmospharischen flussigphaseenchemie. Statushollоguium luftverunreinigung und waldschaden.* Minister fur Umwelt, Raumplanung und Umwelt, Nordrhein-Westfalen, Dusseldorf, 106.

Koch, F. A., & Oldham, W. K. (1985). Oxidation–reduction potential: A tool for monitoring control and optimization of biological nutrient removal systems. *Water Sci. Technol.*, *17*, 11, 259.

Kok, G. L., Holler, T. B., Lopez, M. B., Nachtrieb, M. A., & Yuan, M. (1978). Chemiluminescent method for determination of hydrogen peroxide in the ambient atmosphere. *Environ. Sci. Technol.*, *12*, 1072.

Kok, G. L., Thompson, K., & Lazrus, A. L. (1986). Derivatization techniques for the determination of peroxides in precipitation. *Anal. Chem.*, *58*, 1192.

Kong, Z., Vanrolleghem, P., & Verstraete, W. (1994). Automated respiration inhibition kinetics analysis (ARIKA). With a respirographic biosensor. *Water Sci. Technol.*, *30*, 275.

Kulys, J., & Kasziauskienne, K. (1980). Yeast BOD sensor. *Biotechnol. Bioeng.*, *22*, 221.

Kunen, S. M., Lazrus, A. L., Kok, G. L., & Meikes, B. G. (1983). Aqueous oxidation of SO_2 by hydrogen peroxide. *J. Geophys. Res.*, *88*, 3671.

Lederer, A. (1913). The relation of the nitrates to putrescibility of sewages. *J. Infect. Dis.*, *13*, 236.

Lederer, A. (1914a). A new method for determining the relative stability of sewage effluent, or polluted river water. *J. Infect. Dis.*, *14*, 482.

Lederer, A. (1914b). A serious fallacy of the "standard" methylene blue putrescibility test. *Am. J. Public Health*, *4*, 421.

Lederer, A. (1915a). Determination of the BOD by the saltpeter method in stockyards, tannery, and corn products wastes. *J. Ind. Eng. Chem.*, *7*, 514.

Lederer, A. (1915b). Notes on the practical application of the saltpeter method for determining the strength of sewages. *Am. J. Public Health*, *5*, 354.

Lee, C. C., et al. (1971a). Adenosine triphosphate in lake sediments. I. Determination. *Soil Sci. Soc. Am. Proc.*, *35*, 82.

Lee, C. C., et al. (1971b). Adenosine triphosphate in lake sediments. II. Origin and significance. *Soil Sci. Soc. Am. Proc.*, *35*, 86.

Legeron, J. P., & Ben Aim, R. (1976). Contribution a la mise au point d'une methode de mesure du taux d'activite totale des boues activees. *Water Res.*, *10*, 273.

Lehmann, M., Chan, C., Lo, A., Lung, M., Tag, K., Kunze, G., & Riedel, K. (1999). Measurement of biodegradable substances using salt-tolerant yeast *Arxula adeninivorans* for a microbial sensor immobilized with poly(carbamoyl)sulfonate (PCS). Part II: Application of the novel biosensor to real samples from coastal and island regions. *Biosens. Bioelectron.*, *14*, 295.

Lelonkiewicz, K. (1972). The TTC Test for the control of effluent purification. *Przegl. Papier*, *28*(4), 129; *Abstr. Bull. Inst. Paper Chem.*, *43*, 4153 (1972).

Lenhard, G. (1967). Determination of protease activity in bottom deposits of sewage stabilization ponds. *Hydrobiologia*, *29*, 67.

Levin, G. V. (1972). Biomass determination: A new technique for activated sludge control. *Water Pollut. Control Res. Ser.*, 17050 EOY. U.S. Dep. Interior.

Li, F., & Tan, T. C. (1994a). Effect of heavy metal ions on the efficacy of a mixed *Bacilli* BOD sensor. *Biosens. Bioelectron.*, *9*, 315.

Li, F., & Tan, T. C. (1994b). Monitoring BOD in the presence of heavy metal ions using a poly(4-vinylpyridine)-coated microbial sensor. *Biosens. Bioelectron.*, *9*, 445.

Li, F., Tan, T. C., & Lee, Y. K. (1994). Effects of pre-conditioning and microbial composition on the sensing effacacy of a BOD biosensor. *Biosens. Bioelectron.*, *9*, 197.

Li, Y. R., & Chu, J. (1991). Study of BOD microbial sensors for wastewater treatment control. *Appl. Biochem. Biotechnol.*, *28–29*, 855.

Logan, B. E., & Kohler, D. (2001). Oxygen mass-transfer coefficients for different sample containers used in the headspace BOD test. *Water Environ. Res.*, *73*, 58.

Logan, B. E., & Patniak, R. (1997). A gas chromatographic-based headspace BOD test. *Water Environ. Res.*, *69*, 206.

Logan, B. E., & Wagenseller, G. A. (1993). The HBOD test: A new method for determining BOD. *Water Environ. Res.*, *65*, 862.

Lynggaard–Jensen, A., & Harremoes, P. (1996). Sensors in wastewater technology. *Water Sci. Technol.*, *33*, 1.

MacLeod, M. (1974). An alarming system for high-solids whitewater. *Pulp Paper*, *48*(11), 116.

Markva, M. (1983). Evaluation of correlations between absorbance at 254 nm and COD of river waters. *Water Res.*, *17*, 231.

Marty, J. L., Olive, D., & Asano, Y. (1997). Measurement of BOD: Correlation between 5-day BOD and commercial BOD biosensor values. *Environ. Technol. Lett.*, *18*, 333.

Moore, W. A., & Ruchhoft, C. C. (1943). Oxidation-reduction studies. II. Oxidation–reduction potentials developed in sewage and sewage activated sludge mixtures. *Sew. Works J.*, *15*, 880.

Mottola, H. A., Simpson, B. E., & Gorin, G. (1970). Absorptiometric determination of hydrogen peroxide in submicrogram amounts with leucocrystal violet and peroxidase as catalyst. *Anal. Chem.*, *42*, 410.

Nelson, P. O., & Lawrence, A. W. (1980). Microbial viability measurements and activated sludge kinetics. *Water Res.*, *14*, 217.

Patniak, R. (1996). Oxygen demand tests and mixed-culture bioreactors using chlorate (ClO_3). as an electron acceptor. Masters thesis, Dep. Chem. Environ. Eng., University of Arizona, Tucson.

Patterson, J. W., et al. (1970). Measurement and significance of adenosine triphosphate in activated sludge. *Environ. Sci. Technol.*, *4*, 569.

Perschke, H., & Broda, E. (1961). Determination of very small amounts of hydrogen peroxide. *Nature*, *190*, 257.

Peters, G. M., et al. (1975). Quantitative determination of the population biological activity of biological sludges used in the activated sludge wastewater treatment process. *Water Res.*, *9*, 901.

Porges, N., Pepinsky, J. B., Jasewicz, L., & Hoover, S. R. (1952). Biochemical oxidation of dairy wastes. III. Failure of sodium nitrate as a source of oxygen. *Sew. Ind. Wastes*, *24*, 874.

Preininger, C., Klimant, I., & Woldbeis, O. S. (1994). Optical fiber sensor for BOD. *Anal. Chem.*, *66*, 1841.

Purnima, D., Anil, K., Abha J., Kumar, S. T., Angamuthu, M., Daya, M. S., & Rita, K. (2008). Quick and reliable estimation of BOD load of beverage industrial wastewater by developing BOD biosensor. *Sens. Actuators B.*, *133*(2), 478–83.

Qian, Z., & Tan, T.C. (1998). Response characteristics of a dead-cell BOD sensor. *Water Res.*, *32*, 801.

Qian, Z. R., & Tan, T. C. (1999). BOD measurement in the presence of heavy metal ions using a thermally-killed *Bacillus subtillis* biosensor. *Water Res.*, *33*, 2923.

Rawson, D. M., Willmer, A. J., & Turner, A. P. F. (1989). Whole-cell biosensors for environmental monitoring. *Biosensors*, *4*, 299.

Riedel, K., Lange, K. P., Stein, H. J., Kuehn, M.; & Scheller, F. (1990). A microbial sensor for BOD. *Water Res.*, *24*, 883.

Riedel, K., Renneberg, R., Kuehn, M., & Scheller, F. (1988). A faster estimation of BOD using microbial sensors. *Appl. Microbiol. Biotechnol.*, *28*, 316.

Rohlich, G. A. (1943). Oxidation–reduction potential measurements in activated sludge and activated sludge–sewage mixtures. *Sew. Works Eng.*, *14*, 348.

Sangeetha, S., Sugandhi, G., Murugesan, M., Madhav, V. M., Berchmans, S., Rajasekar, R., Rajasekar, S., Jeyakumar, D., & Rao, G. P. (1996). *Torulopsis candida*-based sensor for the estimation of BOD and its evaluation. *Electroanalysis*, *8*, 698.

Sayler, G. S., et al. (1984). Predictive assessment of priority pollutant bio-oxidation kinetics in activated sludge. *Environ. Prog.*, *3*, 153.

Sekoulov, I., & Heinrich, D. (1981). The continuous oxygen uptake rate measurement and its applicability as an activated sludge control parameter. *Water Sci. Technol.*, *13*, 205.

Smith, R.-K. (1995). *Water and wastewater laboratory techniques*. Water Environment Federation.

Smith, R.-K. (1999a). *Handbook of environmental analysis* (4th ed.). Genium Publishing.

Smith, R.-K. (1999b). *Lectures on wastewater analysis and interpretation*. Genium Publishing.

Strand, S. E., & Carlson, D. A. (1984). Rapid BOD measurement for municipal wastewater samples using a biofilm electrode. *J. Water Pollut. Control. Fed.*, *56*, 464.

Stuetz, R. M., Fenner, R. A., & Engin, G. (1999a). Assessment of odours from sewage treatment works by an electronic nose, H_2S analysis and olfactometry. *Water Res.*, *33*, 453.

Stuetz, R. M., Fenner, R. A., & Engin, G. (1999b). Characterization of wastewater using an electronic nose. *Water Res.*, *33*, 442.

Tan, T. C., Li, F., Neoh, K. G., & Lee, Y. K. (1992). Microbial membrane-modified dissolved oxygen probe for rapid BOD measurement. *Sensors Actuators B: Chem.*, *B8*(2), 167.

Tan, T. C., & Qian, Z. R. (1996). Activity of the enzyme system in thermally killed *Bacillus* cells. *Enzyme Microbiol. Technol.*, *19*, 150.

Tan, T. C., & Qian, Z. R. (1997). Dead *Bacillus subtilis* cells for sensing BOD of water and wastewaters. *Enzyme Microbiol. Technol.*, *19*, 150.

Tan, T. C., & Wu, C. (1999). BOD sensors using multi-species living or thermally killed cells of a BODSEED microbial culture. *Sensors Actuators B: Chem.*, *54*, 252.

Tanaka, H., Nakamura, E., Minamiyama, Y., & Toyoda, T. (1994). BOD biosensor for secondary effluent from wastewater treatment plants. *Water Sci. Technol.*, *30*(4), 215.

Theriault, E. J. (1927). The oxygen demand of polluted waters. I. A critical review. II. The rate of deoxygenation. *U.S. Public Health Serv. Bull.*, 173.

van Baalen, C., & Marler, J. E. (1966). Occurrence of hydrogen peroxide in sea water. *Nature*, *211*, 951.

Vanrolleghem, P., Kong, Z., Rombouts, G., & Verstraete, W. (1994). An on-line respirographic biosensor for the characterization of load and toxicity of wastewaters. *J. Chem. Tech. Biotechnol.*, *59*, 321.

Vlekke, G. J. F. M., et al. (1988). Biological phosphate removal from wastewater with oxygen or nitrate in sequencing reactors. *Environ. Technol. Lett.*, *9*, 791.

Werker, A. G., & Hall, E. R. (2000). Assessment of microbial fatty acid analysis for quantifying mixed culture biomass during batch growth. *Water Environ. Res.*, *72*, 484.

Yamane, M., Kawahara, A., Wakida, S., Takeda, S., & Higashi, K. (1991a). Automated BOD measurement in environmental water. 1. Development of rapid bod measuring system with microbial sensor. *Chem. Express.*, *6*, 615; *Chemical Abstracts*, *115*, 166197 (1991).

Yamane, M., Kawahara, A., Wakida, S., Takeda, S., & Higashi, K. (1991b). Automated BOD measurement of environmental water. 2. Field monitoring of BOD with rapid BOD measuring system. *Chem. Express.*, *6*, 707; *Chemical Abstracts*, *115*, 189370 (1991).

Yang, Z., Suzuki, H., Sasoki, S., & Karube, I. (1996). Disposable sensor for BOD. *Appl. Microbiol. Biotechnol.*, *46*, 10.

Zepp, R. G., Skurlatov, Y. I., & Rittmiller, L. F. (1988). Effects of aquatic humic substances on analysis for hydrogen peroxide using peroxidase-catalyzed oxidations of triarylmethanes or *p*-hydroxyohenoxyphenylacetic acid. *Environ. Technol. Lett.*, *9*, 287.

10

Biodegradation and Toxicity

Most persons who use and are familiar with the dilution bottle biochemical oxygen demand (BOD) test are well acquainted with the interpretation of the test results as a measure of the strength (waste load) of the sample, that is, how much time and effort is going to be required to get the pollutant burden of the treatment facility influent under the regulatory limits listed in the permit for the facility effluent. As the active agents in the test are biological organisms, any discussion of the BOD test is automatically going to be centered on the biodegradability of the waste, although this sometimes gets overlooked when the test results do not agree with some preconceived notion of what the results should be. The lack of correlation between the BOD results and other measures of wastewater strength, such as volatile suspended solids, chemical oxygen demand (COD), total organic carbon, and total oxygen demand (TOD), will eventually bring the term "toxicity" into the discussion after analyst technique issues have been eliminated as the causative agents for the disagreement.

In contrast to the familiar usages of BOD data for treatment facility control and water quality monitoring, there is a separate field of research and study that utilizes the BOD test as a characteristic probe of a combination of properties exhibited by pure compounds and industrial products that can indicate the potential for adverse effect on the environment. These combined properties are lumped together into the term "biodegradability." Within this context, biodegradable and toxic are the two opposing ends of a single measurement scale. Rather than being absolute and unrelated characteristics on the biodegradability scale, the one slowly changes into the other.

This chapter presents general information and examples of formal techniques for determination of biodegradation and toxicity. It then presents a summary of toxicity as it complicates determination of BOD and a technique for obtaining BOD results despite the presence of sample toxicity.

1.0 BIODEGRADATION AND TOXICITY

Although earlier studies generally recognized that BOD is a measure of the biodegradability of the wastewater, the first explicit application of the test to determine the potential for biodegradation of a pure compound was in 1929 (Symons & Buswell, 1929). In this study, the 5- to 30-day BOD of the test compound was determined and then compared with the theoretical BOD (ThBOD) derived from permanganate COD. The biodegradability of the test compound was expressed as percent BOD/ThBOD. The biodegradability results were peptone (82.5%), urea (100%), sodium oleate (76%), sodium palmitate (78%), lactose (75%), starch (83%), and cellulose (71%). This initial study was followed by a further study examining the BOD exerted by a variety of pure sodium soaps (Symons & Buswell, 1932). Studies on the biological decomposition of aromatic wastes (Aitken, 1936) and phenol (Akano & Tyo, 1939; Baziakina, 1940) followed. The toxic properties of wastes from nonferrous metallury processes toward microorganisms were examined (Kalabina et al., 1944), as was the toxicity of tri- and hexavalent chromium (Kreiger & Moore, 1949). These humble beginnings were reviewed (Rudolfs et al., 1950), which set the stage for a tremendous increase in the number of reports, leading eventually to development of biodegradation studies as a field apart from wastewater treatment.

Beginning in the 1950s and continuing on through the 1970s, the degradation of individual industrial products by the mixed microorganism cultures routinely found in wastewater resource recovery facilities (WRRFs) was a popular area of study. Various classes of materials that were studied included anionic sufactants and other synthetic components of detergent formulations (Bogan & Sawyer, 1954, 1955; Brink & Meyers, 1966; Malaney & Sheets, 1957; Sawyer & Bogan, 1954, 1955; Sawyer et al., 1956; Sheets & Malaney, 1956), non-ionic surfactants (Garrison & Matson, 1964), formaldehyde (Gellman & Heukelekian, 1950), fats (Jezeski et al., 1950), alcohols (McKinney & Jeris, 1955; Scope & Kenna, 1950), aromatics (McKinney et al., 1956), alkanes and other petrochemicals (Pitter, 1976; Price et al., 1974), aminocarboxylic acid lactams (Rogovskaya, 1954), volatile organic acids (Sato & Akiyama, 1972; Scope & Kenna, 1950), amines (Oberton &

Stack, 1957; Scope & Kenna, 1950), benzidine and other aromatic amines (Baird et al., 1977), textile industry chemicals (Gallotti, 1971; Gottsching & Luttgen, 1978; Luttgen & Gottsching, 1979), and carbohydrates (Ghosh & Pohland, 1972; Varma & Lindley, 1966). Occasionally, these studies are performed to trace sources of particular components of waste streams in industries and in publicly owned treatment works collection systems (Hale, 1953). Concurrent with these studies on the biodegradability of organic materials, a number of groups were examining the toxic effects of metals on microorganisms (Ingols, 1954, 1955; Kalabina et al., 1944; Kreiger & Moore, 1949; Mowat, 1976). Two books including information on these and other studies on metal toxicity toward microorganisms are available and should be consulted for more information (Beveridge & Doyle, 1989; Hughes & Poole, 1989). The majority of these studies used the dilution bottle BOD technique as the major tool for data generation, with lesser attention given to respirometer-based methods. Heukelekian and Rand (1955) collected BOD results on pure organic compounds for the Federation of Sewage and Industrial Wastes Associations (predecessor of the current Water Environment Federation) and published the summary. Table 10.1 is based for the most part on their work.

The table presents BOD results on pure compounds. These data are derived from both respirometric and dilution bottle BOD tests with incubation times ranging from 5 to 10 days, although some results represent incubation times up to 80 days. In some instances, acclimated seeds were used. The higher values are more frequently derived from extended incubation times, but this is not always the case. Although respirometric methods are commonly thought to give higher values, in general, these data do not support this conclusion.

Three conclusions can be drawn from perusal of the data in Table 10.1. First, there is a wide range of reported results for some of the compounds, one example being phenol. An average BOD is presented based on exclusion of the extreme (outlier) values when multiple data are available. This gives a value that is more consistent within a class of compounds such as the phenolics. The presence of outlier values is generally not indicative of "analyst mistakes" and should not be interpreted in this manner. Instead, it is an indication that certain microbial populations exist that are either well adapted or particularly inept with regard to consumption of the substrate. For compounds in which only one value is presented, the value is essentially worthless as an expectation of results, particularly when the value is an exception to other data within the class of compounds. The second conclusion is that there are materials, the dyes for example, that are inert to the BOD test.

The third conclusion is that minor structural changes in a compound can markedly change its biodegradability. Take, for example, the isomeric

TABLE 10.1. Biochemical oxygen demand ranges for pure compounds (Heukelekian & Rand, 1955, except as marked).

Class	Compound	BOD range, g/g	Average, g/g
Acids and salts	Formic acid	0.02–0.27	0.20
	Sodium formate	0.04–0.096	0.068
	Calcium formate	0.20	—
	Acetic acid	0.34–0.88	0.62
	Sodium acetate	0.11–0.67	0.34
	Calcium acetate	0.43	—
	Dichloroacetic acid	0	—
	Propionic acid	0.36–1.10	0.63
	Sodium propionate	0.36–0.61	0.50
	Butyric acid	0.34–1.40	0.84
	Sodium butyrate	0.41	—
	Sodium isobutyrate	0.43	—
	Valeric acid	1.06–1.35	1.28
	Sodium valerate	0.48	—
	Sodium isovalerate	0.57	—
	Pivalic acid***	0.21	—
	Caproic acid	1.46	—
	Sodium caproate	0.69	—
	Glucono δ-lactone	0.52	—
	Caprylic acid	1.46	—
	Sodium heptylate	0.33–0.82	0.53
	Sodium caprylate	0.76	—
	Capric acid****	1.19–1.40	—
	Nonylic acid	0.59	—
	Palmitic acid	1.07	—
	Sodium palmitate	0.40–1.02	0.43
	Stearic acid	0.786	—
	Sodium stearate	0.45–1.72	1.06
	Glycolic acid	0.175	—
	Oxalic acid	0.10–0.14	0.12
	Sodium oxalate	0.08–0.16	0.11
	Lactic acid	0.63–0.88	0.72
	Succinic acid	0.21–0.64	0.39
	Fumaric acid	0.57–0.70	0.62

(continued)

TABLE 10.1. Biochemical oxygen demand ranges for pure compounds (Heukelekian & Rand, 1955, except as marked) (*continued*).

Class	Compound	BOD range, g/g	Average, g/g
	Malic acid	0.34–0.57	0.46
	Maleic acid/anhydride	0.38–0.60	0.46
	Tartaric acid	0.30	—
	Mucic acid	0.245	—
	Citric acid	0.40	—
	Sodium citrate	0.32–0.39	0.37
	Undecylenic acid	0.537	—
	Sodium oleate	0.47–1.43	0.93
	Benzoic acid	1.34–1.40	1.38
	Sodium benzoate	1.13–1.36	1.24
	Benzoyl benzoic acid	0.001	—
	tert–Butylbenzoic acid***	0.26	—
	Salicylic acid	0.95	—
	Mandelicacid	0.31	—
	Phthalic acid/anhydride	0.72–1.44	1.06
	Anthranilic acid	1.32	—
	Sulfanilic acid	1.11	—
	Toluene sulfonic acid	0	—
	Anthraquinone sulfonic acid	0	—
	Gallic acid	0.08	—
	Tannic acid	0.31–0.46	0.38
	Methacrylic acid	0.89	—
	Methyl methacrylic acid	0.14	—
	EDTA	0.01	—
	Uric acid	0.30	—
	n-Dodecenylsuccinic anhydride***	1.02–1.82	—
	n-Noneylsuccinic anhydride***	0.25–1.30	—
Alcohols			
	Methyl alcohol	0.13–1.12	0.84
	Ethyl alcohol	0.93–1.67	1.30
	n-Propyl alcohol	0.47–1.5	0.98
	Isopropyl alcohol	0.075–1.72	1.40
	Allyl alcohol***	0.20–1.79	—

(*continued*)

TABLE 10.1. Biochemical oxygen demand ranges for pure compounds (Heukelekian & Rand, 1955, except as marked) (*continued*).

Class	Compound	BOD range, g/g	Average, g/g
	3-Chloro-1-propanol***	0.06–0.43	—
	3-Chloro-1,2-propanediol***	0.01	—
	Diacetone alcohol***	0.07–0.68	—
	1,3-Dichloro-2-propanol***	0.01	—
	Glycerol	0.617–0.98	0.76
	n-Butyl alcohol***	1.71	—
	sec-Butyl alcohol***	2.15	—
	tert-Butyl alcohol***	0.02–1.63	—
	Amyl alcohol	1.10–1.62	1.45
	tert-Butyl alcohol**	0	—
	4-Methyl-2-pentanol***	2.37	—
	4-Methyl-2,4-pentanediol***	0.02	—
	Cyclohexanol	0.08	—
	Octyl alcohol	1.09	—
	n-Nonyl alcohol***	0.49	—
	Ethylene glycol	0.16–1.10	0.67
	Ethylene chlorohydrin***	0.03–0.50	—
	Diethylene glycol***	0.05–0.10	—
	Triethylene glycol	0.03–0.50	—
	Tetraethylene glycol	0.50	—
	Polyethylene glycol	0.03–0.08	0.06
	Ethylene glycol monomethyl ether***	0.12–1.1	—
	Ethylene glycol monoethyl ether	1.03–1.58	—
	Ethylene glycol monobutyl ether***	0.71–1.68	—
	Ethylene glycol monoethyl ether acetate***	0.74	—
	Ethylene glycol monomethyl ether acetate***	0.49	—
	Diethylene glycol monoethyl ether***	0.20–0.58	—
	Diethylene glycol monobutyl ether***	0.25	—
	Diethylene glycol monomethyl ether***	0.12	—
	Triethylene glycol monoethyl ether***	0.05	—
	Ethylene glycol mon-i-propyl ether***	0.18	—

(*continued*)

TABLE 10.1. Biochemical oxygen demand ranges for pure compounds (Heukelekian & Rand, 1955, except as marked) (*continued*).

Class	Compound	BOD range, g/g	Average, g/g
	1,2-Propanediol***	1.08	—
	1,3-Propanediol***	0.18–0.27	—
	Dipropylene glycol***	0.09	—
	1,5-Pentanediol	2.0	—
	Hexylene glycol***	0.02	—
	Methyl phenyl carbitol	1.0	—
	Benzyl alcohol	1.55	—
Aldehydes/ ketones			
	Formaldehyde	0–1.10	0.69
	35% Formalin	0.37	—
	Acetaldehyde	1.27	—
	Butyraldehyde	1.06–1.62	1.43
	Acetoldol	0.90	—
	Crotonaldehyde	1.30	—
	Furfural	0.28–0.77	0.52
	Benzaldehyde	1.50	—
	Acetone	0–1.85	0.88
	Methyl ethyl ketone	2.14	—
	Diethyl ketone	1.00–1.98	1.49
	Methyl isobutyl ketone	0.12–2.14	0.95
	Methyl *n*-amyl ketone***	0.50–2.20	—
	4-Metyl-2-pentanone***	2.06	—
	4-Methoxy-4-methyl-2-pentanone***	0.11	—
	Methyl phenylketone	0.518–1.40	0.96
	Phorone***	0.19	—
	1,4-Naphthoquinone	0.81	—
	Acrolein***	0	—
	Methylal**	0	—
Amines/amides			
	Ethanolamine	0.78–1.10	0.89
	Diethanolamine***	0.03–1.17	—
	Triethanolamine**	0.08–0.45	—

(*continued*)

TABLE 10.1. Biochemical oxygen demand ranges for pure compounds (Heukelekian & Rand, 1955, except as marked) (*continued*).

Class	Compound	BOD range, g/g	Average, g/g
	Aniline	1.47–2.26	1.68
	N,*N*-Dimethylaniline	0.25	—
	N-Ethylaniline	0.048	—
	N,*N*-Diethylaniline	0	—
	Melamine	0	—
	Butylamine	1.25	—
	Hexamethylene tetramine	0.015–0.026	0.02
	Toluidine	0.24–1.63	1.50
	Morpholine	0–0.02	0.01
	Diamino pyridine	0	—
	Xylidine	0	—
	Naphthylamine	0.89	—
	Acetanilide	1.20	—
	N-Acetyl ethanolamine	1.10	—
	N-Acetyl morpholine	0	—
	Acylonitrile	0.70	—
	Formamide	0	—
	Nitrobenzene	0	—
	o-Bromonitrobenzene***	0.10–0.26	—
	Allyl amine***	0	—
	Diallyl amine***	0.06	—
	Isopropylamine***	0.06–0.73	—
	Diisopropylamine***	0.02–0.81	—
	Acrylamide***	0.05–0.92	—
	Skatol	1.51	—
Carbohydrates			
	Mannitol	0.68	—
	Glucose	0.50–1.01	0.68
	Sucrose	0.49–0.76	0.67
	Lactose	0.45–0.72	0.58
	Starch	0.22–1.25	0.57
	Dextrin	0.40–0.50	0.45
	Cellulose	0.08–1.1	0.63

(*continued*)

TABLE 10.1. Biochemical oxygen demand ranges for pure compounds (Heukelekian & Rand, 1955, except as marked) (*continued*).

Class	Compound	BOD range, g/g	Average, g/g
	Methyl cellulose	0.008	—
	Carboxymethylcellulose	0.3	—
	Lignin	0	—
Dyes			
	Methyl orange	0	—
	Aniline yellow	0	—
	Tropeolin	0	—
	Congo red	0	—
	Malachite green	0	—
	Methyl violet	0	—
	Phloxin	0	—
	Bromothymol blue	0	—
	Phenol red	0	—
Esters			
	Ethyl formate	0.50	—
	Ethyl acetate	0.29–1.14	0.83
	Butyl acetate	0.15–0.52	0.34
	Amyl acetate	0.31–0.88	0.60
	Dibutyl phthalate	0.43	—
	Isopropyl acetate	0.26	—
Ethers			
	Diethyl ether	0–0.03	0.02
	Ethylene glycol diethyl ether	0.10	—
	Diethylene glycol diethyl ether	0.10	—
	Polyethylene glycol diacetate	0.20	—
	Dioxane	0	—
	Isoamyl ether	0	—
	Ethylene oxide***	0.06	—
	Propylene oxide***	0.17–0.20	—
	Allyl glycidyl ether***	0.06	—
	Phenyl glycidyl ether***	0.14	—
	Diisopropyl ether***	0.19	—
	2-Ethylhexyl glycidyl ether***	0.14	—

(*continued*)

TABLE 10.1. Biochemical oxygen demand ranges for pure compounds (Heukelekian & Rand, 1955, except as marked) (*continued*).

Class	Compound	BOD range, g/g	Average, g/g
	Glycerin allyl ether***	0.01	—
	Pentaerythritol triallyl ether***	0.43	—
N-Heterocycles			
	Pyridine	0–1.47	1.31
	Quinoline	1.71	—
	Indole	2.07	—
	n-Butyl imidazole***	0.09	—
Phenolics			
	Phenol	0.05–2.1	1.67
	ortho Cresol	1.64	—
	meta Cresol	1.70	—
	Xylenol	0.10–1.50	1.16
	Pyrocatechol	0.69	—
	Resorcinol	1.15	—
	Hydroquinone	0.48	—
	Pyrogallol	0.016	—
	Phloroglucinol	0.47	—
	Guiacol	1.40	—
	Naphthol	1.69–1.79	1.74
	Oxyquinoline	0	—
Proteins			
	Casein	0.25–1.1 7	0.80
	Gelatin	0.55–0.92	0.74
	Peptone	0.15–0.88	0.52
Amino acids			
	Glycine	0.52–0.55	0.54
	Alanine	0.94	—
	Aspartic acid	0.58	—
	Glutamic acid	0.42–0.64	0.50
Hydrocarbons			
	Benzene***	2.18	—
	Toulene***	2.15	—
	Styrene***	1.29–2.45	—

(*continued*)

TABLE 10.1. Biochemical oxygen demand ranges for pure compounds (Heukelekian & Rand, 1955, except as marked) (*continued*).

Class	Compound	BOD range, g/g	Average, g/g
	o-Xylene***	1.64–1.80	—
	m-Xylene***	2.53	—
	p-Xylene***	1.40–2.35	—
	p-t-Butyl-toluene***	0.06–0.19	—
	Allyl chloride***	0.23–0.42	—
	Allyl bromide***	0.59–0.78	—
	1-Chloro-3-bromopropane***	0.02	—
	1,3-Dichloropropane***	0.19	—
	Methallyl chloride***	0.81	—

**Hatfield (1957).
***Bridie et al. (1979).
****Gaffney & Heukelekian (1958).

alcohols *n*-amyl alcohol, *sec*amyl alcohol, and *tert*-amyl alcohol. These alcohols have the same formula, $C_5H_{12}O$, yet differ structurally by the number of carbons attached to the carbon atom bearing the alcohol function. *n*-Amyl alcohol is a primary alcohol with two hydrogens and one carbon attached to the alcohol-bearing carbon and exhibits an average 5-day BOD (BOD_5) of 1.54 mg/mg substrate with a theoretical BOD of 2.84 mg/mg. *tert*-Amyl alcohol has three carbons attached to the alcohol-bearing carbon and is completely refractory to biodegradation. *sec*-Amyl alcohol exhibits a BOD_5 midway between the two and is partially degradable. These tests were conducted with an unadapted seed (Hatfield, 1957). Adapted seeds in general perform better (Bridie et al., 1979) but there are some compounds that are outside the catabolism ability of microorganisms.

One of two approaches to determining the biodegradability/toxicity of a compound is generally taken. Biodegradation can be determined by placing a limited amount of the test material into a suitable medium containing microorganisms and supporting nutrients, then monitoring the disappearance of the test compound as time passes. The BOD dilution bottle technique and related shake flask technique are manifestations of this approach. The river die-away test, in which a portion of test compound is spiked into a bottle of river water, and then the disappearance of the test compound monitored, is another. In some cases, examinations similar to these are conducted in the natural setting, with samples being withdrawn periodically for analysis of residual pollutants. This in situ biodegradation is called natural attenuation, and is a recognized technique for remediation of pollution (Smith, 1999a). The progress of natural attenuation is followed by charting the disappearance of the target pollutants over time from the samples. The tests used can be specific for the pollutants, such as gas chromatography-mass spectrometry, or nonselective general analytical procedures may be used. One example of the latter is evaluation of the biodegradation of pollutants in groundwater through repetitive measurement of dissolved organic carbon (Gron et al., 1992). Batch reactors containing activated sludge have been used to allow lengthy acclimation toward the test material, then additional amounts of the substrate chemical are added and analyzed for compound assimilation (Luttgen & Gottsching, 1979).

The other approach is to test for evidence of microorganism metabolism, such as respiration, growth, or reproduction. Bacterial oxygen consumption measured in respirometers or dilution bottles is a common technique. Selected references are listed in Table 10.2. The results of biodegradability evaluations using the BOD dilution bottle technique have been compared with those using respirometry as the method (Heukelekian, 1958). For rapidly degradable materials, the respirometry techniques are generally preferable. The dilution bottle techniques were at one time more suitable for longer-term studies, although modern electrolytic respirometers and respirometric biosensors are replacing the bottles in this application.

The oxygen-uptake rate of pure cultures of the protozoan *Tetrahymena pyriformis* has been noted as particularly sensitive to heavy metal toxicity (Slabbert, 1980). The respiration of batch bioreactors offers both short-term (Brouzes et al., 1978) and long-term evaluations. Compound inhibition of the oxygen production of photosynthetic organisms is another aspect of respiration-based degradability determinations (Knopp, 1961). Carbon dioxide evolution is used as another metabolic activity indicator (Means & Anderson, 1981). The light output from luminescent bacteria is used as a monitoring method (Bulich, 1982; Dutka & Kwan, 1981; Sweet et al., 1997; Toussaint et al., 1995) and has evolved into commercially available test kits (Microtox®, Azur Environmental, Carlsbad, California). *Standard Method* 8050 B describes the use of bacterial bioluminescence for toxicity determination using Microtox. The changing millivolt values in a culture as a reflection of the reduction-oxidation level have been used as a metabolic state probe (Shaikh et al., 1985). A redox biosensor has been used in a similar application (Schowanek et al., 1987).

Biodegradability is quantitatively assessed by comparing the metabolism metric to some standard such as the theoretical maximum expected for the metric for the substance or to the value of the metric obtained from testing an equivalent amount of a known biodegradable material. An example of the former method is computation of the biorefractory index (Hart &

TABLE 10.2. Some examples of the use of BOD and respirometry for determination of biodegradation rates and toxicity.

Reference	Comment
Sierp & Fransemeier, 1934	Sierp respirometer used to study biodegradation of phenol
Tang & Liu, 1937	Effect of organic compounds on respiration fungi
Ruchhoft et al., 1940	Effect of acid mine water on BOD of wastewater
Heukelekian, 1947a, 1947b	Effect of industrial wastes on BOD
Levine & Novak, 1949	Fatty acid effects on oxygen uptake of fungi
Rigbi, 1951	Effect of hydrogen sulfide on BOD of wastewater
McKinney & Jeris, 1955	Metabolism of low-weight alcohols by activated sludge respirometry
Mills & Stack, 1955	Organic chemical oxidations by respirometry
McKinney et al., 1956	Warburg respirometry of aromatic compound degradation
Malaney et al., 1959	Warburg studies of toxic metal acclimation
Malaney, 1960	Degradation of aniline determined with Warburg respirometer
Winter, 1962	Warburg respirometer used to measure biodegradation of surfactants
Lamb et al., 1964	Respirometry of industrial wastes
Dozank, 1965, 1966	Effects of ionic detergents on BOD by Warburg respirometer
Schulze & Hoogerhyde, 1967	Differential manometric determination of biodegradability
Malaney & Gerhold, 1969	Oxidation of organic compounds by activated sludge respirometry
Mills et al., 1969	Effects of feeding regime on BOD of steer excreta
Nay et al., 1974	Biodegradation of trinitrotoluene by dynamic BOD
Verstraete et al., 1974	Toxicity evaluations made with Sapromat manometric respirometer
Bjorndal et al., 1975	Online oxygen uptake rate based toxicity monitor
Hisset et al., 1975	Compared several respirometer methods for determination of biodegradation of agricultural wastes
Ziglio et al., 1975	Warburg respirometry to determine degration of organic compounds
Crandall, 1976	Autorespirator for biodegradation studies
Gibbs, 1976	Oil biodegradation measured with a polarographic respirometer
Stratten & Witte, 1976	Biochemical oxygen demand used for toxicity evaluation
Bormann et al., 1978	Compared oxygen uptake rate of mixed culture with added sample or standard nutrient solution to determine toxicity

(continued)

TABLE 10.2. Some examples of the use of BOD and respirometry for determination of biodegradation rates and toxicity (*continued*).

Reference	Comment
Ajmal et al., 1980	Pharmaceutical waste effects on BOD respiration
Bauer et al., 1981	40-minute oxygen uptake rate based determination of pollutant effects on mixed bacterial culture
Larson & Perry, 1981	Electrolytic respirometer to measure biodegradation rates
Minier et al., 1981	Online respirometric determination of pharmaceutical biodegradability
Taylor et al., 1981	Warburg respirometer to measure phthalic acid ester degradation rates
Busch, 1982	Toxicity measured using BOD bottle bacterial test system
Sommerville, 1985	Physiological aspects of petrochemical biodegradation measured by respirometry
Volskay & Grady, 1988	Organisation for Economic Cooperation and Development activated sludge respiration inhibition test used to determine toxicity of Resource Conservation and Recovery Act priority pollutants
Ivanov et al., 1989	Effects of thiosulfate and sulfide on biodegradation of surfactants
Brown et al., 1990	Biodegradation kinetics of phenolics from electrolytic respirometer
Ros et al., 1990	Respirometric determinations of biodegradation rates
Volskay & Grady, 1990	Respiration inhibition kinetics analysis
Govind et al., 1991	Anaerobic respirometer used to determine degradation of organics
Aichinger et al., 1992	Biodegradation rates determined by respirometry limited by solubility of substrate
Tabak et al., 1992	Respirometry used to develop structure–biodegradation relationships
Vanrolleghem et al., 1994	Respirometric biosensor used to determine toxicity
Kong, Aerewijck et al., 1996; Kong, Vanrolleghem et al., 1996	Respirometric biosensor used to determine toxicity
Smets et al., 1996	Biodegradation load response individual chemicals measured by respirometric assay

Helfgott, 1975). The ultimate BOD of the material, determined from incubation times of 28 days or longer, is divided by the ultimate oxygen demand. The ultimate oxygen demand is computed from the result of a TOD test plus the oxygen equivalents required for theoretical oxidation of the total organic nitrogen (TON). The TON is obtained from the total Kjeldahl nitrogen minus the ammonia-nitrogen value of the substance. Examples presented as biorefractory substances arc histidine and lysine, which give index values of approximately 0.5.

Relating the biodegradability of the test substrate to that of a known easily degradable material is embodied in Organisation for Economic Cooperation and Development (OECD) test methods 301D, 301E, and 301F, in which glucose or sodium acetate are used as representative known biodegradable materials (OECD, 1998). This type of metric is also used to evaluate toxicity (Hermann, 1959). In this case, the toxicity of the test material, measured by the inhibition of BOD or respiration rate, is compared with the effects of a known toxicant. Mercuric chloride is commonly used as the known toxicant at a standard dose of 2.0 mg/L.

Organism-free indicators of biodegradability have been proposed. The persulfate oxidizability of pure compounds has been used to predict the bioavailability of selected polyaromatic hydrocarbons (PAH) to soil and sediment organisms (Cuypers et al., 2000). Reaction of the sample with persulfate in a ratio of 1 g soil organic matter to 12 g persulfate, at 70 °C for 3 hours, followed by comparison of the total PAH with recoverable PAH after digestion, gives a good correlation with 21-day biodegradation by soil microorganisms.

The approaches have been compared many times. Warburg respirometry has been compared with the river die-away test, and the latter test is regarded as very inconsistent (Garrison & Matson, 1964; Wylie et al., 1982). The dilution bottle test performed at 25 °C for 28 days is preferred over the river die-away test (Bunch & Chambers, 1967). Results from BOD dilution bottle (monitoring either dissolved oxygen [DO] or dissolved organic carbon [DOC]), shake flask (monitoring DOC), activated sludge bioreactor (monitoring DOC or the specific compound), carbon dioxide evolution, and the Glendhill test (which monitors carbon dioxide evolution and DOC decrease) on five tested compounds were compared and widely divergent biodegradability evaluations obtained (Means & Anderson, 1981).

The quenching of bioluminescence by toxic compounds and effluents, as embodied in the popular Microtox test kit, has been subjected to comparative testing many times (Sweet et al., 1997; Toussaint et al., 1995). In one study, comparing Microtox with three other procedures, each procedure was found to display unique sensitivity patterns such that use of all the procedures gave a more complete evaluation than was possible from use of any single procedure (Dutka & Kwan, 1981). Other researchers found that, when a suite of test compounds comprising three quaternary amines, two phenols, and chloroaniline was tested by Microtox, BOD, OECD Method

209, biodegradation inhibition, and nitrification procedures, the bioluminescence test was the most consistent (Reynolds et al., 1987). Nitrification, respiration, bacterial growth, and Microtox were compared on industrial effluents, and the bioluminescent procedure was found to be the most sensitive (Robinson, 1988).

One facet of the biodegradation/toxicity phenomenon that has intrigued researchers is the frequently made observation that the level of toxicity noted in pure compound studies is not always seen when the toxic compound is administered in a mixture with other substrates. Particularly when the culture is allowed time to acclimate to the mixed substrate, the organisms seem to be able to ignore the presence of the toxic material and concentrate on assimilation of the readily biodegradable components in the medium. This effect has been noted for mixed sugar substrates (Ghosh & Pohland, 1972, and Varma & Lindley, 1966) and for chlorinated organic toxicants (Kim & Maier, 1986). The effect is called co-metabolism.

To a large extent, the observation of co-metabolism arises from the presence of one of two general processes. The first process regards the biochemical mode of action of the toxicant. Toxicants that fundamentally interfere with the functioning of the bacterial cell membrane or metabolism, such as surfactants, serve as unilateral toxicants with their effects not ameliorated by the presence of any readily biodegradable substrates. Toxicants that do not function unilaterally, but instead inhibit the functioning of the individual cells that are attempting to metabolize it, can give rise to an adapted subpopulation of bacteria that is less affected by the toxicant. The presence of alternate food sources in the reactor maintains the bacterial population while adaption is occurring.

The second general mechanism of co-metabolism occurs in mixed populations in which only one particular species is able to convert the toxicant to a more biodegradable form, but other species are needed to subsequently complete the mineralization of the metabolite. The most studied example of this process involves the metabolism of the pesticide 1,1,1-trichloro-2,2-bis(4-chlorophenyl)-ethane (DDT). There are a few bacterial species that can metabolize DDT to an oxidized intermediate. For example, *p*-chlorophenyl acetic acid is formed from DDT by *Hydrogenomonas*. This metabolite is then completely mineralized by other species such as *Arthrobacter, Psuedomonas, Aspergillus,* or *Penicillium* to complete the process (Pfaender & Alexander, 1972; Subba-Rao & Alexander, 1985). A variant on this mechanism occurs when the first metabolite of the substrate is more toxic to the species performing the initial degradation than the original substrate. Other microorganism species in the culture that can successfully degrade the toxic metabolite serve to protect the susceptible microbes (Fogel et al., 1986; Kobayashi & Rittman, 1982).

Most studies of co-metabolism have been performed on anaerobic systems; however, there have been several performed under aerobic conditions. Some studies have found that selected substrates, such as polychlorinated biphenyls, are degraded under both anaerobic and aerobic conditions through co-metabolic mechanisms (Adriaens & Focht, 1990; Castro, 1993; Castro & Belser, 1990; Fava, et al., 1991; Sugiura, 1992). Many co-metabolic degradation studies have been performed on soil matrices, and respirometric techniques have played an important role in determining the mechanisms (Graves et al., 1991; Haines et al., 1996).

There was initially no standard for evaluation of the biodegradability of a tested material, which meant that results from different laboratories were sometimes contradictory. Sometimes, there was only a qualitative evaluation to whether a material is biodegradable or toxic, and even up to very recently, when standardized procedures and quantitative assessments are widely available, evaluations based on the BOD/COD or total organic carbon (TOC)/BOD are reported (Tuhkanen et al., 1997; Versino et al., 1977; Zhang & Wu, 1992). Even the studies that generated quantitative data generally lacked a common relative standard, such as comparison of results to those obtained from duplicate determination of known degradable materials like glucose as a quality control on the measurement system. The conditions of the tests were not standardized with respect to the composition or prior acclimation of the microorganism seed, or the duration of the test.

Efforts to produce standardized test procedures and evaluations began in the late 1960s. Phenol was used as a comparison standard to evaluate the biodegradation of phenolic compounds in the dilution bottle test, in which the amount of residual compound after 28-day incubation was the metric (Bunch & Chambers, 1967). Other workers began to use known biodegradable standards like glucose or sodium acetate.

The lists of the most frequently used currently available methods are presented in Tables 10.3, 10.4, and 10.5. The closed bottle test for ready biodegradability is the procedure that is closest in technique to the dilution bottle BOD test, with the major difference being a 20- to 28-day incubation time. The closed bottle test is considered to be the most rigorous of the tests for ready biodegradability. Materials that are not readily biodegradable rarely or seldom give false-positive results. However, one of the shortfalls of the test is that materials that are toxic to the test inoculum will generate results indicating that the material is not biodegradable, at least not readily so. The tests for inherent biodegradability include a lengthy acclimation period for the test culture that can extend up to 6 months before conducting the quantitative evaluation.

One of two different quantitation techniques is used in standardized biodegradability evaluations. The first method for evaluation starts with

TABLE 10.3. U.S. Environmental Protection Agency test procedures in 40 CFR for data supporting Toxic Substances Control Act submissions.

U.S. EPA number	Test
796.3100	Aerobic aquatic biodegradation
796.3140	Anaerobic biodegradability of organic chemicals
796.3180	Ready biodegradability: modified AFNOR[a] test
796.3200	Ready biodegradability: closed bottle test
796.3220	Ready biodegradability: modified MITI[b] test (I)
796.3240	Ready biodegradability: modified OECD[c] screening test
796.3260	Ready biodegradability: modified Sturm test
796.3300	Simulation test—aerobic wastewater treatment: coupled units test
796.3340	Inherent biodegradability: modified SCAS[d] test
796.3360	Inherent biodegradability: modified Zahn–Wellens test
796.3400	Inherent biodegradability in soil

[a]Association Francaise de Normalisar.
[b]Ministry of International Trade and Industry (Japan).
[c]Organisation for Economic Cooperation and Development.
[d]Semicontinuous activated sludge.

TABLE 10.4. U.S. Environmental Protection Agency Office of Prevention, Pesticides, and Toxic Substances harmonized test guidelines for Federal Insecticide, Fungicide, and Rodenticide Act and Toxic Substances Control Act applications.

U.S. EPA number	Test
835.3100	Aerobic aquatic biodegradation
835.3110	Ready biodegradability
835.3120	Sealed-vessel carbon dioxide production test
835.3160	Biodegradability in seawater
835.3170	Shake flask die-away test
835.3180	Sediment/water microcosm biodegradation test
835.3200	Zahn–Wellens/EMPA[a] test
835.3210	Modified SCAS[b] test
835.3220	Porous pot test
835.3300	Soil biodegradation
835.3400	Anaerobic biodegradability of organic chemicals
835.5045	Modified SCAS test for insoluble and volatile chemicals
835.5154	Anaerobic biodegradation in the subsurface

[a]Swiss Federal Laboratories for Materials Testing and Research.
[b]Semicontinuous activated sludge.

TABLE 10.5. Organisation for Economic Cooperation and Development degradation and toxicity methods.

OECD number	Test
209	Activated sludge, respiration inhibition test
301A	Ready biodegradability: DOC die-away test
301B	Ready biodegradability: CO_2 evolution test
301C	Ready biodegradability: modified MITI[a] test
301D	Ready biodegradability: closed bottle test
301E	Ready biodegradability: modified OECD[b] screening test
301F	Ready biodegradability: manometric respirometry test
302A	Inherent biodegradability: modified SCAS[c] test
302B	Inherent biodegradability: Zahn–Wellens/EMPA[d] test
302C	Inherent biodegradability: modified MITI test
303A	Simulation test—aerobic wastewater treatment: coupled units test
304A	Inherent biodegradability in soil

[a]Ministry of International Trade and Industry (Japan).
[b]Organisation for Economic Cooperation and Development.
[c]Semicontinuous activated sludge.
[d]Swiss Federal Laboratories for Materials Testing and Research.

calculation of the theoretical BOD of the test compound. The BOD value is then obtained from addition of the substance to a seeded nutrient solution. Finally, the BOD result is compared with the theoretical BOD; the calculation of theoretical BOD is covered in the theory chapter. If the BOD result is greater than 40% of the theoretical result within a 20-day incubation time, then the material is regarded as readily biodegradable (Fischer, 1971). In cases where the tested material is a mixture rather than a single pure compound, the COD or TOC of the material can be substituted for the theoretical BOD value (Fischer et al., 1975; Fischer & Gerike, 1975). A variation on this method is to determine the BOD of glucose, sodium acetate, or some other readily degradable material at the same time and under the same seed conditions as the test substance. If the BOD of the tested sample or the rate of oxygen usage is at least 60%, the amount of carbon dioxide evolution is at least 60%, or the decrease in dissolved organic carbon is at least 70% of the glucose value, then the material is regarded as biodegradable (U.S. EPA Method 835.3110).

Problems associated with the BOD approach commonly stem from lack of test substance solubility. Insoluble solid test substances can be homogeneously dispersed in the medium; however, microbial availability can be a significant concern. In some cases, the substance is dissolved in a water-miscible solvent such as methanol, then dispersed in the medium. The BOD of the solvent must be separately determined and factored out from the total BOD to arrive at a BOD resulting from the test substance.

Even when using a standardized method, the individual components in a mixed product may affect the overall biodegradability evaluation. Some components may be readily biodegradable, while others require lengthy acclimation of the culture (Vaishnau & Babeu, 1986). This may result in a mixture passing the minimum requirements for biodegradability, for instance, 70% decrease in the DOC (U.S. EPA Method 835.3110), but none of one or more components is degraded. There is also the occasional finding that the DOC does not achieve the 70% threshold and the substance or mixture may be declared nonbiodegradable, while an alternate, more selective probe such as thin-layer chromatography or high-performance liquid chromatography indicates that the original compound has been transformed into another, more slowly degradable substance (Lindgaard-Jorgensen & Nyholm, 1988). A mixture may be composed of substances that are individually biodegradable, whereas together they are refractory. A relatively minor amount of a toxic compound in the test substance can serve to passivate all the organisms in the culture. The presence of tetraalkyl ammonium and benzyltrialkyl ammonium cationic surfactant disinfectants is notorious for creating nonbiodegradable results from the BOD test (Smith, 2000; Urano & Saito, 1985). Long-chain alcohols and surfactants in general are able to skew biodegradation test results through bacterial cell membrane disruption (Therien et al., 1984).

The other technique to establish biodegradability is to add a known portion of the test substance to a microorganism culture, then after incubation, chemically determine the residual substance. Although this may seem to be a more definitive method, in practice, it frequently presents more difficulties than the BOD-like method. The first difficulty is the availability of a chemical test method and associated instrumentation for the determination of the substance in the incubation medium. The second difficulty is that steps must be taken to ensure than the test chemical is utilized by the microbes rather than simply adsorbed to the biomass. This frequently occurs, especially when the substance is not water soluble.

An extreme example of the difficulties than can be encountered is the determination of the biodegradability of bis(2-ethylhexyl)phthalate (DEHP, from the alternate name, diethylhexyl phthalate). This compound is used as an additive in plastics as well as being an industrial solvent and keeper. It is in widespread use, being found in soft plastic items such as tubing, children's toys, and blood transfusion bags. It is a nonpolar organic chemical that is insoluble in water to any significant degree.

A survey of the literature reveals two exactly opposite views as to the biodegradability of DEHP. One position holds that it is significantly refractory to biological treatment by both aerobic and anaerobic systems (Burns & Roe Industrial Services Corporation, 1982; Iannone et al., 1984; Johnson et al., 1984; Shelton et al., 1984; Stubin et al., 1996). All of these studies base their conclusions on chemical analysis of residual DEHP after incubation or biological treatment. The other position, which is supported by many laboratory studies, holds that it is easily degradable.

Among the studies reporting the successful biodegradation of DEHP, several are significant. Using ^{14}C-labeled DEHP, it is found to be rapidly mineralized in soil, probably by bacteria, with little uptake by plants (Schmitzer et al., 1988). Using pure and mixed cultures of bacteria, phthalate esters are found to be biodegradable (Keyser et al., 1976). Biodegradation of DEHP is rapid in both river water and activated sludge, as determined by both gas chromatography–flame-ionization detection determination of residuals and carbon dioxide evolution (Saeger & Tucker, 1976). A study using activated sludge as a source of microorganisms found 12 phthalates, including DEHP, to be very rapidly degraded (O'Grady et al., 1985). Using Warburg respirometry rather than a chemical technique for detection of the substrate, rapid degradation of phthalate esters, including DEHP, is found in marine environments (Taylor et al., 1981). Other studies found DEHP and other phthalate esters to be nontoxic up to the solubility limits of the compound (Adams et al., 1995; Defoe et al., 1990; Kwan, 1988; Rhodes et al., 1995). Still another study found that the microbial degradation of phthalate esters is so quick that it hampers evaluation of alternate mechanisms of mineralization (Russell et al., 1985). There are other reports of DEHP being readily biodegradable, and the microorganisms that have been found to accomplish the degradation include Gram-negative and Gram-positive bacteria and yeasts (Group, 1986; O'Conner et al., 1989; Sugatt et al., 1984; Taylor et al., 1981; Warns, 1987).

The number-one reason behind these divergent views is found in the laboratory work that generates the data on which the conclusions are based. There are several factors that lead to unreliable laboratory results. Methylene chloride and DEHP are the two most common laboratory contaminants detected in analysis (Berger et al., 1996; Smith, 1999a, 1999b, 2000). Quality control techniques have evolved that help to monitor ambient laboratory contaminant levels; however, they must be applied and attention paid to the evaluation of the results to avoid falsely reporting that the analyte is present in the sample when, in fact, it is not. In testimony of this fact, two literature reports went so far as to state that the biodegradation of DEHP and phthalic acid is technically difficult to assess in their respective studies because of laboratory contamination problems (Thuren, 1986; Wylie et al., 1982).

Another laboratory factor is the solubility of the compound. Organic compounds often have limited solubility in water, and DEHP is soluble to only the extent of 0.36 mg/L (Aichinger et al., 1992; Defoe et al., 1990; Howard et al., 1985). At concentrations greater than this, micelles (microdroplets) of the compound in water are formed. Bacteria are for the most part incapable of attacking micelles (Vaishnau et al., 1987). The biodegradation rate for butyl benzyl phthalate, di-*n*-octyl phthalate, DEHP, and several polynuclear aromatic hydrocarbons is totally controlled by the dissolution rate, a direct contributor to the bioavailability of the substrate (Aichinger et al., 1992). Thus, DEHP added to the medium at levels greater than the solubility can be recovered from the sample, and determination of no-effect levels of toxicity is impossible (Adams et al., 1995). One of the studies reporting a lack of biodegradability of DEHP used a spike level of 20 mg/L, a level substantially above the solubility limit, then went so far as to mention that the recovery of DEHP from the inoculated cultures was often higher than the spike level, indication of the complicating influence of sample contamination (Shelton et al., 1984).

A complicating factor when degradation studies are run at very low dilutions of the substrate involves another aspect of co-metabolism. At high concentrations of substrate, significant toxic or antagonistic effects are easy to detect, as is induction of specific metabolic enzyme systems. However, at low substrate concentrations, these effects may be masked through insufficient substrate stimulus to the bacteria.

2.0 METHODS FOR DETERMINING BIOCHEMICAL OXYGEN DEMAND IN THE PRESENCE OF TOXICITY

Sample toxicity can cause problems in the BOD test. In some cases of gross toxicity, the seed organisms are completely suppressed and no BOD is exhibited at all. In less extreme cases, if toxicity is even noticed, it is generally evidenced by an increasing BOD as the dilution increases (Heukelekian, 1947a). If the analyst continues to perform the test on more and more dilute samples, at some point, the toxicant may present such a small effect that the BOD of the sample can be determined. However, working against this effect is the simultaneous dilution of the BOD, and if the inherent BOD is low to begin with, this technique may fail to work.

Application of the dilution approach is at best a qualitative process, requiring keen observation, and often being guided by intuition. And, frequently, low levels of toxicity in the sample are overlooked. There are better approaches.

One approach is to perform a toxicity test by one of the approved U.S. EPA procedures, such as the survival and reproduction test with the microcrustacean *Ceriodaphnia dubia* or the survival and growth test with larval fathead minnows, *Pimephales promelas* (Methods 1000.0 and 1002.0, U.S. EPA, 1994; Smith, 1999a). These are somewhat involved procedures and not conducive to occasional use. Additionally, they do not use bacteria as the test subjects. A variation on this idea is to use a microbially based toxicity test kit such as Microtox. Evaluation of toxicity by these tests is subject to consideration of the sensitivity of the subject organisms. For instance, the organophosphate pesticides chloropyrifos and diazinon are deadly at low part per trillion concentrations to the water flea, *Ceriodaphnia dubia*, while activated sludge is unaffected. Ammonia is very toxic to eukaryotic organisms, while the prokaryotes that are the predominant life forms in activated sludge can tolerate up to 50 mg/L ammonia.

Another approach is to mix diluted portions of the sample with seed and nutrient agar, then prepare a pour plate. After 24 to 48 hours of incubation, comparison of the plate with one prepared from an uninhibited sample can reveal toxicity through suppressed growth or colony morphological aberrations. This is a qualitative approach and further requires an experienced microbiological observer, although skill can be developed by simply performing the test often. This is not a bad approach, and if the same seed, such as a commercial preparation, is always used, sample-caused aberrations can be reliably noticed. This procedure is very easy to perform and can serve as a general screening procedure for new samples received by the laboratory.

Thus, there is a difference in interpretation between toxicity as applied to environmental pollutants and human health problems and toxicity as applied to the results of a BOD test. There may be absolutely no connection between these considerations. However, toxicity as encountered in the BOD test does have a practical side to analysts, which is that it makes obtaining BOD results more difficult. There are techniques to circumvent BOD toxicity problems.

A somewhat involved procedure offers quantitative data that will detect almost all samples that exhibit toxicity while still generating BOD_5 results. The quantitative data are represented by plots of DO versus time for a variety of treatment regimes. The technique is based on the idea that bacterial toxicants interact in a nonreversible fashion with the bacteria and thus can be removed from the sample by using sacrificial bacteria. The procedure is adapted from a presentation in a pulp and paper industry publication (Bordon & Woodcock, 1949).

The BOD of the sample is estimated through use of a COD or TOC test, then a dilution series is prepared in at least triplicate, labeled A, B, and C, that will adequately cover the estimated BOD range. A glucose–glutamic acid (GGA) in duplicate and a set of dilution water bottles in at least triplicate for determination of seed demand are also prepared. A dilution water blank can be added to the schedule and read daily as an added control to determine cross-bottle contamination. On the first day each bottle except the blank is seeded, the DO read, and the bottles placed in the incubator. At the end of 24 hours of incubation, each sample bottle is read with a DO probe, then the bottles in series A and B, one of the GGA bottles, and two of the dilution water bottles are reseeded immediately after reading using a long-needle syringe to add the seed to the very bottom of the bottle. If settled sediment is present in the bottle, the bottle contents need to be homogenized by inversion while the DO probe is in place, and then the probe removed and the sample immediately reseeded and stoppered before the sediment settles again. The volume of the added seed must be sufficient to replace the volume displaced by the DO probe, and care must be taken to prevent reaeration. The lost volume in the series C bottles is replaced using marbles. All the bottles are returned to the incubator.

At the end of the next day—the 48-hour total incubation time—all bottles are read and reseeding is again performed on all the bottles reseeded at the 24-hour incubation time. On the third day—72-hour total incubation time—all the bottles are read; however, only the bottles in the A series samples and seed demand bottles are reseeded. Bottles in the B series have the lost volume replaced by marbles, same as the bottles in the C series. The procedure of the third day is continued through a 5- to 7-day total incubation period, although it can be extended to 20 or more days. Table 10.6 summarizes the scheduled treatments.

At the end of the test, the following data have been collected. The daily DO results from the C series bottles represent the respiration of the sample in the standard BOD procedure. The C series seed demand and GGA are the standard controls for the BOD test. The data from series B represent 2 days' worth of readdition of organisms. Frequently, the first seed added to the sample will bind all the toxicant, and the first reseeding adds organisms that are metabolically uninhibited. The reseeding event at 48 hours should add sufficient organisms to overcome any residual effects from toxicant bound by the 24-hour reseeding. This is indicated by the plot of DO versus time quickly falling from the inhibited point at day 1 to become approximately congruent with the C series GGA control line (Figure 10.1). The B series seed demand allows calculation of the BOD_5 between the readings at day 1 and day 6 for the B sample dilutions that meet DO depletion and residual requirements. Series A samples, seed demand, and GGA controls are just in case the amount of toxicant in the sample is really excessive. If the B series data are adequate, this will be indicated by the plot of the A data quickly falling below the line of the B data and the series C GGA control, then leveling off. This behavior is attributed to a vast overabundance of viable

TABLE 10.6. Summary of scheduled treatments (C series GGA control and dilution water blank not included).

Day	A Sample	A Seed demand	B Sample	B Seed demand	C Sample	C Seed demand	A GGA
0	Read and seed	Read and seed	Read and seed	Read and seed	Read and seed	Read and seed	Read and seed
1	Read and seed	Read and seed	Read and seed	Read and seed	Read	Read	Read and seed
2	Read and seed	Read and seed	Read and seed	Read and seed	Read	Read	Read and seed
3	Read and seed	Read and seed	Read	Read	Read	Read	Read and seed
4	Read and seed	Read and seed	Read	Read	Read	Read	Read and seed
5	Read and seed	Read and seed	Read	Read	Read	Read	Read and seed
6	Read and seed	Read and seed	Read	Read	Read	Read	Read and seed
7	Read	Read	Read	Read	Read	Read	Read

microorganisms in the A bottles and a quicker depletion of the available food. An approximate BOD_5 can be calculated from the data represented by series A; however, it is intended more as a guide to how much seed needs to be added initially at day 0 in a repeat of a series B experiment. It may

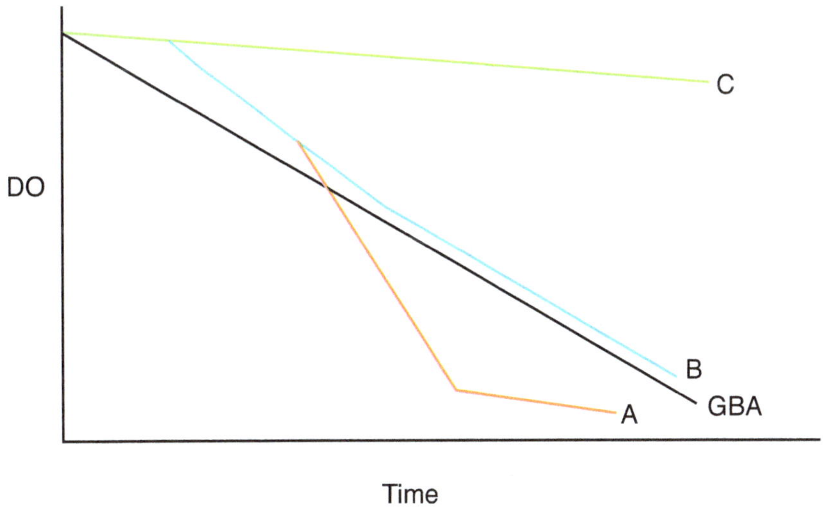

FIGURE 10.1. Idealized plot of DO behavior for different treatment schedules in the presence of toxicity.

indicate that a triple or quadruple dose of sacrificial seed is needed to absorb all the toxicant in the sample, saving time in the long run.

A simple technique that may provide a BOD value in the presence of an organic toxicant is addition of powdered activated carbon (PAC) to the BOD bottle. The PAC functions through adsorption of the organic toxicant and metabolic byproducts, and slowing their release into the liquid, resulting in lessened effect on the microorganisms. Powdered activated carbon has been demonstrated to moderate the effect of phenol in bioreactors (Schultz & Keinath, 1984). Addition of PAC is determined to be working by observation of an increase in BOD. Respirometry can be effectively used to determine the optimal dosage of PAC for use in the BOD test.

Toxicity resulting from the presence of metal ions can be removed from the sample by filtration of the sample through an ion-exchange resin before BOD analysis. If ion-exchange treatment is used, it is imperative to check the possible contribution of leachable materials from the resin to the BOD of the sample. This is achieved by passing a reagent water blank through the resin and assaying the water as a sample. Dilution water should not be used as the blank, as the nutrient mineral composition of the dilution water will be altered by the resin.

3.0 CONCLUSION

Both biodegradation and toxicity affect the performance of the BOD test. Thus, it behooves a technician to develop skill in interpreting individual test results in terms of these factors. A detailed procedure is presented for determination of BOD in the presence of sample toxicity.

Aside from the importance of knowledge of these concepts as they influence test results, from both remediation and prevention points of view, biodegradation and toxicity studies are quickly becoming an integral part of the worldwide regulatory framework for environmental protection (Kaiser, 1998). For approximately 50 years, the "Annual Literature Review," was published as part of *Water Environment Research* by the Water Environment Federation, has included summaries of recent biodegradation and toxicity studies of chemicals and substances that have been performed on both marine and freshwater species. The tables presented in these reviews are extensive, summarizing data from hundreds of references. Many of the references report results based on BOD-like procedures for pure and mixed cultures. Without any exaggeration, biodegradation and toxicity studies comprise the greatest use of the BOD procedure outside the generation of wastewater compliance monitoring data. Even for the selfish reason as job security, technicians should become proficient in these techniques.

4.0 REFERENCES

Adams, W. J., Biddinger, G. R., Robillard, K. A., & Gorsuch, J. W. (1995). A summary of the acute toxicity of 14 phthalate esters to representative aquatic organisms. *Environ. Toxicol. Chem.*, *14*, 1569.

Adriaens, P., & Focht, D. D. (1990). Continuous coculture degradation of selected biphenyl congeners by *Acinetobacter* spp. in aerobic reactor system. *Environ. Sci. Technol.*, *24*, 1042.

Aichinger, G., Grady, C. P., Jr., & Tabak, H. H. (1992). Application of respirometric biodegradability testing protocol to slightly soluble organic compounds. *Water Environ. Res.*, *64*, 890.

Aitken, E. V. (1936). A study of the biological utilization of an aromatic waste product. Unpublished thesis in sanitary engineering, Johns Hopkins University.

Ajmal, M., et al. (1980). Detrimental effects of pharmaceutical industrial waste on microorganisms. *Water, Air, Soil Pollut.*, *13*, 447.

Akano, R., & Tyo, Z. (1939). Biological decomposition of phenol by activated sludge, with consideration of the method of decomposition. *Mitt. Med. Akad. Kioto*, *27*, 389, 430.

Baird, R. B., Carmona, L., & Jenkins, R. L. (1977). Behavior of benzidine and other aromatic amines in aerobic wastewater treatment. *J. Water Pollut. Control Fed.*, *49*, 1609.

Bauer, N. J., et al. (1981). A simple, rapid bioassay for detecting effects of pollutants on bacteria. *Bull. Environ. Contam. Toxicol.*, *27*, 577.

Baziakina, N. A. (1940). Biochemical oxidation of phenol. *Vodosnabzh. Sanit. Tekh.* (USSR), *15*, 38.

Berger, W., McCarty, H., & Smith, R.-K. (1996). *Environmental laboratory data evaluation.* Genium Publishing.

Beveridge, T. J., & Doyle, R. J. (1989). *Metal ions and bacteria.* Wiley.

Bjorndal, H., et al. (1975). Continuous monitoring of acute toxicity in wastewater. *Nordforsch, Miljoevardssekr.*, *9*, 393; *Chemical Abstracts*, *86*, 8299 (1977).

Bogan, R. H., & Sawyer, C. N. (1954). Biochemical degradation of synthetic detergents. I. Preliminary studies. *Sew. Ind. Wastes*, *26*, 1069.

Bogan, R. H., & Sawyer, C. N. (1955). Biochemical degradation of synthetic detergents. II. Studies on the relation between chemical structure and biochemical oxidation. *Sew. Ind. Wastes*, *27*, 917.

Bordon, G. C., & Woodcock, I. (1949). Reseeding BOD bottles: A means of detecting toxic interferences in the BOD test. *Tappi J.*, *32*, 516.

Bormann, F., et al. (1978). Automatic control of differently charged wastewater before or in clarification devices. *Ger. Offen., 2*, 634.

Bridie, A. L., Wolff, C. J. M., & Winter, M. (1979). BOD and COD of some petrochemicals. *Water Res., 13*, 627.

Brink, R. H., & Meyers, J. A. (1966). Anionic surfactant biodegradability studies by Warburg respirometry. *J. Am. Oil Chem. Soc., 43*, 449.

Brouzes, P. H., et al. (1978). Rapid appraisal of the effects of various substances on microorganisms. *Prog. Water Technol., 10*, 715.

Brown, S. C., Grady, C. P. L., Jr., & Tabak, H. H. (1990). Biodegradation kinetics of substituted phenolics: Demonstration of a protocol based on electrolytic respirometry. *Water Res., 24*, 853.

Bulich, A. A. (1982). A practical and reliable method for monitoring the toxicity of aquatic samples. *Process Biochem., 17*, 45.

Bunch, R. L., & Chambers, C. W. (1967). A biodegradability test for organic compounds. *J. Water Pollut. Control Fed., 39*, 181.

Burns & Roe Industrial Services Corporation. (1982). *Fate of priority pollutants in publicly owned treatment works*. Final Rep., Vols. I and II, EPA-440/1-82-303. U.S. EPA.

Busch, A. W. (1982). Bioassay technique for relative toxicity in water pollution control. *J. Water Pollut. Control Fed., 54*, 1152.

Castro, C. E. (1993). Biodehalogenation: The kinetics and rates of the microbial cleavage of carbon–halogen bonds. *Environ. Toxicol. Chem., 12*, 1609.

Castro, C. E., & Belser, N. O. (1990). Biodehalogenation: Oxidative and reductive metabolism of 1,1,2-trichloroethane by *Pseudomonas putida:* Biogeneration of vinyl chloride. *Environ. Toxicol. Chem., 9*, 707.

Crandall, P. G. (1976). The development of instrumentation for evaluating biodegradation. *Dissertation Abstr. Intl., 37*, 1392B.

Cuypers, C., Grotenhuis, T., Joziasse, J., & Rulkens, W. (2000). Rapid persulfate oxidation predicts PAH bioavailability in soils and sediments. *Environ. Sci. Technol., 34*, 2057.

Defoe, D. L., Holcombe, G. W., Hammermeister, D. E., & Biesinger, K. E. (1990). Solubility and toxicity of eight phthalate esters to four aquatic organisms. *Environ. Toxicol. Chem., 9*, 623.

Dozank, W. (1965). Influence of nonionic detergent (alphenol 710) on BOD process occurring in sewage. *Rocz. Panstw., 16*, 363; *Chemical Abstracts, 64*, 1805 (1966).

Dozank, W. (1966). Effect of nonionic detergents (olbrotol 18 and olcol 18). on BOD in sewage. *Rocz. Panstw., 17*, 167; *Chemical Abstracts, 65*, 10319 (1966).

Dutka, B. J., & Kwan, K. K. (1981). Comparison of three microbial toxicity screening tests with the microtox test. *Bull. Environ. Contam. Toxicol.*, *27*, 753.

Fava, F., Zappoli, S., Marchetti, L., & Morselli, L. (1991). Biodegradation of chlorinated biphenyls (fenclor 42) in batch cultures with mixed and pure aerobic cultures. *Chemosphere*, *22*, 3.

Fischer, W. K. (1971). Control and valuation of biodegradability of nonionic surfactants. II. Control of the biodegradation of nonionic surfactant in the closed flask test. *Tenside*, *8*, 182; *Chemical Abstracts*, *75*, 99372 (1971).

Fischer, W. K., & Gerike, P. (1975). Biodegradability determinations via unspecific analysis (COD, DOC) in coupled units of the OECD confirmatory test. II. Results. *Water Res.*, *9*, 1139.

Fischer, W. K., et al. (1975). Biodegradability determinations via unspecific analysis (COD, DOC) in coupled units of the OECD confirmatory test—I. The test. *Water Res.*, *9*, 1131.

Fogel, M. M., Taddeo, A. R., & Fogel, S. (1986). Biodegradation of chlorinated ethenes by a methane-utilizing mixed culture. *Appl. Environ. Microbiol.*, *51*, 720.

Gaffney, P. E., & Heukelekian, H. (1958). Oxygen demand of the lower fatty acids. *Sew. Ind. Wastes*, *30*, 673.

Gallotti, L. (1971). Water pollution by the textile industry with particular reference to biodegradability. Part 2. *Tinctoria*, *68*(7), 251.

Garrison, L. J., & Matson, R. D. (1964). Comparison of Warburg respirometry and die-away studies of the degradability of select non-ionic surface-active agents. *J. Am. Oil Chem. Soc.*, *41*, 799.

Gellman, I., & Heukelekian, H. (1950). Biological oxidation of formaldehyde. *Sew. Ind. Wastes*, *22*, 1321.

Ghosh, S., & Pohland, F. G. (1972). Kinetics of assimilation of multiple substrates in dispersed growth systems. *Water Res.*, *6*(1), 99.

Gibbs, C. F. (1976). A continuously recording polarographic respirometer and its use in oil biodegradation studies. *Water Res.*, *10*, 443.

Gottsching, L., & Luttgen, W. (1978). Organic additives as a source of water contamination in papermaking. *Papier* (Germany), *32*(10A), V46.

Govind, R., et al. (1991). Fate and effects of semivolatile organic pollutants during anaerobic digestion of sludge. *Water Res.*, *25*, 54.

Graves, D., Lang, C., & Leavitt, M. (1991). Respirometric analysis of the biodegradation of organic contaminants in soil and water. *Appl. Biochem. Biotechnol.*, *28*, 29.

Gron, C., et al. (1992). Biodegradability of dissolved organic carbon in groundwater from an unconfined aquifer. *Sci. Total Environ.*, *118*, 241.

Group, E. F., Jr. (1986). Environmental fate and aquatic toxicity studies on phthalate esters. *Environ. Health Perspect.*, *65*, 337.

Haines, J. R., Holder, E. L., & Venosa, A. D. (1996). Respirometry as a tool to assess effects of environmental variables in bioremediation research and field testing. *Proc. 51st Ind. Waste Conf., Purdue University.* Ann Arbor Press, 463.

Hale, F. (1953). Sources of BOD in brewery wastes. *Sew. Ind. Wastes*, *25*, 1187.

Hart, F. L., & Helfgott, T. (1975). Bio-refractory index for organics in water. *Water Res.*, *9*, 1055.

Hatfield, R. (1957). Biological oxidation of some organic compounds. *Ind. Eng. Chem.*, *49*, 192.

Hermann, E. R. (1959). A toxicity index for industrial wastes. *Ind. Eng. Chem.*, *51*, 84.

Heukelekian, H. (1947a). Use of the dilution method for determining the effect of industrial wastes on deoxygenation. *Sew. Works J.*, *19*, 612.

Heukelekian, H. (1947b). Use of the direct method of oxygen utilization in waste treatment studies. *Sew. Works J.*, *19*, 875.

Heukelekian, H. (1958). Method for studying the toxicity of industrial wastes. *Water Sew. Works*, *95*, 285.

Heukelekian, H., & Rand, M. C. (1955). BOD of pure organic compounds. *Sew. Ind. Wastes*, *27*, 1040.

Hisset, R., Evans, M. R., & Baines, S. (1975). The use of respirometric methods for assessing the biodegradability of different components of agricultural wastes. *Prog. Water Technol.*, *7*(2), 13.

Howard, P. H., Bannerjee, S., & Robillard, K. A. (1985). Measurement of water solubilities, octanol/water partition coefficients, and vapor pressure of commercial phthalate esters. *Environ. Toxicol. Chem.*, *4*, 653.

Hughes, M. N., & Poole, R. K. (1989). *Metals and microorganisms.* Chapman-Hall.

Iannone, J., Papi, M., & Papamichael, F. (1984). Organic priority pollutants in New York City wastewater: Their sources and impacts. Paper presented at 57th Annu. Conf. Water Pollut. Control Fed.

Ingols, R. S. (1954). Toxicity of mercuric chloride, chromic sulfate, and sodium chromate in the dilution BOD test. *Sew. Ind. Wastes*, *26*, 536.

Ingols, R. S. (1955). Experience with solid platinum electrodes in the determination of dissolved oxygen. II. Georgia Institute of Technology studies. *Sew. Ind. Wastes*, *27*, 7.

Ivanov, Y. G., et al. (1989). Effect of thiosulfate and of sulfide on biochemical oxidation of surface active substances. *Fibre Chem.*, *20*, 306.

Jezeski, J. J., Halvorson, H. O., & Macy, H. (1950). The action of microorganisms on fats. I. Oxygen uptake by bacteria in the presence of lipid substrates. *J. Bacteriol.*, *59*, 645.

Johnson, B. T., et al. (1984). Environmental and chemical factors influencing the biodegradation of phthalic acid esters in freshwater sediments. *Environ. Pollut. Ser. B*, *8*, 101.

Kaiser, K. L. E. (1998). Review of biodegradability tests for the purpose of developing regulations. *Water Qual. Res. J. Can.*, *33*(2), 185.

Kalabina, M. M., et al. (1944). Effect of the toxic substances of waste water of non-ferrous metallurgy on microorganisms and biochemical processes associated with self-purification of reservoirs. *Gig. Sanit.* (Russia), *9*, 10; *Chem Abstr.*, *41*, 6007 (1947).

Keyser, P., Pujar, B. G., Eaton, R. W., & Ribbons, D. W. (1976). Biodegradation of the phthalates and their esters by bacteria. *Environ. Health Perspect.*, *18*, 159.

Kim, C. J., & Maier, W. J. (1986). Acclimation and biodegradation of chlorinated organic compounds in the presence of alternate substrates. *J. Water Pollut. Control Fed.*, *58*, 157.

Knopp, H. (1961). The A–Z test, a new process for toxicological testing of waste waters. basis and description of the method. *Deutsch Gewasserkundl. Mitt.*, *5*, 66.

Kobayashi, H., & Rittman, B. E. (1982). Microbial removal of hazardous organic compounds. *Environ. Sci. Technol.*, *16*, 170A.

Kong, Z., Aerewijck, M., & Werstraete, W. (1996). On-line stBOD measurement and toxicity control of wastewater with a respirographic biosensor [letter]. *Environ. Technol.*, *17*, 399.

Kong, Z., Vanrolleghem, P., Willems, P., & Verstraete, W. (1996). Simultaneous determination of inhibition kinetics of carbon oxidation and nitrification with a respirometer. *Water Res.*, *30*, 825.

Kreiger, H. S., & Moore, W. A. (1949). Determination of the toxicity of trivalent and hexavalent chromium on oxygen utilization of sewage. U.S. Public Health Serv., Jan. 1949 memo.

Kwan, K. K. (1988). Synthetic activated sludge techniques for toxicity assessment of chemicals and environmental samples. *Toxicity Assess.*, *3*, 93.

Lamb, J. C., III, Westgarth, W.C., Rodgers, J. L., & Vernimmen, A. P. (1964). A technique for evaluating the biological treatability of industrial wastes. *J. Water Pollut. Control Fed.*, *36*, 1263.

Larson, R. J., & Perry, R. L. (1981). Use of the electrolytic respirometer to measure biodegradation in natural waters. *Water Res.*, *15*, 697.

Levine, S., & Novak, M. (1949). The effect of fatty acids on the oxygen uptake of *Blastomyces dermatiditis*. *J. Bacteriol.*, *57*, 93.

Lindgaard-Jorgensen, P., & Nyholm, N. (1988). Characterization of the biodegradability of complex wastes. *Chemosphere*, *17*, 2073.

Luttgen, W., & Gottsching, L. (1979). Biodegradability of organic additives in a batch test. *Papier*, *33*, 133.

Malaney, G. W. (1960). Oxidation abilities of aniline-acclimated activated sludge. *J. Water Pollut. Control Fed.*, *32*, 1300.

Malaney, G. W., & Gerhold, R. M. (1969). Structural determinants in the oxidation of organic compounds by activated sludge. *J. Water Pollut. Control Fed.*, *41*, 18.

Malaney, G. W., & Sheets, W. D. (1957). Detergent builders and BOD. *Sew. Ind. Wastes*, *29*, 263.

Malaney, G. W., Sheets, W. D., & Quillin, P. (1959). toxic effects of metallic ions on sewage microorganisms. *Sew. Ind. Wastes*, *31*, 1309.

McKinney, R. E., & Jeris, J. S. (1955). Metabolism of low weight alcohols by activated sludge. *Sew. Ind. Wastes*, *27*, 728.

McKinney, R. E., Tomlinson, H. D., & Wilcox, R. L. (1956). Metabolism of aromatic compounds by activated sludge. *Sew. Ind. Wastes*, *28*, 547.

Means, J. L., & Anderson, S. J. (1981). Comparison of five different methods for measuring biodegradability in aqueous environments. *Water, Air, Soil Pollut.*, *16*, 301.

Mills, E. J., Jr., & Stack, V. T., Jr. (1955). Suggested procedure for evaluation of biological oxidation of organic chemicals. *Sew. Ind. Wastes*, *27*, 1061.

Mills, K. C., Parker, B. F., & Ross, I. J. (1969). Effects of feed consumption on BOD of steer excreta. *Trans. Am. Soc. Agric. Eng.*, *12*(1), 133.

Minier, A., et al. (1981). On line control of pharmaceutical influent loads to a biological wastewater treatment plant. *Water Sci. Technol.*, *13*, 523.

Mowat, A. (1976). Measurement of metal toxicity by BOD. *J. Water Pollut. Control Fed.*, *48*, 853.

Nay, M. W., Jr., et al. (1974). Biological treatability of trinitrotoluene manufacturing wastewater. *J. Water Pollut. Control Fed.*, *46*, 485.

Oberton, A. E., & Stack, V.T., Jr. (1957). BOD of organic chemicals. *Sew. Ind. Wastes*, *29*, 1267.

O'Conner, O. A., et al. (1989). Toxicity and degradation of phthalic acid esters under methanogenic conditions. *Environ. Toxicol. Chem.*, *8*, 569.

O'Grady, D. P., et al. (1985). Activated sludge biodegradation of 12 commercial phthalate esters. *Appl. Environ. Microbiol.*, *49*, 443.

Organisation for Economic Cooperation and Development. (1998). *Guidelines for testing of chemicals.* 10th Suppl.

Pfaender, F. K., & Alexander, M. (1972). Extensive microbial degradation of DDT in vitro and DDT metabolism by natural communities *J. Agric. Food Chem.*, *20*, 842.

Pitter, P. (1976). Determination of the biological degradability of organic substances. *Water Res.*, *10*, 231.

Price, K. S., et al. (1974). Brine shrimp bioassy and seawater BOD of P. *J. Water Pollut. Control Fed.*, *46*, 63.

Reynolds, L., Blok, J., de Morsier, A., Gerike, P., Wellens, H., & Bontinck, W. J. (1987). Evaluation of the toxicity of substances to be assessed for biodegradability. *Chemosphere*, *16*, 2259.

Rhodes, J. E., Adams, W. J., Biddinger, G. R, Robillard, K. A., & Gorsuch, J. W. (1995). Chronic toxicity of 14 phthalate esters to *daphnia magna* and rainbow trout (*Oncorhynchus mykiss*). *Environ. Toxicol. Chem.*, *14*, 1967.

Rigbi, M. (1951). The effect of ammonia nitrogen and hydrogen sulfide on the BOD of sewage. *Bull. Res. Council Isr.*, *1*, 1.

Robinson, R. J. (1988). Assessment of methods for evaluating the treatability of sewage and the effects of industrial discharges. *Toxicity Assess.*, *3*, 17.

Rogovskaya, T. I. (1954). Influence of aminocarboxylic acid lactams on microorganisms in water and biochemical processes in self-purification of water. *Microbiology* (Moscow), *23*, 464.

Ros, M., Zagorc-Koncan, J., Dular, M., & Pompe, M. (1990). Respirometric determination of wastewater biodegradation in biological treatment plants. *Z. Wasser Abwasser Forsch.* (Germany), *23*, 236.

Ruchhoft, C. C., Ettinger, M. B., & Walker, W. W. (1940). Biochemical oxidation in acid water containing sewage. *Ind. Eng. Chem.*, *32*, 1394.

Rudolfs, W., Barnes, G. E., Edwards, G. P., Heukelekian, H., Hurwitz, E., Renn, C. E., Steinberg, S., & Vaughan, W. F. (1950). Review of literature on toxic materials affecting sewage treatment processes, streams, and BOD determinations. *Sew. Ind. Wastes*, *22*, 1157.

Russell, D. J., et al. (1985). The effect of biodegradation on the determination of some chemodynamic properties of phthalate esters. *J. Environ. Sci. Health*, *A20*, 927.

Saeger, V. W., & Tucker, E. S. (1976). Biodegradation of phthalic acid esters in river water and activated sludge. *Appl. Environ. Microbiol.*, *31*, 29.

Sato, T., & Akiyama, T. (1972). Biochemical oxidation of volatile organic acids by activated sludge. *Water Res.*, *6*, 1059.

Sawyer, C. N., & Bogan, R. H. (1954). Biochemical degradation of synthetic detergents. *Sew. Ind. Wastes*, *26*, 1069.

Sawyer, C. N., & Bogan, R. H. (1955). Biochemical degradation of synthetic detergents. II. Studies on the relation between chemical structure and biochemical oxidation. *Sew. Ind. Wastes*, *27*, 917.

Sawyer, C. N., Bogan, R. H., & Simpson, J. R. (1956). Biochemical behavior of synthetic detergents. *Ind. Eng. Chem.*, *48*, 236.

Schmitzer, J. L., et al. (1988). Fate of bis(2-ethylhexyl)phthalate in laboratory and outdoor soil–plant systems. *J. Agric. Food Chem.*, *36*, 210.

Schowanek, D., et al. (1987). The Rodtox biosensor for rapid monitoring of BOD and toxicity of wastewaters. *Meded. Fac. Landbouwwet. Rijksuniv. Gent.*, *52*(4B), 1757; *Chemical Abstracts 109*, 11448 (1988).

Schultz, J. R., & Keinath, T. M. (1984). Organic removal mechanisms in biophysical treatment systems. *Water Sci. Technol.*, *17*, 1043.

Schulze, K. L., & Hoogerhyde, T. C. (1967). Biodegradability as measured by differential manometer. *Dev. Ind. Microbiol.*, *8*, 284.

Scope, H. G., & Kenna, M. (1950). Effects of organic compounds on BOD. *Sew. Ind. Wastes Eng.*, *21*, 467.

Shaikh, A., et al. (1985). Redox potential and oxygen diffusion rate as parameters for monitoring biodegradation of some organic wastes in soil. *Nucl. Chem. Waste Manage.*, *5*, 337.

Shelton, D. R., Boyd, S. A., & Tiedje, J. M. (1984). Anaerobic biodegradation of phthalic acid esters in sludge. *Environ. Sci. Technol.*, *18*, 93.

Sheets, W. D., & Malaney, G. W. (1956). Effect of synthetic detergents on the BOD test. *Sew. Ind. Wastes*, *28*, 10.

Sierp, F., & Fransemeier, F. (1934). Biological decomposition of phenol. *Wasser Abwasser* (Germany), *8*, 85.

Slabbert, J. L. (1980). A bioassay technique for rapid detection of toxicity in water. *Water Rep.*, *12*, 3.

Smets, B. F., Fehniger, S. M., & Grady, C. P. L., Jr. (1996). Development of a respirometric assay to measure the transient load response of activated sludge to individual organic chemicals. *Water Sci. Technol.*, *33*, 6, 49.

Smith, R.-K. (1999a). *Handbook of environmental analysis* (4th ed.). Genium Publishing.

Smith, R.-K. (1999b). *Lectures on wastewater analysis and interpretation.* Genium Publishing.

Smith, R.-K. (2000). *Interpretation of Organic Data.* Genium Publishing.

Sommerville, H. J. (1985). Physiological aspects of biotreatment of petrochemical wastes. *Conserv. Recycl.*, *8*(1), 73.

Stratten, G., & Witte, H. (1976). Proposal for the assessment of the BOD of sewage from various sources. *Z. Wasser Abwasser Forsch.* (Germany), *9*, 38.

Stubin, A. I., Brosnan, T. M., Porter, K. D., Jimenez, L., & Lochan, H. (1996). Organic priority pollutants in New York City municipal wastewaters: 1989–1993. *Water Environ. Res.*, *68*, 1037.

Subba-Rao, R. V., & Alexander, M. (1985). Bacterial and fungal cometabolism of 1,1,1-trichloro-2,2-bis(4-chlorophenyl)ethane (DDT) and its breakdown products. *Appl. Environ. Microbiol.*, *49*, 509.

Sugatt, R. H., O'Grady, D. P., Banerjee, S., Howard, P. H., & Glenhill, W. E. (1984). Shake flask biodegradability of 14 commerical phthalate esters. *Appl. Environ. Microbiol.*, *47*, 601.

Sugiura, K. (1992). Microbial degradation of polychlorinated biphenyls in aquatic environments. *Chemosphere*, *24*, 881.

Sweet, L. I., Travers, D. F., & Meier, P. G. (1997). Chronic toxicity evaluation of wastewater treatment plant effluents with bioluminescent bacteria: A comparison with invertebrates and fish. *Environ. Toxicol. Chem.*, *16*, 2187.

Symons, G. E., & Buswell, A. M. (1929). BOD of certain substances. *Ind. Eng. Chem. Anal. Educ.*, *1*, 161.

Symons, G. E., & Buswell, A. M. (1932). Preparation and BOD of pure sodium soaps. *Ind. Eng. Chem.*, *24*, 460.

Tabak, H. H., Gao, C., Desai, S., & Govind, R. (1992). Development of predictive structure-biodegradation relationship models with the use of respirometrically generated biokinetic data. *Water Sci. Technol.*, *26*, 763.

Tang, P. S., & Liu, C. Y. (1937). Studies on the kinetics of cell respiration. V. Effect of certain organic compounds on the rate of O_2 consumption and of anaerobic CO_2 production by *Saccharomyces wanching*. *J. Cell. Comp. Physiol.*, *10*, 487.

Taylor, B. F., Curry, R. W., & Corcoran, E. F. (1981). Potential for biodegradation of phthallic acid esters in marine regions. *Appl. Environ. Microbiol.*, *42*, 590.

Therien, N., Le Clave, P., & Jones, P. (1984). A respirometric study of the influence of aliphatic alcohols on activated sludges. *Water Res.*, *18*, 905.

Thuren, A. (1986). Determination of phthalates in aquatic environments. *Bull. Environ. Contam. Toxicol.*, *36*, 33.

Toussaint, M. W., Shedd, T. R., Van der Schalie, W. H., & Leather, G. R. (1995). A comparison of standard acute toxicity tests. *Environ. Toxicol. Chem.*, *14*, 907.

Tuhkanen, T., Naukkarinen, M., Blackburn, S., & Tanskanen, H. (1997). Ozonation of pulp mill effluent prior to activated sludge treatment. *Environ. Technol.*, *18*, 1045.

U.S. Environmental Protection Agency. (1994). *Short-term methods for estimating the chronic toxicity of effluents and receiving waters to freshwater organisms* (3rd ed.). EPA-600-4-91-002.

Urano, K., & Saito, M. (1985). Biodegradability of surfactants and inhibition of surfactants to biodegradation of other pollutants. *Chemosphere*, *14*, 1333.

Vaishnau, D. D., & Babeu, L. (1986). Occurrence and rates of chemical biodegradation in Superior Harbor water. *J. Great Lakes Res.*, *12*, 184.

Vaishnau, D. D., Boethling, R. S., & Babeu, L. (1987). Quantitative structure-biodegradability relationships for alcohols, and alicyclic compounds. *Chemosphere*, *16*, 695.

Vanrolleghem, P., Kong, Z., Rombouts, G., & Verstraete, W. (1994). An on-line respirographic biosensor for the characterization of load and toxicity of wastewaters. *J. Chem. Tech. Biotechnol.*, *59*, 321.

Varma, M. M., & Lindley, H. H. (1966). BOD of pure carbohydrates and its relation to the chemical structure. *Water Sew. Works*, *113*, 102.

Versino, C., Margara, A., Sarzanini, C., & Bianco, P. (1977). Characterization of urban primary sludge. Note III: Biochemical aspects. *Inst. Chim. Fis., Univ. Torino*, *4*, 75; *Chemical Abstracts*, *86*, 145221 (1977).

Verstraete, W., et al. (1974). Three-step measurement by the sapromat to evaluate the BOD_5, the Mineral imbalance, and the toxicity of water samples. *Water Res.*, *8*, 1077.

Volskay, V. T., & Grady, C. P. L. (1988). Toxicity of selected RCRA compounds to activated sludge microorganisms. *J. Water Pollut. Control Fed.*, *60*, 1850.

Volskay, V. T., & Grady, C. P. L. (1990). Respiration inhibition kinetics analysis. *Water Res.*, *24*, 863.

Warns, T. J. (1987). Diethylhexylphthalate as an environmental contaminant: A review. *Sci. Total Environ.*, 66, 1.

Winter, W. (1962). Biochemical decomposition in sewage purification. *Wasserwirtsch.-Wassertech.* (Germany), *12*, 265; *Chemical Abstracts*, *58*, 7711 (1963).

Wylie, G. D., Jones, J. R., & Johnson, B. T. (1982). Evaluation of the river die-away biodegradation test. *J. Water Pollut. Control Fed.*, *54*, 1231.

Zhang, Z., & Wu, Y. (1992). A study of correlation between biodegradability of wastewater and efficiency of biological oxidation pond treatment. *Water Treat.*, *7*, 243.

Ziglio, G., et al. (1975). Quanţitative loads and biodegradability of wastewaters from food processing industries. *Igiene Mod.*, *68*, 259; *Chemical Abstracts*, *84*, 169153 (1976).

11

Sediment Oxygen Demand

The idea of studying the oxygen demand of solids has been around almost as long as biochemical oxygen demand (BOD) studies on wastewater effluents have been reported. Two types of solids are most frequently brought to mind when waste treatment is discussed. The first type consists of the residual solids from activated sludge purification processes. These sludges are for the most part composed of living organisms (biomass) that have associated to form floc particles. The oxygen demand that they exert is an integral part of the general discussion of BOD that has been presented in the other chapters of this book.

The second type of solids is the bottom sediment beds of lakes, rivers, streams, estuaries, and oceans that are the recipients of effluents from treatment facilities. These beds are not simply gouged-out, sterile, inert tracks in the earth that define and confine the water body. Instead, the lining layers of rock, gravel, sand, clay, and organic decomposition detritus provide a home and shelter to a very complex ecosystem of plants, bacteria, protists, and higher animal life-forms that has developed in the interface between the bulk of the earth and the bulk of the water. All the normal functions of living ecosystems are exhibited by the sediments. Nutrient cycling and respiration are observed as the organisms feed, grow, reproduce, and compete among themselves. Sediments, instead of being anoxic areas completely isolated from oxygen, have long been recognized as important active participants in the oxygen flux in streams and other water bodies (Eliassen et al., 1950; Fair et al., 1941; Jansa & Akerlindh, 1941; Moore & Ruchhoft, 1942; Ruchhoft & Moore, 1940; Rudolfs, 1932; Theriault & McNamee, 1930; Velz, 1949,

1958), with contributions of 25% to 40% to the total oxygen flux being reported (Hopkinson, 1985). Beginning in the 1960s, intensive studies of sediment oxygen demand (SOD) have resulted in a body of knowledge, particularly useful in the management of watersheds and water quality planning, that is significantly different from that of BOD as it is normally considered in the context of waste treatment operations.

This chapter presents an introduction to the concepts and techniques of SOD. It is intended to show the differences and some of the similarities between these two areas of research rather than be a definitive treatise on SOD. There are several major compilations of information about SOD for those desiring to delve into this area in greater detail (Cerco et al., 1992; Hatcher, 1986; Lichtenberg et al., 1988; Price et al., 1992).

1.0 SEDIMENT OXYGEN DEMAND TECHNIQUES

The third edition of *Standard Methods* (APHA, 1917) contained a BOD procedure for river muds and wastewater sludge without including any references. A 1% stock solution of the river mud is prepared in oxygen-saturated tap or distilled water, then the stock solution used to prepare 1:10 and 1:100 dilutions in bottles for BOD testing. The dissolved oxygen (DO) in the dilution water, determined by the Winkler titration, is used as the initial DO for the test. A 5-day incubation at 20 °C is recommended, followed by determination of the final DO in the settled supernatent. The oxygen demand is reported in terms of the percentage of dried mud or sludge. Interestingly, the BOD procedure for wastewaters and effluents, performed by the sodium nitrate method, recommended a 10-day incubation.

To this day, techniques for SOD measurement based on a dilution bottle have changed only slightly from the method outlined in the third edition of *Standard Methods* (Cross & Summerfelt, 1987; Rosa & Burns, 1987; Wang, 1980, 1981). Most of the changes have been a refinement in the composition of the dilution water and a move toward DO probe measurements rather than manual Winkler-like titrations.

Manometric respirometers like the Warburg apparatus and the more sophisticated open and closed continuous electrometric respirometers have been used to study sediments and sludges (Baillie, 1986; Barcelona, 1983; Fillos & Molof, 1972; Gunnison et al., 1983; Hanes & Irvine, 1968; Ward, 1984; Whittemore, 1978, 1979). These techniques allow oxygen consumption as a function of time to be easily measured, in contrast to the information derived from the one-point-in-time technique commonly practiced with the dilution bottle. The respirometric techniques also allows for measurement of the SOD of an undisturbed core of sediment.

For the most part, use of laboratory-based dilution bottle and laboratory-based respirometry techniques for SOD studies continued up until the 1960s, when in situ measurement devices were developed. Early experiments with in situ devices demonstrated that laboratory studies of SOD gave significantly different (higher) results than those obtained from field studies (Pamatmat & Fenton, 1968; Stein & Denison, 1967; Whittemore, 1982). In contrast to studies on the time-defined total SOD value exhibited by the sediment, experiments that focused on the uptake rate tended to give higher results with in situ measurements (Edberg & Hofster, 1973; Hall & Berkas, 1988). A distinct advantage stemming from the controlled environment of the laboratory is that tests conducted there tend to give more consistent and reproducible results from a data quality control viewpoint than in situ measurements (Bowman & Delfino, 1980).

One of the major characteristics of the dilution bottle technique as applied to SOD is that the sample of mud or sludge is thoroughly mixed, frequently during sampling and always during analysis. Bottom sediments are known to be distinctly layered in relatively quiescent water bodies, rather than being homogeneous. Particularly in lakes and oceans, the layering has been used as a means of dating substrata in the sediments, leading, for instance, to timelines of environmental pollution deposition extending back in some cases to thousands of years. Lead deposition rates have been studied from the present back to pre-Roman times based on the lead concentrations found at increasing sediment depths in lakes, and these observations have been related to the level of industrialization as cultures waxed and waned. Situations in which there is a general lack of sediment layering are generally associated with rapidly flowing streams and rivers that are characterized by continual bottom scouring and a relative lack of biological activity in the sediments. In quiescent waters with stratified sediments, the oxygen utilization profile is observed to change with sediment depth, the shallow layers of the sediment bed being relatively aerobic, and the deeper layers tending to be strictly anaerobic (Jorgensen & Revsbech, 1985). Oxygen utilization rate studies of intact sediment layers using oxygen microelectrodes surprisingly indicate that the maximum oxygen uptake rate occurs at the oxic–anoxic interface (Revsbech et al., 1986) rather than at the surface of the sediment. Studies of oxygen utilization by sediments in water bodies have tended toward needing measurements in either undisturbed situations, in situ measurements, or toward careful sample acquisition of unmixed sediment cores that can be dissected allowing study of individual layers in the laboratory (Howes et al., 1984, 1986; Truax et al., 1995). Recent studies that utilize the traditional laboratory-based dilution bottle or respirometers as measurement techniques, and where layering has been ignored, tend to label the results as homogenized, for example, a 6-day homogenized biochemical oxygen demand ($HBOD_6$) (Ward, 1984). The reader should be aware that the acronym HBOD is also used to mean headspace BOD (Logan & Wagenseller, 1993) and the usages should not be confused. Some studies directed at the effect of flooded soils have intentionally needed to homogenize the solids (Gunnison et al., 1983). A diagram of a SOD chamber for laboratory use that maintains the integrity of the sediment core is illustrated in Figure 11.1.

In situ measurements of SOD arose in the 1960s and have become the dominant technique in sediment research. A flooded watertight chamber resting on the sediment is equipped with a DO probe, and the oxygen concentration of the overlying water is measured over time. Glass jars were commonly used in the early days of SOD work (Odum, 1957; Pamatmat & Fenton, 1968; Teal, 1957) and are even encountered in much later work (Drabkova, 1983; Lindeboom & Sandee, 1984; Lindeboom et al., 1984, 1985). Other designs for isolating a small portion of sediment and overlying water have been described, such as domes (Costa-Pierce et al., 1984; Veenstra & Nolen, 1991), cylinders (Brewer et al., 1977; Edberg & Hofster, 1973; Hall & Berkas, 1988; Uchrin & Ahlert, 1985), rectangles and cubes (James, 1974; Truax et al., 1995), cubes capped by a pyramidal top (Markert et al., 1983), triangular pyramids (Davis et al., 1987), and others. The effects of the different shapes on SOD determinations have been compared (Snodgrass & Fay, 1987). Construction materials have ranged from glass to the currently popular engineering plastics like plexiglass, fiberglass, acrylic, rigid polyvinyl chloride, and others. Stainless steel is normally avoided as a construction material, as it has been shown to catalytically lower DO levels

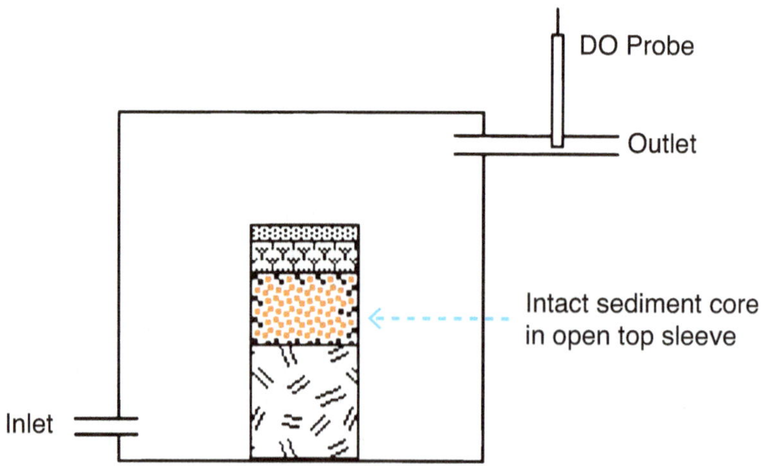

FIGURE 11.1. Schematic of an SOD chamber for laboratory use in water quality models.

(Cramer, 1989). Chambers have been constructed of opaque or transparent/translucent materials or painted over to study the effects of light on oxygen uptake (Nowicki & Nixon, 1985; Uchrin & Ahlert, 1985; Veenstra & Nolen, 1991). Rigid chambers have been universally used for SOD studies rather than flexible designs like the flow-through bags employed for in situ toxicity studies of natural waters. Some chambers are constructed to simply rest on the surface of the sediment, often with a wide flange around the base of the chamber to prevent tipping or excessive settling into soft sediments. Other chambers are designed to core deeply into the sediment to prevent diffusion from sediment outside the isolation area. Figure 11.2 illustrates a schematic for a generic in situ SOD chamber. One design that was developed to study benthic films on the surface of the sediment consists of a polyvinyl cylinder that is driven into the sediment as a permanent fixture, with an upper flange flush with the surface of the sediment that allows attachment of a dome respirometer over the embedded cylinder. After placement of the cylinder, the biofilm is allowed to grow on the surface, and then periodically the dome is attached and SOD readings are taken (Boyle & Scott, 1984).

Static measurements on the sediment and overlying water isolated within the chamber give information on the rate of oxygen consumption in the isolated sediment and water sample, but the rate itself is found to vary with the DO concentration (Davis et al., 1987; Fillos & Molof, 1972) and the degree of agitation of the overlying water. Further, the studies are limited

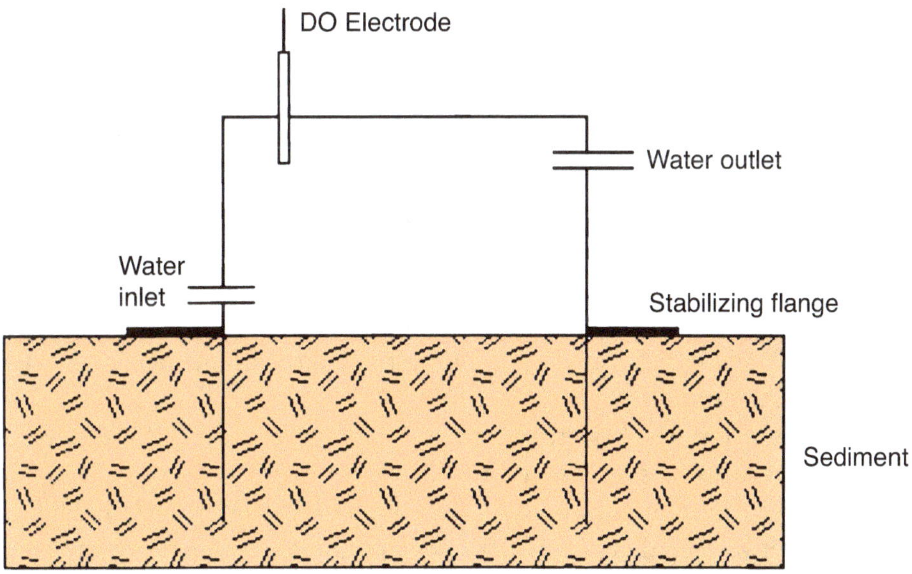

FIGURE 11.2. Generic schematic of an in situ SOD chamber.

to short-term measurements because of the limited solubility of oxygen in water and relatively rapid depletion. These considerations led to the use of respirometer-like devices in which the DO level is kept constant through replenishment, and the rate of replenishment becomes the measured parameter (Chiaro & Burke, 1980).

There are a number of different designs for oxygen replenishment in the SOD chamber, applicable variously to either or both in situ and laboratory-based devices. An elaborate sealed setup consisting of three or more interconnected tanks, only one of which contains the sediment, with pumps used for recirculation of the water and a DO probe placed somewhere in the recirculation plumbing, is limited to laboratory use but has been employed for numerous studies. The large mass of water present in the system compared to the amount of sediment allows oxygen depletion to be monitored without reareation. The SOD is calculated based on the initial and final DO values of the water multiplied by the amount of water in the system (Baity, 1938; Fair et al., 1941; Fillos & Molof, 1972). Filters can be integrated into the water flow to remove sediment-produced materials that may inhibit the SOD measurements resulting from increasing oxygen demand by the water (Davis et al., 1987; Fillos & Molof, 1972; Flint, 1984; Martin & Bella, 1971; Ogunrombi & Dobbins, 1970; Snodgrass, 1987).

If reaeration or reoxygenation is included in the experimental design, a DO probe is normally placed at both the inlet and the outlet to the sediment-containing tank to monitor DO consumption as the difference between the two readings. The oxygen consumption by the sediment is then calculated based on the difference in readings integrated over time and the water flow rate (Shelton & Hunter, 1974). Monitoring the inlet and outlet DO has been used for in situ devices and is especially useful when the naturally surrounding water at the site is pumped through the chamber. The most reliable readings are obtained when the rate of flow through the chamber is adjusted to create no more than a 30% difference between the ambient and outlet DO concentrations (Wu, 1990).

A novel concept for maintenance or quantitative adjustment of the DO concentration within the chamber for SOD measurements uses controlled addition of hydrogen peroxide and catalase (Hickey, 1985, 1988a, 1988b, 1988c). This technique was used to study the contributions of wastewater fungus biofilms to SOD.

The water that overlies the sediment is not sterile in the natural setting. The microorganisms present in natural waters exert an oxygen demand separate from that of the sediment (Campbell & Rigler, 1986). This is termed water oxygen demand (WOD) or hypolimnic oxygen demand (Rosa & Burns, 1987). Various approaches to limiting the measured oxygen demand to that resulting only from the sediment have included preparation and use of a sterile synthetic water, sterilization of a volume of natural water, and separate measurement of the WOD. The sterile synthetic water that is used is normally the standard BOD dilution water (Wang, 1980). Sterilization of natural water is most commonly accomplished by filtration through a 0.45-μm or smaller pore-sized membrane (Brewer et al., 1977; Sommaruga, 1991).

The BOD dilution bottle techniques are widely used to determine the oxygen demand resulting from the overlying water column (Cross & Summerfelt, 1987; Uchrin & Ahlert, 1985). Another technique is to fill duplicates of the chambers used for SOD with samples of the overlying water without sediment and use the oxygen demand of the water column chambers as a control (Truax et al., 1995).

Overlying water that is quiescent affects both the measured SOD and the oxygen uptake rate (Martin & Bella, 1971). Movement of the water is necessary to achieve reliable measurements, but the degree of agitation must be controlled so that there is no resuspension of the sediment. Doubling the velocity of the overlying water roughly doubles the measured SOD (Whittemore, 1978). A leveling of the SOD enhancement because of flow rate occurs at 0.12 m/s (0.4 ft/s) velocity (Whittemore, 1979). The use of stirrers within the chamber (Uchrin & Ahlert, 1985) or recirculation of the overlying water achieves the desired water movement; however, in many cases, only a qualitative evaluation of the degree of mixing within the chamber has been determined; that is, complete without any dead spots, rather than any quantitative surface or volume flow measurement.

The effect of temperature on SOD measurements has been evaluated several times (Edberg & Hofster, 1973; Hargrave, 1969; Reynolds et al., 1973; Whittemore, 1981). The overall conclusion is that within reasonable temperature ranges, the SOD increases with increasing temperature; however, the imprecision of the SOD measurement itself, which runs approximately 40% relative standard deviation (Whittemore, 1979, 1981), renders any temperature effect insignificant.

The measurement of oxygen demand and oxygen uptake rates are not the only probes of the biological activity of sediments. Particularly when anaerobic portions of the sediment are the focus of the study, monitoring of methane, ammonia, hydrogen sulfide, and volatile fatty acid fluxes has been found to be more suitable (DiToro et al., 1990; Molongoski & Klug, 1980). The activity of the electron transport system has been examined as a surrogate to oxygen utilization. Dehydrogenase and protease activities have been examined several times (Lenhard, 1967; Lenhard & DuPlooy, 1965; Pamatmat & Bhagwat, 1973). Good correlations of the triphenyl tetrazolium chloride assays to SOD ($r > 0.8$) are reported for both freshwater and marine environments (Trevors, 1984a, 1984b).

2.0 SEDIMENT OXYGEN DEMAND CONCEPTS

Sediment oxygen demand is a different concept than BOD. On the one hand, BOD tends to focus on the food content of the sample as the limiting factor in microorganism growth and respiration, although it should most correctly be viewed as a measure of the biological response to the presence of food rather than a quantitative measure of the amount of food (Clark, 1974; Eliassen, 1944). On the other hand, SOD studies tend to focus on the oxygen flux into and out of the sediment in relation to the oxygen balance in the whole water body. This difference in interpretation is most vividly brought out in the work of one marine researcher, in which he concludes that because of the rapid increase in bacterial numbers in the flasks containing seawater samples, the dilution bottle BOD test does not give a true evaluation of the biochemical consumption of oxygen in the natural environment (Kriss, 1963). The term "benthic oxygen demand" has been suggested to describe oxygen flux that is solely due to microsize and larger organisms; however, this term includes considerations of life that exists in both the sediment and in the overlying water (Belanger, 1981). Further, the abbreviation for benthic oxygen demand (BOD) is readily confused with that of BOD. An alternate term, benthic oxygen flux, has been presented as more descriptive of the interaction between the sediment and the overlying water column.

Water body sediments are composed of inorganic solids, organic detritus, and a variety of life-forms. These components are not independent, but interact to influence the oxygen flux. A major effort has been made in SOD studies to elucidate the interactions (Brewer et al., 1977; Dechev et al., 1977; DiToro et al., 1990; Wang, 1980). The type of interaction varies with the depth of the sediment and the depth of the overlying water. The uppermost layer of the sediment that is in closest contact with the overlying water has the relatively highest exposure to oxygen, while underlying layers are progressively more anaerobic in character. One study arrives at a figure of 85% of the SOD in a saltwater marsh is accomplished in the uppermost 2 cm of the intact sediment (Howes, et al, 1984). The transition from relative aerobic to strictly anaerobic with sediment depth means that the dominant microbiological populations change with depth. Sediments from relatively shallow waters also have upper layers that contain photosynthetic organisms that are lacking in the sediment depths.

Some studies have concluded that the chemical reactions of the inorganic component are the major factors in oxygen consumption by sediments

(Barcelona, 1983; Dechev et al., 1977). Significant inorganic contributors include the oxidation–reduction cycles of carbon, sulfur, nitrogen, iron, and manganese (Wang, 1980). In the performance of the experiments to determine the relative contributions of chemical fractions, the living organism demands are inhibited by addition of toxicants like phenol, formalin solution (formaldehyde), antibiotics (streptomycin–penicillin), and sodium chloride, with phenol being preferred (Colley, 1989; Wang, 1980). Sulfide demand is inhibited by addition of zinc salts to precipitate the sulfide. Iron(II) oxygen demand is estimated by addition of known amounts of iron(II) salts to the biologically inhibited sediment in a BOD dilution bottle with extrapolation back to the original demand resulting from iron(II) (Wang, 1980).

Other studies have stated that the biological factors are the most important (Brewer et al., 1977). Competing influences on the biological use of oxygen are between aerobic, anaerobic (Anderson & Hargrave, 1984), and photosynthetic organisms (Baillie, 1986). The carbon cycling between dissolved organic carbon, biomass, methane, and carbon dioxide, as well as the nitrogen cycling between ammonia and nitrate, are biologically mediated and use up to 30% of the SOD (Adams et al., 1982; Barcelona, 1983), although at least one author (Cooper, 1984) ascribes 85% of the total SOD as being due to the effect of benthic nitrogeneous oxygen consumption. Others have presented evidence that the sulfur cycle is biologically linked to all carbon mineralization (Blackburn et al., 1988). Still others hold that 94% of carbon mineralization is to the result of oxygen-supported respiration (Carlton et al., 1989).

Within the living portion of sediment, researchers fractionate the SOD into subcategories. One fractionation looks at the contributions by heterotrophic and autotrophic organisms. The heterotrophs are found to consume oxygen and dissolved organic carbon (DOC), whereas the autotrophs contribute to increases in the DOC on a diel cycle (Kuserk et al., 1984). The biological portion of SOD is also fractionated according to the rate of substrate removal—rapid versus slowly degradable—and physical state of the substrate—soluble versus fine particulate and colloidal (Harremoes, 1982; Hvitved-Jacobsen, 1982).

Overall, quantitative assignment of portions of SOD to various factors is muddied by the wide variety of techniques that have been employed and by the different sediment settings that have been studied. In a qualitative sense, the following factors are found to affect the oxygen demand of sediments (Frisk, 1982; Frisk & Kyla-Harakka, 1982). First, only the topmost layers of the sediment are in contact with and utilize oxygen to any large degree (Sweerts et al., 1986). Second, the sublayers of the sediment are anaerobic and are biologically inhibited by the intrusion of oxygen through diffusion and burrowing by life-forms (biotubation), which changes the millivolt potential and alters the chemistry of the microenvironment (Mortimer, 1971; Sweerts et al., 1986; Walker & Snodgrass, 1986). The anaerobic layers also contribute to the available substrate of the aerobic portions of the sediment through diffusional release of carbon dioxide, methane, and ammonia to the upper layers (Adams et al., 1982). One consequence of the anaerobic state of the lower sediment layers is that the reduced inorganics tend to have greater solubility than the related oxidized inorganics and are capable of exerting a disproportionately large COD when mixed with oxygen-bearing water because of the solubilization and immediate reaction of reduced forms of sulfur and iron. Third, the oxygen demand of the relatively aerobic layers of the sediment is to the result of metabolism by microorganisms and higher life forms. Fourth, the level of oxygen demand by the aerobic sediment-bound life is affected by the concentration of food substrates and of oxygen, both of which are mostly controlled by fresh inputs from the flowing overlying water, although there is some cycling of substrates within the sediment and some production of oxygen by photosynthetic organisms. Fifth, the overlying water is also biologically active and there is interchange of substrate between the water and the sediment, in the forms of fresh substrate entering the system and decomposition of existing biomass. Considering the overlying water and the sediment as an intact ecological system, only approximately 25% to 40% of the total oxygen respiration is attributable to the sediment (Hopkinson, 1985; Snodgrass & Ng, 1985a, 1985b).

3.0 CONCLUSION

The SOD is a significant component in the overall oxygen cycling within a water body. The SOD can be simplistically broken down into contributions from chemical and biological factors; however, current models indicate that there are complex interactions between the anaerobic and aerobic layers within the sediment and the overlying water column. Some of these interactions include cycling of nutrients, carbon sources, and electron acceptors, including oxygen.

The field of SOD research is hampered to some extent by the lack of standardized methods for quantitative evaluation. The SOD literature is filled with conclusions that on the surface are apparent contradictions, while in reality are simply artifacts of inconsistent technique. The increasing use of SOD values within water quality models to produce regulatory discharge limits for industries and municipalities places a monetary incentive on the regulated community to obtain input values for these models that are representative of the actual water body into which they discharge, rather than some approximate number that can best be described as being extracted from the air. The books edited by Cerco et al. (1992), Hatcher (1986), and Price et al. (1992) are particularly useful sources of information on the use of SOD values.

4.0 REFERENCES

Adams, D. D., Matisoff, G., & Snodgrass, W. J. (1982). Flux of reduced chemical constituents (Fe^{2+}, Mn^{2+}, NH_4^+, and CH_4) and SOD in Lake Erie. *Hydrobiologia*, *92*, 405.

American Public Health Association. (1917). *Standard methods for the examination of water and sewage* (3rd ed.).

Anderson, F. O., & Hargrave, B. T. (1984). Effects of *Spartina* detritus on aerobic/anaerobic benthic metabolism in an intertidal sediment. *Mar. Ecol. Prog. Ser.*, *16*, 161.

Baillie, P. W. (1986). Oxygenation of intertidal estuarine sediments by benthic microalgal photosynthesis. *Est. Coast Shelf Sci.*, *22*, 143.

Baity, H. G. (1938). Studies of sewage sludge. *Sew. Works. J.*, *10*, 539.

Barcelona, M. J. (1983). SOD fractionation, kinetics and reduced chemical substances. *Water Res.*, *17*, 1081.

Belanger, T. V. (1981). Benthic oxygen demand in Lake Apopka, Florida. *Water Res.*, *15*, 267.

Blackburn, T. H., et al. (1988). C- and N-mineralization in the sediments of earthen marine fishponds. *Mar. Ecol. Prog. Ser.*, *41*, 177.

Bowman, G. T., & Delfino, J. J. (1980). SOD techniques: A review and comparison of laboratory and in situ systems. *Water Res.*, *14*, 491.

Boyle, J. D., & Scott, J. A. (1984). The role of benthic films in the oxygen balance in an east devon river. *Water Res.*, *18*, 1089.

Brewer, W. S., Abernathy, A. R., & Paynter, M. J. B. (1977). Oxygen consumption by freshwater sediments. *Water Res.*, *11*, 471.

Campbell, P. J., & Rigler, F. H. (1986). Effect of ambient oxygen concentration on measurements of sediment oxygen consumption. *Can. J. Fish Aquat. Sci.*, *43*, 1340.

Carlton, R. G., et al. (1989). Relative values of oxygen, nitrate, and sulfate to terminal microbial processes in the sediments of lake superior. *J. Great Lakes Res.*, *15*, 133.

Cerco, C., Gunnison, D., & Price, S. B. (1992). *Proc. U.S. Army Corps Eng. Workshop Sediment Oxygen Demand.*

Chiaro, P. S., & Burke, D. A. (1980). SOD and nutrient release. *J. Environ. Eng.*, *106*, 177.

Clark, D. W. (1974). BOD: A re-evaluation. *Water Sew. Works*, *121*, 68.

Colley, R. (1989). Oxygen demand of the sbehind the Dat Altenworth, River Danube, Austria. *Acta Hydrochim. Hydrobiol.*, *17*, 9.

Cooper, A. B. (1984). Activities of benthic nitrifiers in streams and their role in oxygen consumption. *Microbiol. Ecol.*, *10*, 317.

Costa-Pierce, B. A., et al. (1984). Correlation of in situ respiration rates and microbial biomass in prawn (*Macrobrachium rosenberg*). *Aquaculture*, *37*, 157.

Cramer, A. (1989). A common artefact in estimates of benthic community respiration caused by the use of stainless steel. *Neth. J. Sea Res.*, *23*, 1.

Cross, T. K., & Summerfelt, R. C. (1987). Oxygen demand of lakes: sediment and water column BOD. *Lake Reservoir Manage. N. Am. Lake Manage. Soc.*, *3*, 109.

Davis, W. S., et al. (1987). Overview of EPA/CLEAR Lake Erie sediment oxygen demand investigations during 1979. *J. Great Lakes Res.*, *13*, 731.

Dechev, G., et al. (1977). Oxygen consumption and oxygen debt in bottom sediment. *Arch. Hydrobiol. Suppl.*, *52*(1), 63.

DiToro, D. M., et al. (1990). Sediment oxygen demand model: Methane and ammonia oxidation. *J. Environ. Eng.*, *116*, 945.

Drabkova, V. G. (1983). Bacterial decomposition of organic matter in lacustrine sediments. *Hydrobiologia*, *103*, 99.

Edberg, N., & Hofster, B. V. (1973). Oxygen uptake of bottom sediments studied in situ and in the laboratory. *Water Res.*, *7*, 1285.

Eliassen, R. (1944). Oxidants and reductants in sewage plant effluents. *Water Sew. Works*, *91*, 6, 219.

Eliassen, R., Grune, W. N., Jaffe, T., & Diachishin, A. N. (1950). BOD of radioactive sludge. *Prog. Rep. U.S. Atomic Energy Comm.*, NYO 1510 (NYU-6), 4; *Public Health Eng. Abstr.*, *30*(9), 18 (1950).

Fair, G. M., Moore, E. W., & Thomas, H. A., Jr. (1941). The natural purification of river muds and pollutional sediments. *Sew. Works J.*, *13*(270), 756, 1209.

Fillos, J., & Molof, A. H. (1972). Effect of benthal deposits on oxygen and nutrient economy of flowing waters. *J. Water Pollut. Control Fed.*, *44*, 644.

Flint, R. W. (1984). Benthic nutrient regeneration in South Texas Coastal Waters. *Est. Coast. Shelf Sci.*, *18*, 221.

Frisk, T. (1982). An oxygen model for Lake Haukivesi. *Hydrobiologia, 86*, 133.

Frisk, T., & Kyla-Harakka, T. (1982). Mathematical methods for predicting DO concentrations and phytoplankton biomass in lakes loaded by waste waters of the pulp and paper industry. *Aqua Fennica, 10*, 21.

Gunnison, D., Chen, R. L., & Brannon, J. M. (1983). Relationship of materials in flooded soils and sediments to the water quality of reservoirs. I: Oxygen consumption rates. *Water Res., 17*, 1609.

Hall, D. C., & Berkas, W. (1988). Comparison of instream and laboratory methods of measuring sediment oxygen demand. *Water Resour. Bull., 24*, 571.

Hanes, N. B., & Irvine, R. L. (1968). New techniques for measuring oxygen uptake rates. *J. Water Pollut. Control Fed., 40*, 223.

Hargrave, B. T. (1969). Similarity of oxygen uptake by benthic communities. *Limnol. Oceanogr., 14*, 801.

Harremoes, P. (1982). Immediate and delayed oxygen depletion in rivers. *Water Res., 16*, 1093.

Hatcher, K. J. (Ed). (1986). *Sediment oxygen demand: Processes, modeling, and measurement*. Inst. Natl. Res., University of Georgia.

Hickey, C. W. (1985). Quantitative addition of DO to in situ benthic chamber systems by use of catalase and hydrogen peroxide. *Appl. Environ. Microbiol., 49*, 462.

Hickey, C. W. (1988a). Benthic chambers for use in rivers: Testing against oxygen balance. *J. Environ. Eng., 114*, 828.

Hickey, C.W. (1988b). Oxygen uptake kinetics and microbial biomass of river sewage fungus biofilms. *Water Res., 22*, 1365.

Hickey, C. W. (1988c). River oxygen uptake and respiratory decay of sewage fungus biofilms. *Water Res., 22*, 1375.

Hopkinson, C. S. (1985). Shallow-water benthic and pelagic metabolism: Evidence of heterotrophy in the nearshore Georgia Bight. *Mar. Biol., 87*, 19.

Howes, B. L., et al. (1984). Carbon flow through oxygen and sulfate reduction pathways in salt marsh sediments. *Limnol. Oceanogr., 29*, 1037.

Howes, B. L., et al. (1986). Factors controlling the growth form of *Spartina alternflora*: Feedbacks between above-ground production, sediment oxidation, nitrogen, and salinity. *J. Ecol., 74*, 881.

Hvitved-Jacobsen, T. (1982). The impact of combined sewer overflows on the DO concentration of a river. *Water Res., 16*, 1009.

James, A. (1974). The measurement of benthal respiration. *Water Res.*, *8*, 955.

Jansa, V., & Akerlindh, G. (1941). A method for computing the oxygen curve in a polluted stream with special reference to the oxygen consumption of sludge deposits. *Sew. Works J.*, *13*, 551.

Jorgensen, B. B., & Revsbech, N. P. (1985). Diffusive boundary layers and the oxygen uptake of sediments and detritus. *Limnol. Oceanogr.*, *30*, 111.

Kriss, A. E. (1963). On the distribution of unstable and stable forms of organic matter in the water mass of the world ocean. *Microbiology* (USSR), *32*, 103; *Water Pollut. Abstr.*, *37*, 1750 (1964).

Kuserk, F. T., et al. (1984). *In situ* measures of dissolved organic carbon flux in a rural stream. *Can. J. Fish. Aquat. Sci.*, *41*, 964.

Lenhard, G. (1967). Determination of protease activity in bottom deposits of sewage stabilization ponds. *Hydrobiologia*, *29*, 67.

Lenhard, G., & DuPlooy, A. (1965). Studies on bottom deposits of the Ural River System. *Hydrobiologia*, *26*, 271.

Lichtenberg, J. J., Winter, J. A., Weber, C. I., & Fradkin L. (Eds.). (1988). *Chemical and biological characterization of municipal sludges, sediments, dredge spoils, and drilling muds.* Am. Soc. Testing Mater., STP 976.

Lindeboom, H. J., & Sandee, A. J. J. (1984). The effect of coastal engineering projects on microgradients and mineralization reactions in sediments. *Water Sci. Techol.*, *16*, 87.

Lindeboom, H. J., et al. (1984). Mineralization of organic carbon on and in the sediment of Lake Grevelingen. *Neth. J. Sea Res.*, *18*, 492.

Lindeboom, H. J., et al. (1985). A new bell jar/microelectrode method to measure changing oxygen fluxes in illuminated sediments with a microalgal cover. *Limnol. Oceanogr.*, *30*, 693.

Logan, B. E., & Wagenseller, G. A. (1993). The HBOD test: A new method for determining BOD. *Water Environ. Res.*, *65*, 862.

Markert, B. E., Tesmer, M. G., & Parker, P. E. (1983). An *in situ* SOD sampler. *Water Res.*, *17*, 603.

Martin, D. C., & Bella, D. A. (1971). Effect of mixing on oxygen uptake rate of estuarine bottom deposits. *J. Water Pollut. Control Fed.*, *43*, 1865.

Molongoski, J. J., & Klug, M. J. (1980). Anaerobic metabolism of particulate organic matter in the sediments of a hypereutrophic lake. *Freshwater Biol.*, *10*, 507.

Moore, W. A., & Ruchhoft, C. C. (1942). Effect of river mud deposits on determination of BOD of supernatent liquor. *Ind. Eng. Chem Anal. Ed.*, *14*, 933; *Sew. Works J.*, *15*, 356 (1943).

Mortimer, C. H. (1971). Chemical exchanges between sediments and water in the Great Lakes: Speculations on probable regulatory mechanisms. *Limnol. Oceanogr.*, *16*, 387.

Nowicki, B. L., & Nixon, S. W. (1985). Benthic community metabolism in a coastal lagoon ecosystem. *Mar. Ecol. Prog. Ser.*, *22*, 21.

Odum, H. T. (1957). Trophic structure and productivity of Silver Springs, Florida. *Ecol. Monogr.*, *27*, 55.

Ogunrombi, J. A., & Dobbins, W. E. (1970). The effects of benthal deposits on the oxygen resources of natural streams. *J. Water Pollut. Control Fed.*, *42*, 538.

Pamatmat, M. M., & Bhagwat, A. M. (1973). Anaerobic metabolism in lake washington sediments. *Limnol. Oceanogr.*, *18*, 611.

Pamatmat, M. M., & Fenton, D. (1968). An instrument for measuring subtidal benthic metabolism in situ. *Limnol. Oceanogr.*, *13*, 537.

Price, C. B., et al. (1992). Sediment oxygen demand and its effects on water quality in USACE projects. *Proc. Water Qual., 9th Seminar.*

Revsbech, N. P., et al. (1986). Oxygen production and consumption in sediments determined at high spatial resolution by computer simulation of oxygen microelectrode data. *Limnol. Oceanogr.*, *31*, 293.

Reynolds, T. D., et al. (1973). Benthic oxygen demands of houston ship channel sediments. TAMU-SG-73, Tex. A&M University.

Rosa, F., & Burns, N. M. (1987). Lake Erie central basin oxygen depletion changes 1929–1980. *J. Great Lakes Res.*, *13*, 684.

Ruchhoft, C. C., & Moore, W. A. (1940). Determination of BOD and dissolved oxygen in river muds and suspensions. *Ind. Eng. Chem. Anal. Educ.*, *12*, 711.

Rudolfs, W. (1932). Relation between BOD and volatile solids of the sludge deposits of the connecticut river. *Sew. Works J.*, *4*, 315.

Shelton, T. B., & Hunter, J. V. (1974). Aerobic decomposition of oil pollutants in sediments. *J. Water Pollut. Control Fed.*, *46*, 2172.

Snodgrass, W. J. (1987). Analysis of models and measurements for sediment oxygen demand in Lake Erie. *J. Great Lakes Res.*, *13*, 738.

Snodgrass, W. J., & Fay, L. A. (1987). Values of SOD measured in the central basin of Lake Erie. *J. Great Lakes Res.*, *13*, 724.

Snodgrass, W. J., & Ng, P. S. (1985a). Biochemical models for the hypolimnetic oxygen depletion in lakes impacted by wastewater discharges. 1. Alternative models. *Arch. Hydrobiol. Suppl.*, *72*(1), 89.

Snodgrass, W. J., & Ng, P. S. (1985b). Biochemical models for the hypolimnetic oxygen depletion in lakes impacted by wastewater discharges. 2. Phytoplankton biomass model. *Arch Hydrobiol. Suppl.*, *72*(2), 220.

Sommaruga, R. (1991). Sediment oxygen demand in man-made Lake Ton-Ton (Uruguay). *Hydrobiologia*, *215*, 215.

Stein, J. E., & Denison, J. G. (1967). In situ benthal oxygen demand of cellulosic fibers. *Advances in Water Pollution Research, Proc. 3rd Int. Conf. Water Pollut. Res.* Water Pollut. Control Fed., 181.

Sweerts, J. P., et al. (1986). Metabolic activities in flocculent surface sediments and underlying sandy littoral sediments. *Limnol. Oceanogr.*, *31*, 330.

Teal, M. (1957). Community metabolism in a temperate cold spring. *Ecol. Monogr.*, *27*, 283.

Theriault, E. J., & McNamee, P. D. (1930). Sludge aeration experiments. I. rate of disapperarance of oxygen in sludge. *J. Ind. Eng. Chem.*, *22*, 1330.

Trevors, J. T. (1984a). Electron transport system activity and its relationship to other measures of activity in an estuarine sediment. *System. Appl. Microbiol.*, *5*, 136.

Trevors, J. T. (1984b). The measurement of electron transport system activity (ETS) in freshwater sediment. *Water Res.*, *18*, 581.

Truax, D. D., Shindala, A., & Sartain, H. (1995). Comparison of two sediment oxygen demand measurement techniques. *J. Environ. Eng.*, *121*, 619.

Uchrin, C. G., & Ahlert, W. K. (1985). In-situ SOD determination in the Passaic River (NJ) during the late summer/early fall 1983. *Water Res.*, *19*, 1141.

Veenstra, J. N., & Nolen, S. L. (1991). In-situ sediment oxygen demand in five southwestern lakes. *Water Res.*, *25*, 351.

Velz, C. J. (1949). Factors influencing self-purification and their relation to pollution abatement. II. Sludge deposits and drought probabilities. *Sew. Works J.*, *21*, 309.

Velz, C. J. (1958). Significance of organic sludge deposits. Oxidation relationships in streams. *U.S. Public Health Serv. Tech. Rep.*, W58-2.

Walker, R. R., & Snodgrass, W. J. (1986). Model for sediment oxygen demand in lakes. *J. Environ. Eng.*, *112*, 25.

Wang, W. (1980). Fractionation of SOD. *Water Res.*, *14*, 603.

Wang, W. (1981). Kinetics of SOD. *Water Res.*, *15*, 475.

Ward, G. H., Jr. (1984). The measurement of benthal oxygen demand in the Houston Ship Channel. *J. Water Pollut. Control Fed.*, *56*, 183.

Whittemore, R. C. (1978). Interfacial velocity effects on the measurement of SOD. *Natl. Council Air Stream Improvement Tech. Bull.*, 317.

Whittemore, R. C. (1979). Further studies of SOD measurement and its variability. *Natl. Council Air Stream Improvement Tech. Bull.*, 321.

Whittemore, R. C. (1981). The effect of temperature on SOD as related to water quality monitoring. *Natl. Council Air Stream Improvement Tech. Bull.*, 351.

Whittemore, R. C. (1982). A review of in-situ and laboratory SOD measurement comparison. *Natl. Council Air Stream Improvement Tech. Bull.*, 386.

Wu, R. S. S. (1990). A respirometer for continuous, in situ measurements of sediment oxygen demand. *Water Res.*, *24*, 391.

12

Chemical Oxygen Demand

Chemical oxygen demand (COD) is a measure of the amount of oxidant, in terms of oxygen, that is required to neutralize a sample of wastewater. The term "neutralize" is used intentionally to imply the complete conversion of all elements in the sample to their respective combustion (oxidation) products. A COD-like test using permanganate as the oxidant is reported (Weems, 1901) as being the starting point by Forchamer in 1848 in the development of the concept of biochemical oxygen demand (BOD) (Foulds & Lundsford, 1968; O'Brien & Clark, 1962).

The idea of using chemical oxidants to neutralize wastewater, as a measure of the strength or load of the sample, has led to the examination of numerous COD procedures using a wide variety of oxidizing reagents. Further, the ability to exactly replicate any COD procedure in any laboratory, regardless of the method used, and generate similar results on the same sample, has led researchers away from reliance on the BOD procedures as means of providing quantitative data for use in treatment facility design and operation. Reproducible data have become a prime concern with the increasing dependence on computerized models for design.

The COD test does not generate the same data as the BOD test. This is because there are few fundamental commonalities between the two other than the word "oxygen," and that the two tests are reported in the same units, mg O_2/L. Thus, there has been an impetus to marry the ease and reproducibility of the COD test with the ability to describe BOD-like aspects of the biological systems used in wastewater treatment. There is a long shelf of procedures that have been developed to fill this need, ranging from rapidly biodegradable soluble COD to nonbiodegradable particulate COD.

This chapter is organized to guide the reader through the complexities of COD as currently understood. It begins with an in-depth description of the standard COD tests and their development. The relationship (or lack of one) between COD and BOD is then discussed, followed by a review of the procedures that use COD to generate BOD-like information on biological systems. Finally, the concepts of total oxygen demand (TOD) and oxygen demand index (ODI) are described.

1.0 STANDARD CHEMICAL OXYGEN DEMAND TESTS

The 20th edition of *Standard Methods* (APHA et al., 1998) presents three standard COD test procedures, all using potassium dichromate as the oxidant. The first (*Standard Methods* 5220B) is the open reflux method. The second is the closed reflux, titrimetric method (*Standard Methods* 5220C), and the third is the closed reflux, colorimetric method (*Standard Methods* 5220D). Similar procedures are present in the methods manuals from Japan and the United Kingdom (Japanese Industrial Standard Committee, 1991; United Kingdom Department of the Environment, 1987).

1.1 Open Reflux Method

The open reflux method treats 50.00 mL of the sample (or diluted sample if the COD is more than 900 mg O_2/L) in a 500-mL flask with 1 g solid mercuric sulfate ($HgSO_4$), followed by 5.0 mL sulfuric acid reagent (concentrated sulfuric acid [H_2SO_4] containing 5.5 g/kg silver sulfate [Ag_2SO_4]) with cooling to prevent loss of volatile materials. Next is added 25.00 mL potassium dichromate solution (0.041 67 *M* $K_2Cr_2O_7$), and a water-cooled reflux condenser is attached to the flask. An additional 70 mL of the sulfuric acid reagent is added with mixing, then the contents of the flask heated to reflux for 2 hours. At the end of the reflux time, the mixture is diluted with reagent water and cooled to room temperature. After cooling, two to three drops of ferroin indicator (1,10-phenanthroline with ferrous sulfate, $FeSO_4$) are added and the contents of the flask titrated with ferrous ammonium sulfate (FAS; 0.25 *M* $Fe[NH_4]_2[SO_4]_2$ in reagent water acidified with sulfuric acid) until the initially blue-green solution changes to a persistent reddish brown. A reagent blank is used along with the determination and the COD calculation is

$$\text{COD, mg O}_2\text{/L} = \frac{[\text{mL FAS}_{\text{blank}} - \text{mL FAS}_{\text{sample}}]M\ \text{FAS}}{\text{mL sample}} \times 8000$$

1.2 Closed Reflux, Titrimetric Method

The sample (2.50 mL) is pipetted into a sturdy screw-capped borosilicate glass test tube of at least 10 mL capacity. Dichromate digestion solution (1.50 mL) (0.016 67 *M* $K_2Cr_2O_7$ containing 33.3 g/L $HgSO_4$) is added, followed by 3.5 mL sulfuric acid reagent (concentrated H_2SO_4 with 5.5 g/kg Ag_2SO_4). The test tube is capped and heated to 150 ± 2 °C for 2 hours. After cooling to room temperature, one to two drops of ferroin indicator is added and the solution titrated with 0.100 *M* FAS solution. The titration end point is indicated by a sharp color change from blue-green to reddish brown. A reagent blank is performed in the same manner, with reagent water being used as a sample with each set of samples. The calculation of the results is the same as is used in the open reflux method.

1.3 Closed Reflux, Colorimetric Method

The sample is processed in the same manner with similar reagents to those described above in the closed reflux, titrimetric method, with the difference being the digestion solution has a higher concentration of dichromate (0.034 73 *M* as compared with 0.016 67 *M*). The determination is completed by measuring the absorption at either 420 or 600 nm and then comparing the absorbance to a five-point calibration curve obtained from processing known standards of potassium hydrogen phthalate. The COD calculation is

$$\text{COD, mg } O_2/\text{L} = \frac{\text{mg } O_2/\text{L from calibration curve}}{\text{mL sample}} \times 1000$$

Quality controls over all three of the tests are quantitatively exerted by determining the COD of blanks and known concentrations of potassium hydrogen phthalate (KHP; $KC_8H_5O_4$) in both laboratory control and matrix spike format (Smith, 1999a, 1999b). Potassium dichromate reacts with KHP according to the equation

$$10K_2Cr_2O_7 + 2KC_8H_5O_4 + 41H_2SO_4 \rightarrow$$
$$16CO_2 + 46H_2O + 10Cr_2(SO_4)_3 + 11K_2SO_4$$

and a 300.0 mg/L solution of KHP exhibits a COD of 353 mg O_2/L (1.176 mg O_2/mg KHP; Pitwell, 1983).

The standard COD tests have the common elements of potassium dichromate, sulfuric acid, silver sulfate, mercuric sulfate, and heating. In the titration procedures, a FAS titration with ferroin indicator is common. Let us examine each of these elements.

Potassium dichromate in sulfuric acid, giving chromic acid, H_2CrO_4, is a powerful oxidant. The half-cell reduction reaction is

$$Cr(VI) + 3e^- \rightarrow Cr(III)$$

This half-cell reaction is coupled with a source of electrons to give a reduction–oxidation (redox) reaction. The most common element in wastewater that is available as an electron source is the carbon in organic materials. Carbon exists in a wide variety of formal oxidation states ranging from C^{4-} (as in methane, CH_4) to C^{4+} (as in carbon dioxide, CO_2). In any organic substance, carbon often exists in several oxidation states. Taking KHP as an example, some of the carbons are in the 3+ oxidation state, whereas others are formally 1− or zero (Smith, 1999b), with the average oxidation state for the carbons in the KHP molecule ($KC_8H_5O_4$) being 0.25+. The oxidation states for potassium and hydrogen are both 1+ and oxygen is 2^-. The oxidation half-cell for each carbon is

$$C^{0.25+} \rightarrow C^{4+} + 3.75\ e^-$$

and for a molecule of KHP is

$$8C^{0.25+} \rightarrow 8C^{4+} + 30e^-$$

Thus, for every molecule of KHP, 10 atoms of chromium(VI) must be reduced to absorb the 30 electrons produced by the oxidation, giving the 1:10 ratio present in the above balanced chemical equation for KHP COD.

Potassium dichromate is not the only oxidant that has been considered for the COD reaction. The original experiment of Forchamer in 1848 (Weems, 1901) examined the reaction of potassium permanganate with wastewater. Since then, many others have used permanganate to determine the chemical oxidizability of polluted water (Batek, 1914; Benson & Hicks, 1931; Burtle & Buswell, 1935, 1937; Chamot, 1908; Forster & Gad, 1938; Fox, 1903; Haupt, 1935; Koshkin & Karasik, 1940; Mallet, 1882; Maturaba, 1938; Northrup, 1954; Okada & Uesugi, 1957; Tidy, 1879; Tiemann & Gartner, 1895; Wyatt, 1893), and it was included as a standard procedure in the first through ninth editions of *Standard Methods* as the "Oxygen Consumed" test. Permanganate COD is still in use even after the publication of a standard dichromate COD in the 10th edition of *Standard Methods* in 1955 (Berka & Zavesky, 1969, 1974a, 1974b; Bevan & Stevenson, 1969; Dahl, 1969; Fleet et al., 1972b; Garcia et al., 1977; Goto et al., 1981; Hornke, 1974; Ito & Yamaguchi, 1980; Korenaga & Ikatsu, 1981; Korenaga et al., 1980; Korenaga et al., 1982; Ludvik, 1963; Ludvik & Siska, 1963; Mukai, 1976; Nasr & MacDonald, 1975; Sawada, 1986; Schumann, 1974; Stones, 1980; Vezzi, 1970).

Potassium permanganate is a strong oxidant, and unlike dichromate, reacts fairly well all by itself with many organic materials. Permanganate contains manganese in the 7+ oxidation state and exhibits two different reduction half-cells that are pH dependent. Under strongly acidic conditions, permanganate can accept five electrons:

$$\mathrm{Mn(VII)} + 5\mathrm{e}^- \rightarrow \mathrm{Mn(II)}$$

However, under neutral to alkaline conditions, the half-cell reduction stops at the 4+ state, forming the insoluble manganese dioxide in a three-electron reduction:

$$\mathrm{Mn(VII)} + 4\mathrm{OH}^- + 3\mathrm{e}^- \rightarrow \mathrm{Mn(IV)O_2} + 2\mathrm{H_2O}$$

Although the early *Standard Method* procedures and most authors describe use of permanganate in a strongly acidic medium, there have been reports of alkaline digestions (Benson & Hicks, 1931; Stamm, 1934; Trommsdorff, 1869), and an alkaline digestion was incorporated into *Standard Methods* by the eighth edition for samples with high chloride content.

The standard technique for the oxygen consumed test is to add 10 mL of 25% sulfuric acid to 100 mL of sample, then add 10 mL of potassium permanganate solution (0.0025 *M*) and heat in a boiling water bath for 30 minutes. At the end of the digestion time, 10 mL of ammonium oxalate solution (0.006 25 *M*) is added, then the excess oxalate ($C_2O_4^{2-}$) determined by back titration with potassium permanganate until a permanent pink color is observed. The reaction of oxalate with permanganate in acidic solution is (Mahin, 1929)

$$5(NH_4)_2C_2O_4 + 2KMnO_4 + 8H_2SO_4 \rightarrow K_2SO_4 + 2MnSO_4 + 5(NH_4)_2SO_4 + 10CO_2 + 8H_2O$$

The reagent concentrations are chosen such that 1.0 mL of the permanganate or oxalate solution is equivalent to 0.1 mg of oxygen equivalents consumed by the 100 mL sample, or to 1.0 mg/L by the undiluted sample. Correction for interferences caused by inorganic ions can be established by performing a direct titration incorporating a room-temperature, 3-minute digestion with permanganate. Another technique for determination of residual permanganate involves a direct titration with hydroquinone using a diphenylamine indicator (Berk & Hoffman, 1963).

It was recognized early on (Fox, 1886, 1903) that oxygen-consumed values are dependent on exactly how the test is performed, in addition to the technical ability of the analyst to perform quantitative analysis. Values from one treatment facility are generally related to values from any other facility in only a qualitative sense because of these uncertainties. The most common variants are the time of heating in boiling water, ranging from 2 minutes (Drown, 1890; Drown & Hazen, 1890) to 10 minutes (Leffman, 1895; Mason, 1910). Because of the accelerated decomposition of permanganate in the presence of chloride (Dupre, 1885), English practice was to conduct the test with the digestion at room temperature with times ranging from 3 minutes to 4 hours (Fowler, 1902; McGowan, 1904; McGowan et al., 1904; Thresh, 1904; Sutton, 1911).

Potassium permanganate possesses a property that makes it less than perfectly desirable as a reagent. Solutions of potassium permanganate are not stable, and decompose to generate the brown coating of manganese dioxide often noted on the inside of bottles used for storing the solution. The decomposition is autocatalytic with the presence of manganese dioxide assisting the reaction. The decomposition is also accelerated by light, requiring storage in the dark or in tinted bottles. This property means that the solution must be restandardized frequently. Normally, the oxalic acid standardization is used. It is not the most convenient procedure to perform, requiring heating during the titration to about 60 °C. Permanganate is more prone to interference from the presence of chloride ion than dichromate COD methods (Mukai, 1976).

Sodium hypoiodide (NaOI) has been used (Cameron, 1944; Douglas, 1978) in a highly alkaline COD procedure at room temperature for 1 hour. Iodine is present in the 1+ oxidation state in hypoiodide and readily accepts two electrons giving iodide ion:

$$I^+ + 2e^- \rightarrow I^-$$

The sodium hypoiodide procedure has the problem that the oxidizing reagent is not available in high purity, unlike most of the other oxidants, and the activity of the reagent must be determined by redox standardization. The advantage of the hypoiodide method is that it is highly aggressive toward nitrogen-containing organics. Oxidized iodine in the form of iodic acid, HIO_3, (Dzyadzio, 1938; Johnson & Halvorson, 1941), with iodine in the 5+ oxidation state, has been used in a COD-like procedure. The sample is reacted with potassium iodate in 65% to 80% sulfuric at 200 °C, then the excess iodate is determined with an iodide-thiosulfate titration. The iodic acid method is reported to be selective for carbon and hydrogen in the wastewater and nonreactive toward nitrogen.

Ferricyanide (Krunchak et al., 1984) has been tried in a COD-like procedure, although the iron(III)–iron(II) reduction is not very powerful. The results were compared with dichromate COD.

Ceriometry (Bezel, 1945; El-Dib & Ramadan, 1966; Golterman & Wisselo, 1981; Klein, 1941; Korenaga et al., 1984) uses the cerium reduction half-cell in strong acid:

$$Ce(IV) + e^- \rightarrow Ce(III)$$

Cerium oxidations are reported not to exhibit interference by chloride up to about 1 *M* concentration, although one of the later papers (Golterman & Wisselo, 1981) recommends the use of mercuric sulfate to remove chloride ion. Cerium has a higher oxidation potential than dichromate but is only about two-thirds as effective an oxidant as dichromate toward organic substrates. The procedure is to heat the sample in an autoclave at 160 °C for 2 hours with excess cerium(IV) sulfate in a mixture of sulfuric, perchloric, and nitric acids, then determine the residual oxidant colorimetrically at 405 nm. Cerium has been used as an extra catalyst in the dichromate-silver COD procedure with a noted improvement in oxidation of aromatic substrates (El-Dib & Ramadan, 1966). Pyridine is inert under these conditions.

Potassium persulfate has been used in a COD-like procedure (Gilmour et al., 1960; Leithe, 1969; Petermann, 1975). It has also been adapted into an in-line COD monitoring instrument in which the persulfate is continuously generated (Davenport & Wynveen, 1977).

Hydrogen peroxide is a commonly employed oxidizing reagent but by itself is not a very powerful oxidant, although it has been tested as a COD reagent (Cooper & Read, 1927). Biological systems have evolved enzymes such as peroxidase and catalase to utilize peroxide. The advantage of using hydrogen peroxide in a COD procedure is that it can be easily converted into hydroxyl radicals that are very strong oxidants. Chemical production of hydroxyl radical is achieved in Fenton's reagent [hydrogen peroxide and iron(II)]. The use of Fenton's reagent in a COD-like reaction has been examined (Medalia, 1951). One of the drawbacks from use of Fenton's reagent is that a high concentration of hydroxyl radicals is difficult to sustain for the period of time needed to quantitatively oxidize the waste.

Ultraviolet (UV) light will also generate hydroxyl radicals from hydrogen peroxide, with the added advantage that it can be continuously applied and the radicals continuously generated during the course of an oxidation. An instrument that performs a UV-catalyzed hydrogen peroxide oxidation has been developed (Guway et al., 1999). The sample is mixed with a known amount of hydrogen peroxide and irradiated in a closed chamber with UV light over a period of 1 hour. Some of the added peroxide is immediately converted to oxygen by microbial catalase already present in the sample (Guway et al., 1998). The oxygen bubbles out of the sample and is measured with a gas meter during the reaction. The sample is then irradiated. At the end of the irradiation time, the oxidized sample is passed through a catalase-filled reactor that converts any remaining peroxide to oxygen, which is also measured. For a given dosing of the sample with peroxide, the sum of the two oxygen measurements is directly related to the amount of hydrogen peroxide (hydroxyl radical) consumed by the oxidizable portion of the waste. This in turn is highly correlated ($r > 0.93$) in many effluents to the COD and 5-day BOD (BOD_5) of the waste.

The various oxidants used in COD methods have been compared many times (Hoover, 1934). The permanganate method is regarded as the least efficient method (Ettinger, 1956; Foulds & Lunsford, 1968; Hornke, 1974; Jenkins et al., 1965; Jones & Dageforde, 1968; Klein, 1941; Korenaga et al., 1984; Larsson et al., 1975; Leszczynski & Skwara, 1966; Moore et al., 1951; Riehl, 1951; Roberts & Sanderson, 1953; Vogl et al., 1968), while either the dichromate–silver or the cerium–COD are variously regarded as the most effective method. The cerium–COD method further has the advantage of being considerably less toxic (Korenaga et al., 1984).

There were numerous early reports of a dichromate-based COD test (Abbott, 1927; Adeny & Dawson, 1926; Popova, 1930). These early experiments mixed the sample with sulfuric acid, then heated it to dryness to drive off chlorides as the volatile hydrogen chloride. After chloride removal, dichromate was added to the sample and it was again reduced in volume by heating. Finally, more acid was added and the mixture heated to reflux. The residual oxidant was determined and the amount consumed in the reaction was related to the oxygen consumed by the sample. *Standard Methods* did not include a dichromate-based COD procedure until the 10th edition in 1955. Previous editions all the way back to the first edition presented COD-like oxygen consumed tests that use potassium permanganate.

Potassium dichromate presents many operational advantages in the COD procedure. It is a primary standard, available in very high purity. The reagent is stable as both a solid and in solution for very long periods of time. The potassium dichromate solution is considered a primary standard and is used to check the titer of the FAS) used in the manual titration with ferroin indicator. The primary standard potassium hydrogen phthalate (KHP) is frequently used in conjunction with dichromate COD methods for quality control purposes. The use of KHP along with dichromate in COD gives an analytical procedure that is quantitatively verified by two primary standards, an analyst's ideal situation.

Iodometric titrations were originally used with dichromate COD (Abbott, 1927; Adeny & Dawson, 1926; Ingols & Murray, 1948; Rhame, 1947) to determine the residual amount of oxidant. Iodide ion added to the cooled digestion solution is oxidized to iodine, which is then determined by thiosulfate titration with starch indicator. This finishing technique still finds use in both a direct titration (Charravari & Som, 1950) and as a determinative final step in auto-analyzer methods. Mohr's salt (FAS), with ferroin indicator is more commonly used for the final determination of residual oxidant. The color transition of ferroin is vivid for visual end point determination. Although this is the visualization method presented in *Standard Methods*, it is not the only technique available. The use of automated titrators for the FAS finish with potentiometric detection (Baughman et al., 1969; Cronholm & Winbladh, 1972; Edwards & Allen, 1984; Matsumae, 1980) or direct coulometric detection (Hoogeveen & Butzelaar, 1972; Li, 1986) are reported to be both faster and more reproducible than visual end point determination. Phenylanthranilic acid has also been used as a visual indicator in the FAS titration (Rozenberg, 1993). Another titration finish for the dichromate COD involves a back titration with permanganate (Singh & Neves, 1974).

Both the oxidized and reduced ions of chromium are highly colored, dichromate being red-orange with maximum absorbance at 420 nm, whereas chromium(III) is green with a usable absorbance at 600 nm, allowing either to be monitored with a spectrophotometer (Chaudhuri et al., 1973; Gaudy & Ramanathan, 1964; Kumar et al., 1986; McNary et al., 1957). The results obtained using the macroscale FAS titration procedure and microscale colorimetric and titration procedures have been compared favorably (Haapala & Starck, 1986; Jones et al., 1985; Klein, 1985; Messenger, 1981; Mobius, 1979, 1980; Schumann, 1984; Trbovic, 1979).

The chromium(VI) salts in the form of sodium or potassium chromate (CrO_4^{2-}) or dichromate ($Cr_2O_7^{2-}$) are strong oxidants in their own right; however, the product of acidification, chromic acid (H_2CrO_4), is a much more effective oxidant in COD reactions. The most common acid used is sulfuric, with strengths ranging from 50% to the concentrated acid, 95% to 97%. The use of 1:1 mixtures of concentrated sulfuric acid and 85% phosphoric acid have been reported (Burtle & Buswell, 1937; Charravari & Som, 1950; Ingols & Murray, 1948; Jeris, 1967; Mohlman & Edwards, 1931; Rhame, 1947) and are particularly recommended for the digestion when the object is determination of evolved carbon dioxide in a total organic carbon procedure.

There are shortcomings to the use of potassium dichromate. It rapidly reacts with chloride to produce chromyl chloride, which has a boiling point of only 117 °C and can volatize out of the reaction mixture. Chromyl chloride, when heated, will decompose, giving an overall oxidation of chloride ion to chlorine that will also volatize from the digestion mixture. Both of these effects give a positive bias to the results by decreasing the amount of chromate remaining in the mixture after the digestion. Dichromate also rapidly reacts with other inorganic ions. Examples are reduced forms of iron, tin, manganese, and other polyvalent cations, plus many simple and complex anions such as chloride, nitrite, sulfide, sulfite, and phosphite. By itself, dichromate, unlike permanganate, is not very effective at oxidizing all organic materials, which probably hampered initial acceptance of the procedure (Moore et al., 1949). Silver was found to be necessary as a catalyst (Muers, 1936). The addition of silver ion to the mixture increases the speed and ease of organic matter oxidation (Moore et al., 1951). This occurs through dichromate oxidation of the silver(I) to silver(II), which is a very powerful oxidant toward organics. For example, there is no oxidation of acetate or glutamic acid by dichromate unless silver is present (Dasgupta & Petersen, 1990). As the silver is reduced by the organic reaction, it is immediately reoxidized by more dichromate.

The requirement for having silver in the reaction presents its own problems. Silver precipitates rapidly in the presence of chloride ion, giving the insoluble silver chloride (Ballinger et al., 1982). Higher concentrations of chloride react further with the precipitated silver chloride to give a soluble, complexed dichloroargenate ion, $(AgCl_2)^-$, which is essentially inert toward oxidation (Smith, 1999b). If any chloride is present in the solution, the catalytic activity of silver is destroyed and the oxidation reaction of many organic substrates for all practical purposes ceases. Thus, there are two mechanisms by which chloride presents interference.

There have been many proposed ways to circumvent the chloride interference. One is to use an excess of silver dichromate as the oxidant (Camps & Arias, 1963). Other proposed techniques add excess silver chloride (Lloyd, 1982) or silver sulfate before the addition of the rest of the reagents (Kincannon & Gaudy, 1966). The solution that is included in *Standard Methods* for dichromate COD is the addition of mercuric ion to remove the chloride as a stable chloromercurate complex, $(HgCl)^+$, $(HgCl_2)^0$, $(HgCl_3)^-$, and $(HgCl_4)^{2-}$. The formation constants for mono- and dichloromercurate, log K = 6.74 and 6.48, respectively, are much larger than those for the silver chloride complexes, log K (AgCl) = 3.04 and log K ($AgCl_2^-$) = 2.00 (Canelli et al., 1976); thus, the order of reagent addition is immaterial. However, most methods call for addition of the mercury before any silver or dichromate is added (Cripps & Jenkins, 1964; Dobbs & Williams, 1963; Jeris, 1967). Mercury effectively sequesters the chloride interference, except in the presence of ammonia, where ammonia reacts with the chloride–mercury complex to displace the chloride back into the reaction mixture and make it available for complexation with silver or oxidation by dichromate (Cotton et al., 1999; Kim, 1989).

The extreme toxicity of mercury spurs continued efforts to find ways that do not use the element for removal of chloride interference. One is to add an excess of silver nitrate to the sample to remove chloride as the insoluble silver chloride (Ballinger et al., 1982). Another calls for addition of up to 8% excess silver sulfate to overcome chloride interference (De Casseres et al., 1984). Chromium(III) salts, such as $KCr(SO_4)_2$, are added to negate chloride interference without addition of mercury (Hejzlar & Kopacek, 1990; Thompson et al., 1986). The chromium(III) was believed to act through formation of a chloride complex; however, later work (Dasgupta & Peterson, 1990) suggested that the effect is the result of lowering of the chromium(VI) oxidation potential. This leads to a decreased ability to oxidize organic compounds, with the end result being that the addition is not completely effective. Reducing the concentration of dichromate to 0.075 *N* is reported to allow elimination of the mercury, by decreasing the oxidation potential of the mixture. The reduced concentration is used with an autoclave procedure at 120 °C for 1 hour to achieve complete reaction (Ryding & Forsberg, 1977). Simply increasing the sulfuric acid concentration is reported to minimize chloride interference (Dasgupta & Peterson, 1990).

An elegant technique for accounting for chloride interference is to trap the escaping chlorine gas in a potassium iodide solution, and then after the reflux is over, titrate the contents of the trap with sodium thiosulfate (Baumann, 1974). The correction is then done mathematically:

$$\text{Corrected COD} = \text{Apparent COD} - \frac{\text{mL thiosulfate} \times N \text{ thiosulfate}}{\text{mL sample}} \times 8000$$

There have been several attempts to develop correction factors for chloride interference. One study reported that 1 mg/L chloride is equivalent to 0.23 mg/L COD, and the results of a Mohr titration ($AgNO_3$) for chloride on the sample can be directly subtracted from the dichromate COD result to give a corrected COD (Charravari & Som, 1950). Another proposed technique calibrated the chloride COD interference through analysis of representative samples with known amounts of chloride in a variation of the method of standard additions (Smith, 1999a, 1999b). A graph was prepared of the apparent COD against the amount of chloride (Burns & Marshall, 1965). When a sample is analyzed, the chloride content is determined along with the COD, then a correction applied to the apparent COD to give the true COD:

$$\text{True COD} = (\text{apparent COD} - D) \times E$$

where D is taken from the interference calibration graph and the constant E is derived by the laboratory for its testing procedure and may be sample-specific. The authors reported a value of $E = 1.20$. Other work (Canelli et al., 1976) suggested that the exerted COD of pure sodium chloride solutions is much greater than the COD of the same concentration of chloride mixed with an organic substrate. The difference was attributed to lower oxidation potential of dichromate in the organic containing sample.

Sulfur dioxide is sometimes an interference in COD determinations. Aeration with a manganous sulfate catalyst has been reported to be an effective way to oxidize the sulfur dioxide (sulfite) to sulfate, before adding the COD reagents (Watts & Adams, 1983).

Hydrogen peroxide is an interference in the COD test because of its ability to function as both an oxidant and a reductant (Flaschka et al., 1980):

$$H_2O_2 + 2H^+ + 2e^- \rightarrow 2H_2O \qquad \text{oxidant}$$

$$H_2O_2 \rightarrow O_2 + 2H^+ + 2e^- \qquad \text{reductant}$$

Hydrogen peroxide consumes dichromate through initial formation of peroxidichromic acid ($H_2Cr_2O_{12}$), giving a transient blue color to the sample, which further reacts with additional dichromate in a reduction reaction, giving an overall false-positive result of 0.25 mg/L COD for every 1 mg/L H_2O_2 (Talinli & Anderson, 1992). The authors suggest determining the actual amount of hydrogen peroxide in the sample from a separate iodometric titration with molybdate catalyst (Klingenberg & Reed, 1967), then correcting the final COD result for the hydrogen peroxide demand of dichromate. The peroxide–dichromate reaction is used intentionally in a postdigestion detector in an autoanalyzer to form a system that is capable of monitoring either dichromate or permanganate COD (Fleet et al., 1972a, 1972b). The excess oxidant from the digestion is reacted with hydrogen peroxide to liberate oxygen, which is detected coulometrically in an electrochemical cell.

The silver-catalyzed dichromate COD reaction does not oxidize all organic compounds. Pyridine, C_5H_5N, the simplest 6-membered aromatic nitrogen heterocycle, is inert to the reaction as are many substituted pyridines, such as nicotinic acid, nicotinamide, vinyl pyridine, and other pyridine carboxylic acids. Aromatic nitrogen heterocycles with two or more rings, such as quinoline and indole, and simple five-membered aromatic nitrogen heterocycles, such as pyrrole and proline, are completely oxidized (Chudoba & Dalesicky, 1973). Other organic molecules are known to be resistant to oxidation, such as benzene, toluene, styrene, acetic acid, lactic acid, and alanine. Dichromate COD of pure compounds have been determined and compared with theoretical values many times (Canelli et al., 1976; Chudoba, 1985; Dobbs & Williams, 1963; Goodwin et al., 1977; Himebaugh & Smith, 1979; Ingols & Murray, 1948; Matsuoka et al., 1987; Medalia, 1951; Moore & Walker, 1956; Moore et al., 1949; Papinsky et al., 1952; Schulze, 1953; Wolff, 1975; Zajic et al., 1970).

One complication is noteworthy with regards to the theoretical COD of nitrogen-containing substrates (Chudoba & Dalesicky, 1973). Under the conditions of the dichromate COD reaction, some of the nitrogen is oxidized to elemental nitrogen (N_2), whereas the rest of the nitrogen is recovered as ammonia (NH_3), which is inert to the conditions of the reaction. The relative proportions of nitrogen and ammonia produced depend on the chemical environment of the nitrogen in the substrate molecule. Thus, uric acid is oxidized by dichromate according to

$$C_4H_4O_2N_4 + 3O_2 + 2H^+ \rightarrow 4CO_2 + 2NH_3 + N_2$$

The reagents used in the dichromate COD are hazardous. Chromium, silver, and mercury are all included on the original U.S. Resource Conservation and Recovery Act (RCRA) list of the eight metals of most environmental concern in defining a waste as hazardous. The reagents are also expensive. The two closed-tube procedures from *Standard Methods* are an expression of the intent to reduce sample size and reagent usage, and thus reduce the generation of hazardous laboratory waste. There have been reports on the recovery and reuse of the toxic metals in the dichromate COD test (Aslam & Walker, 1982; Holm, 1996).

Many technical adaptations that use the COD reaction to minimize use of hazardous reagents have been reported. A semiautomated procedure that performs the oxidation in a small test tube at 150 °C for 2 hours, then determines the residual dichromate in a flow-through system has been described (Jirka & Carter, 1975). Automated analyzers using either air-segmented (Fleet et al., 1972a, 1972b; Morfaux & Sarris, 1974; Morfaux et al., 1972; Schierjott et al., 1974; Tian & Wu, 1992) or continuous-flow injection techniques (Appleton et al., 1986; Korenaga, 1982; Korenaga & Ikatsu, 1982; Spohn et al., 1991) have been reported for the silver catalyzed dichromate COD. Flow injection analyzer methods using permanganate have been reported using colorimetric determination (Korenaga, 1980; Korenaga et al., 1980) and amperometric determination (Goto et al., 1981).

A variety of detectors have been used with these automated and semiautomated techniques, with colorimetric detection at 600 or 420 nm of the oxidized–reduced chromium being the most common (Adelman, 1968; Hey et al., 1969; Jirka & Carter, 1975). Electrochemical determination of oxidized iodine produced from excess oxidant after merging the digestate with potassium iodide solution has been used in an automated apparatus (Novic et al., 1988). A coulometric detector has been used with an autoanalyzer (Mitev et al., 1978), as has a potentiometric cell (Geisler et al., 1974).

An automated microscale kinetic COD technique that monitors the disappearance of dichromate through an optical fiber has been developed (Dasgupta & Peterson, 1990). The standard dichromate reagents are added from reservoirs to 0.46 mL of the sample in a microreactor that is then heated to 140 °C for 3 minutes. An advantage to this technique is that the kinetics of the reaction can be studied as it is taking place.

The last important aspect of the wet chemical COD test is heating. Basic chemical theory indicates that as the temperature of the reaction is raised, the reaction progresses faster. In the COD, this is true. Temperatures ranging from 60 to 180 °C have been reported. The concentrations of the reagents in the macroscale titration procedure are designed to achieve a reflux temperature in the upper part of this range, 145 to 150 °C (Moore et al., 1949). Increasing the sulfuric acid concentration achieves better oxidation (Chian & DeWalle, 1975; Dasgupta & Peterson, 1990; Geisler et al., 1974; Harkness et al., 1972); however, in the macroscale reflux, this also increases the temperature, which can lead to dichromate decomposition. Dichromate has been reported to decompose at temperatures greater than 165 °C; thus, control of the reaction temperature at 165 $\pm$ 1 °C is recommended to achieve maximum efficiency (Wells, 1970). In one report (Ingols & Murray,

1948), the high temperature decomposition of dichromate was described as resulting from the oxidation of water by dichromate to form hydrogen peroxide. Some experimental justification for this conclusion was presented.

One concern associated with the open reflux of COD reaction mixtures is the potential loss of volatile organic components from the sample before oxidation. Most practitioners of the technique use a water-cooled condenser with the apparatus, although use of a passive air-cooled condenser has been advocated (TeKippe & Zehnpfennig, 1971). The use of recirculating refrigeration units for cooling the condenser fluid will minimize the loss of volatile components (Smith, 1995). Current trends in performance of COD have been toward dosed-vessel heating, which completely sidesteps the issue of volatile component loss.

The closed-tube microscale techniques use a heated block at 150 °C. These higher temperatures for 2 hours are used to ensure complex oxidation, although temperatures on the low end of the scale have been reported (Canelli et al., 1976). Very brief heating times, on the order of 10 to 15 minutes, have been reported to be able to achieve up to 88.5% oxidation for selected samples (Ishcherikova & Drozdova, 1973; Ishcherikova et al., 1975). A 30-minute heating time has been reported as sufficient for other samples (Jeris, 1967). Autoclaving the samples at 120 °C for 30 minutes to 1 hour is described as sufficient in studies of the COD reactions of pure compounds (Ryding & Forsberg, 1977; Schulze, 1953). A quick method for achieving the needed temperature and complete reaction uses a microwave oven operated at 700-W power for 7 minutes (Jardim & Rohwedder, 1989). With certain samples, temperatures as low as 60 °C have been reported as optimum (Gonzalez, 1986). In studies of KHP oxidation, complete rapid oxidation at any temperature from 110 to 170 °C is found (Morfaux & Sarris, 1974; Pitwell, 1983).

Digestion temperature and completion of reaction are key considerations when autoanalyzers are being used for COD determinations. Most autoanalyzers actually function in a kinetic mode, meaning that the chemical reaction does not go to completion. Instead, the reaction is stopped at some point before completion, say for instance 25%. The reason the technique works for many analytes like orthophosphate and nitrate is that the analyte and standards are in only one chemical form, and all samples exhibit approximately the same rate of reaction. It does not matter in these situations whether the reaction results are read at 5%, 25%, or 95% completion. As long as the autoanalyzer is calibrated with a complete range of standards and samples run under exactly the same conditions, the results are acceptable.

These requirements are typically not met in autoanalyzer COD procedures, regardless of which oxidant is being used. The foremost reason is that the chemical characteristics of the actual materials being oxidized are not the same in every sample. Some samples are rich in easy-to-oxidize materials like sugars and proteins, while others contain a majority of slow oxidizers like hydrocarbons, fatty acids, and industrial chemicals. The KHP standard used to calibrate the system is easy to oxidize under almost any temperature conditions (Korenaga, 1982; Morfaux & Sarris, 1974; Pitwell, 1983). Building a calibration based on an easily oxidized material like KHP leads to a sense of false confidence that all materials are as easy to oxidize as KHP, when in fact, falsely low COD values may be obtained on actual samples. A system that achieves, for example, 50% reaction completion on KHP, may actually deliver less than 10% completion on other samples, even under exactly the same conditions of reactant concentration, temperature, and time of reaction.

The reaction completion issue has been directly addressed by incorporation of a microwave heating unit in a flow injection autoanalyzer (FIA) system for COD determination. A microwave oven containing a Teflon reaction coil wound around a polytetraflouroethylene-carbon block, allows heating for 3 minutes and gives substantially better oxidation. A debubbling unit has to be added to the system to counteract the production of bubbles by the oven in the tubing (Balconi et al., 1992).

Other heating techniques that have potential application to automated analyzers have been reported. In one, the sample is mixed with dichromate and silver in 95% sulfuric acid then percolated over 3 to 5 minutes down through a quartz sand-packed column heated to 170 °C. The analysis is completed by FAS titration with phenylanthranilic acid as the visual indicator. Comparison of this digestion technique to the standard reflux indicates about the same results for easily oxidized compounds, but a substantial improvement for difficult materials. One difficult compound, ε caprolactam, gives 88% oxidation by the heated column method, while standard reflux only gives 38% recovery (Rozenberg, 1993). Another potentially useful technique is a UV light-assisted oxidation (Topalian & Scntheimer, 1979).

A COD autoanalyzer system is useful in treatment facility process monitoring situations in which the influent to the facility is fairly consistent in a chemical sense, and repetitive analysis is being performed on essentially the same sample. In this case, the instrument can be calibrated with KHP standards, and then the instrumental results on the real sample compared with the COD results obtained from a more rigorous manual COD technique. Normally, a statistically valid correction factor in the form of a multiplier with a low standard deviation (less than 5%) or a linear equation with a regression coefficient value greater than 0.95 can be found to relate instrument results to the rigorous results. Under these conditions of repetitive analysis of a consistent sample, reliable quantitative results can be produced. But this does not apply to the situation in a commercial environmental laboratory, where every sample is essentially unique, rendering autoanalyzer COD procedures unreliable.

A study exhaustively examining alternate wet chemical COD oxidants is available (Boyles, 1997a, 1997b; Miller et al., 2001). The oxidants examined included permanganate, persulfate, cerate, periodate, iodate, bromate, perbromate, hypochlorite, perchlorate, ferrate, bismuthate, hydrogen peroxide, ozone, vanadate, and manganese(III). From these and other reagent and condition combinations, manganese(III) was selected for commercial development.

The reagent is prepared by controlled oxidation of manganese(II) with potassium permanganate

$$2KMnO_4 + 8MnSO_4 + 8H_2SO_4 \rightarrow 5Mn_2(SO_4)_3 + K_2SO_4 + 8H_2O$$

The addition of manganese(II) salts in strong sulfuric acid helps preserve the manganese(III) reagent by interfering with the disproportionation reaction of manganese(III) to manganese(II) and manganese(IV).

The manganese(III) COD is performed in a capped vial with heating at 150 °C for 1 hour, then the excess manganese(III) determined spectrophotometrically at 510 nm (U.S. Patent no. 5,556,787). Chloride still presents an interference in the procedure; however, it can be removed by pretreatment of the sample with sodium bismuthate at room temperature. Comparison of the manganese(III) COD with the standard dichromate COD indicates the manganese(III) procedure gives approximately 10% lower results.

As an alternative to either the titration or spectrophotometric finish of the dichromate COD test, single sweep polarography has been examined to determine residual dichromate (Dan et al., 2000). In direct comparison with titration COD on real wastewater samples, a less than 1.5% relative error was reported.

A novel direction in COD determination is oriented toward development of COD sensors. One of the first reported consists of a thin electrochemical cell (0.1-mm thick, about 45 μL volume) with copper working electrode and a potentiostat with platinum counter electrode and silver–silver chloride reference probe. Operated at 650 mV and 25 °C, it provides data in 3 to 10 minutes that correlate well ($r = 0.946$) with wet chemical dichromate COD on glucose, glycine, glutamic acid, phenol, hemicellulose, lignin, tannic acid, and humic acid standards (Lee et al., 1999). Chemical oxygen demand analysis using a copper oxide electrode in alkaline solution is also described by Silva (2009). Determination of metal complexation capacity by electrochemical complexometeric titration using differential pulse anodic stripping voltammetry is reported by Mohan et al. (2000). The complexation capacity results correlated well with standard COD measurements. Another approach uses a photocatalytic titanium dioxide element coupled with an oxygen electrode (Kim et al., 2000). The membrane of the oxygen electrode is impregnated with titanium dioxide. The sample is stirred to achieve dissolved oxygen (DO) stabilization, then irradiated with UV light (365 nm). Titanium dioxide serves as a photocatalyst to form superoxide radical ion, which quickly oxidizes organic materials in the sample. The decrease in DO concentration is calibrated to COD values of known standards. A novel electrode consisting of carbon nanotubes-polystyrene composite with Ni, NiCu, CoO, and CuO/AgO nanoparticles was constructed and tested on wastewater to obtain COD values (Gutiérrez-Capitán et al., 2015).

2.0 RELATION OF CHEMICAL OXYGEN DEMAND TO BIOCHEMICAL OXYGEN DEMAND

There have been many reports of efforts to correlate results from COD and similar tests to BOD results in an effort to avoid the 5-day time period commonly used for the BOD test. As presented in Table 12.1, these efforts have ranged from correlations on specific waste streams to attempts to develop universal relationships. As a general summary of all these efforts, useful correlations on specific waste streams are frequently obtained, whereas a universal relationship is impossible to derive.

The basic reason why a universal relationship is impossible is that the tests are fundamentally different measures of unrelated characteristics of the water. On the one hand, as detailed in other chapters in this book, the BOD test measures the biological response of microorganisms to the presence of substrates in the water that can be used as energy and food. On the other hand, COD and the other similar tests measure a defined chemical property of the sample. There is no a priori reason why these characteristics should be the same. And, in fact, the reported successful correlations between COD-like procedures and BOD should appropriately be regarded as mere coincidence, reflecting an effluent that is qualitatively and quantitatively very consistent over time more than any other factor. The extreme example that supports this viewpoint is the correlation that is reported between sodium content and BOD in paper mill effluents (Fowler et al., 1974). Sodium, while a necessary mineral for microorganism growth, is never the limiting growth factor in wastewater, and any correlation with BOD is purely accidental.

Within the environmental laboratory community, there is the widespread belief that for any sample as an estimation the BOD is about two-thirds of the COD, and the total organic carbon (TOC) is about one-half the BOD. This guidance arises from the values obtained on KHP, a material frequently

TABLE 12.1. Efforts to relate COD and other tests to BOD.

Reference	Relation
Burtle & Buswell, 1935	$KMnO_4$ COD to BOD
Forster & Gad, 1938	$KMnO_4$ COD to BOD
Klein, 1941	Mixed oxidant COD to BOD
Rhame, 1947, 1950	COD to BOD
Lewin, 1949	COD to BOD
Charravari & Som, 1950	Dichromate COD to BOD_5 and BOD_{20} on wastewater
Chelsley et al., 1951	COD to BOD on industrial effluents
Moore & Ruchhoft, 1951	COD to BOD
Riehl, 1951	COD to BOD
Papinsky et al., 1952	COD to BOD
Kaufman, 1954	COD to BOD in treated industrial effluent
Northrup, 1954	$KMnO_4$ COD to BOD in textile wastes
Ettinger, 1956	Dichromate COD to BOD for industrial wastes
Carpenter, 1966	COD to BOD of kraft mill effluents
Vogl et al., 1968	$KMnO_4$ and dichromate COD to BOD
Bevan & Stevenson, 1969	$KMnO_4$ COD to BOD_5 of sugar mill wastes
Branch, 1969	Correlated COD, TOC, and BOD of paper mill wastes
Dahl, 1969	$KMnO_4$ COD to BOD of slaughterhouse wastes
Krogh, 1970	COD and ignition loss to BOD_7 of slaughterhouse waste
Vezzi, 1970	$KMnO_4$ COD to BOD in surface water
Zajic et al., 1970	COD to BOD of parafinine hydrocarbons
Davis, 1971	COD, TOC, and TOD to BOD
Delcommune, 1971	COD to BOD of malt and brewery wastes
South et al., 1971	COD to BOD of pine industry effluents
McKeown et al., 1972	COD, TOC, and TOD to BOD of paper mill effluents
Shriver & Young, 1972	COD and ODI to BOD
Bochinski, 1973	In general examined BOD, COD, TOC, and TOD
McDonald, 1973	Johnson COD machine to BOD
McGhee et al., 1972; McGhee, Christenson & Bonneau, 1973; McGhee, Torrens & Smaus, 1973; McGhee et al., 1976	COD to BOD and ultimate BOD of feedlot runoff
Fowler et al., 1974	Sodium and COD to BOD for kraft pulp effluent
Hornke, 1974	COD to BOD on paper mill wastes
Hwang et al., 1974	COD and TOC to BOD for board mill effluent

(*continued*)

TABLE 12.1. Efforts to relate COD and other tests to BOD (*continued*).

Reference	Relation
O'Herron, 1974	COD, TOC, and TOD to BOD
Reynolds & Goellner, 1974	ODI to BOD_5
Rhame, 1974	COD to first stage BOD
Schumann, 1974	COD to BOD of brewery wastewater
Sorensen, 1974	COD to BOD of dairy effluent
Ishcherikoa et al., 1975	COD to BOD
Janardan & Schaffer, 1975	COD to BOD_5
Larsson et al., 1975	COD to BOD of polymers
Reinefeld et al., 1975	COD to BOD_5 of sugar factory wastes
Chian & DeWalle, 1975	COD and TOC to BOD of landfill leachates
Hayes et al., 1976	COD and ODI to BOD on kraft mill effluents
Virarghavan, 1976	COD and SOC to BOD
Garcia et al., 1977	COD, TOD, TOC, and $KMnO_4$ ODI to BOD of paper mill effluents
Voorn & Marlowe, 1977	COD, TOD, and TOC to BOD_5
Douglas, 1978	NaOI COD to BOD_5 for kraft mill waste, excellent correlation
Strom & Solyom, 1978	COD and TOC to BOD_7 and BOD_{20} for dairy effluent
Uchida et al., 1978	COD, TOD, and TOC to BOD in seafood processing wastewater
Zimina & Zaleshina, 1978	COD to BOD_5 in paper mill effluent
Jamieson & Stewart, 1979	COD to BOD of wool scour liquor
Bucker, 1980	COD to BOD of dairy wastes
Chernigovskii & Kurenkova, 1980	COD to BOD of yeast manufacturing effluent
Crespi-Roswell & Cegarra Sanchez, 1980	COD to BOD of dye waters
Stones, 1980, 1985	COD, nitrogenous oxygen demand, and TOD to BOD
Vtill & Spiegel, 1980	COD, TOC, and BOD interrelationships in industrial waste
Christensen & Nielsen, 1983	COD to BOD of landfill leachates
Evans et al., 1983	COD to BOD of piggery slurry
Earle et al., 1984	COD to BOD for swine waste
Belen'kaya et al., 1986	$KMnO_4$ and dichromate COD results combined in BOD_5 and BOD_{20} of brewery wastes
Belen'kaya et al., 1987	COD to BOD of brewery waste

(*continued*)

TABLE 12.1. Efforts to relate COD and other tests to BOD (*continued*).

Reference	Relation
Matsuoka et al., 1987	COD to BOD of fatty acids
Rao & Nyns, 1987	COD and TOC to BOD
Rein & Bello, 1988	COD to BOD of beetsugar wastewater
Timofeeva & Beim, 1989	COD to BOD of kraft mill wastewater
Kemeny & Bannerjee, 1997	COD to BOD in bleached kraft mill
Christian et al., 2017	COD, TOC, and fluorescence spectroscopy to BOD

used as a substrate in performance evaluation blind samples and as a quality control check solution. For a 300-mg/L solution of KHP, the COD is 353 mg O_2/L, the BOD_5 is 240 mg O_2/L, and the TOC is 141 mg/L (Smith, 1999a). Thus, for assessing data from performance evaluation samples that contain KHP as the test chemical, the estimation is useful.

However, the guidance that holds for KHP-based samples can not be reliably transferred to real samples, even though urban legends and published reports abound that COD is equivalent to 20-day BOD and 5-day BOD is equivalent to 0.68 × COD (Charravari & Som, 1950; Ingols & Murray, 1948; Rhame, 1947). In other studies, a TOC/COD ratio was found to be constant at 1:1, whereas BOD/TOD varied from 1:3 to 1:9 (Bochinski, 1973). (As described below, TOD is frequently equal to or slightly greater than COD.) An evaluation of the treated effluent from a paper mill using the kraft process examined COD results that were corrected for lignin content. No relation between corrected COD and BOD was found. The author suggested that biological treatment creates biologically refractive materials that exert a COD (Carpenter, 1966). For landfill leachate, the magnitude of BOD/COD ratios was taken to indicate landfill age and stability (Chian & DeWalle, 1975). Another study found that when BOD is much less than anticipated for a given COD on a landfill leachate, the leachate should be regarded as not easily biodegradable (Christensen & Nielsen, 1983; Popadopoulos et al., 1998). For slaughterhouse wastes, BOD values are moderately correlated to $KMnO_4$ COD, with $r = 0.74$ to 0.84 and wide variations in 95% confidence limits (Dahl, 1969). An examination of a variety of waste streams from a brewery found BOD/COD ratios of 0.42 for wort, 0.67 for spent malt, 0.71 for barley soak, and 0.83 for yeast (Delcommune, 1971). Ultimate BOD was found to be approximately equal to COD for feedlot runoff (McGhee et al., 1976). In kraft pulp mills, a BOD/COD ratio of 0.6 to 0.7 was found (Timofeeva & Beim, 1989). The BOD_5/COD ratio has been used as an index of biodegradability for dye waters. A ratio greater than 0.20 indicates material that is rapidly biodegradable. A ratio between

0.20 and 0.10 indicates probably biodegradable, and a ratio less than 0.10 indicates poorly biodegradable (Crespi-Roswell & Cegarra Sanchez, 1980).

In a study of 14 fatty acids, the BOD/COD ratio for the pure compounds was found to range from a low of 0.019 to a high of 0.36, a considerable measure away from the 0.68 ratio for KHP (Matsuoka et al., 1987). Another study found BOD_5/COD ratios on a variety of effluents to range from less than 0.01 to 0.9 (Riehl, 1951).

In a study of board mill effluents (Hwang et al., 1974), a BOD_5-to-COD correlation was found of the form

$$BOD_5 = 0.0285\ (COD) - 0.73$$

with a correlation coefficient $r = 0.9935$ for one of the waste streams. Three other waste streams in the facility all produced equations of similar form with equally high correlation coefficients; however, the individual constants in the equations are very different.

Part of the substrate that is initially assimilated by bacteria is converted to substances needed for food gathering, growth, and reproduction. A portion of these bacterial products are inert to further biodegradation and reutilization by other bacteria, although the products are for the most part readily oxidized by chemical means used in COD determinations. This phenomenon means that the COD tends to double-count the biodegradability of the effluent as compared with the influent and overestimate what can actually be consumed by bacteria. This results in the widespread observation that facility efficiency computed from COD determinations is generally worse than that computed from BOD tests (Krogh, 1970; Sykes, 1982).

3.0 FRACTIONATION OF CHEMICAL OXYGEN DEMAND

While recognizing that there is no general relation between BOD and COD, still there is great interest in quantitative characterization of the biological oxidation of pollutants in wastewater with the intent of developing predictive models of effluent quality. Unfortunately, effluent quality is most frequently defined by a regulatory agency as some maximum limit of allowable BOD_5; thus, modelers have a need to express final results in terms of discharged BOD_5. At the same time, the BOD test is notorious for having very poor precision; that is, a wide spread of results. Data that are used as inputs to models need to have some degree of consistency rather than being a different value depending on who performed the test on what day of the week. Thus, the data precision that is easily achieved in COD-based chemical procedures is very attractive to those engineers and scientists who model wastewater facilities.

There has been a great amount of effort put into breaking down total COD into smaller categories that can be more closely identified with the biological processes in treatment facilities. The first of these efforts determined the COD before and after an 8-hour aeration and presented it as equivalent to BOD (Hiser & Busch, 1964; Mullis & Schroeder, 1971). Later, the term ΔCOD was defined for this before and after evaluation:

$$\Delta COD = \frac{\text{Total COD}_{\text{init}}}{\text{Total COD}_{\text{Fin}}}$$

where some sort of biological treatment process occurred between the initial and final COD measurements. The ΔCOD was presented as a more precise and easily measured value than BOD_5, while still purporting to have the same information content as BOD_5 (Gaudy & Gaudy, 1972). The idea of ΔCOD is part of every effort and definition since then, although the intervening processes between the initial and final COD measurements have become quite sophisticated.

A breakdown of the contents of an influent is presented in Figure 12.1. This should be compared with a breakdown of the total influent COD presented in Figure 12.2 (Orhon & Cokgor, 1997). These tables are not equivalent, although they contain some common elements.

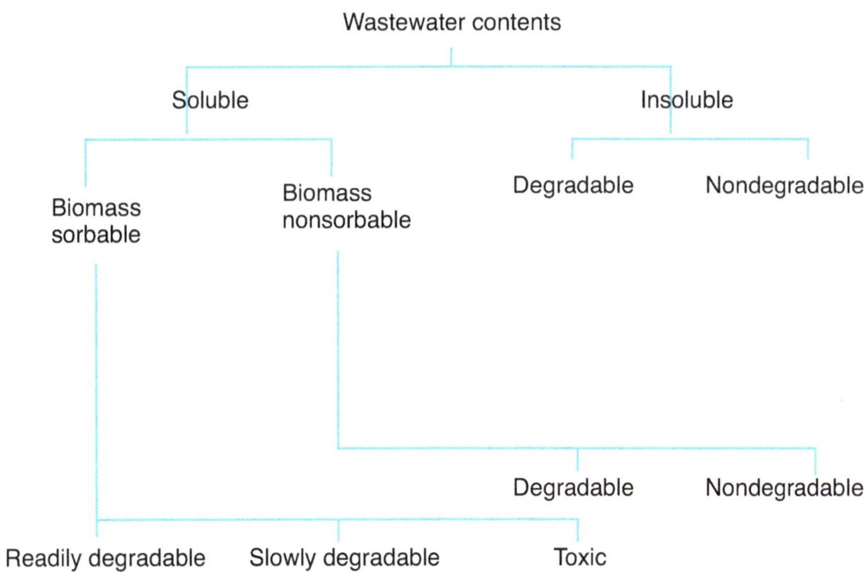

FIGURE 12.1. Breakdown of contents of an influent.

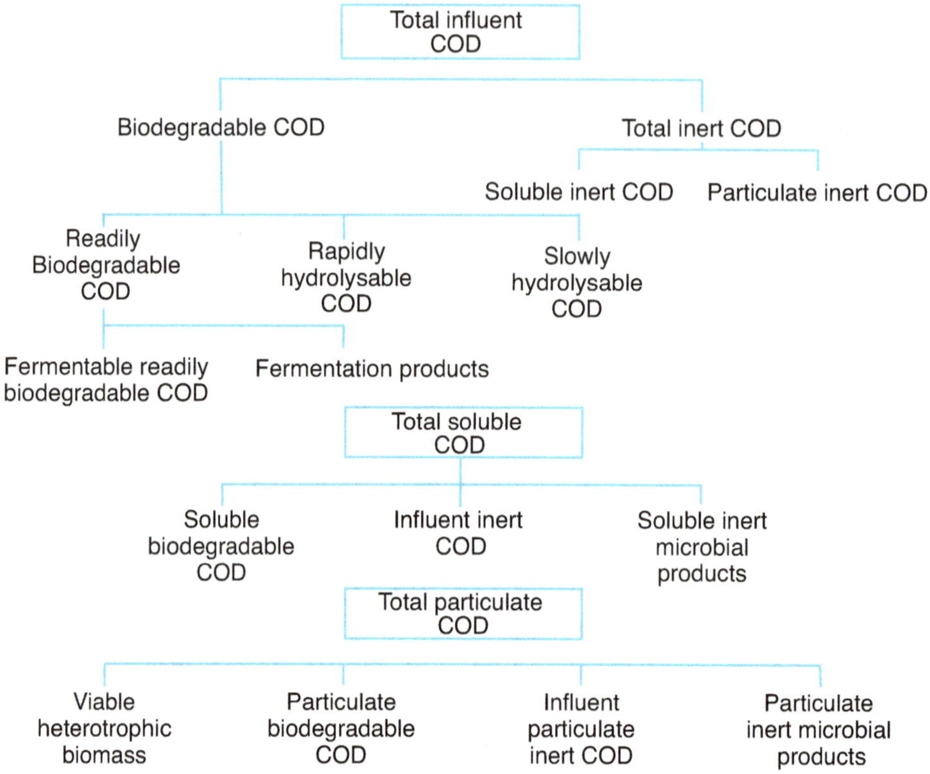

FIGURE 12.2. Breakdown of total influent COD.

3.1 Total Influent Chemical Oxygen Demand

The total influent COD combines the contributions of both the particulate and the soluble materials in the sample that can react as COD. In many interpretations, total influent COD is equated to the organic carbon content of the sample (Orhon & Cokgor, 1997) and does not include consideration of inorganic substances, although in reality, these do contribute to the determination. The COD procedures presented in *Standard Methods* can be performed on samples that contain solids; however, the presence of solids in the sample precludes the use of autoanalyzer and other techniques in which the small-bore tubing that is used in the sample introduction mechanism and throughout the other parts of the instrument can be plugged. Another consideration is that the digestion used must be chemically and physically rigorous so as to be able to include the slowly reacting components of the sample in the completed test.

3.2 Total Residual Chemical Oxygen Demand

Total residual COD is defined as the sum of that portion of the original influent COD that remains after exhaustive biodegradation of the wastewater (inert) plus any residual COD that is produced as a result of the biological treatment. Note that this definition does not say the inert COD is equivalent to the residual COD that is left after biodegradation. Exhaustive biodegradation is commonly regarded as requiring from 10 to 20 days, although in certain cases, it can require a much longer time. This concept of a time limit for biodegradation results from linking the extent of biodegradation to the maximum retention times that can be achieved in facility operations under normal operating conditions. The use of bioreactors with up to 613 days of holding time has demonstrated that up to 90% of what is initially residual COD can eventually be degraded (Gaudy & Blachly, 1985). Thus, the concept of residual COD is more of a kinetic idea, rather than an absolute characteristic.

3.3 Soluble and Particulate Chemical Oxygen Demand

Soluble COD is defined as the part of the total COD that is dissolved in the water. Particulate COD is that part of the total COD that is not dissolved in the water. The reason for the division into soluble and particulate is that the soluble portion is the only form able to pass through the cell wall of the bacteria, allowing utilization by the cells, while the particulate fraction must be broken down into soluble form if it is to be utilized at all. Soluble COD in effluent water is found to comprise 40% to 60% of the total COD with apparent molecular weights, based on membrane fractionation, ranging from less than 165 to greater than 1800, although this upper limit may be an exaggeration. Soluble COD is nominally negatively charged and approximately 95% can be absorbed on activated carbon, indicating a high organic content (Keller et al., 1978). The soluble COD can be frequently correlated with oxygen uptake rates (Huang & Cheng, 1984).

The filtration of samples before BOD and COD determinations is recommended as the general procedure to produce a more reliable indicator of effluent quality (Boutin et al., 1979; Collins & Tenney, 1976; Friedman et al., 1988; Gottsching et al., 1977; Rogers, 1983). Vacuum suction devices used in sampling can increase the proportion of sediments in the sample and skew the resulting values upward unless the samples are filtered (Haider & Haider, 1998). Sedimentation and centrifugation (Friedman et al., 1988; Rickert & Hunter, 1971; Takamatsu et al., 1970) have been used to separate these fractions for determination, although the more commonly employed technique is filtration. The type and nominal pore size of the filter has a large influence on the results (Smith, 1995, 1999a, 1999b). The separation of bacteria from protozoa is achieved with a membrane filter having a 5- to 8-μm pore size (Tezuka, 1968). The frequently used glass fiber filters (Whatman GF/C) and 0.45-μm membrane filters (Ekama et al., 1986; Novak et al., 1995) are designed to retain large floc particles and bacteria, respectively, and do not provide a true division into COD fractions that are immediately usable by the bacteria. Particulates that are slightly smaller than bacteria will pass these devices, and should not be considered as soluble, especially with respect to bacterial assimilation. Researchers will often determine only the total and soluble COD (i.e., filtered), then calculate the particulate COD as the difference between the two measurements.

3.4 Soluble Inert Chemical Oxygen Demand

Soluble inert COD is defined as those soluble substances in the wastewater that pass through the treatment system unaffected by the microorganisms in the system. One technique for soluble inert COD includes filtration of the mixed liquor from reactors with sludge ages from 10 to 20 days, then determination of the COD (Ekama, et al., 1986). Another aerates the filtered mixed liquor from a reactor with sludge age greater than 10 days before determination of the COD (Henze et al., 1987). Problems identified with these approaches include consideration that some of the inert substances are initially in the influent; however, others can be biochemically manufactured by the organisms (Boero et al., 1991; Chudoba, 1985; Daigger & Grady, 1977; Germirli et al., 1991; Kappeler & Gujer, 1992; Orhon et al., 1994; Sykes, 1982), and that the approaches actually measure "soluble" residual COD. One treatment is to express the amount of microbial waste as a derivative fraction of the microbial mass and cellular growth rate (Daigger & Grady, 1977). Measurement techniques have been developed for differentiating the two components of residual COD. One technique to differentiate initial from microbially produced inert COD is to perform batch aeration experiments on the waste itself and additional samples of the waste spiked with known amounts of glucose, a material that has a value of zero for initially inert COD. A regression is prepared of the final inert COD results, with the intercept equal to the initial amount of inert COD in the waste. Another technique performs a batch aeration experiment on the waste and another batch experiment on an equivalent amount of glucose. The residual COD from the glucose is subtracted from the residual COD of the waste to give the initial inert COD of the waste (Germirli et al., 1991). The latest methods establish a reactor with a minimal amount of adapted biological seed and varying amounts of filtered wastewater, then monitor the COD of aliquots from the reactor over a set period of time or until a steady state is reached. The microbial product portion of the residual is related to the initial amount of biodegradable COD (Ince et al., 1998; Orhon & Cokgor, 1997). Regardless of source, these inert substances contribute to a residual COD that is detected in the effluent from the facility.

3.5 Particulate Inert Chemical Oxygen Demand

Particulate inert COD is defined as those insoluble substances in the wastewater that are unaffected by the microorganisms in the system. They are generally removed from the water as part of the solids-handling mechanisms and do not contribute to residual effluent COD. Methods have been developed for the determination of particulate inert COD (Ekama et al., 1986; Henze et al., 1987; Kappeler & Gujer, 1992; Orhon et al., 1994) and are essentially the same as those used for soluble inert COD, with the exception that the wastewater is not filtered initially. These methods for determination of initial particulate inert COD have been criticized (Orhon et al., 1999b) and a technique using three aerated batch reactors proposed. The first reactor is charged with raw wastewater, the second with 0.45-μm filtered wastewater, and the third with a glucose solution exhibiting the same total COD as the filtered wastewater. The reactors are operated until a constant residual COD is obtained, taking from 20 to 40 days. Using the total initial COD and the total residual COD from the three reactors, it is possible to calculate the initial inert COD fractions, both soluble and particulate, and the microbial contribution to residual COD, expressed as a fraction of the influent degradable COD.

3.6 Biodegradable Chemical Oxygen Demand

Biodegradable COD is the portion of the organic carbon in the waste stream that can be utilized by the microorganisms as a food and energy source. The experimental determination of this fraction harks back to the work of Hiser & Busch (1964), although recent techniques have relied more on respirometry (Sierra-Alvarez et al., 1990; Ubay, 1998; van den Eynde et al., 1984) than dilution bottle BOD methods. The biodegradable fraction is divided into two components, readily and slowly biodegradable, based on an order of magnitude rate difference in utilization (Dold & Marais, 1986; Hao & Li, 1987). The slowly biodegradable soluble COD and BOD are often considerably larger than the rapidly biodegradable fractions, with up to 91% of the total degradable COD being ascribed to the slow component by some (Hao & Li, 1987; Ubay, 1998). Other authors report a much lower percentage. Part of the confusion is the result of the intermediate fate of the rapidly assimilated COD, part of it being quickly converted to excreted bacterial products (utilization-associated products), such as slime and extracellular hydrolysis enzymes, while another part is converted to biomass that can become waste products on cell death or protozoan predation (biomass-associated products) (Ibrahim & Abasaeed, 1995). Portions of these products are biodegradable by other bacteria. The slowly biodegradable component is also known as a hydrolyzable COD fraction.

3.7 Readily Biodegradable Chemical Oxygen Demand

Readily biodegradable COD consists of soluble materials that can be quickly consumed by the microorganisms, and is generally believed to consist of simple carbohydrates and volatile fatty acids. The readily biodegradable fraction is further divided into fermentable readily biodegradable COD and fermentation products (Henze et al., 1994; Kruit et al., 1994). Readily biodegradable COD can be determined from respirometer studies (Cokgor et al., 1998; Ekama et al., 1986). The oxygen uptake between the beginning of the experiment is followed until it either changes rate, indicating the presence of a slowly biodegradable component, or it returns to the endogenous metabolism rate. The readily biodegradable COD (S_{S1}) is calculated as:

$$S_{S1} = \frac{DO_2}{1 - Y_H}$$

where Y_H is the heterotrophic yield, also determined through respirometer studies (Sollfrank & Gujer, 1991). Another technique for determination of readily biodegradable COD (Mamais et al., 1993) adds zinc hydroxide to a sample of wastewater that has been filtered through a 0.45-μm membrane filter. The pH is adjusted to 10.5, then the sample is allowed to settle. Determination of the COD of the supernatant gives the readily biodegradable fraction. The zinc flocculation treatment removes the colloidal portion of the sample that passed through the filter, which is equated with the hydrolyzable portion of total biodegradable COD.

3.8 Hydrolysable Chemical Oxygen Demand

Originally characterized as a slowly biodegradable particulate fraction (Dold & Marais, 1986), this fraction is now regarded as containing both particulate and soluble components (Orhon & Cokgor, 1997). Hydrolysable COD is further broken into two components, the rapidly hydrolysable and slowly hydrolysable COD (Henze, 1992; Nowak et al., 1999; Orhon et al., 1999a). The rapidly hydrolysable portion is regarded as all soluble and can be equated with the slowly biodegradable soluble COO (Ibrahim & Abasaeed, 1995; Novak et al., 1995). A portion of the very slowly hydrolysable COD is identified as part of the heterotrophic biomass that is produced in the reactor (Nowak et al., 1999).

These hydrolysable fractions are normally calculated based on mass balance; however, it is reasonable that they should be determinable from extended oxygen uptake studies. Examination of results from these type of studies has indicated that part of what is nominally identified as the endogenous rate of metabolism actually is attributed to slow extracellular hydrolysis. The slowly hydrolysable COD fraction is thought to comprise or be derived from 59% to 83% of the total influent COD (Nowak et al., 1999; Orhon et al., 1999a; Vanrolleghem et al., 1999).

3.9 Biodegradable Volatile Suspended Solids

Biodegradable volatile suspended solids (BVSS) are used to determine the portion of the biomass in a system that can be further degraded (Balmat, 1957; Bulloch et al., 1996; Droste & Sanchez, 1986). This characterization is used in active biomass studies and models. Suspended solids are those materials that will not pass through a glass-fiber filter or sintered glass disk. The retained solids are then subjected to COD determination (Paulsen, 1965).

3.10 Chemically Hydrolysable Chemical Oxygen Demand

This fraction is added to the list, even though it does not appear in the literature as such, because of the number of studies that have reported the phenomenon. It arises when a raw or treated (stabilized) wastewater is further subjected to chemical oxidation with chlorine, ozone, chlorine dioxide, and so forth. The most common observation is that the chemical treatment breaks down initially inert organic materials, often large natural and synthetic polymers, to smaller fragments that are amenable to biodegradation. This effect has been noted many times after ozonation (Paraskeva et al., 1998).

4.0 TOTAL OXYGEN DEMAND AND ALTERNATE TESTS TO CHEMICAL OXYGEN DEMAND

Related to COD are TOD and ODI tests. The development of the TOD analyzer grew out of the work on TOC determinations (Clifford, 1968; Goldstein et al., 1968). The TOD relies on a high-temperature oxidation of the sample, and measures oxygen use rather than carbon oxidation. The rigor of the oxidation produces values at least as high as conventional dichromate COD and succeeds for many COD-resistant compounds. It has the restriction that particulates cannot be present in the samples. The ODI is a very mild procedure that is designed to chemically generate values that are closer to those of BOD.

In the standard TOD instrument, a water sample is oxidized in a nitrogen–oxygen gas stream at 900 °C over a platinum catalyst. The instrument is designed around two properties of zirconium oxide. The first is the ability of zirconium oxide to transport oxygen when a current is applied across the material. Current control is used to accurately dose a continuous flow of purified nitrogen carrier gas with a preset amount of oxygen, the atmosphere being the source of the oxygen. The second property is the generation of a measurable potential across zirconium oxide when two parallel surfaces of the ceramic are in contact with different concentrations of oxygen. Both of these properties are observable and efficient only at high temperatures. Two cylinders of zirconium oxide ceramic are positioned in airtight series connected by a higher-temperature oxidation furnace, the whole assembly contained in a 600 °C oven. The carrier gas is dosed with oxygen in the first cylinder, then the carrier reacts with the injected sample in the combustion furnace at 900 °C over platinum. The amount of oxygen used in the combustion is measured in the monitoring cell.

The potential in millivolts that exists between the inside surface of the zirconium cylinder and the outside of the monitoring cell, which is in contact with atmospheric oxygen, can be used as the output metric. This metric requires calibration. A metric based on feedback to the dosing cell and measurement of the amount of current needed to maintain a constant potential in the monitoring cell gives an absolute readout that does not require calibration.

The test requires 2 minutes and gives values approximately 10% higher than standard wet COD for most samples (Voorn, 1977; Voorn & Lee-Frampton, 1976; Voorn & Mitchell, 1976). It is not affected by chloride presence (Butzelaar & Hoogeveen, 1975). Results from the COD and TOD procedures have been compared many times (Davis, 1971; Garcia et al., 1977; Kita, 1980; Lueck et al., 1981; McKeown et al., 1972; Schulze, 1976; Stones, 1980; Voorn & Marlowe, 1977). Commercial instruments for measurement of TOD are available from a number of manufacturers.

An alternate TOD technique has been described. The sample is reduced in an oxygen-free atmosphere in an 875 °C platinum-catalyzed combustion tube, then reacted with carbon dioxide carrier gas. The oxygen deficiency in the reduced sample is satisfied by removal of oxygen from the carbon dioxide to form carbon monoxide. The amount of carbon monoxide is measured by infrared. The result has been described as a 2-minute automated vapor phase redox COD (Stenger & van Hall, 1967, 1968).

A combined TOD–TOC analyzer has been described. Using a 200-ppm oxygen-dosed nitrogen carrier gas, the sample is combusted in a 900 °C furnace over platinum gauze pellets. The carbon dioxide produced from the oxidation is then passed through a nondirect infrared cell for TOC, followed by an in-series oxygen detector for the TOD measurement. Inorganic carbon is determined from a valve-controlled parallel system that passes the sample through a 150 °C furnace over phosphoric acid-wetted, quartz chip packing (Nakajima et al., 1982).

Oxygen demand index consists of performing a COD with mild heating at 100 °C for 20 to 60 minutes, then comparing the absorbance or titration result to the COD of a readily oxidized material such as glucose run at the same time under the same conditions (Westerhold, 1965a, 1965b, cited in Young, 1977). The numerous studies comparing ODI results to COD and BOD_5 results have revealed that ODI can be related to COD with a fair degree of confidence; however, the correlation to BOD_5 is much less reliable (Bochinski, 1973; Chen & Wang, 1983; Hayes et al., 1976; Reynolds and Goellner, 1974; Shriver & Young, 1972).

There have been attempts to provide surrogate measurements that can be used to predict COD, especially for use in continuous online monitoring situations. The UV 254-nm absorbance of samples has been related to COD (Mrkva, 1983). Correlations ranged from 0.766 to 0.989, with most of the variation being exhibited by low-level samples. Turbidity-free samples give consistent correlations, whereas turbid samples are not determinable (Yeun & Viraraghavan, 1987). Online multiwavelength fluorometry tests that probe the extent of protein hydrolysis of the sample have been correlated to COD (Tartakovsky et al., 1996). The tests consist of excitation at 290 and 400 nm, with recording emission at 340 and 530 nm, respectively. These efforts have demonstrated that correlations can be developed for effluents that are very qualitatively and quantitatively consistent, but general correlations are lacking.

Ziglio et al. (2001) have described a respirometric technique for the characterization of readily biodegradable COD. The method takes only 30 minutes per sample after calibration. A microbial fuel cell with a bacteria-covered anode is reported to have current area related to COD when coupled with a neural network (Feng & Harper, 2013).

5.0 CONCLUSION

While BOD is by far the more commonly monitored parameter listed in effluent permits, COD is vastly preferred as a tool by researchers and modelers for characterizing the strength of a waste. Part of the attraction of COD is the ease and reproducibility of the measurement along with the reduced laboratory time needed for obtaining results. Commercially available kits for COD determinations are widely used, adding to its popularity. These qualities have led to experimental designs that use the COD test to produce results that mimic those of the BOD test.

But COD is not without its own set of fundamentally undesirable qualities, the foremost being the generation of hazardous wastes in the most common techniques. There are modified COD procedures available that minimize waste generation; however, the long-term solution lies not with the COD test. Instead, instrumental procedures that avoid waste generation, such as total oxygen demand and total organic carbon analyzers, hold the most promise for the future, while not sacrificing any of the good qualities of the COD test.

6.0 REFERENCES

Abbott, W. E. (1927). Estimation of organic matter in sewage and effluents. *Ind. Eng. Chem.*, *19*, 919.

Adelman, M. H. (1968). Automated measurement of COD. *Water Wastes Eng.*, *5*, 52.

Adeny, W. E., & Dawson, R. B. (1926). The determination of organic matter in water by potassium dichromate and sulfuric acid. *Sci. Proc. Royal Dublin Soc.*, *18*, 199.

American Public Health Association, American Water Works Association, and Federation of Sewage and Industrial Wastes Association. (1955). *Standard methods for the examination of water and wastewater* (10th ed.). American Public Health Association.

American Public Health Association, American Water Works Association, & Water Environment Federation. (1998). *Standard methods for the examination of water and wastewater* (20th ed.). American Public Health Association.

Appleton, J. M. H., Tyson, J., & Mounce, R. P. (1986). The rapid determination of COD in wastewaters and effluents by flow injection analysis. *Anal. Chim. Acta* (Netherlands), *179*, 269.

Aslam, S., & Walker, O. L. (1982). Recycling of mercury and silver from COD Tests. *J. Water Pollut. Control Fed.*, *54*, 1148.

Balconi, M. L., Borgarello, M., Ferraroli, R., & Ralini, F. (1992). COD determination in well and river waters by flow-injection analysis using a microwave oven during the oxidation step. *Anal. Chim. Acta* (Netherlands), *261*, 295.

Ballinger, D., et al. (1982). Determination of COD of wastewaters without use of mercury salts. *Analyst*, *107*, 1047.

Balmat, J. L. (1957). Biochemical oxidation of various particulate fractions of sewage. *Sew. Ind. Wastes*, *29*, 757; *Pub. Health Eng. Abstr.*, *38*(1–2), 15 (1958).

Batek, A. (1914). Determination of oxygen consumed for water and sewage by the Kubel-Tiemann method. *Chem. Ztg.*, *38*, 1129; *Chemical Abstracts*, *9*, 678 (1915).

Baughman, G. L., Butler, B. T., & Sanders, W. M., III. (1969). Automatic potentiometric titration of low COD. *Water Sew. Works*, *116*, 359.

Baumann, F. L. (1974). Dichromate reflux COD. A proposed method for chloride correction in highly wastes. *Anal. Chem.*, *46*, 1336.

Belen'kaya, S. L., et al. (1987). Rapid method for the determination of BOD in wastewaters of breweries. *Ferment. Spirt. Prom-st.*, *4*, 13; *Chemical Abstracts*, *107*, 241991 (1987).

Belen'kaya, S. L., Popova, Y. I., & Kalitvyanskaya, V. I. (1986). Standardized method for the determination of pollutants in wastewater from breweries. *Pishch. Prom-st. 1*, 41; *Chemical Abstracts*, *105*, 11667 (1986).

Benson, H. K., & Hicks, J. F. G., Jr. (1931). Proposed modification of oxygen consumed method for determination of sea water pollution. *Ind. Eng. Chem., Anal. Ed.*, *3*, 30.

Berk, A., & Hofmann, P. (1963). Estimation of DO and determination of permanganate oxidizability in waters. *Chem. Prumysl.*, *13*, 287; *Chemical Abstracts*, *59*, 9675 (1963).

Berka, A., & Zavesky, Z. (1969). Determination of organic compounds by permanganate oxidation: Oxidizability of phenolic waters. *Chem. Listy*. *63*, 1148.

Berka, A., & Zavesky, Z. (1974a). Determination of organic substances by oxidation with potassium permanganate. XV. *Mikrochim. Acta*, 135.

Berka, A., & Zavesky, Z. (1974b). Determination of organic substances by oxidation with potassium permanganate. XVI. *Mikrochim. Acta*, 493.

Bevan, D., & Stevenson, J. C. (1969). Modified permanganate value test for the estimation of carbonaceous substances in sugar mill wastes. *Proc. Queensland Soc. Sugar Cane Technol.*, *36*, 317.

Bezel, L. I. (1945). Application of cerium sulfate for the determination of oxidizable material in water. *J. Appl. Chem. USSR*, *18*, 361; *Chemical Abstracts*, *40*, 3548 (1946).

Bochinski, J. (1973). Measuring oxygen demand: Take your pick: TC, BOD, COD, TOC, and TOD. *Pollut. Eng.*, *5*(1), 45.

Boero, G. L., Eckenfelder, W. W., & Bowers, A. R. (1991). Soluble microbial product formation in biological systems. *Water Sci. Technol.*, *23*, 1067.

Boutin, P., et al. (1979). Parameters for monitoring processed effluents. *Tech. Eau Assainissement* (Belgium), 9, 390; *Chemical Abstracts*, *91*, 216211 (1979).

Boyles, W. (1997a). Manganese III COD. Technical Paper 17057, Hach Company.

Boyles, W. (1997b). The science of chemical oxygen demand. *Tech. Inf. Ser.*, Booklet No. 9, Hach Company.

Branch, C. T. (1969). Correlation between BOD, COD, and TOC in paper mill wastes. M.S. thesis, Mississippi State University; *Abstr. Bull. Inst. Paper Chem.*, *42*, 11409 (1972).

Bucker, J. (1980). Effluent purification plants for dairies: Flow metering-sampling-testing. *Molkereitechnik*, 48.

Bulloch, C., Bicho, P., & Zhang, Y. (1996). Solid COD method for determination of biomass in waste waters. *Water Res.*, *30*, 1280.

Burns, E. R., & Marshall, C. (1965). Correction for chloride interference in the COD Test. *J. Water Pollut. Control Fed.*, *37*, 1716.

Burtle, J., & Buswell, A. M. (1935). A comparison of permanganate oxygen demand, BOD, and direct absorption of oxygen. *Sew. Works J.*, *7*, 839.

Burtle, J., & Buswell, A. M. (1937). Oxygen demand studies. *Sew. Works J.*, *9*, 224.

Butzelaar, P. F., & Hoogeveen, L. P. J. (1975). Measuring the oxygen demand of water: A new method. *Water Serv.*, *79*, 984.

Cameron, W. M. (1944). Notes on oxygen absorbed tests. *Inst. Sew. Purif. J.*, 204.

Camps, J. M., & Arias, E. (1963). Determination of organic carbon in waters of Castellon de la Plana. *Invest. Pesquera*, *23*, 125; *Chemical Abstracts*, *59*, 13686 (1963).

Canelli, E., Mitchell, D. G., & Paine, R.W. (1976). An improved determination of COD in water and wastes by a simplified acid dichromate digestion. *Water Res.*, *10*, 351.

Carpenter, W. L. (1966). COD and BOD relationships of raw and biologically treated kraft mill effluents. *Natl. Council Air Stream Improvement Tech Bull.*, *193*.

Chamot, E. M. (1908). The determination of oxygen consumed in waters. *Am. J. Public Hyg.*, *18*; *Chemical Abstracts*, *2*, 2966 (1908).

Charravari, S. C., & Som, A. B. (1950). Comparative studies of BOD values of sewage with that obtained by chemical oxidation. *J. Indian Chem. Soc., Ind. News Ed.*, *13*, 171; *Chemical Abstracts*, *45*, 6327 (1951).

Chaudhuri, N., Nyogi, S., De., A., & Basu, A. (1973). A colorimetric technique suggested for COD determination. *J. Water Pollut. Control. Fed.*, *45*, 537.

Chelsley, K. G., White, R. W., & Hogg, R. V. (1951). Correlation of BOD and chemical tests on industrial effluents. *Tappi J.*, *34*, 86.

Chen, T., & Wang, X. (1983). Discussion of methods for determining oxygen demand index (ODI). *Huanjing Kexue*, *4*, 60; *Chemical Abstracts*, *99*, 10523 (1983).

Chernigovskii, N. N., & Kurenkova, I. (1980). Monitoring the waste waters of yeast works. *Khlebopek. Konditer. Prom-st.*, *5*, 44; *Chemical Abstracts*, *94*, 52279 (1981).

Chian, E. S. K., & DeWalle, F. B. (1975). Compilation of methodology for measuring pollution parameters of landfill leachate. EPA-600/3-75-011, U.S. Environmental Protection Agency.

Christensen, P. H., & Nielsen, C. W. (1983). Leaching from land disposal municipal composts: 1. Organic matter. *Waste Manage. Res.*, *1*(1), 83.

Christian, E., Batista, J. R., Gerrity, D. (2017). Use of COD, TOC, and fluorescence spectroscopy to estimate BOD in wastewater. *Water Environ. Res.*, *89*, 168–77.

Chudoba, J. (1985). Quantitative estimation in COD units of refractory organic compounds produced by activated sludge microorganisms. *Water Res.*, *19*, 37.

Chudoba, J., & Dalesicky, J. (1973). COD of some nitrogenous heterocyclic compounds. *Water Res.*, *7*, 663.

Clifford, D. A. (1968). Automatic measurement of total oxygen demand. *Proc. 23rd Ind. Waste Conf., Purdue University*; *Chemical Abstracts*, *71*, 94541 (1969).

Cokgor, E. U., Sozen, S., Orhon, D., & Henze, M. (1998). Respirometric analysis of activated sludge behavior. I. Assessment of the readily biodegradable substrate. *Water Res.*, *32*, 461.

Collins, J., & Tenney, R. S. (1976). Fishery waste effluents: A method to determine relationships between COD and residue. *Fish. Bull.*, *74*, 725.

Cooper, E. A., & Read, W. H. (1927). The chemical oxidation of the constituents of sewage: The Action of hydrogen peroxide. *J. Soc. Chem. Ind.*, *46*, 155T.

Cotton, F. A., Wilkinson, G., Murillo, C. A., & Bachmann, M. (1999). *Advanced inorganic chemistry* (6th ed.). Wiley & Sons.

Crespi-Rosell, M., & Cegarra Sanchez, J. (1980). Purification of the waste waters coming from dyeing with carriers. *Biol. Inst. Invest. Text. Coop. Ind., 77*, 41; *Chemical Abstracts, 94*, 180046 (1981).

Cripps, J. M., & Jenkins, D. (1964). A COD method suitable for the analysis of highly saline waters. *J. Water Pollut. Control Fed., 36*, 1240.

Cronholm, M., & Winbladh, E. (1972). Control of COD of wastewater with the Axel Johnson Institute KS-meter. *Vatten, 28*, 367; *Chem Abstr., 78*, 101765 (1973).

Dahl, O. (1969). Determining organic constituents in slaughterhouse effluents. *Fleischwirtschaft, 49*, 1064.

Daigger, G. T., & Grady, C. P. L., Jr. (1977). A model for the bio-oxidation process based on product formation concepts. *Water Res., 11*, 1049.

Dan, D. Z., Dou, F. L., Xiu, D. J., & Qin, Y. Q. (2000). Chemical oxygen demand determination in environmental waters by mixed-acid digestion and single sweep polarography. *Anal. Chim. Acta, 420*, 1, 39.

Dasgupta, P. K., & Peterson, K. (1990). Kinetic approach to measurement of COD with an automated micro batch analyzer. *Anal. Chem., 62*, 395.

Davenport, R. J., & Wynveen, R. A. (1977). Fundamental studies for development of electrochemical COD and TOC analyzers. *Army Med. Res. Dev. Command*, 17-76-C-6077.

Davis, E. M. (1971). BOD vs. COD vs. TOC vs. TOD. *Water Wastes Eng. 8*(2), 32.

De Casseres, K., et al. (1984). Determination of COD by a mercury-free open tube method. *Water Pollut. Control, 83*, 416.

Delcommune, T. (1971). Pollution of malt factory and brewery waste waters. *Bull. Assoc. Roy. Anciens Etud. Brass. Univ. Louvain, 67*, 45; *Chemical Abstracts, 75*, 121148 (1971).

Dobbs, R. A., & Williams, R. T. (1963). Elimination of chloride interference in the COD test. *Anal. Chem., 35*, 1064.

Dold, P. L., & Marais, G. v. R. (1986). Evaluation of the general activated sludge model proposed by the IAWPRC Task Group. *Water Sci. Technol., 18*, 6, 63.

Douglas, T. B. (1978). A test and technique for predicting BOD_5 values of kraft mill waste. *Tappi J., 61*, 4, 53.

Droste, R. L., & Sanchez, W. A. (1986). Modeling active mass in aerobic sludge digestions. *Biotechnol. Bioeng., 28*, 1699.

Drown, T. M. (1890). The chemical examination of waters and the interpretation of analysis. I. Examination of water supplies. *Examinations of Water Supplies by the Mass. State Board of Health, 1887–1890*, 519.

Drown, T. M., & Hazen, A. (1890). A report of the chemical work done at the Lawrence Experiment Station. II. Purification of water and sewage. *Examinations of Water Supplies by the Mass. State Board of Health, 1887–1890*, 707.

Dupre, A. (1885). Some observations on the permanganate test in water analysis. *Analyst*, *10*, 118.

Dzyadzio, A. M. (1938). The true oxygen consumption of sewage and a method for its determination. *Vodosnabzh. Sanit. Tekh.* (Russia), *8*, 117.

Earle, J. F. K., et al. (1984). Role of purple sulfur bacteria in swine waste reclamation. *Agric. Wastes*, *10*, 297.

Edwards. S. J., & Allen, M. (1984). Rapid and precise method for the determination of the COD of factory effluent samples. *Analyst*, *109*, 671.

Ekama, G. A., Dold, P. L., & Marais, G. v. R. (1986). Procedures for determining influent COD fractions and the maximum specific growth rate of heterotrophs in activated sludge systems. *Water Sci. Technol.*, *18*(6), 91.

El-Dib, M. A., & Ramadan, F. M. (1966). Dichromate versus ceric sulphate in COD determinations. *J. Sanit. Eng. Div., Proc. Am. Soc. Civ. Eng.*, *92*, 97.

Ettinger, M. B. (1956). Evaluation of methods currently used to determine BOD, COD, and odor for industrial wastes. *Sew. Ind. Wastes*, *28*, 1116.

Evans, M. R., et al. (1983). The effect of temperature and residence time on aerobic treatment of piggery slurry: Degradation of carbonaceous compounds. *Agric. Wastes*, *5*(1), 25.

Feng, Y., & Harper, W. F. (2013). Precise water quality monitoring with microbial fuel cells and artificial neural networks. *Proceedings of the Water Environment Federation*. WEF.

Flaschka, H. A., Barnard, A. J., & Sturrock, P. E. (1980). *Quantitative Analytical Chemistry* (2nd ed.). Willard Grant Press.

Fleet, B., Ho, A. Y. W., & Tenygl, J. (1972a). Automated determination of permanganate and dichromate using a porous catalytic oxygen probe. *Anal. Chem.*, *44*, 2156.

Fleet, B., Ho, A. Y. W., & Tenygl, J. (1972b). A fully automated method for determination of COD. *Analyst*, *97*, 321.

Forster, A. S., & Gad, G. (1938). The BOD and the potassium permanganate consumption of organic compounds. *Kleine Mitt. Mitglied. Ver. Wasser-, Boden-, u. Lufthyg.*, *14*, 214.

Foulds, J. M., & Lundsford, J. V. (1968). An analysis of the COD method. *Water Sew. Works*, *115*, 112.

Fowler, D. E., et al. (1974). A study of the relationship between residual soda and water extractable components of vacuum drum washed kraft pulp and of repulped corregated container effluent characteristics. *Natl. Council Air Stream Improvement Tech. Bull., 277.*

Fowler, G. J. (1902). *Sewage works analysis.* Wiley & Sons.

Fox, C. B. (1886). *Sanitary examinations of water, air, and food.* J. and A. Churchill.

Fox, C. B. (1903). The composition of sewage in relation to problems of disposal. *Tech. Quart., 16*, 132.

Friedman, A. A., Smith, J. E., Desantis, J., Ptak, T., & Ganley, R. C. (1988). Characterization of residues from wet air oxidation of anaerobic sludges. *J. Water Pollut. Control Fed., 60*, 1971.

Garcia, M. D., Caminos Muruzabal, R. M., & Garcia Vellejo, M. C. (1977). Control of COD in pulp and paper mill effluents. *Invest. Tec. Papel, 14*(53), 643; *Abstr. Bull. Inst. Paper Chem., 48*, 9348 (1978).

Gaudy, A. F., Jr., & Blachly, T. R. (1985). A study on the biodegradability of residual COD. *J. Water Pollut. Control Fed., 57*, 332.

Gaudy, A. F., Jr., & Gaudy, E. T. (1972). COD gets nod over BOD test. *Ind. Water Eng., 9*(5), 32.

Gaudy, A. F., Jr., & Ramanathan, M. (1964). A colorimetric method for determining COD. *J. Water Pollut. Control Fed., 36*, 1479.

Geisler, C., et al. (1974). New COD analysis arrives. *Water Wastes Eng., 11*, 26.

Germirli, F., Orhon, D., & Artan, D. (1991). Assessment of the initial inert soluble COD in industrial wastewaters. *Water Sci. Technol., 23*, 1077.

Gilmour, C. M., Merryfield, F., Burgess, F., Punkerson, L., & Carswell, K. (1960). Persulfate oxidizable carbon and BOD as a measure of organic pollution in water. *Purdue Univ. Eng. Bull. Ext. Ser., 106*, 143; *Chemical Abstracts, 55*, 26323 (1961).

Goldstein, A. L., Katz, W. E., Meller, F. H., & Murdock, D. M. (1968). *Total oxygen demand: A new automatic instrumental method for measuring pollution and loading on oxidation process.* Ionics Corporation.

Golterman, H. L., & Wisselo, A. G. (1981). Ceriometry, a combined method for COD and DO (with a discussion on the precision of the Winkler technique). *Hydrobiologia, 77*, 37.

Gonzalez, J. F. (1986). Semi-micro determination of COD on fish filleting wastewater. *Environ. Technol. Lett., 7*, 269.

Goodwin, A. E., Cabbiness, D. E., & Mottola, H. A. (1977). Some rate considerations on COD determinations. *Water, Air, Soil Pollut.*, *8*, 468.

Goto, M., et al. (1981). Continuous monitoring system of COD at micro scale based on electrochemical measurement. *Bunseki Kagaku*, *30*, 403; *Chemical Abstracts*, *95*, 120406 (1981).

Gottsching, L., et al. (1977). Determination of COD, BOD, & TOC in the filtrate from suspensions of pulp, paper, and wastepaper. *Papier (Darmstadt)* (Germany), *31*(1), 1.

Gutiérrez-Capitán, M., Baldi, A., Gómez, R., García, V., Jiménez-Jorquera, C., Fernández-Sánchez, C. (2015). Electrochemical nanocomposite-derived sensor for the analysis of chemical oxygen demand in urban wastewaters. *Anal. Chem.*, *87*(4), 2152-60.

Guway, A. J., Buckland, H., Hawkes, F. R., & Hawkes, D. L. (1998). Active biomass in activated sludge: Comparison of respirometry with catalase activity measured using an on-line monitor. *Water Res.*, *32*, 3705.

Guway, A. J., Farley, L. A., Cunnah, P., Hawkes, F. R., Chase, M., & Buckland, H. (1999). An automated instrument for monitoring oxygen demand in polluted waters. *Water Res.*, *33*, 3142.

Haapala, K., & Starck, B. (1986). COD_{Cr-} results obtained by the closed tube method and the standard procedure are in agreement. *Vatten* (Sweden), *42*, 311.

Haider, R., & Haider, S. (1998). Overestimated COD values for raw wastewaters extracted by vacuum sampling devices due to sedimentation effects. *Water Sci. Technol.*, *37*(12), 293.

Hao, O. J., & Li, C. T. (1987). Effect of slowly biodegradable organics on kinetic coefficients. *J. Environ. Eng.*, *113*, 205.

Harkness, N., Hey, A. E, & Willetts, D. G. (1972). Automatic methods of analysis, instrumentation, and mechanical aids in the laboratory. *Water Pollut. Control*, *71*, 261.

Haupt, H. (1935). The permanganate number in the judging of wastes from pulp factories. *Vom Wasser* (Germany), *10*, 60; *Chemical Abstracts*, *31*, 8775 (1937).

Hayes, E. R., Munroe, V. G., & Spurr, J. (1976). Using ODI, COD, and BOD tests to characterize kraft mill effluents. *Water Pollut. Control*, *114*(9), 22.

Hejzlar, J., & Kopacek, J. (1990). Determination of low COD values in water by the dichromate semi-micro method. *Analyst*, *115*, 1463.

Henze, M. (1992). Characterization of wastewater for modelling of activated sludge process. *Water Sci. Technol.*, *25*(6), 1.

Henze, M., Grady, C. P. L., Jr., Gujer, W., Marais, G. v. R., & Matsuo, T. (1987). Activated Sludge Model No. 1. *Sci. Tech. Rep. No. 1, Int. Assoc. Water Pollut. Res. Control.*

Henze, M., Gujer, W., Mino, T., Matsuo, T., Wenzel, M., & Marais, G. v. R. (1994). Activated Sludge Model No. 2. *Sci. Tech. Rep. No. 3, Int. Assoc. Water Pollut. Res. Control.*

Hey, A. E., Green, A., & Harkness, N. (1969). Automated system for the determination of COD. *Water Res.*, *3*, 873.

Himebaugh, R. R., & Smith, M. J. (1979). Semi-micro tube method for COD. *Anal. Chem.*, *51*, 1085.

Hiser, L. L., & Busch, A. W. (1964). An 8-hour BOD test using mass culture aeration and COD. *J. Water Pollut. Control Fed.*, *36*, 505.

Holm, T. R. (1996). Treatment of spent COD solutions for safe disposal. *Illinois State Water Survey.*

Hoogeveen, L. P. J., & Butzelaar, P. F. (1972). Detector for measuring the COD. *German Offen.* (patent) 2220976; *Chemical Abstracts*, *78*, 33763 (1973).

Hoover, C. R. (1934). Comparison tests of oxygen consumed methods. *Fifth Biennial Report, Connecticut State Water Comm.*, 49.

Hornke, R. (1974). Informational capacity and interpretation of wastewater indices, particularly potassium permanganate, COD, and BOD. *Wochenbl. Papierfabr.*, *102*(6), 199; *Abstr. Bull. Inst. Paper Chem.*, *45*, 6080 (1975).

Huang, J. Y. C., & Cheng, M. D. (1984). Measurement and new applications of oxygen uptake rates in activated sludge processes. *J. Water Pollut. Control Fed.*, *56*, 259.

Hwang, C. P., Good, G., & Davis, E. (1974). Technique cuts BOD and TOC calculations to 5 minutes. *Water Pollut. Control*, *112*(3), 28.

Ibrahim, G., & Abasaeed, A. E. (1995). Modeling of sequencing batch reactors. *Water Res.*, *29*, 1761.

Ince, O., Babuna, F. G., Kassapgil, B., & Anderson, G. K. (1998). Experimental determination of the inert soluble COD fraction of a brewery wastewater under anaerobic conditions. *Environ. Technol.*, *19*, 437.

Ingols, R. S., & Murray, P. E. (1948). An oxygen consumed test for sewage. *Water Sew. Works*, *95*, 113.

Ishcherikova, G. A., & Drozdova, N. P. (1973). Rapid determination of COD. *Gidroliz. Lesokhim. Promst.* (Russia), *6*, 25; *Abstr. Bull. Inst. Paper Chem.*, *44*, 8510 (1974).

Ishcherikova, G. A., et al. (1975). Rapid determination of the COD of effluents. *Gidroliz. Lesokhim. Promst.* (Russia), *7*, 25; *Abstr. Bull. Inst. Paper Chem.*, *46*, 10326 (1976).

Ito, S., & Yamaguchi, T. (1980). Correlation between sample volume and titrant volume in COD determination by acidic permanganate method at 100C. *Bunseki Kagaku Shimpo Sosetsu*, *29*, 27; *Chemical Abstracts*, *93*, 53553 (1980).

Jamieson, R. G., & Stewart, R. G. (1979). Part III: The relation between the components of wool-scour liquor and its BOD_5 and COD. *J. Text. Inst.*, *22*, 253.

Janardan, K. G., & Schaffer, D. (1975). Oxygen demand parameters: Correlation of BOD_5 with COD. *Water Sew. Works*, *122*(10), 62.

Japanese Industrial Standard Committee. (1991). *Testing Methods for Industrial Wastewater*. Method JIS K0101.

Jardim, W. F., & Rohwedder, J. J. R. (1989). COD using microwave digestion. *Water Res.*, *23*, 1069.

Jenkins, S. H., Harkness, N., Hewitt, P. J., Snaddon, X. V. M., Ellerker, R., Divito, B., & Dee, H. J. (1965). Some analytical methods used in examination of sewage and trade wastes *J. Inst. Sew. Purif.*, 533.

Jeris, J. S. (1967). A rapid COD test. *Water Wastes Eng.*, *4*(5), 89.

Jirka, A. M., & Carter, M. J. (1975). Micro semi-automated analysis of surface and wastewaters for COD. *Anal. Chem.*, *47*, 1397.

Johnson, D. W., & Halvorson, H. O. (1941). A chemical method for the determination of BOD. *J. Bacteriol.*, *42*, 145; *J. Am. Water Works Assoc.*, *34*, 956 (1942).

Jones, B. M., Sakaj, R. H., & Daughton, C. G. (1985). Comparison of microcolorimetric and macrotitrimetric methods for COD of oil shale wastewaters. *Anal. Chem.*, *57*, 2334.

Jones, R. H., & Dageforde, A. F. (1968). Application of a high sensitivity total organic carbon analyzer. *Proc. Instr. Soc. Am.*, *6*, 43.

Kappeler, J., & Gujer, W. (1992). Estimation of kinetic parameters of heterotrophic biomass under aerobic conditions and characterization of wastewater for activated sludge modelling. *Water Sci. Technol.*, *25*(6), 125.

Kaufman, S. (1954). The BOD–COD relationship in a treated industrial effluent. *Water Sew. Works*, *10*, 362.

Keller, J. V., Leckie, J. O., & McCarty, P. L. (1978). Investigation of soluble organic nitrogen compounds in municipal secondary effluent. *J. Water Pollut. Control Fed.*, *50*, 2522.

Kemeny, T. E., & Bannerjee, S. (1997). Relationships among effluent constituents in bleached kraft pulp mills. *Water Res.*, *31*, 1589.

Kim, B. R. (1989). Effect of ammonia on COD analysis. *J. Water Pollut. Control Fed.*, *61*, 614.

Kim, Y.-C., Lee, K.-H., Sasaki, S., Hashimoto, K., Ikebukuro, K., & Karube, I. (2000). Photocatalytic sensor for COD determination based on oxygen electrode. *Anal. Chem.*, *72*, 3379.

Kincannon, D. F., & Gaudy, A. F., Jr. (1966). Some effects of high salt concentrations on activated sludge. *J. Water Pollut. Control Fed.*, *38*, 1148.

Kita, A. (1980). Regulation of pollutants in total amount of effluent, and use of TOD automatic analyzer. *Yuasa Jiho*, *49*, 61; *Chemical Abstracts*, *93*, 155352.

Klein, H. (1985). The COD cuvette test: Area of usage and proof of confidence. *WLB, Wasser Luft. Betr.*, *29*, 16; *Chemical Abstracts*, *103*, 109661 (1985).

Klein, L. J. (1941). The strength of sewage: some comparative results. *Proc. Inst. Sew. Purif.*, 174.

Klingenberg, J. J., & Reed, K. R. (1967). *Introduction to Quantitative Chemistry.* Reinhold, 304.

Korenaga, T. (1980). Apparatus for measuring COD based on flow injection analysis. *Bunseki Kagaku Shimpo Sosetsu*, *29*, 222.

Korenaga, T. (1982). The continuous determination of filtered COD with potassium dichromate by means of flow injection analysis. *Bull. Chem. Soc. Japan*, *55*, 1033.

Korenaga, T., & Ikatsu, H. (1981). Continuous flow injection analysis of aqueous environmental samples for COD. *Analyst*, *106*, 653.

Korenaga, T., & Ikatsu, H. (1982). The determination of COD in wastewaters with dichromate by flow injection analysis. *Anal. Chim. Acta*, *141*, 301.

Korenaga, T., et al. (1980). Semiautomated determination of COD in wastewater samples. *Mem. Sch. Eng., Okayama University*, *14*(2), 119; *Chemical Abstracts*, *93*, 191459 (1980).

Korenaga, T., et al. (1984). Evaluation of three flow injection analysis methods for the determination of COD. *Mem. Sch. Eng., Okayama University*, *19*, 53.

Korenaga, T., Ikatsu, H., & Moriwake, T. (1982). The oxidation behavior of various organic compounds on the determination of COD by means of flow injection analysis with acidic permanganate. *Bull. Chem. Soc. Japan*, *55*, 2622.

Koshkin, M. L., & Karasik, R. M. (1940). An improved method for determining the oxygen consumption of a potable water. *Med. Exptl.*, *4*, 9; *Chemical Abstracts*, *36*, 4241 (1942).

Krogh, O. (1970). Analysis of organic materials in slaughterhouse sewage water. *Vatten*, *26*, 1, 9; *Chemical Abstracts*, *76*, 62953 (1972).

Kruit, J., Boley, F., Jacobs, L. J. A. M., & Nouda, T. W. N. (1994). Prediction of the O_2 conditions in the selector. *Water Sci. Technol.*, *29*(7), 229.

Krunchak, V. G., et al. (1984). Redoxometric method of monitoring the biological treatment of wastewater from antibiotics production. *Khim. Tekhnol.* (Russia). *Vody.*, *6*(3), 242; *Chemical Abstracts*, *101*, 136315 (1984).

Kumar, S., Shukla, H., & Viswanathan, L. (1986). A colorimetric method for the determination of COD. *Int. Sugar J.*, *88*, 76; *Chemical Abstracts*, *105*, 11686 (1986).

Larsson, S. A., Ostman, B. A.-L., & Back, E. L. (1975). Biochemical and chemical oxygen demand of carbohydrates with different degrees of polymerization. *Environ. Sci. Technol.*, *9*(2), 160.

Lee, K.-H., Ishikawa, T., McNiven, S., Nomura, Y., Sasaki, S., Arikawa, Y., & Karube, I. (1999). COD sensor employing a thin layer electrochemical cell. *Anal. Chim. Acta (Neth.)*, *386*, 211.

Leffman, H. (1895). *Examination of water* (3rd ed.). P. Blakiston's Son & Co.

Leithe, W. (1969). New procedure for determination of the COD in wastewater. *Gas-Wasserfach*, *110*, 1233; *Chemical Abstracts*, *72*, 6069 (1970).

Leszczynski, C., & Skwara, Z. (1966). Comparison of various techniques for the determination of cod in waste waters from cellulose and paper manufacturing industry. *Przegl. Papier.* (Poland), 22, 213.

Lewin, V. H. (1949). Routine estimation of BOD. *Surveyor*, *108*, 281.

Li, M. (1986). COD-I coulometer and its application. *Lihua Jianyan, Huaxue Fence*, 22, 99.

Lloyd, A. (1982). Simplified procedure for the determination of COD using silver chloride to suppress chloride interference. *Analyst*, *107*, 1316.

Ludvik, J. (1963). Oxygen demand of tannery wastes. I. Reaction conditions in permanganate number: four hour determinations. *Kozarstvi*, *13*, 187; *Chemical Abstracts*, *60*, 3865 (1964).

Ludvik, J., & Siska, S. (1963). Chemical oxygen demand in tannery waste waters. II. Some problems of optimum oxidant consumption in the $KMnO_4$: four hour test. *Kozarstvi*, *13*, 263; *Chemical Abstracts*, *60*, 10377 (1964).

Lueck, D. E., et al. (1981). Total oxygen demand: Recent advances in the automatic oxygen demand analysis technique. *ISA Trans.*, *20*, 67; *Anal. Instrum.*, *18*, 59 (1980).

Mahin, E. G. (1929). *Introduction to quantitative analysis.* McGraw-Hill.

Mallet, J. W. (1882). Water analysis. *Annu. Rep., Natl. Board Health*, 189.

Mamais, D., Jenkins, D., & Pitt, P. (1993). A rapid physical–chemical method for the determination of readily biodegradable soluble COD in municipal wastewater. *Water Res.*, *27*, 195.

Mason, W. P. (1910). *Examination of water* (4th ed.). Wiley & Sons.

Matsumae, K. (1980). Automatic COD meter. *Kankyo Gijutsu*, *9*, 324; *Chemical Abstracts*, *93*, 209617 (1980).

Matsuoka, C., et al. (1987). COD and BOD of 14 fatty acids determined by the LIS K0102 methods. *Kenkyu Kiyo-Tokushima Bunri Daigaku*, *34*, 193.

Maturaba, T. (1938). Sanitation studies on the automatic purification of river waters. I. The determination of BOD of water. *Mitt. Med. Akad. Kioto*, *23*, 193.

McDonald, D. (1973). Squeezing five days BOD determination down to an hour's work. *Process Eng.*, *12*, 68.

McGhee, T. J., Backer, D. S., & O'Neal, M. V. (1976). Biological treatment of feedlot runoff. *J. Water Pollut. Control Fed.*, *48*, 153.

McGhee, T. J., Christenson, L. R., & Bonneau, W. F. (1973). Laboratory studies on feedlot runoff. *J. Sanit. Eng. Div., Proc. Am. Soc. Civ. Eng.*, *99*, 883.

McGhee, T. J., Torrens, R. L., & Smaus, R. J. (1972). BOD determination on feedlot runoff. *Water Sew. Works*, *119*, 58.

McGhee, T. J., Torrens, R. L., & Smaus, R. J. (1973). Aerobic treatment of feedlot runoff. *J. Water Pollut. Control Fed.*, *45*, 1865.

McGowan, G. (1904). Chemical report (section on chemical standards of purity for sewage effluents). *Royal Comm., Fourth Rep.*, 4,(5), 37.

McGowan, G., Floris, R. B., & Finlow, R. S. (1904). Methods of analysis as applied to sewage and sewage effluents. *Royal Comm., Fourth Rep.*, *4*(5). 11.

McKeown, J. J., et al. (1972). Oxidation analysis of mill effluents. *Natl. Council Air Stream Improvement Tech. Bull. 256.*

McNary, R. R., Dougherty, M. H., & Wolford, R. W. (1957). Determinattion of COD of citrus waste waters. *Sew. Ind. Wastes*, *29*, 894.

Medalia, A. I. (1951). Test for traces of organic matter in water. *Anal. Chem.*, *23*, 1318.

Messenger, A. L. (1981). Comparison of sealed digestion chamber and standard method COD tests. *J. Water Pollut. Control Fed.*, *53*, 232.

Miller, D. G., Brayton, S. V., & Boyles, W. T. (2001). COD analysis of wastewater using trivalent manganese oxidant with chloride removal by sodium bismuthate pretreatment. *Water Environ. Res.*, *73*, 63.

Mitev, S., et al. (1978). Coulometric determination of the COD of wastewater. *God. Vissh. Khim.-Tekhnol. Inst. Burgas. Bulg.*, *13*, 189; *Chemical Abstracts*, *93*, 137792 (1980).

Mobius, C. H. (1979). COD Determinations, new ISO draft. *Wochenbl. Papierfabr* (Germany), *107*, 537.

Mobius, C. H. (1980). Photometric COD determinations with cuvette test systems in comparison with the referee method. *Wochenbl. Papierfabr.* (Germany), *108*, 481.

Mohan, S. V., Mohan, S. K., & Reddy, S. J. (2000). Determination of organic matter in water using organo-metal complexes. *Wat. Res.*, *34*, 3761.

Mohlman, F. W., & Edwards, G. P. (1931). Determination of carbon in sewage and industrial wastes. *Ind. Eng. Chem. Anal. Ed.*, *3*, 119.

Moore, W. A., Kroner, R. C., & Ruchhoft, C. C. (1949). Dichromate reflux method for determination of oxygen demand. *Anal. Chem.*, *21*, 953.

Moore, W. A., Ludzack, F. J., & Ruchhoft, C. C. (1951). A comparative study of oxygen consumed methods. *Anal. Chem.*, *31*, 1297; *Sew. Ind. Wastes*, *24*, 925 (1952).

Moore, W. A., & Ruchhoft, C. C. (1951). Chemical oxygen consumed and its relationship to BOD. *Sew. Ind. Wastes*, *23*, 705.

Moore, W. A., & Walker, W. W. (1956). Determination of low COD of surface waters by dichromate oxidation. *Anal. Chem.*, *28*, 164.

Morfaux, J. N., & Sarris, J. (1974). Automated determination of COD. Critical study of the oxidation of pure substances. *Trib. CEBEDEAU* (Belgium), 27, 97.

Morfaux, J. N., Sarris, J., & Catroux, G. (1972). Relation between COD and organic carbon in waste waters. Automated method of simultaneous determination of these parameters. *Ind. Aliment Agric.*, *89*, 1579; *Chemical Abstracts*, *79*, 45420 (1973).

Mrkva, M. (1983). Evaluation of correlations between absorbance at 254 nm and COD of river waters. *Water Res.*, *17*, 231.

Muers, M. M. (1936). The biological purification of whey solutions. *J. Soc. Chem. Ind.*, *55*, 71T.

Mukai, T. (1976). Determination of COD II. Effect of chloride ions on alkaline cod with potassium permanganate. *Jpn. Analyst*, *25*, 219.

Mullis, M. K., & Schroeder, E. D. (1971). A rapid BOD test suitable for operational control. *J. Water Pollut. Control Fed.*, *43*, 209.

Nakajima, K., Ishizuka, T., & Sunahara, H. (1982). The role of elemental oxygen demand in total oxygen demand for water pollutants. *Water Res.*, *16*, 1003.

Nasr, M. S., & MacDonald, D. G. (1975). Measuring effluent COD with rapid permanganate test. *Pulp Paper Mag. Can.*, *76*(3), 91.

Northrup, H. J. (1954). A comparison of the oxygen consumed from permanganate and BOD tests in the study of textile wastes. *Am. Dyestuff Rep.*, *43*, 130.

Novak, L., Larrea, L., & Wanner, J. (1995). Mathematical model for soluble carbonaceous substrate biosorption. *Water Sci. Technol.*, *31*(2), 67.

Novic, M., Pihlar, B., & Duler, M. (1988). Use of flow injection analysis based on iodometry for automation of DO (Winkler method) and COD (dichromate method) determinations. *Fresenius Z. Anal. Chem. (Germany)*, *332*, 750.

Nowak, O., Svardal, K., Franz, A., & Kuhn, V. (1999). Degradation of particulate organic matter: A comparison of different model concepts. *Water Sci. Technol.*, *39*(1), 119.

O'Brien, W. J., & Clark, J. W. (1962). The historical development of the biochemical oxygen demand test. *Bull. No. 20, Eng. Exp. Stn., New Mexico State University*.

O'Herron, R. J. (1974). *Literature survey of instrumental measurements of BOD for control applications: 1960–1973*. EPA-670/4-74-001, Environ. Protection Technol. Ser.

Okada, S., & Uesugi, K. (1957). Chemical oxygen demand of polluted water. *Himeji Kogyo Daigaku Kenkyu Hokolu*, 7, 74; *Public Health Eng. Abstr.*, *38*(4), 24 (1958).

Orhon, D., Artan, N., & Ates, E. A. (1994). A description of three methods for the determination of the initial inert particulate COD of wastewater. *J. Chem. Technol. Biotechnol.*, *61*, 73.

Orhon, D., & Cokgor, E. U. (1997). COD fractionation in wastewater characterization: The state of the art. *J. Chem. Tech. Biotechnol.*, *68*(3), 283.

Orhon, D., Cokgor, E. U., & Sozen, S. (1999a). Experimental basis for the hydrolysis of slowly biodegradable substrate in different wastewaters. *Water Sci. Technol.*, *39*(1), 87.

Orhon, D., Cokgor, E. U., & Sozen, S. (1999b). The effect of residual microbial products on the assessment of the particulate inert COD in wastewaters. *Water Res.*, *33*, 3191.

Papinsky, J. B., Proges, N., & Hoover, S. R. (1952). Application of a rapid COD test for determining organic pollutants. *Sew. Ind. Wastes*, *24*, 922.

Paraskeva, P., Lambert, S., & Graham, N. (1998). Influence of ozonation conditions on the treatability of secondary effluents. *Ozone: Sci. Eng.*, *20*(2), 133.

Paulsen, B. B. (1965). Determination of suspended organic solids by filtration and wet dichromate oxidation. *Vatten Hygiene*, *21*, 9.

Petermann, E. P. (1975). Modified method for COD determination. *Wochenbl. Papierfabr.*, *103*(9), 321; *Abstr. Bull. Inst. Paper Chem.*, *46*, 3890 (1975); *Chemical Abstracts*, *85*, 151813 (1975).

Pitwell, L. R. (1983). Standard COD. *Chem. Br.*, *19*, 907.

Popadopoulos, A., Fatta, D., & Loizidou, K. C. (1998). Treatment of stabilized landfill leachate by physico-chemical and bio-oxidation processes. *J. Environ. Sci. Health A*, *33*, 651.

Popova, E. S. (1930). Potassium dichromate for estimating the oxidizability of sewage. *Trav. Comm. Recherches Epurat Eaux Eqout*, *12*, 143; *Chemical Abstracts*, *24* (1930).

Rao, C. U., & Nyns, E. J. (1987). Correlations among TOC, COD, and BOD of wastewaters. *J. Inst. Public Health Eng. Ind.*, *4*, 14.

Rein, D. A., & Bello, T. (1988). *Changes in the BOD/COD relationship during processing of beet-sugar wastewater.* Georgia Inst. Technol. Educ. Ext Serv.

Reinefeld, V. E., et al. (1975). Analytical investigations of sugar factory waste waters, particularly the COD of storage pond water. *Zucker*, *28*(4), 165.

Reynolds, J. F., & Goellner, K. A. (1974). Statistical evaluation of BOD_5 versus ODI. *Water Sew. Works*, *121*(1), 31.

Rhame, G. A. (1947). Determination of BOD values by chemical oxidation. *Water Sew. Works*, *94*, 192.

Rhame, G. A. (1950). BOD or COD: A commentary. *Water Sew. Works*, *97*, 179.

Rhame, G. A. (1974). Rationalization of the COD–BOD relationship. *Water Sew. Works*, *121*, 68.

Rickert, D. A., & Hunter, J. V. (1971). General nature of soluble and particulate organics in sewage and secondary effluent. *Water Res.*, *5*, 421.

Riehl, M. L. (1951). Evaluation of chemical oxygen consumed tests. *Sew. Ind. Wastes*, *23*, 1398.

Roberts, H. V., & Sanderson, W. W. (1953). An evaluation of the oxygen consumed test. *Sew. Ind. Wastes*, *25*, 793.

Rogers, D. (1983). Deep-tank aeration/flotation, clarification adds a new treatment dimension. *Ind. Wastes*, *29*(1), 10.

Rozenberg, M. (1993). The fast method of the COD determination. *Anal. Lett.*, *26*, 2025.

Ryding, S. O., & Forsberg, A. (1977). A mercury-free accelerated method for determining the COD of large numbers of water samples by autoclaving them under pressure with acid-dichromate. *Water Res.*, *11*, 801.

Sawada, S. (1986). A simplified method of COD determination. *Bull. Kyoto Univ. Educ., Ser. B*, *69*, 29.

Schierjott, G., Bleier, H., & Malissa, H. (1974). Simultaneous continuous determination of COD and of organic carbon. *Vom Wasser* (Germany), *39*, 1; *Chemical Abstracts*, *81*, 29238 (1974).

Schulze, N. F. (1953). Assay of certain organic materials by dichromate oxidation. *Anal. Chem.*, *25*, 1762.

Schulze, R. (1976). TOD analysis by automatic instruments. A. *Wasser Abwasser Forsch.*, *9*, 11; *Chemical Abstracts*, *85*, 112487 (1976).

Schumann, G. (1974). New methods for the measurement and evaluation of wastewater. *Branntweinwirtschaft*, *114*(12), 257; *Chemical Abstracts*, *82*, 21352 (1975).

Schumann, G. (1984). Determination of the COD value of brewery wastewater. Comparison of the values determined by the DIN (German Industrial Standard). Method with those of the COD cuvette test system. *Brauwelt*, *124*, 1566; *Chem Abstr.*, *102*, 12061 (1985).

Shriver, L. E., & Young, J. C. (1972). Oxygen demand index as a rapid estimate of BOD. *J. Water Pollut. Control Fed.*, *44*, 2140.

Sierra-Alvarez, R., et al. (1990). The continuous anaerobic treatment of pulping wastewaters. *J. Ferment. Bioeng.*, *70*, 119.

Silva, C. R., Conceicao, C. D. D., Bonifacio, V. G., Filho, O. F., Teixeira, M. F. S. (2009). Determination of the chemical oxygen demand (COD) using a copper electrode: A clean alternative method. *J Solid State Electrochem.*, *13*, 665.

Singh, T., & Neves, J. (1974). Simplify your COD testing. *Water Wastes Eng.*, *11*, 39.

Smith, R.-K. (1995). *Water and wastewater laboratory techniques.* Water Environment Federation.

Smith, R.-K. (1999a). *Handbook of environmental analysis* (4th ed.). Genium Publishing.

Smith, R.-K. (1999b). *Lectures on wastewater analysis and interpretation.* Genium Publishing.

Sollfrank, U., & Gujer, W. (1991). Characteristics of domestic wastewater for mathematical modelling of the activated sludge process. *Water Sci. Technol.*, *23*, 1057.

Sorensen, P. E. (1974). Correlation between parameters of analyses for dairy effluent. *Ann. Bull. Int. Dairy Fed.*, *77*, 35.

South, W. D., et al. (1971). New approaches to in-plant load control and monitoring. *Natl. Council Air Stream Improvement Tech. Bull*, *248*.

Spohn, U., et al. (1991). On-line monitoring of the COD with flow injection analyzer (FIA). *BioEngineering*, *7*(3), 20.

Stamm, H. (1934). Ein neues verfahren fur titrationen alkalisher permanganatlosung. *Angew. Chemie*, *47*, 791.

Stenger, V. A., & van Hall, C. E. (1967). Rapid method for the determination of COD. *Anal. Chem.*, *39*, 206.

Stenger, V. A., & van Hall, C. E. (1968). Analyses of municipal and chemical waste waters by an instrumental method for COD determination. *J. Water Pollut. Control Fed.*, *40*, 1755.

Stones, T. (1980). The oxygen demand of sewage. *Effluent Water Treat. J.*, *20*, 437.

Stones, T. (1985). Anomalies arising from arbitrary measures of the strength of sewage. *Effluent Water Treat. J.*, *25*, 103.

Strom, A., & Solyom, P. (1978). Origin and composition of effluent in a Swedish liquid milk dairy. *Nordisk Mejeriindustri*, *5*(2), 41.

Sutton, F. (1911). *Volumetric analysis* (10th ed.). P. Blakiston's Son & Co.

Sykes, R. M. (1982). Hydraulic regime and activated sludge performance. *J. Environ. Eng.*, *108*, 286.

Takamatsu, T., Hashimoto, I., & Sioya, S. (1970). Model identification of wet air oxidation process thermal decomposition. *Water Res.*, *4*(1), 33.

Talinli, I., & Anderson, G. K. (1992). Interference of hydrogen peroxide on the standard COD test. *Water Res.*, *26*, 107.

Tartakovsky, B., Lishman, L. A., & Legge, R. L. (1996). Application of a multi-wavelength fluorometry for monitoring wastewater treatment process dynamics. *Water Res.*, *30*, 2941.

TeKippe, R. J., & Zehnpfennig, R. G. (1971). Improved COD apparatus. *J. Water Pollut. Control Fed.*, *43*, 515.

Tezuka, Y. (1968). A method for estimating bacterial respiration in natural water. *Jpn. J. Ecol.*, *18*, 60.

Thompson, K. C., Mendham, D., Best, D., & de Casseras, K. D. (1986). Simple method for minimizing the effect of chloride on the COD test without the use of mercury salts. *Analyst*, *111*, 483.

Thresh, J. C. (1904). *The examination of waters and water supplies*. P. Blakiston's Son & Co.

Tian, L., & Wu, S. (1992). Determination of chemical oxygen in aqueous environmental samples by segmented flow-injection analysis. *Anal. Chim. Acta* (Netherlands), *261*, 301.

Tidy, C. M. (1879). The process for determining the organic purity of potable waters. *J. Chem. Soc.*, *35*, 46.

Tiemann, F., & Gartner, A. (1895). *Handbuch der Wasser* (4th ed.). F. Vieweg und Sohn.

Timofeeva, S. S., & Beim, A. M. (1989). BOD and COD correlation functions and prospects for their use in analytical monitoring of kraft pulp mill waste water. *Sov. J. Water Chem. Technol.*, *11*, 52.

Topalian, P., & Scntheimer, H. (1979). Determination of COD using UV oxidation. *Gewaesser, Wasser, Abwasser* (Germany), *39*, 285; *Chemical Abstracts*, *92*, 134834 (1980).

Trbovic, M. (1979). Experience with the Dr. Lange cuvette test system for COD determination with photometric evaluations. *Papiermacher*, *29*(9), 134.

Trommsdorff, H. (1869). Bestimmung de organischen. *Zeit. Anal. Chem.*, *8*, 344.

Ubay, C. E. (1998). Respirometric analysis of activated sludge behavoir assessment of the readily biodegradable substrate. *Water Res.*, *32*, 461.

Uchida, T., et al. (1978). Pollution Index on the organic waste water marine processing waste water. *Shizuuka-Ken Kogyo Shikenjo Hokulu*, *21*, 43; *Chemical Abstracts*, *88*, 125979 (1978).

United Kingdom Department of the Environment. (1987). COD (dichromate value) of polluted and waste waters (2nd ed.). *Methods Exam, Waters Assoc. Mater.*

van den Eynde, E., et al. (1984). Plug flow simulating and completely mixed reactors with a premixing tank, in the control of filamentous bulking. *Appl. Microbiol. Biotechnol., 19*, 4, 288.

Vanrolleghem, P. A., Spanjers, H., Petesen, B., Ginestet, P., & Takacs, I. (1999). Estimating (combinations of) ASM1 parameters and components by respirometry. *Water Sci. Technol., 39*(1), 195.

Vezzi, S. (1970). Relationship between BOD, oxidizability, and stability tests as indexes of the degree of pollution of surface waters. *Rass. Chim., 22*(5), 149; *Chemical Abstracts, 75*, 9670 (1971).

Virarghavan, T. (1976). Correlation of BOD, COD, and soluble organic carbon. *J. Water Pollut. Control Fed., 48*, 2213.

Vogl, K., Schumann, G., & Kolb, H. (1968). Comparison of various methods of effluent examination. *J. Agric. Sci., 19*, 257.

Voorn, G. (1977). Monitoring continuously the oxygen demand of industrial effluent and surface waters. *Water Waste Treat. J., 20*(10), 14.

Voorn, G., & Lee-Frampton, J. (1976). New meter makes quick work of TOD measurement. *Water Pollut. Control., 114*(8), 30.

Voorn, G., & Marlowe, J. S. (1977). Metering total oxygen demand. *Effluent Water Treat J., 17*(6), 301.

Voorn, G., & Mitchell, P. (1976). A new instrumental method for measuring the oxygen demand of water. *Prog. Water Technol., 8*, 423.

Vtill, D. R., & Spiegel, S. J. (1980). Characterization of industrial wastes by evaluating BOD, COD, and TOC. *J. Water Pollut. Control Fed., 52*, 2704.

Watts, R. J., & Adams, V. D. (1983). The elimination of sulfur dioxide interference in the low level COD analysis. *Water Res., 17*, 715.

Weems, J. B. (1901). The evolution of the oxygen absorption test in water analysis. *Proc. 13th Annu. Meeting Iowa Eng. Soc.*; *Rev. Am. Chem. Res., 8*, 323 (1902).

Wells, W. N. (1970). Evaluation of the Jeris rapid COD test. *Water Sew. Works, 117*, 123.

Westerhold, A. F. (1965a). *Digester*, 22, 1, 4 (Quoted in Young, 1977).

Westerhold, A. F. (1965b). *Digester*, 22, 2, 18 (Quoted in Young, 1977).

Wolff, C. J. M (1975). COD determinations of volatile compounds. *Water Res., 9*, 1015.

Wyatt, F. (1893). Tests for purity of drinking water. *Eng. Mining J., 56*, 168.

Yeun, A., & Viraraghavan, T. (1987). Relationship between COD and UV absorbance at 254 nm of wastewaters and surface waters. *Asian Environ., 9*, 4.

Young, J. C. (1977). *Biochemical oxygen demand: Measurement and application*. Unpublished.

Zajic, J., et al. (1970). BOD and COD analyses of paraffinic hydrocarbons. *J. Am. Water Works Assoc.*, *62*, 784.

Ziglio, G., Andreottola, G., Foladori, P., & Ragazzi, M. (2001). Experimental validation of a single-OUR Method for wastewater RBCOD characterization. *Water Sci. Technol.*, *43*(11), 119.

Zimina, G. N., & Zaleshina, G. M. (1978). Rapid method for determining 5-Day BOD. *Tsellyuloza Bumaga Karton Ref. Inform.*, *23*, 4; *Abstr. Bull. Inst. Paper Chem.*, *49*, 524.

13

Total Organic Carbon and Organic Matter

Carbon is present in samples in one of three forms: elemental (charcoal), inorganic (carbon dioxide, carbon monoxide, carbonates, and bicarbonates), and organic. Elemental carbon is essentially inert to biological organisms and is not considered further here.

Inorganic carbon is most frequently encountered as either carbonate or bicarbonate minerals with a variety of cations, calcium and magnesium being the most common. Some organisms can use carbonates and bicarbonates as a primary carbon source, whereas plants that perform photosynthesis use carbon dioxide as the sole source of carbon. The amount of carbon dioxide that can be present in water is a function of the pH, with higher pH values favoring increased amounts stored in the form of bicarbonate and carbonate. Determination of the inorganic carbon fraction in samples may be of interest in certain situations, but in the majority of cases, inorganic carbon presents an interference to organic carbon analysis. Fortunately, the inorganic carbonate and bicarbonate salts have the common property that acidification converts them to carbon dioxide, which is a gas only slightly soluble in water. As most samples have much greater amounts of inorganic carbon as compared with organic carbon, acidification serves as a very easy means of removing the inorganic carbon (Smith, 1999a, 1999b).

Most bacterially usable substrates in wastewater contain carbon in the organic form. This means that total organic carbon (TOC) is directly related to the potential amount of food present in the waste, and significantly different than the indirect measures of chemical oxygen demand (COD) and biochemical oxygen demand (BOD), which are oriented toward the process of utilization. Determination of TOC and organic matter is a logical extension of the concept of biological purification of wastewater, but it is a relatively new parameter in the waste treatment arena as compared with BOD and COD.

Most of the early work on TOC determinations actually took place within the context of soils analysis for agricultural purposes. Wet chemical methods for TOC developed and refined in the 1920s through 1960s are still in use in soils laboratories. Extensive use of TOC data in wastewater facility design and operation and in marine and oceanographic research did not occur until after the invention and commercialization of automated TOC instruments in the 1960s. Detailed reviews of the early agricultural work on wet chemical TOC are available (Nelson & Sommers, 1982; Tiessen & Moir, 1993) and the salient points are included in the discussion below.

1.0 WET CHEMICAL METHODS FOR TOTAL ORGANIC CARBON

The first experiments concerning wet chemical determination of TOC were reported in 1927 (Schollenberber, 1927, 1931). The sample is heated with potassium dichromate and sulfuric acid, then the excess dichromate determined with ferrous ammonium sulfate, Mohr's salt [FAS, $Fe(NH_4)_2(SO_4)_2$], with a diphenylamine visual indicator. The consumed dichromate is related to the amount of carbon in the sample based on the chemical equation

$$2Cr_2O_7^{2-} + 3C^o + 16H^+ \rightarrow 4Cr^{3+} + 3CO_2 + 8H_2O$$

where organic carbon is assumed to be, on average, in the zero valence state. This procedure is very similar to that used for the first determinations of dichromate COD (Abbott, 1927; Adeny & Dawson, 1926; Popova, 1930), with the difference being the calculation and report of the final result is in terms of carbon rather than oxygen.

A mixture of hydrogen peroxide and dichromate–sulfuric acid has been suggested as a digestion reagent (Degtijareff, 1930). This gives high values of carbon resulting from the oxidation of hydrogen peroxide by dichromate (Walkley & Black, 1934) if the method of determination is based on consumed or remaining dichromate in the sample solution.

Modification of the procedure to trap the evolved carbon dioxide and determine it, rather than the consumed oxidant, results in elimination of many of the interferences (Heck, 1929). The first reported application of a TOC method (Mohlman & Edwards, 1931) to wastewater used dichromate in a heated sulfuric acid–phosphoric acid digestion mixture and carbon dioxide absorption in a barium hydroxide trap. The carbon dioxide reacts with barium hydroxide to form the insoluble barium carbonate, which is filtered from the trap contents after the completion of the digestion. The amount of carbon dioxide absorbed is determined by the difference between the amount of the barium hydroxide in the trap solution before and after the digestion, with barium hydroxide measured by a simple acid–base titration. The reflux temperature of the digestion mixture is 210 °C, considerably greater than the temperature of the 50% sulfuric acid (150 °C) that is used in COD determinations, where the intent is to prevent decomposition of dichromate.

One interferent that is not eliminated by trapping the carbon dioxide is chloride. Chloride combines with dichromate on mixing the digestion solution with the sample to form chromyl chloride:

$$3H_2SO_4 + K_2Cr_2O_7 + 4KCl \rightarrow 2CrO_2Cl_2 + 3K_2SO_4 + 3H_2O$$

Though quite stable at room temperature, chromyl chloride decomposes with heating, generating chlorine, the reaction becoming quantitative around 190 °C. Both chromyl chloride (BP 117 °C) and chlorine are volatile and can readily pass out of the digestion flask. In the absorption tube, both compounds react as acids with the barium hydroxide, lowering the titer of the trap solution and thereby giving a false-positive bias to the amount of carbon dioxide evolved from the digestion.

Another shortcoming of this simple procedure is that easily digested substrates, such as carbohydrates, generate a mixture of carbon monoxide and carbon dioxide (van Slyke, 1954). Although carbohydrates are not frequently encountered in agricultural soils, they can be significant components of the TOC in a wastewater. Added copper or manganese salts have been used to improve the yield of carbon dioxide. Complete oxidation to carbon dioxide can be ensured by using a digestion mixture of fuming sulfuric acid, phosphoric acid, potassium iodate, and chromic oxide, CrO_3 (Archer, 1954; McCready & Hassid, 1942; Van Slyke & Folch, 1940). A recipe for use on wastewater samples that avoids fuming sulfuric acid and dichromate but includes periodic acid in the digestion reagent with a sodium hydroxide trap for carbon dioxide has been described (Iwasaki, Hirayama, Ohmori, & Kubomura, 1966; Iwasaki, Hirayama, Ohmori, &Ono, 1966). An alternate wastewater digestion consists of potassium persulfate, sulfuric acid, and silver sulfate at 200 to 250 °C for 1 hour with trapping of evolved carbon dioxide in potassium hydroxide (Ohmori et al., 1968).

An optimized method widely used in agricultural laboratories for determination of TOC by carbon dioxide trapping is described by Allison (1960) and repeated in detail in the standard manuals for soil analysis (Nelson & Sommers, 1982; Tiessen & Moir, 1993). This procedure includes silver sulfate and potassium iodide traps to remove chlorine–chromyl chloride interference with the carbon dioxide absorption. Silver arsenite filled traps have also been examined for removal of interferences arising from chlorides (Pickhardt et al., 1955; Weber & Morris, 1964). Sulfur interferences can be removed as sulfate in a barium chloride trap and nitrogen oxides removed in a 190 °C trap filled with silver metal and lead oxide (Weber & Morris, 1964).

Sample pretreatment at room temperature with sulfurous acid, H_2SO_3, and ferrous sulfate to remove inorganic carbon before digestion has been described several times (Nelson & Sommers, 1982; Tiessen & Moir, 1993) for wet and dry digestion procedures. The excess sulfurous acid can be removed from the sample by mild heating under a hood, with evolution of sulfur dioxide; however, this may lead to loss of some volatile organic components. The sulfurous acid pretreatment has been applied to solids for removal of inorganic carbon (Heron et al., 1997). Acid pretreatment of solids needs to be performed in such a manner as to avoid the use of solutions that are removed by filtration rather than by evaporation. The use of liquid acidification followed by filtration has been found to reduce the organic carbon content of solid samples through partial solubilization of the organic fraction in the liquid (Roberts et al., 1973).

The contents of the trap have been assayed in a number of ways. The bicarbonates formed in sodium or potassium hydroxide traps are weaker bases than the hydroxide and lead to lower phenolphthalein titers in a direct acid–base titration. Alternatively, conductimetric titrations can be performed on the contents of sodium hydroxide traps (Krey & Szekielda, 1965; Szekielda & Krey, 1965). Barium chloride is added to the trap to precipitate barium carbonate, then an acid–base titration is performed on the soluble contents of the trap to determine residual hydroxide (Iwasaki, Hirayama, Ohmori, & Kubomura, 1966; Iwasaki, Hirayama, Ohmori, & Ono, 1966). Excess barium has been added to the trap to precipitate barium carbonate, then the excess barium remaining in the supernatant is determined by back-titration in a complexometric procedure (Ohmori et al., 1968). When a barium hydroxide trap is used, the pH changes as barium carbonate is precipitated. The pH of the trap can be used as a feedback to an electrolysis unit, which coulometrically reestablishes the pH, the amount of current required being the analytical metric (Lyutsarev, 1969). The residual barium hydroxide in the trap can be directly determined by acid–base titration after filtration (Mohlman & Edwards, 1931; Weber & Morris, 1964).

A novel approach is to assay the contents of the sodium hydroxide trap using a TOC instrument set in the inorganic carbon mode of analysis (Suzuki et al., 1983). This is used as a very sensitive technique for sediment TOC determination.

An important application of benchtop wet chemical methods for TOC digestion is when isotopic distributions are studied. Carbon-13 is the most frequently studied isotope, and recovery of the evolved carbon dioxide for isotopic fractionation is a necessary step in the analysis (Lindenbaum et al., 1948; Tan & Edmond, 1993; Westerhausen et al., 1993). Mass spectrometry of the evolved carbon dioxide released when the contents of the trap are acidified is used to determine the $^{12}C/^{13}C$ ratio. Carbon-14 has also been used as a tracer, and TOC measurements combined with mass spectrometer and radioactivity assessments allow determination of carbon flows in soil and sediment biomass (Dobbins & Pfaender, 1988).

2.0 INSTRUMENTAL TOTAL ORGANIC CARBON METHODS

The widespread use of TOC data in the wastewater field did not occur until after the development and commercialization of TOC instruments (Busch, 1967). The instruments present advantages by automating the procedure, thereby providing increased precision in the determination. Often an automatic sampler is used to introduce aqueous samples sequentially into the instrument. Figures of 1% relative standard deviation for precision are commonplace for instrumental techniques, particularly when autosamplers are used (Croll, 1974).

The instruments perform two main functions in the analysis. First, there is a release of the carbon from the sample (digestion), and second, there is a determination of the released carbon (detection). There are many different instruments available reflecting the many possible combinations of digestion and detection that have been invented. Further, there are a variety of ways that have been developed to deal with inorganic carbon. The 20th edition of *Standard Methods* (APHA, et al., 1998) approaches the plethora of available choices in the TOC description (Method 5310) by dividing the instruments into three general types based on the digestion technique: high-temperature combustion (5310B), persulfate–UV or heated persulfate oxidation (5310C), and wet-oxidation (5310D); and then instructing the user to essentially follow the manufacturer's directions.

The total carbon, TOC, and inorganic carbon fractions are determined in a variety of ways. Some instruments split the injected sample into two portions, one being processed with acid for inorganic carbon, while the other portion is routed through the digestion unit to give a total carbon result. The TOC is the difference between the two values. Other instruments use a serial operation, in which inorganic carbon is first stripped with acid treatment, then the residual sample is introduced into the digestion unit. This latter design gives discrete values for inorganic carbon and organic carbon, with the total carbon calculated by adding the values. When volatile organic carbon, also called purgeable organic carbon (POC), is a concern, the first configuration is generally the more reliable.

The high-temperature combustion method oxidizes the sample at temperatures from 550 °C to greater than 950 °C in the presence of an oxygen source to convert organic carbon to carbon dioxide (Blackmore & Voshel, 1967; Busch, 1966; Katz et al., 1954; Schaffer et al., 1965; van Hall et al., 1963, 1965; Williams, 1967). Copper oxide is a very effective oxidant at these temperatures. One technique uses an inert carrier gas (nitrogen) and performs the combustion in a copper oxide-packed combustion tube at elevated temperatures, where oxygen to form carbon dioxide comes from the copper oxide (Cropper et al., 1967a, 1967b; West, 1964). Another design vaporizes the sample on copper oxide wire in a 900 °C combustion chamber in a helium carrier gas (Dobbs et al., 1967). Oxygen gas as carrier is used as the oxygen source in furnaces set at temperatures ranging from 600 °C (Grunnett & Hansen, 1972) to 900 °C (Teal et al., 1967) to 950 °C (Goulden, 1976). At the higher temperatures, no catalyst is needed, and an inert asbestos packing with oxygen carrier gas has been successful (Emery et al., 1971; Schaffer et al., 1965). When oxygen gas is used in conjunction with copper oxide in the furnace, the solid is regenerated as it is used (Menzel & Vaccaro, 1964; Russell, 1977). Cobalt(III) oxide has also been used as the regenerated catalyst at 800 °C (Gacs & Payer, 1989; Schaffer et al., 1965; Suzuki et al., 1983).

The UV–persulfate method mixes the sample with a potassium peroxydisulfate reagent, then exposes the flowing mixture to intense UV light. The UV light breaks the oxygen–oxygen bond in the peroxide to generate sulfate radical anion, a very powerful oxidant. Carbon in the sample is converted to carbon dioxide (Aiken, 1992; Armstrong & Tibbits, 1968; Armstrong et al., 1966; Beattie et al., 1961; Bramer et al., 1966; Cauwet, 1984; Collins & Williams, 1977; Dobbs et al., 1967; Enrhardt, 1969; Godec et al., 1992; Goulden & Brooksbank, 1975; Gravelet-Blondin et al., 1980; Jones &

Dageford, 1968; Montgomery & Thom, 1967; Oake, 1981; van Steenderen & Lin, 1981). Various designs allow for either continuous-flow operation, in which the sample and persulfate are mixed before passage into the reactor, or batch-wise injection of sample and reagent directly to the UV reactor.

The wet oxidation instrumental technique typically mixes the sample with a potassium persulfate solution, then heats it to approximately 130 °C for periods of time ranging from minutes up to 4 hours, which converts carbon in the sample to carbon dioxide. The *Standard Methods* (APHA et al., 1998) technique, Method 5310D, describes sealing the sample and solid potassium persulfate in an ampule and performing the oxidation in an autoclave (Hughes et al., 1974; Williams, 1969; Wilson, 1961, 1963). After digestion, the ampule contents are assayed for carbon dioxide. This technique has been applied to digestion of solids and sediments many times (Menzel & Vaccaro, 1964; Mills & Quinn, 1979). The use of a potassium persulfate–sodium hydroxide digestion mixture has been described (Fiedler et al., 1975). A silver-catalyzed peroxysulfate mixture heated to 95 °C has been successfully used for digestions (Baldwin & McAtee, 1974; Goulden & Brooksbank, 1975; van Steenderen et al., 1979), and is described in a patent (Kramer et al., 1972) for a continuous-flow online instrument. A digestion solution containing sulfuric acid, silver dichromate, and potassium dichromate has been described, with the evolved carbon dioxide swept in an oxygen carrier gas through a silver permanganate trap before carbon dioxide detection (Krey & Szekielda, 1965).

A novel digestion system makes use of the titanium dioxide photo-catalyzed near UV–oxygen gas oxidation of carbon-containing organics to carbon dioxide (Matthews et al., 1990). The titanium dioxide is coated on the inside of a borosilicate glass reactor wound around a 20 W black light with maximum energy output at 365 nm. Complete conversion of organic carbon to carbon dioxide is accomplished with a 10-minute illumination time. Borosilicate glass is transparent to near UV; thus, the use of the quartz glass reaction cell required in the UV–persulfate technique is avoided. The pH-neutral and nonheated characteristics of the procedure eliminate the need for inorganic carbon suppression procedures.

Detection of the evolved carbon dioxide can take many forms. One is the nondispersive infrared detector (Blackmore & Voshel, 1967; Van Hall & Stenger, 1964; Van Hall et al., 1963). Carbon dioxide is a greenhouse gas, absorbing many wavelengths of infrared (IR) radiation and converting it to heat. The absorbed radiation can be directly monitored with a photocell. In another technique, the heat produced by the carbon dioxide can be detected by monitoring the temperature of the carrier gas in the IR cell and comparing it with the temperature of a parallel stream of carrier gas that is also illuminated. The heat evolved causes an expansion of the gas in the IR cell; thus, constructing the cell as a pressure transducer gives another measure of the response of carbon dioxide in the carrier (Emery et al., 1971).

Other detection techniques are possible when an absorbing solution is used to trap the evolved carbon dioxide. As carbon dioxide is absorbed by a flowing sodium hydroxide solution, the conductance changes, which is monitored downstream from the point of mixing the effluent from the digestion unit with the detection solution (Enrhardt, 1969; Krey & Szekielda, 1965). The change in conductance is calibrated against the response obtained from standard solutions.

A modification of the flowing detector system uses a trap solution of anhydrous ethanol containing 2 *M* 3-methoxypropyl amine, which reacts with the carbon dioxide to form a carbamate (Gacs & Payer, 1989). The conductivity changes in the flowing solution are monitored to create the analytical signal. The detector is reported to have a repeatable sensitivity of 0.05 mg carbon/L.

A U.S. patent describes conductivity detection used in an online instrument, in which the effluent gases from a high-temperature combustion oven with oxygen gas carrier are condensed to form a self-trap (Lossberg, 1967).

Gas chromatography can be used as a separation–detection system for total carbon (TC) and TOC analysis. The oldest application used a heated copper oxide bed in the injector with a helium carrier (West, 1964). The thermal conductivity detector used is not a very sensitive technique for carbon dioxide assay, and modifications to the procedure needed to be made. Carbon dioxide is reduced easily to methane in a hydrogen atmosphere over nickel or copper catalysts at elevated temperatures from 300 °C (Cropper et al., 1967a, 1967b) to 400 °C (Dobbs et al., 1967; Stainton, 1973) to 800 °C (Cauwet, 1984). The methane in the hydrogen carrier gas is then detected by burning in a flame-ionization detector (FID). The FID is quite versatile, and serves as a general detector for almost all compounds containing hydrogen bonded to carbon. It is a mass-sensitive detector, generating one signal for each ion created in the flame (Smith, 1999a & 1999b). These characteristics are adapted to a carbon analyzer that uses a temperature-controlled oven to selectively volatize organic molecules into a nitrogen carrier gas that transports the compounds to the detector (Eggertsen & Stross, 1972). The signal from the FID is continuously collected and integrated as the oven temperature is increased. At oven temperatures up to 150 °C, volatile organic materials in the sample are detected in the flame, while a semivolatile fraction is determined between 150 and 550 °C.

An FID has been used to monitor the effluent from a high-temperature pyrolysis unit (Lysyj and Nelson, 1971; Lysyj et al., 1968) in a very fast TOC technique. Modification of the pyrolysis unit to achieve reductive conditions generates methane directly as the product, which is determined by FID. The TC–TOC procedure is complete in less than 5 minutes per sample (Takahashi et al., 1972). Modifying the 900 °C pyrolysis unit to serve as an injector to a gas chromatograph with a Porapak-Q column and helium carrier, then transporting the column effluent to a mass spectrometer as a detector, produces an instrument that can simultaneously monitor TOC and total organic nitrogen in the sample (Kakajima, 1986).

A dissolved organic carbon (DOC) analyzer that uses liquid chromatography with a detector for carbon dioxide or DOC has been described (Gloor & Leidner, 1979; Gloor et al., 1981). The effluent from the chromatograph enters an 800 to 900 °C heated ceramic tube with a copper oxide on platinum gauze packing. A carbon dioxide–free air carrier gas transfers the oxidation products to an indirect IR detector. The use of a gel-permeation chromatography column in the chromatograph allows a molecular size fractionation of the DOC.

An air-gap electrode has been described as a detector for continuous monitoring of carbon dioxide from a digestion unit (Fiedler et al., 1975).

Instrumental methods for TOC analysis have been adapted to soils and sediments. In one design, the sample is added to a platinum or quartz boat, which is placed in a furnace with a flowing oxygen carrier gas. High-temperature combustion (900 °C) converts the inorganic and organic carbon to carbon dioxide for determination. A very reproducible technique involves initially drying and grinding the solid, then forming a homogeneous suspension in water before addition of a 30-µL aliquot to a platinum boat with a manganese(IV) oxide catalyst and oxygen carrier for reductive pyrolysis at 850 °C (McQuaker & Fung, 1975). Removal of inorganic carbon can be accomplished by pretreatment of the sample with acids such as hydrochloric, phosphoric, or sulfurous. The instrument design may preclude the use of hydrochloric acid. Several reviews and comparisons of instrument methods for TOC analysis of soils and sediment are available (Charles & Simmons, 1986; Powell et al., 1989), with the dry, high-temperature combustion methods generally preferred.

3.0 TOTAL ORGANIC CARBON–BIOCHEMICAL OXYGEN DEMAND RELATIONSHIP

For any pure substance, the TOC can be exactly calculated based on the molecular formula. The calculation is

$$\text{TOC, mg}\frac{\text{C}}{\text{mg}}\text{substance} = \frac{\text{No. carbons} \times 12}{\text{Molecular weight of compound}}$$

One distinct advantage derived from this characteristic of the test is that any pure substance can be used as a standard. For potassium hydrogen phthalate, KHP ($C_8H_5O_4K$, Mwt 204.23), the TOC is 0.470 mg C/mg KHP. Commonly, KHP is used as a standard for the calibration of TOC procedures and for preparation of quality control solutions (Smith, 1999a, 1999b). With regard to its use as a quality control for digestion efficiency, KHP is reported to be too easily digested by most procedures (van Steenderen & Lin, 1981). Reliance on only KHP recoveries can lead to a false sense of confidence that the test is performing at an optimum level.

The first difference between TOC and BOD is in the reporting units. The TOC is reported in milligrams carbon per liter, whereas BOD is reported in milligrams oxygen per liter. The reporting unit difference consists of more than a simple ratio of atomic weights, but must also include the oxidation reaction relationship. The reaction for the oxidation of carbon is

$$C + O_2 \rightarrow CO_2$$

which indicates the ratio is one atom of carbon per molecule of oxygen: 12/32 or 0.375. No organic molecules consist of just carbon, and the other elements in the molecules can consume oxygen. In Chapter 6 on theory, the theoretical oxygen demand of glucose, $C_6H_{12}O_6$, is described and a value of 1.07 mg O_2/mg glucose derived. The TOC result can be calculated based on the amount of carbon in the molecule, giving 0.40 mg C/mg glucose. The TOC/BOD ratio is, therefore, 0.374, approximately the same value as the ratio of units. However, the alkane decane, $C_{10}H_{22}$, gives a calculated TOC of 0.845 mg C/mg and BOD of 3.49 mg O_2/mg, for a ratio of only 0.242. Looking at real results for a common substance such as KHP, the TOC is 141 mg/L and the BOD is 240 mg/L, for a 300-mg/L KHP standard solution (Smith, 1999a, 1999b). These values give a ratio of 0.588. Comparing these results, it is possible to draw the conclusion that for reduced organic molecules, the TOC/BOD ratio will be low. For carbohydrates that contain significant amounts of oxygen, the ratio will be higher, approximately 0.37, and for more oxidized molecules, such as functionalized aromatics, the ratio goes still higher. These examples demonstrate that there is no a priori reason for a simple mathematical relationship between TOC and BOD, and a simple unit conversion will not work in most situations.

Most of the substrates that are usable by bacteria are organic compounds, and they contain carbon. It can, therefore, be said that TOC accounts for all the biologically usable food. However, not all organic compounds are amenable to bacterial utilization, and thus TOC tends to overestimate the amount of food available (Maier & McConnell, 1974). The rate at which biological organisms consume the substrate is also an important factor leading to a lack of correlation between TOC and BOD.

Overall, then, there is little to no relationship between TOC and BOD, or TOC and COD. This has been repeatedly reported from observations on real wastewater effluents. In a study of the effluent from seven pulp and paper mills, 1131 samples collected over 40 days demonstrated no general relationship between TOC and BOD (Howard & DeWitt, 1971). Another extensive study verified a lack of correlation and, in general, questioned the value of TOC determinations (Jones, 1972). Variations in the COD/BOD ratio in spent liquor effluents from a pulp plant were indicative of a lack of a direct TOC to BOD conversion (Leithe, 1975). Nonetheless, for waste streams that are very consistent from one day to the next, simple usable relationships have been successfully developed. Examples are list in Table 13.1. Other examples are listed in the table in Chapter 12 on COD.

TABLE 13.1. Examples of TOC/BOD/COD relationships.

Reference	Relationship
Schaffer et al., 1965	TOC/BOD and TOC/COD
Ford, 1968	General TOC/BOD and TOC/COD ratios
Ford et al., 1971	TOC/BOD in petrochemical and refinery wastewaters
Lysyj & Nelson, 1971	TOC to BOD
D'Alessandro & Characklis, 1972	Soluble TOC to soluble BOD progression
Grunnett & Hansen, 1972	Comparison of TOC, BOD, and COD
Robbins et al., 1972	TOC to BOD of swine waste effluents
Anonymous, 1973	TOC to BOD general relationship
Chandler et al., 1976	TOC to BOD and COD of domestic effluent
Rider, 1976	Total carbon to COD
Whitehead, 1976	TOC to BOD and COD in poultry processing wastewater
Kehoe, 1977	TOC to BOD and COD
Bories et al., 1978	TOC to BOD of winery waste
Cooper et al., 1980	TOC to BOD of leather tannery waste
Basei, 1984	TOC used to estimate effluent COD
Heckershoff & Weismann, 1986a, 1986b	Comparison of DOC, BOD, and COD continuous monitoring
Wilson, 1993	TOC to BOD in vegetable pickling wastewater
Murren, 1995	TOC to BOD and COD in textile effluents

Measurement of TOC is an excellent process control parameter by itself. The analysis time is short, providing immediate feedback to operators. Tracking TOC from day to day for a particular effluent generates the same quality, if not better, information as that provided by BOD or COD. The analytical sensitivity and precision of TOC, particularly for low load levels, gives a more consistent parameter than BOD or COD (Arin, 1974; Maier & Swain, 1978). The changes between influent and effluent TOC are used to monitor plant efficiency and kinetic reaction rates (Wilson, 1997). The TOC has been used as the primary control parameter for monitoring membrane separation processes in the textile dying industry (Lei et al., 1998) and is preferred over BOD and COD for effluent containing dyestuffs and dye-fixing reagents (Ohmori et al., 1980). The TOC is the preferred parameter for monitoring effluents from coke plant and blast furnace blowdown effluents (Melcer et al., 1984). The TOC is related to food-to-microorganism ratios and sludge age in activated sludge treatment processes (Siber & Eckenfelder, 1980). The TOC is used as an indicator of groundwater contamination stemming from oil refinery operations (Spruill, 1988). An online TOC analyzer is preferred for water resource recovery facilities (Toth, 1976).

4.0 FRACTIONATION OF TOTAL ORGANIC CARBON

The concepts of organic carbon and organic matter are often used interchangeably, although organic matter is reported in milligrams total mass and organic carbon is in milligrams carbon. There are other differences. Organic matter is frequently determined by the weight difference of a dried sample before and after firing at 550 °C in an oven in the presence of oxygen, the idea being that the organic fraction of the sample is converted to carbon dioxide in the furnace. This time-honored crude technique includes contributions from relatively volatile inorganic salts, loss of water from the more thermally stable hydrates, and carbon dioxide from decomposition of certain carbonates, thus, generally overestimating the amount of semi- and nonvolatile organic matter in the sample. The initial drying step before furnace treatment can lead to loss of volatile and some semivolatile organic components, depending on the temperature and duration of the sample preparation.

The organic carbon determinations generally give faster and reliable values for the organic carbon content of the sample, but accurate conversion from mass units in terms of carbon to those of total organic matter require knowledge of the percent carbon in the sample, which changes from one sample to the next. Pure carbohydrates, such as glucose, exhibit approximately 40% carbon, whereas petroleum-derived materials display a range of values from pure aliphatic values of approximately 84% (decane) to pure aromatic values of 94% (phenanthrene) and higher. Sludge biomass, formulated as $C_8H_{15}O_5N$ (Burkhead & Waddell, 1971) to $C_5H_7NO_2$ (Porges et al., 1956) to C_4H_8ON (Burkhead & Waddell, 1971), exhibits 47% to 53% to 56% carbon, respectively. The empirical formula for algae (Milojevic, 1967) also gives 53% carbon. Depending on the actual substrate or sample, a percentage carbon for use as a conversion factor can be easily determined by measurement of TOC on a weighed dried portion of the sample. The extent of loss of volatile components from the dried sample can be determined by comparison of the aqueous TOC and solid TOC, giving a more reliable conversion factor. There are also technique modifications that can be used to ensure inclusion of all the volatile organic components in the aqueous TOC measurement (Barcelona, 1984). Performing the digestion in a sealed vessel will generally include the contributions from volatile components, but they can be lost in the accounting if a purge is used in conjunction with removal of carbon dioxide after acidification in the inorganic carbon (IC) determination before digestion in instruments that perform serial inorganic carbon and TOC procedures. Instruments that calculate TOC as the difference between total carbon and inorganic carbon do not exhibit this problem.

The commonly encountered primary subcategories of TOC are the dissolved component and the biodegradable component. The dissolved fraction is variously labeled dissolved organic carbon (DOC) and dissolved organic matter (DOM). The biodegradable fraction is sometimes referred to as biodegradable organic matter (BOM) and biodegradable organic carbon (BDOC); however, the even more restrictive category of assimilable organic carbon (AOC) has been developed, which is based on a bacterial plating test rather than a chemical carbon determination.

The concept of the term dissolved can be qualitative or quantitative, but most uses of the term are actually referring to an operationally defined process, such as filtered or nonfiltered. A rigorous application of the term "dissolved" implies that each individual molecule of the substance is completely surrounded by a shell of water molecules such that there is little to no interactions between the molecules of the substance. This is absolutely true only for solutions approaching infinite dilution, as most organic molecules tend to be hydrophobic rather than hydrophilic and, anthropomorphically, they prefer their own company to that of the water molecules. Thus, the presence of clusters of organic molecules in water tends to be the rule. The clusters may be small or they may be large, giving rise in extreme cases to micellular emulsions or colloidal dispersions in the water. Most emulsions and dispersions will pass through the commonly employed filtering media, so even though samples may be filtered, the filtrate is not the strictly defined dissolved fraction.

This may seem to be nit-picking, but it is an important consideration when the intent of the process is determining the portion of the organic carbon that is readily available to bacteria and the portion that is not. Bacteria can only assimilate truly dissolved molecules; otherwise, some prior manipulation of the potential food must occur before consumption. These considerations have led to further characterization of the filtrate (dissolved fraction) into slowly and rapidly hydrolyzable components of biodegradable organic matter (Carlson & Amy, 1997).

For most aqueous samples, the plumbing of the TOC instrument will only allow determination of aqueous samples that do not contain particulates. The particulates can be filtered from the samples before instrument analysis, and the organic carbon fraction retained by the filter determined as particulate organic carbon, POC (Menzel & Vaccaro, 1964; Whitehouse et al., 1989). The particulates are burned off the glass filter at 550 °C in an oxygen atmosphere, and the evolved carbon dioxide is trapped in sodium hydroxide and determined conductimetrically (Szekielda & Krey, 1965). The abbreviation POC, meaning particulate organic carbon, should not be confused with purgeable organic carbon, which is also encountered (Smith, 1999a). The commonly used glass fiber filters and 0.45-μm membrane filters (Srinanathakumar & Amirtharajah, 1983) for isolation of the particulates do not give the same size-based fractionation, and further, there is some indication that the use of glass fiber filters can lead to retention of some purportedly dissolved carbon (Abdel-Moati, 1990).

A simple technique for determination of true dissolved organic carbon using filters has been reported. The sample is filtered through a sequence of silver membrane filters of 1.2- and 0.2-μm pore size before being cross-flow filtered with a 10 000 nominal molecular weight ultrafilter. The TOC analysis of the permeate from the ultrafiltration gives the true DOC, and TOC analysis of the filtrate that passed the 1.2-μm membrane gives a measure of DOC plus any organic carbon associated with colloids. The contribution from colloids is the difference between the two measurements (Whitehouse et al., 1989).

Another technique for DOC uses flocculation of the sample with alum, followed by settling before analysis of the clarified supernatant for TOC (McLachlin, 1981). The flocculation procedure captures not only particulate TOC but also some of the dissolved material that is charged; thus, results from this technique are not comparable with those derived from a simple filtration.

Molecular size distributions of organic carbon have been determined through use of gel-permeation columns (Gjessing & Lee, 1967; Leidner et al., 1984; Reinhard, 1984). Ultrafiltration has been used to achieve molecular weight separations of DOC (Ogura, 1974). Ultrafiltration has also been used on acidic, basic, and neutral fractions of DOC to further characterize the molecular weight distributions of these chemical classes (Tadaanier et al., 2000). In general, these studies have employed manual collection of fractions, which are then subjected to further analysis for TOC and other parameters.

Techniques to determine the biodegradable fraction of TOC and organic matter range from straightforward batch experiments to use of the AOC test. The batch tests determine TC or TOC on the sample, then add the sample to a bioreactor for periods of time ranging from 10 to 28 days, followed by another determination of TC or TOC. The difference in the values is ascribed to the biodegradable portion, BDOC, while the residual TOC is described as biorefractory or inert (Ciapparone et al., 1997; Frias et al., 1992; Joret et al., 1989; Kaplan & Newbold, 1955; Khan, Babcock, Suffett, & Stenstrom, 1998; Khan, Babcock, Viriyavejakul, Suffett, & Stenstrom, 1998; Khan et al., 1999). The biodegradablility of particulate TOC is determined in much the same manner (Tyteca, 1981). Integrating the flux of bacterial mortality during the batch test has been proposed as an alternate to TOC for determination of biodegradable organic matter (Servias et al., 1987).

The key factor that has been found to influence BDOC results is the state of acclimation of the seed to the waste (Werner, 1984). Using an acclimated seed, preferably isolated from the sample source, gives the best results. In a logical extension of this idea, the operational efficiency of biological treatment facilities can be determined and tracked through comparison of influent and effluent TOC, where the whole facility is used as the bioreactor (Wilson, 1997). The hydraulic retention time can be used to correlate individual influent and effluent results for the most reliable tracking.

The duration of the BDOC test is shortened to approximately 3 hours when dynamic flow-through reactors are used, consisting of a glass bead-filled column with attached biomass (Kaplan & Newbold, 1995; Lucena et al., 1990; Ribas et al., 1991). The growth of the biomass within the reactor is reported to take up to 6 months before optimum TOC removal is obtained. Absorption versus assimilation of the TOC by the biomass can be ascertained through addition of sodium azide to the flowing water of one of a pair of duplicate reactors (Kaplan & Newbold, 1995).

Assimilable organic carbon has been included in *Standard Methods* since the 18th edition (APHA et al., 1992) as Method 9217. The idea behind the test is that the amount of organic carbon is the limiting growth factor in the medium, and that the number of colony-forming units (cfu) is directly related to the concentration of usable organic carbon. The sample is dispensed into nine 45-mL vials and pasteurized in a 70 °C water bath for 30 minutes, then each vial is inoculated with 500 cfu each of *Pseudomonas fluorescens* strain P-17 and *Spirillum* strain NOX. The vials are incubated for 9 days in the dark at 15 °C. On days 7, 8, and 9, three vials are removed from the incubator, and decade-dilution pour plates in duplicate are prepared from each vial. The plates are incubated at 23 °C for 3 to 5 days, then counted. If the counts indicate that the bacteria in the vials have reached the stationary phase, the AOC is calculated based on the yield factor: 4.1×10^6 cfu/µg acetate-C for P-17, and 2.9×10^6 cfu/µg oxylate-C for NOX (Kaplan & Bott, 1990; Vahala et al., 1998; van der Kooij et al., 1982, 1989).

The relationship between BDOC and AOC results is tenuous at best. In a comparative study of five different techniques for BDOC, the major difference being a range in incubation times from 10 to 21 days, all gave relatively equivalent results. However, all of the BDOC methods gave significantly higher results than the standard AOC method (Frias et al., 1995).

5.0 PHOTOMETRIC ESTIMATION OF TOTAL ORGANIC CARBON AND ORGANIC MATTER

Quick methods for estimating TOC and DOM have been sought for application in online continuous monitoring and for field use in lake evaluations. Particularly in lakes and other water bodies, the largest contributors to TOC and organic matter are the leaf and wood detritus degradation products humic acid, fulvic acid, and assorted lignins. All of these have strong UV-absorbing chromophores that are used as surrogate probes for organic matter content. Absorbance of UV at 254 nm is a standard estimation procedure for natural streams (De Haan et al., 1982); however, each water body needs a unique correlation, as a general relationship is lacking. When time is taken to develop a correlation, the results can be impressive. Fluorescence in particular can serve as an excellent surrogate probe for molecular size distributions of DOC in lakes affected by humic acids (De Haan & De Boer, 1987). A similar approach uses ultraviolet spectrophotometry (Chevalier et al., 2002) and compares the result to historical results with known TOC, COD, and BOD values.

Turbidity is a recognized problem in the correlation, and can be overcome by simultaneous measurement of the absorbance at 546 nm (Mattson et al., 1974). The flow rate of streams is a complicating factor in TOC estimations because of decreased natural settling as the flow rate increases. The TOC can be estimated from the absorbance at 250 nm, with differential flowrates taken into account by preparing three different correlations for low-, medium-, and high-flow regimes (Reid et al., 1980). The applicability of UV absorbance TOC to wastewater has been studied, with the normal finding that strong correlations exist for effluents with very consistent quality (Dobbs et al., 1972).

Other photometric probes have included the use of the IR absorbance of –OH groups and –CH groups as the quantitative metric for establishing correlations with TOC (Braunlich, 1973). This patent describes an online instrument that uses attenuated total reflectance from two alternately excited laser diodes as the analytical metric.

Fluorescent measurement with excitation at 365 nm and wide-range detection at 400 to 600 nm has proven to be correlated to filtered TOC (Smart et al., 1976). Individual compounds display unique synchronous fluorescent spectra that are related to the concentration of the organic matter in the sample. The spectrum derived from the sample can be used to trace the presence of the compound back to the effluent source (Cabaniss & Shuman, 1987).

Thermal lensing has been adapted as a spectrophotometric method for determination of DOM. According to the authors, turbidity adjustment is not necessary (Power & Langford, 1988).

An online colorimetric TOC analysis is demonstrated to be very useful for process control at three wastewater treatment facilities (Fair et al., 2003). Recent efforts have produced descriptions of a neural network (Fan & Boshnakov, 2010) to construct a soft measurement approach to monitor the effluent COD and BOD in wastewater treatment processes. Data derived from a soft sensor setup are shown to be useful for process control (Haimi et al., 2013). Based on a microspectrometer, an automatic, miniature, and multiparameter online water quality monitoring system was designed and implemented by Xie et al. (2016). The system was capable of real-time process control. Online optical monitoring variables were deemed a potential tool to be used to maintain stable process control (Tomperi et al., 2017). Degradation of synthetic wastewater in two microbial fuel cells hydraulically connected in series was optimized using a perturbation-observation maximum power point algorithm (Recio-Garrido et al., 2017). Developing a model-based soft sensor further improved the technique.

6.0 CONCLUSION

Determination of TOC is faster, more precise, and considerably less labor-intensive than either COD or BOD measurements. It also has the increasingly important advantage of minimizing the amount of hazardous waste generated from the analysis, in some instruments minimizing this factor to zero. Further, in most instances the technique can be adapted to experimental designs that provide exactly the same information as that derived from

COD or BOD. These considerations should lead one to choose TOC as the preferred instrument for the majority of effluent biological characterization and monitoring applications within the wastewater field.

7.0 REFERENCES

Abbott, W. E. (1927). Estimation of organic matter in sewage and effluents. *Ind. Eng. Chem.*, *19*, 919.

Abdel-Moati, A. R. (1990). Adsorption of dissolved organic carbon (DOC) on glass fibre filters during particulate organic carbon (POC) determination. *Water Res.*, *24*, 763.

Adeny, W. E., & Dawson, R. B. (1926). The determination of organic matter in water by potassium dichromate and sulfuric acid. *Sci. Proc. Royal Dublin Soc.*, *18*, 199.

Aiken, G. R. (1992). Chloride interference in the analysis of dissolved organic carbon by the wet oxidation method. *Environ. Sci. Technol.*, *26*, 2435.

Allison, L. E. (1960). Wet-combustion apparatus and procedure for organic and inorganic carbon in soil. *Soil Sci. Soc. Am. Proc.*, *24*, 36.

American Public Health Association, American Water Works Association, & Water Environment Federation (1992). *Standard methods for the examination of water and wastewater* (18th ed.). American Public Health Association.

American Public Health Association, American Water Works Association, & Water Environment Federation (1998). *Standard methods for the examination of water and wastewater* (20th ed.). American Public Health Association.

Anonymous (1973). Brave new world of TOC and TOD. *Ind. Water Eng.*, *10*(5), 13.

Archer, E. E. (1954). Semimicro wet combustion method for the determination of carbon. *Analyst*, *79*, 30.

Arin, M. L. (1974). Monitoring with carbon analyzers. *Environ. Sci. Technol.*, *8*, 898.

Armstrong, F. A. L., & Tibbits, S. (1968). Combustion of organic matter in sea water for nitrogen, phosphorus, and carbon determination photochemical. *J. Mar. Biol. Assoc. U.K.*, *48*, 143.

Armstrong, F. A. L., Williams, P. M., & Strickland, J. D. H. (1966). Photooxidation of organic matter in sea water by ultraviolet radiation, analytical and other applications. *Nature*, *211*, 481.

Baldwin, J. M., & McAtee, R. E. (1974). Determination of organic carbon in water with a silver-catalyzed peroxydisulfate wet chemical oxidation method. *Microchem. J.*, *19*, 179.

Barcelona, M. J. (1984). TOC determinations in ground water. *Ground Water*, *22*, 18.

Basei, R. (1984). On-line measurement of organic carbon in wastewater. *Hydrocarbon Process.*, *63*(10), 42.

Beattie, J., Bricker, C., & Garvin, D. (1961). Photolytic determination of trace amounts of organic matter in water. *Anal. Chem.*, *33*, 1890.

Blackmore, R. H., & Voshel, D. (1967). Rapid determination of TOC in sewage. *Water Sew. Works*, *114*, 398.

Bories, A., et al. (1978). Characterization of the organic load of winery waste water by TOC. *Trib. Cebedeau*, *31*, 411; *Chemical Abstracts*, *89*, 135107 (1978).

Bramer, H. C., Walsch, M. J., & Caruso, S. C. (1966). Instrument for monitoring trace organic compounds in water. *Water Sew. Works*, *113*, 275.

Braunlich, G. (1973). Continuous determination of organic substances in waste water by infrared photometry. *Ger. Offen.* (patent), 2164670; *Chemical Abstracts*, *79*, 83238 (1973).

Burkhead, C. E., & Waddell, S. L. (1971). Composition studies of activated sludges. *Proc. 24th Ind. Waste Conf, Purdue University*, 135, 576.

Busch, A. W. (1966). Energy, total carbon, and oxygen demand. *Water Resour. Res.*, *2*, 59.

Busch, A. W. (1967). Carbon analysis as a concept and procedure in water pollution technology. *Analyzer*, *8*(1), 3.

Cabaniss, S. E., & Shuman, M. S. (1987). Synchronous fluorescence spectra of natural waters: tracing sources of dissolved organic matter. *Mar. Chem.*, *21*(1), 37.

Carlson, K., & Amy, G. (1997). The formation of filter-removable biodegradable organic matter during ozonation. *Ozone: Sci. Eng.*, *19*(2), 179.

Cauwet, G. (1984). Automatic determination of dissolved organic carbon in seawater in the sub-ppm range. *Marine Chem.*, *14*, 297.

Chandler, R. L., O'Shaughnessy, J. C., & Blanc, F. C. (1976). Pollution monitoring with TOC analysis. *J. Water Pollut. Control Fed.*, *48*, 2791.

Charles, M. J., & Simmons, M. S. (1986). Methods for the determination of carbon in soils and sediments. A review. *Analyst*, *111*, 385.

Chevalier, L. R., Irwin, C. N., Craddock, J. N (2002). Evaluation of InSpectra UV analyzer for measuring conventional water and wastewater parameters. *Adv. Environ. Rese.*, *6*, 369.

Ciapparone, L. A., Diehl, A. C., & Speitel, G. E. (1997). Ozonation and BDOC Removal: Effect on water quality. *J. Am. Water Works Assoc.*, *89*(2), 84.

Collins, K. J., & Williams, P. J. L. (1977). An automated photochemical method for the determination of dissolved organic carbon in sea and estuarine waters. *Mar. Chem.*, *5*, 123.

Cooper, J. E., et al. (1980). Use of TOC as an alternative to COD for monitoring the biooxidation of tannery unhairing waste. *J. Am. Leather Chem. Assoc.*, *75*, 331.

Croll. B. T. (1974). Determination of organic carbon in water. *Anal. Chem.*, *46*, 1341.

Cropper, F. R., Heinekey, D. M., & Westwell, A. (1967a). The determination of total organic matter (carbon content in aqueous media). I. Organic matter in aqueous plant streams. *Analyst*, *92*, 436.

Cropper, F. R., Heinekey, D. M., & Westwell, A. (1967b). The determination of total organic matter (carbon content in aqueous media). II. Involatile organic matter in demineralized water. *Analyst*, *92*, 443.

D'Alessandro, P. L., & Characklis, W. G. (1972). Simple measurement technique for soluble BOD progression. *Water Sew. Works*, *119*(9), 106.

Degtijareff, W. T. (1930). Determining soil organic matter by means of hydrogen peroxide and chromic acid. *Soil Sci.*, *29*, 239.

De Haan, H., & De Boer, T. (1987). Applicability of light absorbance and fluorescence as measures of concentration and molecular size of DOC in humic Lake Tjeukemeer. *Water Res.*, *21*, 731.

De Haan, H., De Boer, T., Kramer, H. A., & Voerman, J. (1982). Applicability of light absorbance as a measure of organic carbon in humic lake water. UV 250 nm correlated with COD of filtered lake water. *Water Res.*, *16*, 1047.

Dobbins, D. C., & Pfaender, F. K. (1988). Methodology for assessing respiration and cellular incorporation of radiolabeled substrates by soil microbial communities. *Microbiol. Ecol.*, *15*, 257.

Dobbs, R. A., et al. (1972). Use of ultraviolet absorbance for monitoring the toc content of water and wastewater. *Water Res.*, *6*, 1173.

Dobbs, R. A., Wise, R. H., & Dean, R. B. (1967). Measurement of organic carbon in water using the hydrogen flame ionization detector. *Anal. Chem.*, *30*, 1255.

Eggertsen, F. T., & Stross, F. H. (1972). Flame detection method for determining organic carbon in water. *Anal. Chem.*, *44*, 709.

Emery, R. M., Welch, E. B., & Christman, R. F. (1971). The total organic carbon analyzer and its application to water research. *J. Water Pollut. Control Fed.*, *43*, 1834.

Enrhardt, M. (1969). Automatic measurement of dissolved organic carbon in seawater. *Deep-Sea Res.*, *16*, 393; *Chemical Abstracts*, *71*, 128533 (1969).

Fair, C., Boyles, W., Miller, M., Kneis, C., Allen, L., Dunkin, B. (2003). Colorimetric TOC analysis as a substitute for COD analysis. *Proceedings of the Water Environment Federation.* WEF.

Fan, L., Boshnakov, K. (2010). Neural-network-based water quality monitoring for wastewater treatment processes. *Proc. 6th Internat. Conf. Nat. Comp.*, *4*, 1746–48.

Fiedler, V., Hansen, E. O., & Ruzicka, J. (1975). Measurements of carbon dioxide with the air-gap electrode. Determination of the total inorganic and total organic carbon contents in water. *Anal. Chim. Acta* (Netherlands), *74*, 423.

Ford, D. L. (1968). TOC as a wastewater parameter. *Public Works*, *99*, 89.

Ford, D. L., Eller, J.M., & Gloyna, E. F. (1971). Analytical parameters of petrochemical and refinery wastewaters. *J. Water Pollut. Control Fed.*, *43*, 1712.

Frias, J., Lucena, F., & Ribas, F. (1992). A method for the measurement of BDOC in waters. *Water Res.*, *26*, 255.

Frias, J., Ribas, F., & Lucena, F. (1995). Comparison of methods for the measurement of biodegradable organic carbon and assimilable organic carbon in water. *Water Res.*, *29*, 2785.

Gacs, I., & Payer, K. (1989). Determination of TOC in water in the form of carbamate by means of a simple denuder/electrolytic conductivity flow cell system. *Anal. Chim. Acta* (Netherlands), *220*, 1.

Gjessing, E., & Lee, G. F. (1967). Fractionation of organic matter in natural waters on Sephadex column. *Environ. Sci. Technol.*, *1*, 631.

Gloor, R., & Leidner, H. (1979). Universal detector for monitoring organic carbon in liquid chromatography. *Anal. Chem.*, *51*, 645.

Gloor, R., et al. (1981). Exclusion chromatography with carbon detection: A tool for further characterization of DOC. *Water Res.*, *15*, 457.

Godec, R., O'Neil, K., & Hutte, R. (1992). New technology for TOC analysis in water. *Ultrapure Water*, 9(9), 17.

Goulden, P. D. (1976). Automated determination of carbon in natural water. *Water Res.*, *10*, 487.

Goulden, P. D., & Brooksbank, P. (1975). Automated determination of dissolved organic carbon in lake water. *Anal. Chem.*, *47*, 1943.

Gravelet-Blondin, L. R., van Vliet, H. R., & Mynhardt, P. A. (1980). An automated method for the determination of dissolved organic carbon in fresh water. *Water SA*, *6*, 138.

Grunnet, K., & Hansen, J. C. (1972). Measurement of organic carbon in water, BOD, COD, and TOC. *Vatten* (Sweden), *28*, 356; *Chemical Abstracts*, *78*, 101764 (1973).

Haimi, H., Mulas, M., Corona, F., & Vahala, R. (2013). Data-derived soft-sensors for biological wastewater treatment plants: An overview. *Env. Modelling & Software, 47,* 88–107.

Heck, A. F. (1929). A method for the determining of total carbon and also for the estimation of carbon dioxide evolved from soils. *Soil Sci.*, *28*, 225.

Heckershoff, H., & Weismann, U. (1986a). A new measuring device for operation, monitoring, and control of activated sludge plants. Part 1: Measurement of substrate concentration. *Korresp. Abwass.* (Germany), *33*, 508; *Chemical Abstracts*, *105*, 65761 (1986).

Heckershoff, H., & Weismann, U. (1986b). A new measuring device for operation, monitoring, and control of activated sludge plants. Part 2: Measurement of the respiration activity of activated sludge and of reaction kinetics. *Korresp. Abwass.* (Germany), *33*, 600; *Chemical Abstracts*, *105*, 158202 (1986).

Heron, G., Barcelona, M. J., Anderson, M. L., & Christensen, T. H. (1997). Determination of nonvolatile organic-carbon in aquifer solids after carbonate removal by sulfurous acid. *Ground Water*, *35*, 1.

Howard, T. E., & DeWitt, J. W. (1971). Effluent characteristics of bleached kraft pulp mills. *Pulp Paper Mag. Can.*, *72*(1), 73.

Hughes, J. L., et al. (1974). Dissolved organic carbon (DOC), an index of organic contamination in ground water near Barstow, Calif. *Ground Water*, *12*, 283.

Iwasaki, I., Hirayama, M., Ohmori, T., & Kubomura, S. (1966). Determination of organic carbon in water by potassium peroxydisulfate oxidation. *Kogyo Yosui* (Japan), *89*, 24; *Chemical Abstracts*, *70*, 6471 (1969).

Iwasaki, I., Hirayama, M., Ohmori, T., & Ono, T. (1966). Determination of organic carbon in polluted water by periodic acid. *Kogyo Yosui* (Japan), *91*, 23; *Chemical Abstracts*, *69*, 89566 (1968).

Jones, R. H. (1972). TOC: How valid is it. *Water Wastes Eng.*, *9*(4), 32.

Jones, R. H., & Dageford, A. F. (1968). Application of a high sensitivity total organic carbon analyzer. *Proc. Instr. Soc. Am.*, *6*, 43.

Joret, J. C., et al. (1989). The measurement of bioeliminable dissolved organic carbon (BDOC): A tool in water treatment. *Water Supply*, *7*, 1.

Kakajima, K. (1986). Simultaneous determination of TOC and total nitrogen in waters by pyrolysis–gas chromatography–mass spectrometry. *Water Res.*, *20*, 233.

Kaplan, L. A, & Bott, T. L. (1990). Measurement of assimable organic carbon in water distribution systems by a simplified bioassay technique. *Proc. Water Qual. Technol. Conf.*, *16*, 475.

Kaplan, L. A., & Newbold, J. D. (1995). Measurement of streamwater biodegradable dissolved organic carbon with a plug-flow bioreactor. *Water Res.*, *29*, 2696.

Katz, J., Abraham, S., & Baker, N. (1954). Analytical procedure using a combined combustion–diffusion vessel: Improved method for combustion of organic compounds in aqueous solution. *Anal. Chem.*, *26*, 1503.

Kehoe, T .J. (1977). Determining TOC in waters. *Environ. Sci. Technol.*, *11*, 137.

Khan, E., Babcock, R. W., Suffett, I. H., & Stenstrom, M. K. (1998). Method development for measuring biodegradable organic carbon in reclaimed and treated wastewaters. *Water Environ. Res.*, *70*, 1025.

Khan, E., Babcock, R. W., Viriyavejakul, S., Suffett, I. H., & Stenstrom, M. K. (1998). Biodegradable dissolved organic carbon for indicating wastewater reclamation plant performance and treated wastewater quality. *Water Environ. Res.*, *70*, 1033.

Khan, E., King, S., Babcock, R. W., Jr., & Stenstrom, M. K. (1999). Factors influencing biodegradable dissolved organic carbon measurement. *J. Environ. Eng.*, *125*, 514.

Kramer, H., et al. (1972). Automatic continuous carbon determination in water. German Offen (patent) 2116148; *Chemical Abstracts*, *78*, 7713 (1973).

Krey, J., & Szekielda, K. H. (1965). Determination of organically-bound carbon in sea water with a new apparatus for the determination of small amounts of CO_2. *Zeits. Anal. Chem.*, *207*, 338.

Lei, L., Hu, X., & Yue, P. (1998). Improved wet oxidation for the treatment of dyeing wastewater concentrate from membrane separation process. *Water Res.*, *32*, 2753.

Leithe, W. (1975). Analytic Characterization of sewer loading by spent liquors of the pulp industry. *Das Papier*, *29*(5), 195.

Leidner, H. A., Fleischmann, T. H., & Hamer, G. (1984). Molecular weight fractionation for the study of complex biodegradation processes. *Anal. Chim. Acta*, *163*, 35.

Lindenbaum, A., Schubert, J., & Armstrong, W. D. (1948). Rapid wet combustion method for carbon determination with particular reference to isotopic carbon. *Anal. Chem.*, *20*, 1120.

Lossberg, L. G. (1967). Apparatus for continuous measurement of organic matter in water. U.S. Patent 3205045.

Lucena, F., Frias, F., & Ribas, F. (1990). A new dynamic approach to the determination of BDOC in water. *Environ. Technol.*, *12*, 343.

Lysyj, I., & Nelson, K. H. (1971). Analysis of biochemical oxygen demand. U.S. Patent 3567391; *Chemical Abstracts*, *74*, 115675 (1971).

Lysyj, I., Nelson, K. H., & Newton, P. R. (1968). Determination of the organic content of natural waters by a rapid method. *Water Res.*, *2*, 233.

Lyutsarev, S. V. (1969). Automatic device for determining the carbon in organic substances in natural waters. *Girdrokhim. Mater. Akad. Nauk.* (Russia), *49*, 207; *Chemical Abstracts*, *71*, 73884 (1969).

Maier, W. J., & McConnell, H. L. (1974). Carbon measurements in water quality monitoring. *J. Water Pollut. Control Fed.*, *46*, 623.

Maier, W. J., & Swain, W. R. (1978). Organic carbon: A nonspecific water quality indicator for Lake Superior. *Water Res.*, *12*, 523.

Matthews, R. W., Abdullah, M., & Low, G. K.-C. (1990). Photocatalytic oxidation for TOC analysis. *Anal. Chim. Acta* (Netherlands), *233*, 171.

Mattson, J. S., et al. (1974). Continuous monitoring of dissolved organic matter by UV-visible photometry. *Limnol. Oceanogr.*, *19*, 530.

McCready, R. M., & Hassid, W. T. (1942). Semimicrodetermination of carbon using the Van Slyke–Folch oxidation mixture. *Ind. Eng. Chem. Anal. Ed.*, *14*, 525.

McLachlin, C. N. S. (1981). UV–DOC characterization of pulp and paper bleach effluents. *Water Res.*, *15*, 983.

McQuaker, N. R., & Fung, T. (1975). Determination of carbonaceous material in sediments by reductive pyrolysis and spectrophotometry. *Anal. Chem.*, *47*, 1435.

Melcer, H., et al. (1984). Combined treatment of coke plant wastewater and blast furnace blowdown water in a coupled biological fluidized bed system. *J. Water Pollut. Control Fed.*, *56*, 192.

Menzel, D. W., & Vaccaro, R. F. (1964). The measurement of dissolved and particulate carbon in seawater. *Limnol. Oceanogr.*, *9*(1), 138.

Mills, G. L., & Quinn, J. G. (1979). Determination of organic carbon in marine sediments by persulfate oxidation. *Chem. Geol.*, *25*, 155.

Milojevic, M. (1967). Contribution of steam-sterilized algae to BOD. *J. Water Pollut. Control Fed.*, *39*, 1201.

Mohlman, F. W., & Edwards, G.P. (1931). Determination of carbon in sewage and industrial wastes. *Ind. Eng. Chem. Anal. Ed.*, *3*, 119.

Montgomery, H. A. C., & Thom, N. S. (1967). The determination of low concentrations of organic carbon in water. *Analyst*, *87*, 689.

Murren, C. (1995). Are you putting money down the drain? *Tech. Text. Int.*, *3*, 24.

Nelson, D. W., & Sommers, L. E. (1982). Total carbon, organic carbon, and organic matter. In *Methods of soil analysis, part 2. Chemical and microbiological properties*, (2nd ed.). American Society of Agronomy, Soil Science Society of America.

Oake, R. J. (1981). A review of photo-oxidation for the determination of total organic carbon in water. *Water Res. Cent., Tech. Rep. TR 160.*

Ogura, N. (1974). Molecular weight fractionation of dissolved organic matter in coastal seawater by ultrafiltration. *Mar. Biol.*, *24*, 305.

Ohmori, T., Hirayama, M., & Iwasaki, I. (1968). Rapid determination of organic carbon in water. *Kogyo Yosui* (Japan), *117*, 30; *Chemical Abstracts*, *69*, 89628 (1968).

Ohmori, M., Kurashima, M., Satoh, E., & Ogawa, H. (1980). Pollution indexes for water quality control of wastewater from dye works by TOC analyzer. *Kenkyu Hokoku-Kanagawa-ken Kogyo Shikensho*, *50*, 75; *Chemical Abstracts*, *94*, 162165 (1981).

Pickhardt, W. P., Oemler, A. N., & Mitchell, J. (1955). Determination of total carbon in organic materials by a wet–dry combustion method. *Anal. Chem.*, *27*, 1784.

Popova, E. S. (1930). Potassium dichromate for estimating the oxidizability of sewage. *Trav. Comm. Recherches Epurat Eaux Eqout*, *12*, 143; *Chemical Abstracts*, *24* (1930).

Porges, N., Jasewicz, L., & Hoover, S. R. (1956). Simple method of estimation of oxygen value of wastes. *Ind. Wastes*, *1*, 95.

Powell, R. M., Bledsoe, B. E., Curtis, G. P., & Johnson, R. L. (1989). Interlaboratory methods comparison for TOC analysis of aquifer materials. *Environ. Sci. Technol.*, *23*, 1246.

Power, J. F., & Langford, C. H. (1988). Optical absorbance of dissolved organic matter in natural water studies using the thermal lens effect. *Anal. Chem.*, *60*, 842.

Recio-Garrido, D., Adekunle, A., Perrier, M., Raghavan, V., Tartakovsky, B. (2017). Wastewater treatment and online chemical oxygen demand estimation in a cascade of microbial fuel cells. *Ind. Eng. Chem. Res., 56*(44), 12471–78.

Reid, J. M. et al. (1980). Observations on the estimation of TOC From UV absorbance for an unpolluted stream. *Water Res., 14*, 525.

Reinhard, M. (1984). Molecular weight distribution of dissolved organic carbon and dissolved organic halogen in advanced treated wastewater. *Environ. Sci. Technol., 18*, 410.

Ribas, F., Frias, J., & Lucena, F. (1991). A new dynamic method for the rapid determination of BDOC in drinking water. *J. Appl. Bacteriol., 71*, 371.

Rider, M. D. (1976). Monitoring organic pollutants continuously by total carbon analyzer. *Ind. Water Eng., 12*, 10.

Robbins, J. W., et al. (1972). TOC determinations on swine waste effluents. *Am. Soc. Agric. Eng., 15*, 105.

Roberts, A. A., et al. (1973). Determination of organic carbon in modern carbonate sediments. *J. Sed. Petrol., 43*, 1157.

Russell, E. R. (1977). Determination of traces of organic material in an aqueous solution. *Anal. Chim. Acta* (Netherlands), *88*, 171.

Schaffer, R. B., Van Hall, C.E ., McDermott, G. N., Barth, D., Stanger, V. A., Sebesta, S. J., & Griggs, S. H. (1965). Application of a carbon analyzer in waste treatment. *J. Water Pollut. Control Fed., 37*, 1545.

Schollenberger, C. J. (1927). A rapid approximate method for determining soil organic matter. *Soil Sci., 24*, 65.

Schollenberger, C. J. (1931). The determination of soil organic matter. *Soil Sci., 31*, 483.

Servias, P., Billen, G., & Hascoet, M. C. (1987). Determination of the biodegradable fraction of dissolved organic matter in waters. *Water Res., 21*, 445.

Siber, S., & Eckenfelder, W. W., Jr. (1980). Effluent quality variation from multicomponent sustrates in the activated sludge process. *Water Res., 14*, 471.

Smart, P. L., Finlayson, B. L., Rylands, W. D., & Ball, C. M. (1976). The relationship of fluorescence to dissolved organic carbon in surface waters. *Water Res., 10*, 805.

Smith, R.-K. (1999a). *Handbook of environmental analysis* (4th ed.). Genium Publishing.

Smith, R.-K. (1999b). *Lectures on wastewater analysis and interpretation.* Genium Publishing.

Spruill, T. B. (1988). Use of TOC as an indicator of contamination from an oil refinery, South-Central Kansas. *Ground Water Monit. Rev.*, *8*(3), 76.

Srinanthakumar, S., & Amirtharajah, A. (1983). Organic carbon decay in Swith biofilm kinetics. *J. Environ. Eng.*, *109*, 102.

Stainton, M. P. (1973). A syringe gas-stripping procedure for gas-chromatographic determination of dissolved inorganic and organic carbon in fresh water and carbonates in sediments. *J. Fish Res. Bd. Can.*, *30*, 1441; *Chemical Abstracts*, *80*, 40847 (1974).

Suzuki, J., Yokoyama, Y., Unno, Y., & Suzuki, S. (1983). A sensitive method for measuring organic carbon in sediment. *Water Res.*, *17*, 431.

Szekielda, K. H., & Krey, J. (1965). Determination of particulate, organically bound carbon in sea water by a new rapid method. *Mikrochim. Ichnoanal. Acta*, *1*, 149; *Chemical Abstracts*, *62*, 11529 (1965).

Tadaanier, C. J., Berry, D. F., & Knocke, W. R. (2000). Dissolved organic matter apparent molecular weight distribution and number-average apparent molecular weight by batch ultrafiltration. *Environ. Sci. Technol.*, *34*, 2348.

Takahashi, Y., Moore, R. T., & Royce, R. J. (1972). Direct determination of organic carbon in water by reductive pyrolysis. *Am. Lab.*, *4*, 31.

Tan, F. C., & Edmond, J. M. (1993). Carbon isotope geochemistry of the Orinoco Basin. *Estuarine Coastal Shelf. Sci.*, *36*, 541.

Teal, J. L., Van Hall, C.E., Stenger, V. A., & Safranko, J. W. (1967). Carbon content of aqueous systems. U.S. Patent 3296435; *Chemical Abstracts*, *66*, 58710 (1967).

Tiessen, H., & Moir, J. O. (1993). Total and organic carbon. In M. R. Carter (Ed.), *Soil sampling and methods of analysis.* Lewis Publishers.

Tomperi, J., Koivuranta, E., Leiviska, K. (2017). Predicting the effluent quality of an industrial wastewater treatment plant by way of optical monitoring. *J. Water Process Eng.*, *16*, 283–89.

Toth, H. (1976). Continuous measurement of organic carbon content. *Oesterr. Abwasser-Rundsch.* (Austria), *20*, 94, *Chemical Abstracts*, *85*, 10016 (1976).

Tyteca, D. (1981). Nonlinear programming model of wastewater treatment plant. *J. Environ. Eng.*, *107*, 747.

Vahala, R., Ala-Peijari, T., Rintala, J., & Laukkanen, R. (1998). Evaluating ozone dose for AOC removal in two-step GAC filters. *Water Sci. Technol.*, *37*(9), 113.

van der Kooij, D., et al. (1982). Determining the concentration of easily assimilable organic carbon in drinking water. *J. Am. Water Works. Assoc.*, *74*, 540.

van der Kooij, D., et al. (1989). The effects of ozonation, biological fand distribution on the concentration of easily assimilable organic carbon (AOC) in drinking water. *Ozone: Sci. Eng.*, *11*, 297.

van Hall, C. E., Barth, D., & Stenger, V. A. (1965). Elimination of carbonates from aqueous solutions prior to organic carbon determinations. *Anal. Chem.*, *37*, 769.

van Hall, C. E., Safranko, J., & Stenger, V. A. (1963). Rapid combustion method for the determination of organic substances in aqueous solutions. *Anal. Chem.*, *35*, 315.

van Hall, C. E., & Stenger, V. A. (1964). Use of infrared analyzer for total carbon determination. *Water Sew. Works*, *3*, 266.

van Slyke, D. D. (1954). Wet carbon combustion and some of its applications. *Anal. Chem.*, *26*, 1706.

van Slyke, D. D., & Folch, J. (1940). Manometric carbon determination. *J. Biol. Chem.*, *136*, 509.

van Steenderen, R. A., Basson, W. D., & van Duuren, F. A. (1979). Automated chemical analysis for measuring microgram levels of organic carbon in potable waters. *Water Res.*, *13*, 539.

van Steenderen, R. A., & Lin, J. (1981). Determination of DOC in water. *Anal. Chem.*, *53*, 2157.

Walkley, A., & Black, I. A. (1934). An examination of the Degtjareff method for determining soil organic matter and a proposed modification of the chromic acid titration method. *Soil Sci.*, *37*, 29.

Weber, W. J., & Morris, J. C. (1964). Determination of carbon in wastewaters by high-temperature wet oxidation. *J. Water Pollut. Control Fed.*, *36*, 573.

Werner, P. (1984). Investigations on the substrate character of organic substances in connection with drinking water treatment. *Z. Bakt. Hyg.*, *180*, 46.

West, D. L. (1964). Determination of total carbon in water by combustion-gas chromatography. *Anal. Chem.*, *36*, 2194.

Westerhausen, L., Poynter, J., Eglinton, G., Erlenkeuser, H., & Sarnthein, M. (1993). Marine and terrigenous origin of organic matter in modern sediments of the equatorial East Atlantic: The $d^{13}C$ and molecular record. *Deep-Sea Res. I*, *40*, 1087: *Chemical Abstracts*, *119*, 256050 (1993).

Whitehead, W. K. (1976). Monitoring poultry processing wastewater with TOC. *Poultry Sci.*, *55*, 679.

Whitehouse, B. G., McDonald, R. W., Iseki, K., Yunker, M. B., & McLaughlin, F. A. (1989). Organic carbon and colloids in the Mackenzie River and Beaufort Sea. *Mar. Chem.*, *26*, 371.

Williams, P. M. (1969). The determination of dissolved carbon in seawater: A comparison of two methods. *Limnol. Oceanogr.*, *14*, 297.

Williams, R. T. (1967). Water-pollution instrumentation. analyzer looks for organic carbon. *Instrum. Technol.*, *14*, 63.

Wilson, F. (1993). Prediction of rotating biological contactor efficiency using TOC. *J. Environ. Eng.*, *119*, 478.

Wilson, F. (1997). Total organic carbon as a predictor of biological wastewater treatment efficiency and kinetic reaction rates. *Water Sci. Technol.*, *35*(8), 119.

Wilson, R. F. (1961). Measurements of organic carbon in sea water. *Limnol. Oceanogr.*, *6*, 259.

Wilson, R. F. (1963). Organic carbon levels in some aquatic ecosystems. *Public Inst. Mar. Sci.*, *9*, 64.

Xie, Y., Wen, Z., Mo, Z., Yu, Z., Kanglin Wei, K. (2016). Implementation of an automatic and miniature on-line multi-parameter water quality monitoring system and experimental determination of chemical oxygen demand and ammonia-nitrogen. *Water Sci. Technol.* 73(3), 697–706.

Index

www.ingramcontent.com/pod-product-compliance
Lightning Source LLC
LaVergne TN
LVHW060556240526
839643LV00003B/23